Soil Chemistry

Soil Chemistry

Edited by Rufus Mitchell

SYRAWOOD
PUBLISHING HOUSE
New York

Published by Syrawood Publishing House,
750 Third Avenue, 9th Floor,
New York, NY 10017, USA
www.syrawoodpublishinghouse.com

Soil Chemistry
Edited by Rufus Mitchell

International Standard Book Number: 978-1-68286-396-1 (Hardback)

Cataloging-in-publication Data

Soil chemistry / edited by Rufus Mitchell.
p. cm.
Includes bibliographical references and index.
ISBN 978-1-68286-396-1
1. Soil chemistry. 2. Plant-soil relationships. 3. Soil ecology. I. Mitchell, Rufus.
S592.5 .S65 2017
631.41--dc23

Printed in China.

TABLE OF CONTENTS

PREFACE

Soil chemistry is the study of the chemical properties of the soil. Chemical changes in the composition of soil layers affect the fertility as well as absorptive capabilities of the soil. This book elucidates the concepts and innovative models around prospective developments with respect to soil chemistry and environmental factors that directly affect chemical characteristics of soil. This book includes some of the vital pieces of work being conducted across the world, on various topics related to this field. For all readers who are interested in soil chemistry, the case studies included in this book will serve as excellent guide to develop a comprehensive understanding.

This book is a comprehensive compilation of works of different researchers from varied parts of the world. It includes valuable experiences of the researchers with the sole objective of providing the readers (learners) with a proper knowledge of the concerned field. This book will be beneficial in evoking inspiration and enhancing the knowledge of the interested readers.

In the end, I would like to extend my heartiest thanks to the authors who worked with great determination on their chapters. I also appreciate the publisher's support in the course of the book. I would also like to deeply acknowledge my family who stood by me as a source of inspiration during the project.

Editor

1

Automated Analysis of Two-Dimensional Positions and Body Lengths of Earthworms (*Oligochaeta*); MimizuTrack

Naomi Kodama[1]*, Toshifumi Kimura[2], Seiichiro Yonemura[1], Satoshi Kaneda[1], Mizue Ohashi[2], Hidetoshi Ikeno[2]

1 National Institute for Agro-Environmental Sciences (NIAES), 3-1-3 Kannondai, Tsukuba, Ibaraki, Japan, **2** School of Human Science and Environment, University of Hyogo, 1-1-12 Shinzaike-honcho, Himeji, Hyogo, Japan

Abstract

Earthworms are important soil macrofauna inhabiting almost all ecosystems. Their biomass is large and their burrowing and ingestion of soils alters soil physicochemical properties. Because of their large biomass, earthworms are regarded as an indicator of "soil heath". However, primarily because the difficulties in quantifying their behavior, the extent of their impact on soil material flow dynamics and soil health is poorly understood. Image data, with the aid of image processing tools, are a powerful tool in quantifying the movements of objects. Image data sets are often very large and time-consuming to analyze, especially when continuously recorded and manually processed. We aimed to develop a system to quantify earthworm movement from video recordings. Our newly developed program successfully tracked the two-dimensional positions of three separate parts of the earthworm and simultaneously output the change in its body length. From the output data, we calculated the velocity of the earthworm's movement. Our program processed the image data three times faster than the manual tracking system. To date, there are no existing systems to quantify earthworm activity from continuously recorded image data. The system developed in this study will reduce input time by a factor of three compared with manual data entry and will reduce errors involved in quantifying large data sets. Furthermore, it will provide more reliable measured values, although the program is still a prototype that needs further testing and improvement. Combined with other techniques, such as measuring metabolic gas emissions from earthworm bodies, this program could provide continuous observations of earthworm behavior in response to environmental variables under laboratory conditions. In the future, this standardized method will be applied to other animals, and the quantified earthworm movement will be incorporated into models of soil material flow dynamics or behavior in response to chemical substances present in the soil.

Editor: Jan M. Hemmi, University of Western Australia, Australia

Funding: This work was supported by Grant for Environmental Research Projects from the Sumitomo Foundation (Contract nr.:123225.) The funders had no role in study design, data collection and analysis, decision to publish, or preparation of the manuscript.

Competing Interests: The authors have declared that no competing interests exist.

* E-mail: naomi.kodama@googlemail.com

These authors contributed equally to this work.

Introduction

Earthworms, the so-called ecosystem engineers, are important soil macrofauna belonging to the subclass *Oligochaeta*. They inhabit various types of soil ranging from 10 to 2000 individuals per meter squared, resulting in a large biomass of between 0.5 to 305 g dry mass m^{-2} [1]. They significantly change soil properties and fertility in the vicinity of the large burrowing areas [2]. Their important role in soil formation has been already emphasized by Darwin [3]. Earthworms are one of the main components of the drilosphere, one of the biological systems regulating mineralization of soil organic matter, where they activate microbes by mixing the litter and soil, breaking down the physical protection of soil organic matter, and developing soil aggregates during gut transit [2]. The changes in soil pore structures and the extent of aggregations also alter soil gas diffusion coefficients physically, which could increase emissions of global warming gases, such as CO_2, CH_4 and N_2O [4]. Because of their behaviors and features, earthworms are regarded as important biological indicators of chemical toxicity in soil ecosystems and many studies have examined the effects on population dynamics of earthworms such as fecundity, reproductive activity, and mortality [5]. Their behaviors affect not only below-ground material flow dynamics, but also indicate "soil health" conditions.

Evaluation of the functional role of earthworms on soil ecosystems has been, however, insufficient because of methodological difficulties [6]. Traditional research methods including descriptive observations of their behavior, gut content analysis, choice tests, and litter bags have increased our understanding of their ecological importance. Recent advances of molecular tools provide information on enzyme activities of microorganisms which interact with earthworms. Isotope labeling techniques have been used to quantify materials exchanged between earthworms and soils or atmosphere, and between earthworms and microorganisms. These new techniques are powerful but still limit our knowledge on the soil biogeochemical processes regulated by earthworms because they are indirect and static methods. In combination with existing methods, additional development of the tools to quantitatively and mechanistically evaluate their movements and exploring factors controlling their feeding behavior

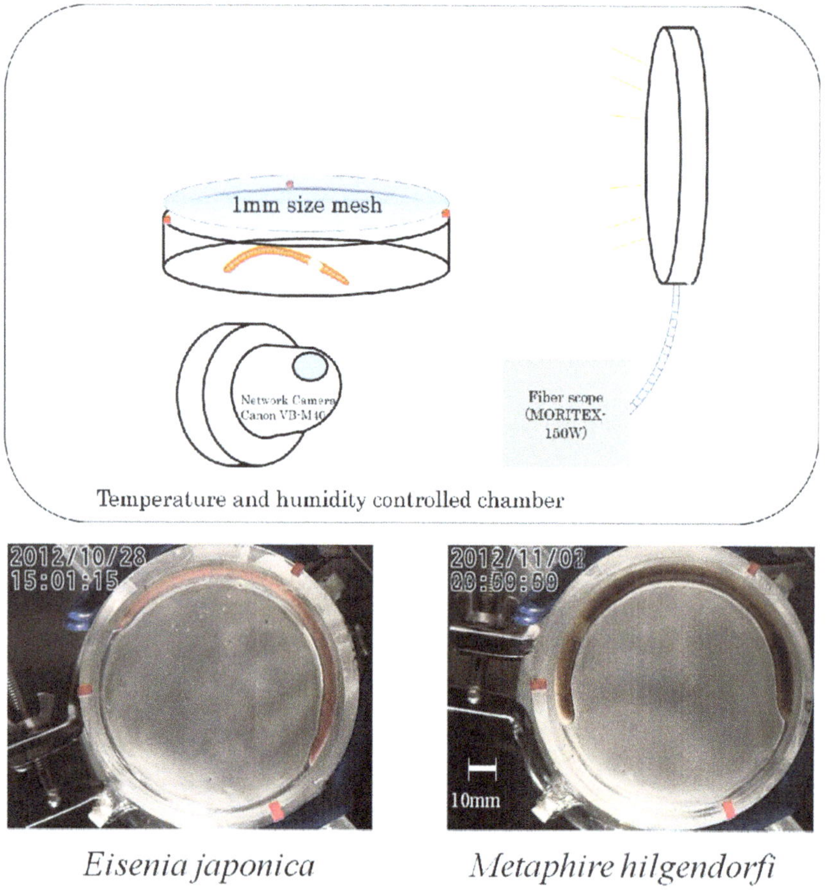

Figure 1. Webcam setup and study animals. The upper diagram shows the system for webcam recording in a controlled chamber. The earthworms used in this study are shown in the bottom panel. The diameter of the glass Petri dish was 85 mm (inside diameter) and the height of the Petri dish was 20 mm.

would be helpful to further understand their ecological roles. Such information will also enable us to establish more realistic models of soil ecosystems because many of the current models do not take the impact of soil animals into account [5]. Thus, observing earthworm feeding behavior, such as their response to some stimuli (such as light, temperature, vibration, and odor cues of food) in vitro without soil will be of importance albeit under very simplified artificial condition. The understanding of their physical activities resulting from their feeding behavior will advance quantification of the earthworm's contribution to whole below-ground biogeochemical processes. To determine earthworm activity, it is first necessary to quantify the extent to which they move and stretch their bodies and relate this to gas emissions from the soil. To confirm the activities, therefore, direct observation is a powerful tool and the combination of conventional methods with

Figure 2. Earthworm body thinning procedure and determination of the head, tail, and central points. The captured image data were used for the latter.

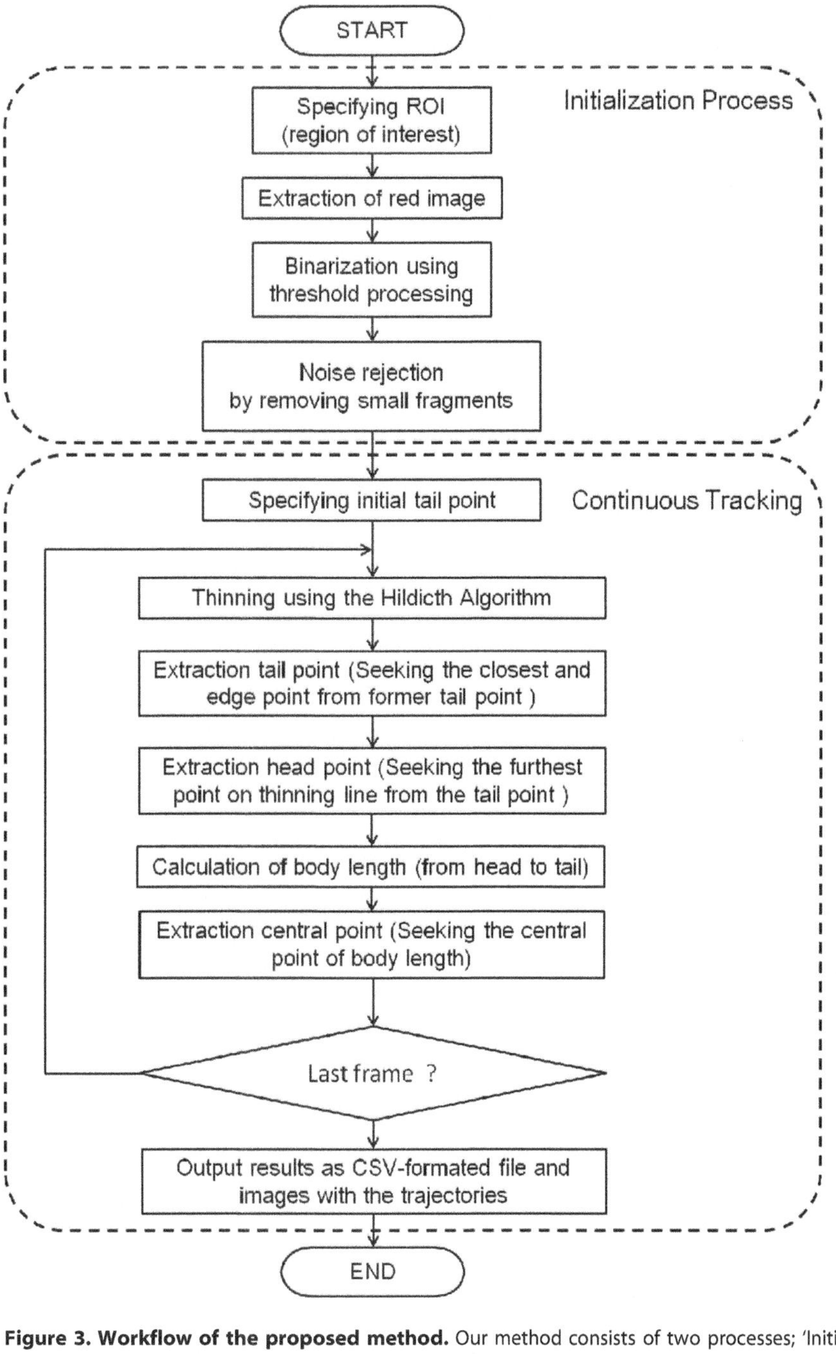

Figure 3. Workflow of the proposed method. Our method consists of two processes; 'Initialization Process' and 'Continuous Tracking'.

direct observation such as using a web camera or scanner will be of great merit in this field.

In addition to the importance of earthworms in ecosystem functions, they are also regarded as an important biological indicator of chemical toxicity in the soil. Ecotoxicology has traditionally used biological indicators to quantify or predict the changes in soil "health" or "quality" [5]. To date many studies have found serious effects of chemicals and heavy metals on the reproductivity, growth rate, and mortality of earthworms ([7] [8,9]), and experimental protocol and earthworm culture techniques in vitro have been developed [10]. However, the observation of earthworms in response to such chemical pollution has been conducted over a relatively long time span with a lower resolution time (on the scale of minutes), and less attention has been paid to the signs in early stages, such as irregular behavior or morphological alterations induced by the uptake of pollutants. However, since the sudden changes in their activity and features might be also tightly connected to the damages and types of the pollutant, observations at high time resolutions (on the scale of seconds) should also be conducted.

To date, automated systems have been used to track movement with high time resolution continuously using a web camera for nematodes, ants, honeybees, and drosophila, and to subsequently analyze the continuous image data [11] [12] [13] [14] [15] [16] [17]. Image analysis is an efficient method for quantifying the tracking data. Previous studies have attempted to track earth-

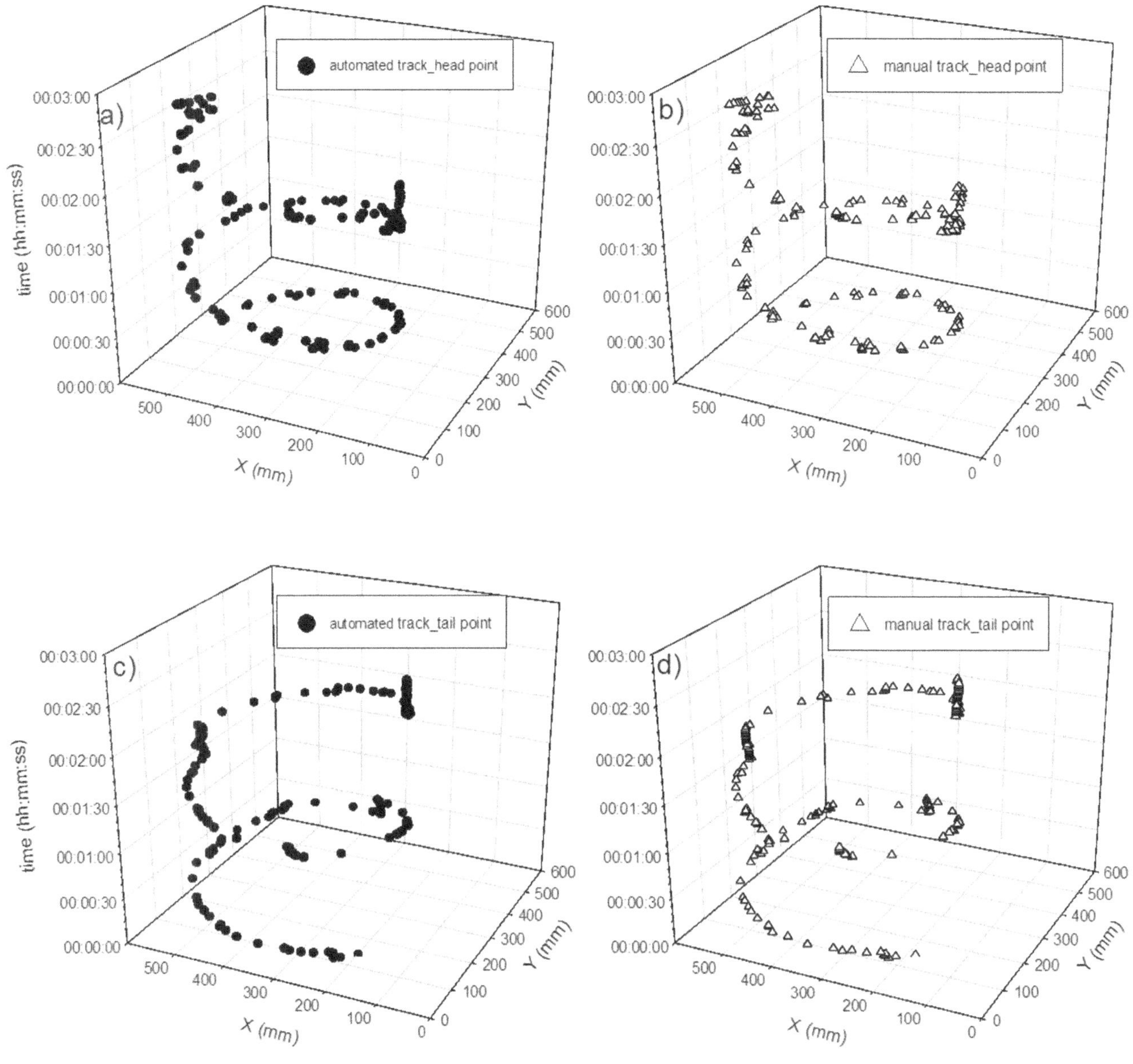

Figure 4. The automatically and manually measured two-dimensional coordinates of *Eisenia japonica* over time. Each panel shows the head point in the upper panel and the tail point in the bottom panel.

worms directly in the soil by means of X-ray or minirhizotron [18] [19]; the time resolution, however, was low to track their dynamic changes over a long time period on a time scale of seconds. Revealing their habits without soil but with a high time resolution would be a first step in investigating their behavior. However, those tools cannot be applied directly to earthworms as some of their traits are different from other animals in terms of the extent to which they can stretch their bodies (their body stretch can reach 1.5 imes that of their default body length) [20].

A number of automated tracking systems for a model worm, *Caenorhabditis elegans*, have not considered changes in the extent of body stretch [20]. The WormScan is indeed a useful tool to observe population dynamics such as fecundity, reproductive activity, or mortality along with the growth rate that occur over a long time span. However, we still lack a program that can track dynamic change in their activities to observe a fast response to chemical pollutants, environmental changes or toxic substances. WormScan's default settings are for the nematode as a target animal. The features of nematodes, such as the movement and body size, vary widely from those of earthworms, therefore it requires intensive modification of the parameters of the program to apply it to earthworms. In addition, a standardized tracking system is required so that reliable comparisons can be made between species or phenotypes when tracking is done by different users. Standardized measurements will reduce misestimation and will output more reliable quantification.

Although existing methods such as WormScan [20] are candidates that may have been applicable in this study, the current WormScan program, which was originally developed for tracking nematodes, was not capable of following the changes in morphology of earthworms because earthworms move more quickly and over a wide area compared with nematodes.

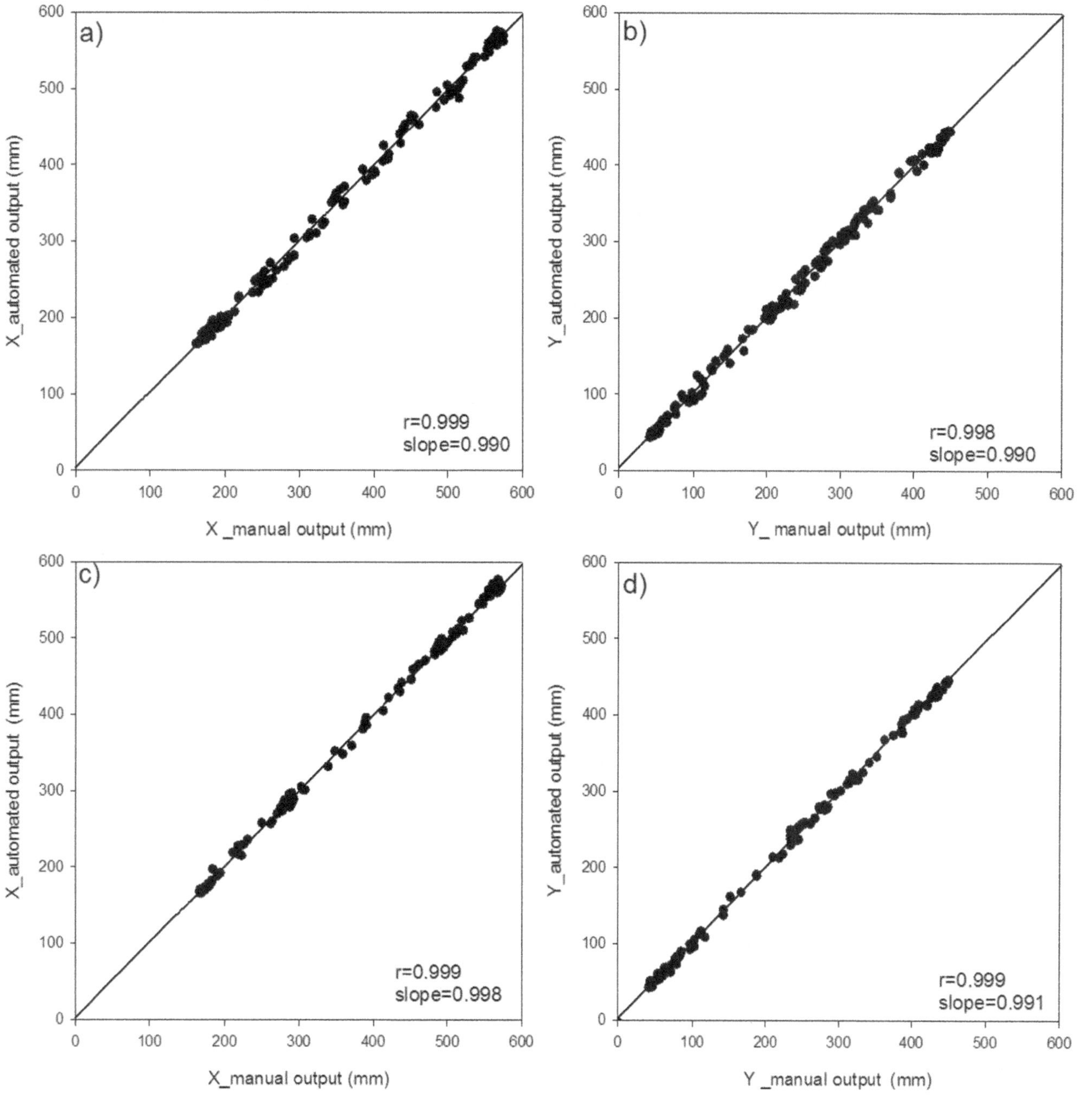

Figure 5. Correlations between the automatically and manually measured two-dimensional coordinates of *Eisenia japonica*. Each panel shows the head point in the upper panel and the tail point in the bottom panel.

WormScan has been developed for observing population dynamics such as reproductive activity, fecundity or mortality which occur over a longer term span (several days to weeks), and the behavior is also relatively static. WormScan, therefore does not suit observations at a continuous and high time resolution, which captures dynamic and rapid movements. This is a remarkable property of earthworms and is an important factor in quantifying their physical activity. We believe that WormScan can track some objects of varying sizes, however we found that earthworms' movements are too fast and dynamic, and WormScan could not cope with this. The observation of dynamic changes would help to understand the ecological background of earthworm behaviors; chemical assays in vitro without soils could also be conducted. The existing application ImageJ (Fiji) (http://rsbweb.nih.gov/ij/ version 1.46 for Windows) can provide a plug-in to manually track the two-dimensional positions and lengths of objects. Even when using this application, analyzing large data sets is labor intensive; therefore, a more reliable and automated quantification method is required for continuous image data.

In the present study, we developed a high-throughput program that tracks two-dimensional coordinates of multiple points and body lengths of objects in continuously acquired image data from earthworms; to the program was able to process approximately 1 frame per second. Following user initialization, the program tracks

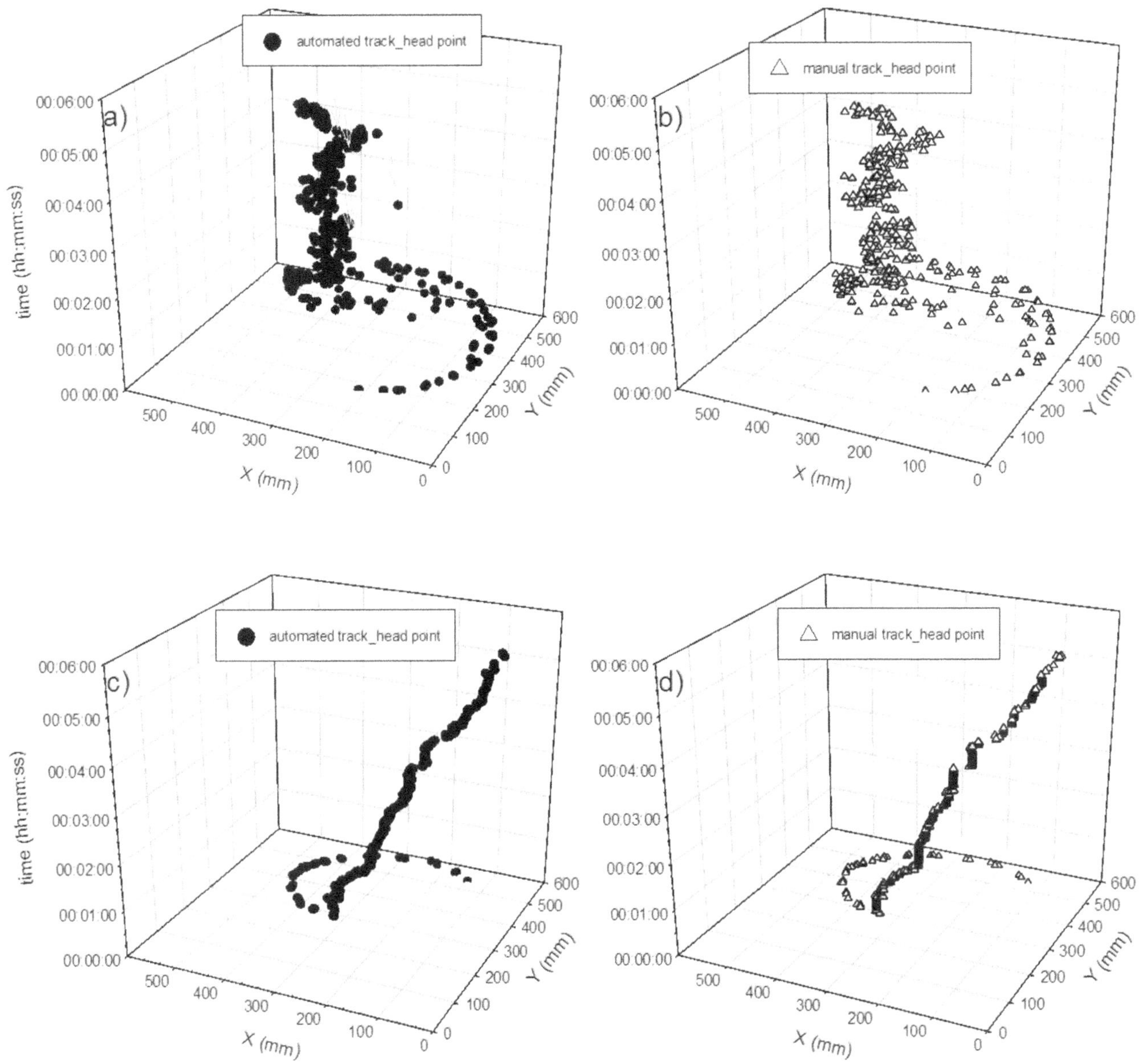

Figure 6. The automatically and manually measured two-dimensional coordinates of *Metaphire hilgendorfi* over time. Each panel shows the head point in the upper panel and the tail point in the bottom panel.

the head, central position, and tail of a single earthworm. This program provides a means to track animals that change their body lengths over time both accurately and efficiently, though still with limited performance. The output was subsequently used to calculate the velocity of those points. Finally, in the future, the quantified behavior could be incorporated into soil material flow dynamic models or applied to measure their reactions to chemical inputs in the soil.

Materials and Methods

Study animals

Two different species of earthworm were used in this study; *Eisenia japonica* [21] with a body size ranging from 4–17 cm [22], and *Metaphire hilgendorfi* [21] with a body size ranging from 9–30 cm [23]. The lifespan of both is usually 6 months from spring to fall. Both species are common in Japanese mountains and fields, and both genera have a wide global distribution [24]. *M. hilgendorfi* is epigeic because it primarily inhabits the litter layer and *E. japonica* is polyhumic-endogeic because it mainly lives in the topsoil layer and has more inorganic content in the gut [25].

Continuous recording of earthworm movement in glass containers (image capture)

Adult earthworms were captured from a field at the National Institute for Agro-Environmental Sciences, Tsukuba, Japan on 28 October, 2012. Earthworms were kept in a plastic box (inner diameter 15 cm, height 6 cm) with approximately 100 g of alluvial soil at 60% of its water holding capacity until the experiment. The weights of *E. japonica* and *M. hilgendorfi* used in the experiment were 1.45 and 4.75 g, respectively. *E. japonica* and *M. hilgendorfi* were transferred from the plastic box to glass containers for network-

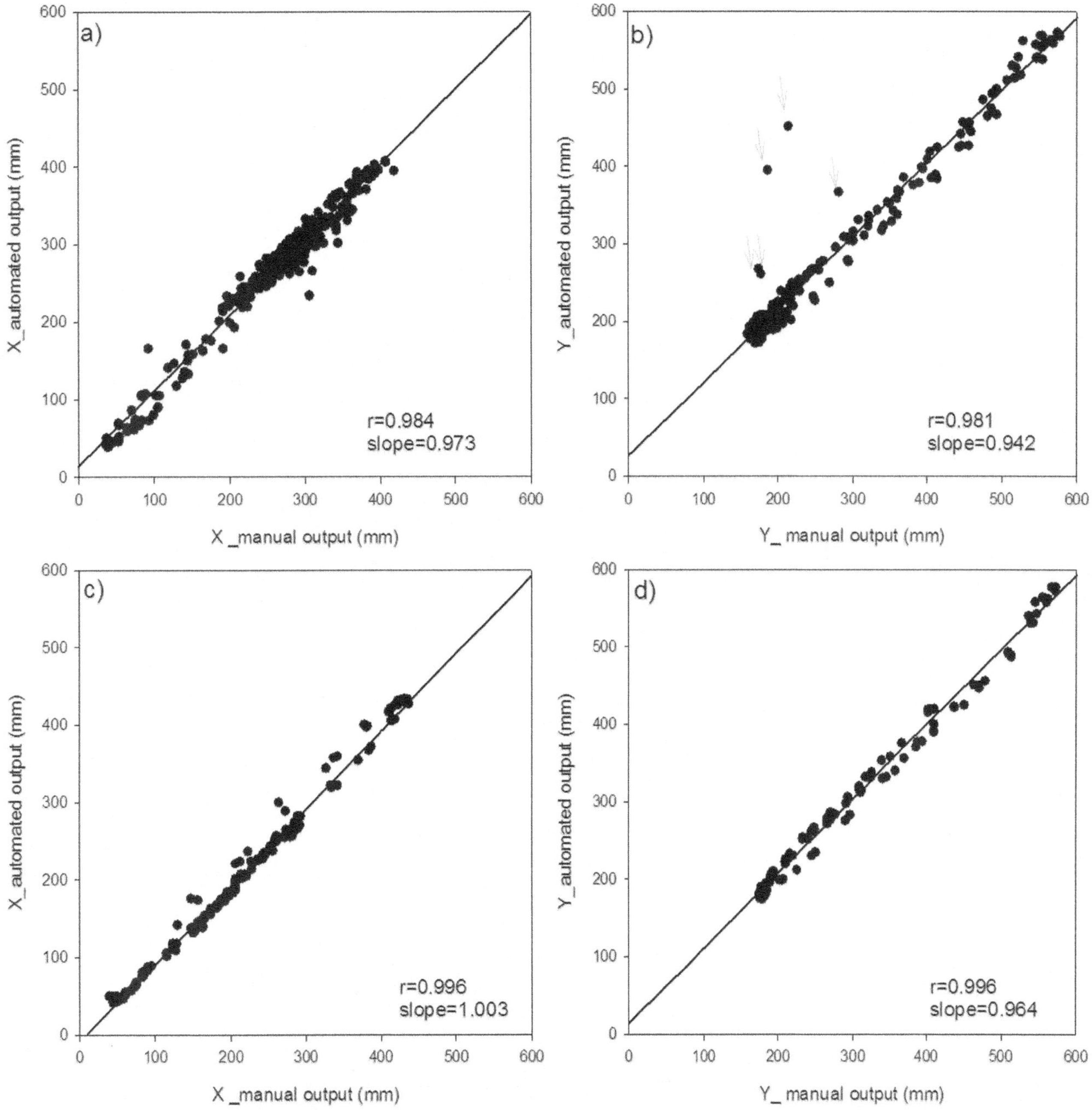

Figure 7. Correlations between the automatically and manually measured two-dimensional coordinates of *Metaphire hilgendorfi*. Each panel shows the head point the in upper panel and the tail point the in bottom panel.

camera recording inside an incubator, where the mean ambient temperature was set at 15°C and the ambient air humidity was saturated at the given temperature (Fig. 1). A fiber scope light was used to maintain constant light conditions (MHAB-150W, Moritex, Tokyo) at approximately 500 lux outside of the container and app. 250 lux inside the container; this was measured with a radiometer (T-10A, Konica Minolta Optics, Inc., Tokyo, Japan). The glass containers were covered with 1-mm mesh plastic cloths and fixed with double-sided cellophane tape to prevent the earthworms from escaping during the experimental period. A small fan (OD2510, Orion Fans, Dallas, USA) was placed on the top of the glass container to ventilate the air inside and wet tissue was put in to keep the humidity constant during the measurement period. A network camera (VB-M40, Canon Inc., Tokyo, Japan) was fixed to the bottom of the glass container with approximately 60 cm of distance between the camera and the glass container (Fig. 1). The network camera was fixed in parallel to the glass container firmly while capturing the image. The position of the camera was calibrated using the ruler to ensure the accurate pixilation of the actual length. The network camera uses a lens that is 40 mm in diameter and can cover 60.4° from the horizontal angle. The captured image data was 640×480 pixels in size. To obtain the coordinates of the glass container, marks were fixed at three positions on the edge of the containers with red tape. The

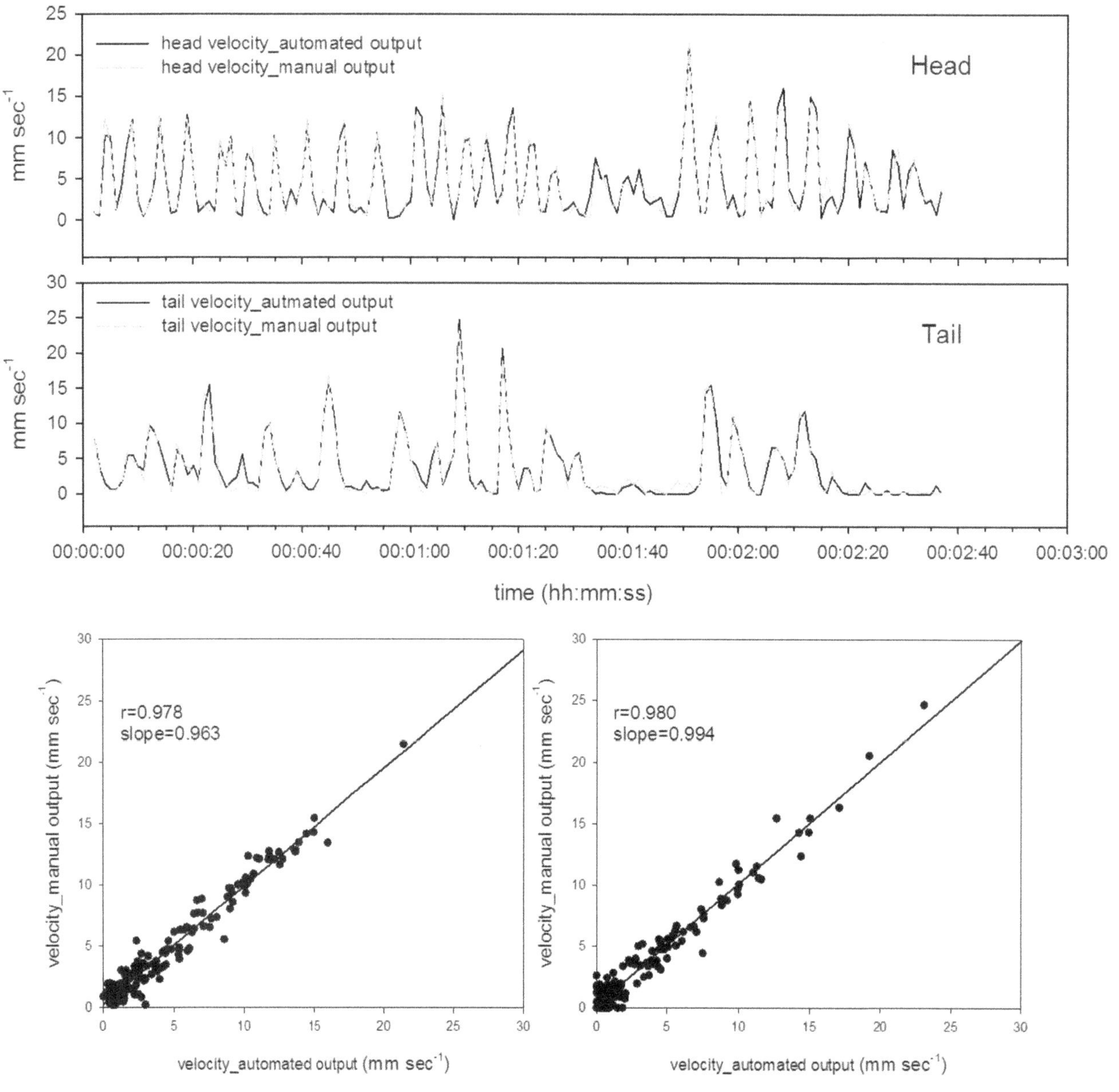

Figure 8. Comparison of the automatically and manually calculated velocity of each point of *Eisenia japonica*. The upper panel shows the calculated velocity of the head point and the central panel shows that of the tail point. Correlations between the automatically and manually calculated velocities of *Eisenia japonica*.

round glass containers were 85 mm in inside diameter and 2.5 mm in thickness. The image data were recorded as a movie using a network camera continuously recording every second and saved as JPEG files. One pixel was equivalent to 0.2 mm in the captured image data.

To estimate local image distortions, checkerboard (8 rows and columns) was located for image capturing system on the same position as the glass container. The length of the diagonal lines were measured using a vernier caliper with a 0.1 mm precision and calculated the errors. The maximum error in mm was 0.2 mm that resulted the precision less than 0.1%.

Image analysis

Two methods were implemented: a manual tracking method and a new automated tracker. First we used a manual tracking method relies on the Fiji implementation of ImageJ (http://rsbweb.nih.gov/ij/version 1.46 for Windows) with an additional plugin, Manual Tracking (http://fiji.sc/wiki/index.php/Manual_Tracking). Currently manual tracking is a standard method to track animal behavior using image data. Second, we then used a newly developed program (MimizuTrack) that was compared with Manual Tracking in Fiji.

Manual tracking

Using the Manual Tracking plug-in on the Fiji implementation of ImageJ, head and tail of the body were tracked separately. The static image data (JPEG) captured at 1 second intervals were converted to movie data, using a function in Fiji to convert from JPEG into.avi formatted data. Each coordinate per one second was tracked to obtain the velocity of the movement. We then

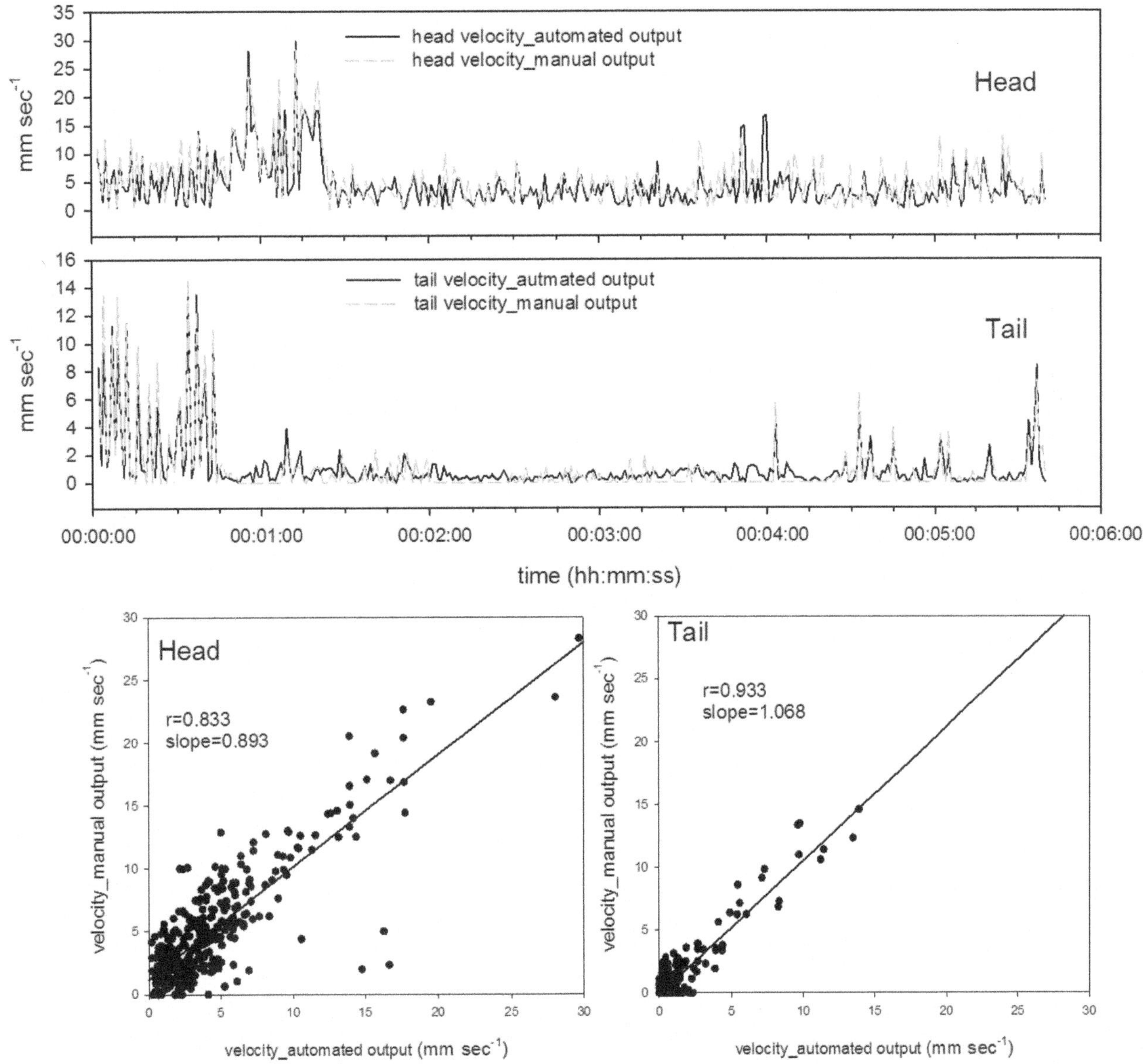

Figure 9. Comparison of the automatically and manually calculated velocity of each point of *Metaphire hilgendorfi.* The upper panel shows the calculated velocity of the head point and the central panel shows that of the tail point. Correlations between the automatically and manually calculated velocities of *Metaphire hilgendorfi*.

handled the output data as ground truth for further evaluation since the data were input by expert users. The output data were used for evaluation of the program described in the next section. Our program was benchmarked against user input only. This was because no automated tracking system currently exists that is directly comparable to our tracker in this specific setting.

Automatic tracking - Initialization Process

The image data obtained were first processed to separate the objects (earthworms from the background). In the original image, the contrast between the earthworms and the background was highest in the red channel so only this component was used for the process. The procedures are done manually as an initial step, then the image data are processed using a program (image shown in Fig. 2); We developed several programs to track moving objects over time, such as honeybees or plant roots in previous studies [12] [26] [27] the source code for which is original, and we have developed a new program (MimizuTrack) for earthworms based on the previous programs. The steps are shown in Fig. 3 and detailed below. The detailed initialization process are described as follows:

- The outline of the inside of the glass container was determined using a tool for drawing around the shape and cutting the outside of the image.
- A color threshold was determined to highlight the color of the earthworm body and split the color channels into three colors (green, blue, and red). The red image was then used for further analysis. Earthworms were assigned a value of 255 and the non-earthworm areas were eliminated by applying a value of 0.

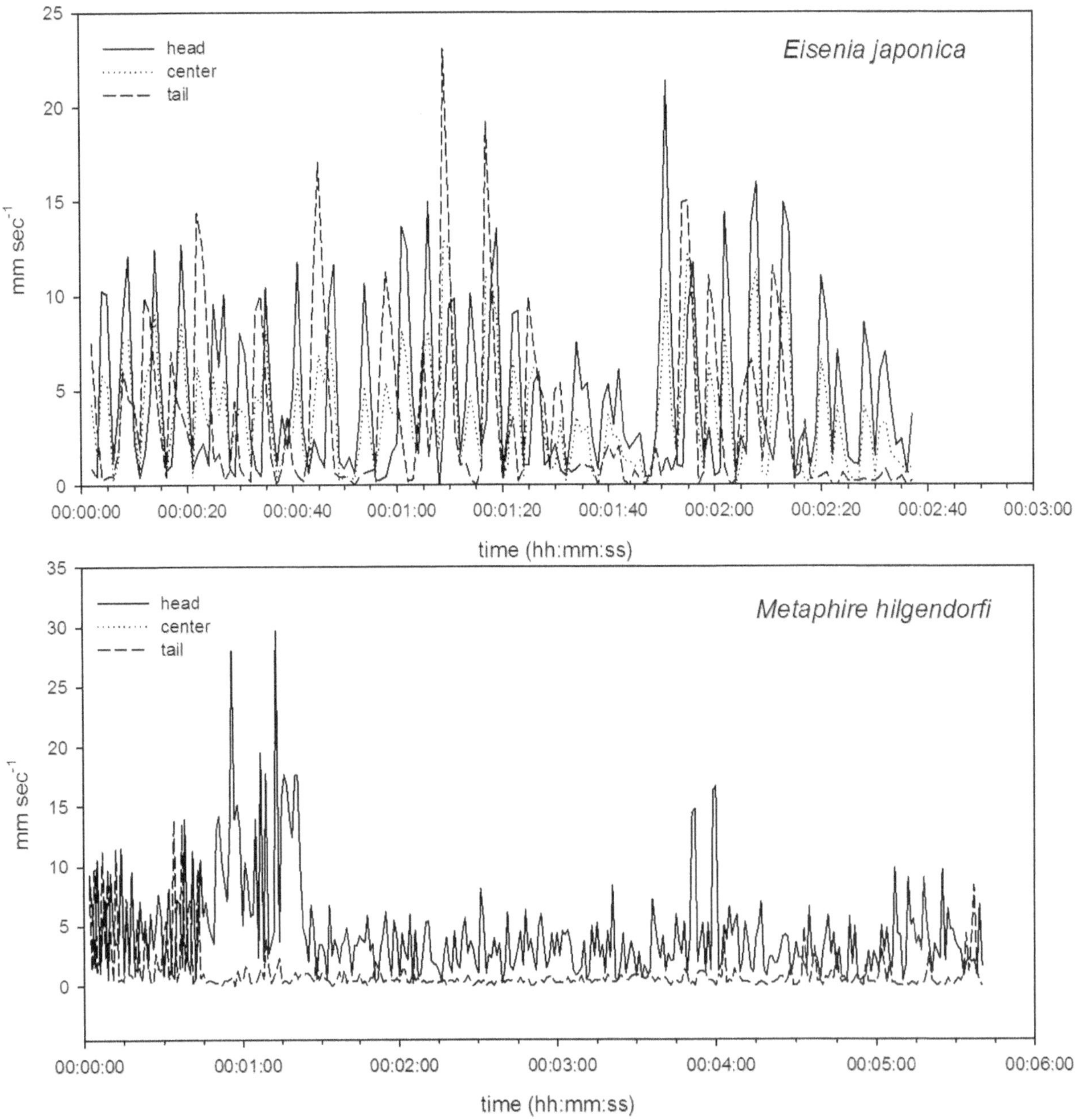

Figure 10. Comparison of velocities among different points (head, central, and tail points). The upper panel shows the velocities of each point of *Eisenia japonica* and the bottom panel shows those of *Metaphire hilgendorfi*.

- A threshold to extract the earthworm body was determined and a binary image with white for the object and black for the background was generated.
- The noise was eliminated by removing small fragments from the background.

Automatic tracking - Continuous Tracking

The rest of the image processing was carried out by our newly developed program. The procedures are shown in Fig. 2. We developed a prototype program using Microsoft Visual Studio 2010 (Visual C++2010) and OpenCV 2.31 on a computer with an Intel Core i5 2.50 GHz (CPU), 16 GB (Memory), 640 GB (HDD), and running Microsoft Windows 7 Enterprise 64-bit (OS). Our developed software is executed as a Win64 console application on Windows.

The tracking method is as follows;

- Thinning the earthworm body areas

To extract feature points, such as the head, tail, and center points, the program was tasked with identifying the end-points (tails) of the line segment corresponding to the worm body. That is as thin as possible while remaining fully connected and centered. Using binary images of the earthworm body, the program was able to obtain three feature points on each frame. First, the Hilditch algorithm is used to reduce the pixel region representing the

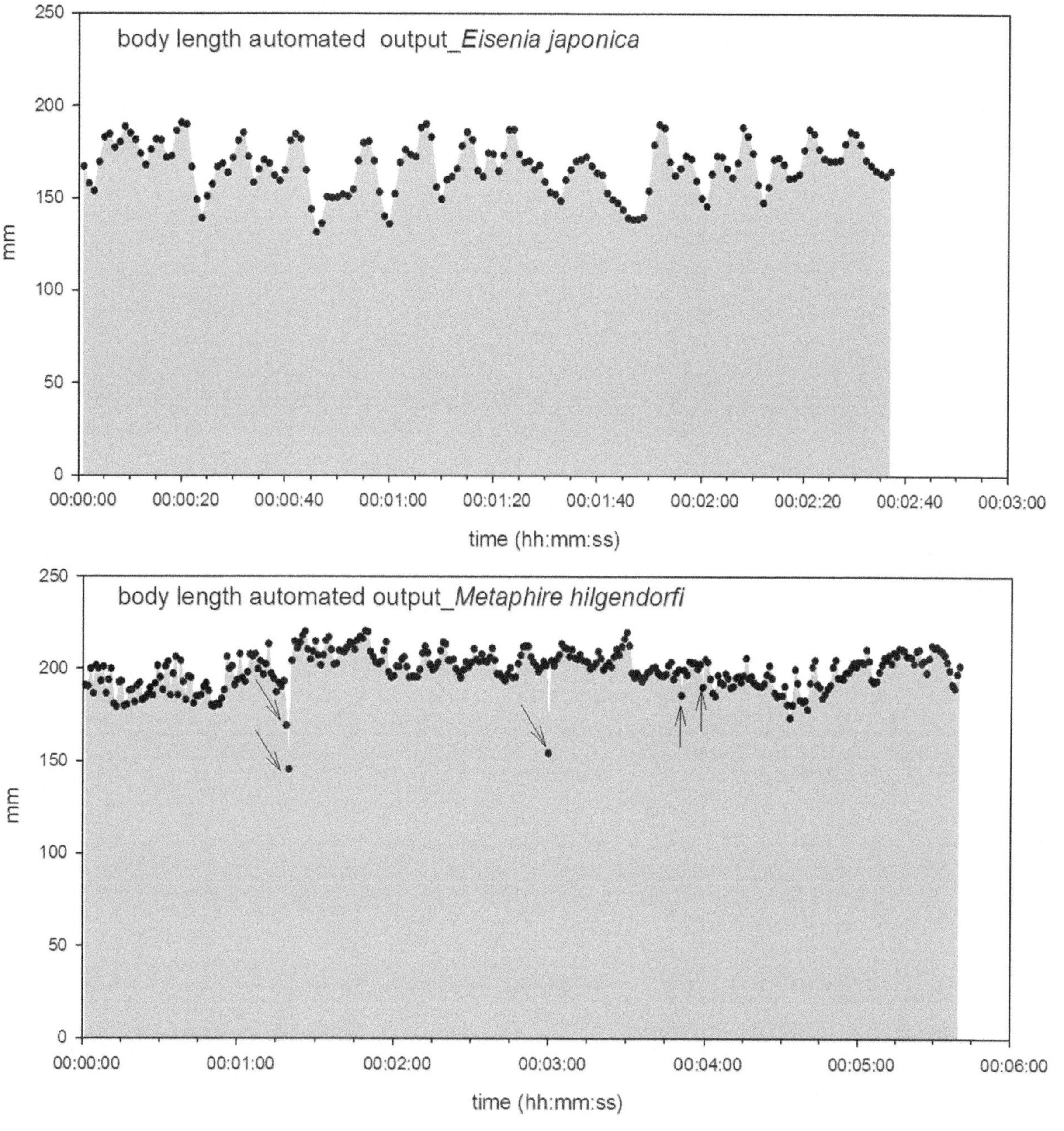

Figure 11. Change in body length over time. The upper panel shows the length of *Eisenia japonica* and the bottom panel shows that of *Metaphire hilgendorfi.*

earthworm to its skeletal form [28]. The end-points of the center line are recognized as candidates of earthworm feature points.

- Extracting the head, tail, and center point of individuals (feature extraction)

The next step was to determine the head, tail, and center point. The program extracted the feature points of the earthworm body by first identifying the head and tail using the different features between the two points (tail points always move slower than the other points), then the center point was determined. To initialize the program, the tail point was determined manually. By doing this, the program automatically calculated four-neighbor distances of the center line from the tail point and extracted the point at the furthest distance. The program defined this as the head point of the earthworm. Then, the program recognized the halfway point between the head and tail as the central point of the body. Using this process, the program could obtain the initial feature points of the earthworm (i.e. head, tail, and center points). From the second frame, the program calculated new feature points using the previous ones. First, the program extracted new tail points based on the fact that the tail points of the earthworm move slower than the other points. Using this feature, the program recognized the edge points as new tail points with the shortest distance from the previous points. Both head and center points were obtained by the same procedure, which were extracted from the initial head and

Table 1. Mean, maximum, and minimum values of automatically and manually calculated velocities of each point (head and tail) of two earthworm species (*E. japonica* and *M. hilgendorfi*).

Species	*velocity*	mean (mm sec^{-1})	max. (mm sec^{-1})	mini. (mm sec^{-1})
E. japonica				
Program	head	4.75±4.36	21.37	0
	tail	3.44±4.38	23.09	0
manual	head	4.81±4.29	21.45	0.21
	tail	3.55±4.44	24.69	0
M. hilgendorfi				
program	head	4.38±4.13	29.71	0.21
	tail	1.03±1.92	13.9	0
manual	head	5.09±4.43	28.29	0
	tail	0.92±2.21	14.58	0

The errors indicate standard deviations over time.

center points. These processes were carried out on all of the images. The program outputs positional data in each frame to a CSV file, allowing trajectories over time to be visualized.

Statistical Analysis

All statistical analyses were performed using SPSS 10.05 (SPSS Inc., Chicago, IL, USA). Relationships between two variables (program output and manual output) were assessed using standard bivariate correlation procedures. The correlation coefficient is shown in the text below and the figures.

Results

Evaluations of automatic output using the manually tracked values

Tracking and extraction of each point. First, we were successful in tracking the three target points, (the head, center, and tail of the body) on a time scale of seconds, continuously followed by body length calculations using the newly developed program. We used a part of the obtained data for the evaluation of the program which shows typical movement of two species of earthworms every 150 seconds and 300 seconds for *E. japonica* and *M. hilgendorfi* respectively (Figs. 4–11). The tracking of the head and tail with the automatic and manual programs over time are shown in Figs. 4 and 6 respectively. All of the regression lines between the automatically and manually tracked coordinates were close to 1 (>0.98), especially in the case of *E. japonica* (Fig. 5). In contrast, some outliers were observed in *M. hilgendorfi* as indicated by the arrows in Fig. 7. There were no significantly different measurements observed in *E. japonica* between different body parts; however, significant differences were observed between the different body parts of *M. hilgendorfi*. In general, the errors in measurement of the head point ($r = 0.981$, $r = 0.984$, X and Y coordinates, respectively) were larger than those for the tail point ($r = 0.996$, $r = 0.996$, X and Y coordinates, respectively). One reason for this was that the head moved more frequently while the tail stayed in the same position. Since the program tracks and determines the feature points by predicting the next steps, the program may fail to track the objects if the objects (in this case, the heads) move more broadly in the image-capturing interval (1 second). In the case of *M. hilgendorfi*, when they looped, coiled, or crossed their bodies, the program faced difficulties in recognizing the edges, and failed to detect the head positions.

Table 2. Mean maximum values of differences between automatically and manually calculated velocities of each point (head and tail) of two earthworm species (*E. japonica* and *M. hilgendorfi*).

Species	mean (mm sec^{-1})
E. japonica	
head	0.07±0.92
tail	0.11±0.88
M. hilgendorfi	
head	0.71±2.49
tail	0.11±0.80

The errors indicate standard deviations over time.

Calculations of velocity of each point and body length. Using their coordinates, the velocity of each point was calculated. The output data obtained with the automated program were further confirmed by comparing them with the manually acquired data (Figs. 9 and 10). The accuracy of each output was satisfactory for the application to quantify earthworm activity over time. Some outliers in head velocity in *M. hilgendorfi* were due to the misestimation of head points by the automated program, in which case the head was attached to a part of the body. Otherwise, the automated program satisfactorily detected the peaks in terms of timing and absolute values. As shown in Table1, the mean velocities of the head and tail of *E. japonica* calculated by the automated program were 4.75±4.36 and 3.44±4.38 mm sec^{-1}, respectively, while those from the manually calculated values were 4.81±4.29 and 3.56±4.44 mm sec^{-1}, respectively. The mean velocities of the head and tail of *M. hilgendorfi* calculated by the automated program were 4.38±4.13 and 1.03±1.92 mm sec^{-1}, respectively, while those from the manually calculated values were 5.09±4.43 and 0.92±2.21 mm sec^{-1}, respectively. The differences between the automated and manually calculated values were 0.07±0.92 and 0.11±0.88 mm sec^{-1} for the head and tail of *E. japonica*, respectively and 0.71±2.49 and 0.11±0.80 mm sec^{-1} for

Table 3. Mean, maximum, and minimum values of body length of two earthworm species (*E. japonica* and *M. hilgendorfi*).

body length	**mean (mm)**	**max. (mm)**	**mini.(mm)**
E. japonica	167.02±13.62	190.73	131.44
M. hilgendorfi	199.28±10.05	220.38	145.33

The errors indicate standard deviations over time.

those of *M. hilgendorfi*, respectively (Table2). The difference was largest in the head of *M. hilgendorfi* but the others matched quite well, with the differences being less than 0.2 mm sec^{-1}. The tail points tended to move slower than the head points according to the calculated velocities, and the head point of *M. hilgendorfi* moved faster than that of *E. japonica* by 1.65±7.07 mm sec^{-1}. The body lengths were calculated from the binary data, which were converted into a line (shown in Table3).

The mean lengths of *E. japonica* and *M. hilgendorfi* were 167.02±13.62 and 199.28±10.05 mm, respectively, while the maximum and minimum lengths of *E. japonica* were 190.73 and 131.44 mm, respectively, and those of *M. hilgendorfi* were 220.38 and 145.33 mm, respectively. Therefore, the stretch from minimum to maximum was 59.19 and 75.05 mm for *E. japonica* and *M. hilgendorfi*, respectively. The extent of the stretchiness was in agreement with previously reported values [1], which indicated that the body could reach 1.5 times longer than its default length. Moreover, our program also generated the change in length over time per second, so that we could observe the frequency of the stretchiness for further applications (Fig. 11).

Discussion

To date, there exists no method to process and extract such high time resolution image data (on a time scale of seconds) capturing three feature points along with tracking elastic objects such as earthworms, which move quickly and widely, although the program is still a prototype at this stage. A number of automated systems have been developed to track insects' movements, and our earthworm tracking system might be similar to that of *Caenorhabditis elegans* (one of the model nematodes) [20,29]. Our program can deal with a time scale of seconds, though there remain some problems with the program. The program could not cope with irregular shapes of earthworms, such as coiling or overlapping. This could be achieved by modifying the program in the future using the Knot Theory, or topological concept. The program also failed to follow the feature points when the head moved more frequently than the general movement. Furthermore, in the case of *M. hilgendorfi*, when they coiled or crossed their bodies, the program faced difficulties in recognizing the edges, and failed to detect the head positions. To overcome such problems, one solution would be to obtain images at higher time resolutions so that the program could keep tracking more frequently moving objects without losing the target points because the higher time resolution is closer to continuous data. Increasing the contrast between the background and the objects for image analysis would also achieve more precise results. Furthermore, there would be alternative methods to detect feature points such as differentiating the shapes or brightness of head or tail, which will further improve the program.

The velocity of each point of the different body parts of *E. japonica* was synchronized; in contrast, those of *M. hilgendorfi* were not. In the case of *M. hilgendorfi*, the head and center points were synchronized but the tail point was not; it tended to stay still on the same coordinate. We recognize that the velocities were under non-soil conditions; therefore, earthworms in the soil will be much slower than our calculations indicate. However, the program picked up the features of each species when observed under the same conditions. We consider our newly developed program to be sufficient to measure earthworm activity; this method can be used to compare dynamic changes in these parameters between species or in response to environmental variables such as temperature, light, and moisture, as well as chemical contaminants.

Image data is powerful, direct information that can be used to quantify physical information of changes in earthworm's activities. The intention of this study is to apply the program to quantify the physical parameters to indicate earthworm's activity from their movement speed or body stretchiness. It will be more beneficial if these data are combined with other data, such as data on gas emissions from the body, enzyme activities using molecular tools, or pathway of elements using isotopic data for mechanistic understanding. The method used in this study is artificial; a small Petri dish without soil as used. However, this study will advance our understanding of the basic behavior of earthworms depending on the alteration of their living environment. This basic behavior can be incorporated in models in terms of material flow belowground or reactions in response to chemical contaminant in soil [5].

Combining different kinds of studies will help to unravel or predict the complex processes in belowground systems [6]. Furthermore, although this system could not be applied to observe their behavior directly in the soil, culturing earthworms in transparent gels or agar medium would be a useful method to observe them three-dimensionally. Furthermore, applying other rapidly developing state of the art technologies, such as MRI or radar, to observe earthworms under soil conditions, it will be theoretically possible to directly observe their behavior in soil, though the time resolution will not be high enough. In fact, radar is already being used for root growth observations in soil.

Conclusions

We demonstrated that our newly developed program successfully tracks the movement of two species of earthworms that have different features in terms of body size, mass, and morphology as well as movement. Based on our data, the program reduced the time required for manual tracking 3-fold and the program could track the movement which is inherent in earthworm behavioral ecology studies. Large continuous image data sets can be immediately processed and the program can determine earthworm body parts along with their changes in elastic body lengths over time. In the future, quantified earthworm behaviors could be incorporated in material flow dynamic models in soil or in models of the response of earthworms to chemical inputs to soil.

The software "MimizuTrack" will be released on INCF Software Center (http://software.incf.org/) under the BSD license. Otherwise the program is open upon request to the following contact person; Assis. Prof. Toshifumi Kimura (email: kimura@shse.u-hyogo.ac.jp).

Acknowledgments

We would like to acknowledge Mrs. Sugawara for her cheerful support in the setup of the measurement systems and taking care of the earthworms.

Author Contributions

Conceived and designed the experiments: NK SY SK. Performed the experiments: NK SY SK. Analyzed the data: NK TK. Contributed reagents/materials/analysis tools: NK TK SY SK MO HI. Wrote the paper: NK TK SY SK MO HI.

References

1. Edwards CA, Bohlen PJ (1996) Biology and ecology of earthworms: Chapman & Hall.
2. Lavelle P, Spain AV (2001) Soil ecology: Kluwer Academic Pub.
3. Darwin C (1892) The formation of vegetable mould, through the action of worms, with observations on their habits: J. Murray.
4. Lubbers IM, van Groenigen KJ, Fonte SJ, Six J, Brussaard L, et al. (2013) Greenhouse-gas emissions from soils increased by earthworms. Nature Climate Change 3 187–194.
5. Bartlett MD, Briones MJI, Neilson R, Schmidt O, Spurgeon D, et al. (2010) A critical review of current methods in earthworm ecology: from individuals to populations. European Journal of Soil Biology 46: 67–73.
6. Curry JP, Schmidt O (2007) The feeding ecology of earthworms - A review. Pedobiologia 50: 463–477.
7. Helling B, Reinecke SA, Reinecke AJ (2000) Effects of the Fungicide Copper Oxychloride on the Growth and Reproduction of *Eisenia fetida* (Oligochaeta). Ecotoxicology and Environmental Safety 46: 108–116.
8. Neuhauser EF, Callahan CA (1990) Growth and reproduction of the earthworm *Eisenia fetida* exposed to sublethal concentrations of organic chemicals. Soil Biology and Biochemistry 22: 175–179.
9. Anderson GL, Boyd WA, Williams P (2001) Assessment of sublethal endpoints for toxicity testing with the nematode Caenorhabditis elegans. Environmental Toxicology and Chemistry 20: 833–838.
10. Yasmin S, D'Souza D (2010) Effects of pesticides on the growth and reproduction of earthworm: a review. Applied and Environmental Soil Science.
11. Feldman A, Balch T (2003) Automatic identification of bee movement. pp. 53–59.
12. Kimura T, Ohashi M, Okada R, Ikeno H (2011) A new approach for the simultaneous tracking of multiple honeybees for analysis of hive behavior. Apidologie 42: 607–617.
13. Mersch DP, Crespi A, Keller L (2013) Tracking Individuals Shows Spatial Fidelity Is a Key Regulator of Ant Social Organization. Science 340: 1090–1093.
14. Rizvandi NB, Pizurica A, Rooms F, Philips W (2008) Skeleton analysis of population images for detection of isolated and overlapped nematode C. elegans.
15. Straw AD, Branson K, Neumann TR, Dickinson MH (2011) Multi-camera real-time three-dimensional tracking of multiple flying animals. Journal of The Royal Society Interface 8: 395–409.
16. Colomb J, Reiter L, Blaszkiewicz J, Wessnitzer J, Brembs B (2012) Open Source Tracking and Analysis of Adult Drosophila Locomotion in Buridan's Paradigm with and without Visual Targets. PLoS ONE 7: e42247.
17. Kimura T, Ohashi M, Crailsheim K, Schmickl T, Okada R, et al. (2014) Development of a New Method to Track Multiple Honey Bees with Complex Behaviors on a Flat Laboratory Arena PLoS ONE.
18. Ligthart TN, Peek GJWC, Taber EJ (1993) A method for the three-dimensional mapping of earthworm burrow systems. Geoderma 57: 129–141.
19. Joschko M, Mueller PC, Kotzke K, Doehring W, Larink O (1993) Earthworm burrow system development assessed by means of X-ray computed tomography. Geoderma 56: 209–221.
20. Mathew MD, Mathew ND, Ebert PR (2012) WormScan: A technique for high-throughput phenotypic analysis of Caenorhabditis elegans. PLoS One 7: e33483.
21. Michaelsen W (1891) Oligochaeten des Naturhistorischen Museums in Hamburg. I [-IV].
22. Yamaguchi E (1965) Allolobophora japonica. In: Okada K editor Illustrated encyclopedia of the fauna of Japan 1 Hokuryukan co ltd: 547.
23. Ishizuka K (2001) Taxonomic Study of the Genus Pheretima s. lat (Oligochaeta, Megascolecidae) from Japan. Bulletin of Seikei University 33: 1–125.
24. Blakemore RJ, Ito MT, Kaneko N (2006) Alien earthworms in the Asia/Pacific region with a checklist of species and the first records of Eukerria saltensis (Oligochaeta: Ocnerodrilidae) and Eiseniella tetraedra (Lumbricidae) from Japan, and Pontoscolex corethrurus (Glossoscolecidae) from Okinawa. Assessment and Control of Biological Invasion Risks: 173–181.
25. Uchida T, Kaneko N, Ito MT, Futagami K, Sasaki T, et al. (2004) Analysis of the feeding ecology of earthworms (Megascolecidae) in Japanese forests using gut content fractionation and 朩エ 15N and 朩エ 13C stable isotope natural abundances. Applied Soil Ecology 27: 153–163.
26. Kimura T, Ohashi M, Crailsheim K, Schmickl T, Odaka R, et al. Tracking of Multiple Honey Bees on a Flat Surface; 2012. IEEE. pp. 36–39.
27. Nakano A, Ikeno H, Kimura T, Sakamoto H, Dannoura M, et al. (2012) Automated analysis of fine-root dynamics using a series of digital images. Journal of Plant Nutrition and Soil Science 175: 775–783.
28. Hilditch CJ (1969) Linear skeleton from square cupboards. Machine intelligence 6: 403–733.
29. Geng W, Cosman P, Berry CC, Feng Z, Schafer WR (2004) Automatic tracking, feature extraction and classification of C. elegans phenotypes. Biomedical Engineering, IEEE Transactions on 51: 1811–1820.

2

Assessment of Bacterial *bph* Gene in Amazonian Dark Earth and Their Adjacent Soils

Maria Julia de Lima Brossi[1]*, Lucas William Mendes[1], Mariana Gomes Germano[2], Amanda Barbosa Lima[1], Siu Mui Tsai[1]

1 Cellular and Molecular Biology Laboratory, Center for Nuclear Energy in Agriculture, University of São Paulo, Piracicaba, SP, Brazil, **2** Brazilian Agricultural Research Corporation, Embrapa Soybean, Londrina, PR, Brazil

Abstract

Amazonian Anthrosols are known to harbour distinct and highly diverse microbial communities. As most of the current assessments of these communities are based on taxonomic profiles, the functional gene structure of these communities, such as those responsible for key steps in the carbon cycle, mostly remain elusive. To gain insights into the diversity of catabolic genes involved in the degradation of hydrocarbons in anthropogenic horizons, we analysed the bacterial *bph* gene community structure, composition and abundance using T-RFLP, 454-pyrosequencing and quantitative PCR essays, respectively. Soil samples were collected in two Brazilian Amazon Dark Earth (ADE) sites and at their corresponding non-anthropogenic adjacent soils (ADJ), under two different land use systems, secondary forest (SF) and manioc cultivation (M). Redundancy analysis of T-RFLP data revealed differences in *bph* gene structure according to both soil type and land use. Chemical properties of ADE soils, such as high organic carbon and organic matter, as well as effective cation exchange capacity and pH, were significantly correlated with the structure of *bph* communities. Also, the taxonomic affiliation of *bph* gene sequences revealed the segregation of community composition according to the soil type. Sequences at ADE sites were mostly affiliated to aromatic hydrocarbon degraders belonging to the genera *Streptomyces*, *Sphingomonas*, *Rhodococcus*, *Mycobacterium*, *Conexibacter* and *Burkholderia*. In both land use sites, shannon's diversity indices based on the *bph* gene data were higher in ADE than ADJ soils. Collectively, our findings provide evidence that specific properties in ADE soils shape the structure and composition of *bph* communities. These results provide a basis for further investigations focusing on the bio-exploration of novel enzymes with potential use in the biotechnology/biodegradation industry.

Editor: Niyaz Ahmed, University of Hyderabad, India

Funding: Funding provided by 'Conselho Nacional de Desenvolvimento Científico e Tecnológico' and the 'Fundação de Amparo à Pesquisa do Estado de São Paulo' (FAPESP/Biota 2011/50914-3). The funders had no role in study design, data collection and analysis, decision to publish, or preparation of the manuscript.

Competing Interests: The authors have declared that no competing interests exist.

* E-mail: majubrossi@gmail.com

Introduction

Amazonian Dark Earth (ADE), locally termed *'Terra Preta de Índio'*, are anthropogenic soil horizons built-up by the Pre-Colombian Indians between 500 and 8,700 years ago. These soil sites were formed by the progressive deposit of materials and organic compounds, such as charcoal, bone, and pottery sheds, which gradually shifted the natural physical and chemical properties of the soil. As a result, relatively infertile Amazon soils were progressively converted into highly fertile spots through processes like increasing the cation exchange capacity and the nutrient content, as well as promoting the stabilization of the soil physical structure [1,2]. Substantial increments of organic material in these sites gradually increased the carbon content, yielding to the formation of soil spots with a high proportion of incompletely combusted biomass (biochar). These spots have been reported to reach up to a 70-fold higher amount of carbon than native soils at adjacent locations (ADJ) [3].

The existence of ADE sites close to their natural ADJ soil locations, which present the same geological history, provides a unique opportunity to investigate the role of biotic and abiotic factors influencing the microbial community assembly and dynamics at these sites. Previous studies revealed that ADE and ADJ sites present differences in microbial community composition, and bacterial diversity has been reported to be higher at ADE sites [4,5,6,7,8]. Most of these studies rely on comparisons between the taxonomic profiles of these communities (i.e., based on the taxonomic bacterial 16S rRNA gene). In this sense, the extent to which the local environment shapes the functional profiles of these communities, and influences their performance, remains mostly elusive [9,10,11,12,13].

Soil is one of the most biodiverse ecosystems on Earth, being able to support communities from multiple trophic levels, which are constantly performing the metabolism of diverse and complex substrates. The extreme spatial and temporal heterogeneity of the soil matrix, paired with the myriad of internal and external feedbacks, are known to determine the structure and function of these communities [14]. Microbes are involved in many ecosystem processes, including biodegradation, decomposition and mineralization, inorganic nutrient cycling, disease causation and suppression, and pollutant removal. Soil disturbances are known to cause shifts in microbial activities, shifting the rate of these processes and triggering impacts on the ecosystem performance [15]. Several environmental factors are known to affect microbial community composition in the soil, including soil temperature, moisture, texture, carbon content, nutrient availability, pH, land use history, seasonality, and the content of incompletely combusted biomass, such as biochar [13,16,17,18,19].

The biochar content is the major physical distinction between ADE and their ADJ soils, which is also known to play an important role in global carbon biogeochemistry [20]. In anthrosols, such as ADE, it is predicted that distinct microbial communities can perform unique processes, such as the retention of high-labile carbon [5]. Despite the unique and specialized capabilities of these soils, functional assessments of their microbial community, particularly those of genes encoding important steps in the carbon cycle, are still scarce. Biodegradation through bacterial activity is one of the most important processes occurring in soils regarding organic matter recycling. This process involves genes acting on key steps in the carbon cycle, for the turnover of more recalcitrant organic carbon, as well as for pollutant degradation in the ecosystem [21]. This process, along with biosynthesis, largely governs the carbon cycle in the environment, which is dependent on microbial enzymatic activities that most often use organic compounds as a primary energy source [22].

The primary step involved in the aerobic microbial degradation of aromatic hydrocarbons is an oxidative attack [23,24], where enzymes named oxygenases are responsible for the insertion of molecular oxygen into aromatic benzene rings [25]. Genes encoding these enzymes have been characterized in *Rhodococcus*, *Acinetobacter*, *Pseudomonas*, *Mycobacterium*, *Burkhoderia*, *inter alia* [25,26]. The α-subunit of oxigenases is known as the catalytic domain involved in the transfer of electrons to oxygen molecules. Due its DNA sequence conservation, this subunit has been currently used as a target gene for the detection of such enzymes in complex communities [21,27,28,29].

In this study, we evaluated the structure, composition and abundance of the bacterial catabolic gene Biphenyl Dioxygenase (*bph*) involved in aromatic hydrocarbon degradation in Amazonian Dark Earth and their adjacent soil locations. We aimed to determine the role of anthropogenic action in the diversity of the *bph* gene in soil bacterial communities. Understanding the diversity of specific bacterial genes in ADE should led to future studies that investigate the microbial ecology of anthropogenic altered soils, especially in regards to their potential source of novel enzymes. Collectively, this study characterized this catabolic gene occurring in Amazonian Dark Earth, and compared the profiles with those obtained from adjacent sites, as well as under distinct land use systems.

Materials and Methods

Ethic statement

No specific permits were required for the described field studies. The locations are not protected. The field studies did not involve endangered or protected species.

Study sites, sample collection and soil chemical analyses

Studied sites are located at the Caldeirão Experimental Station of Amazon Brazilian Agricultural Research Corporation (Embrapa) in Iranduba County in the Brazilian Central Amazon (03°26′00" S, 60°23′00" W). A detailed description of the soil sampling locations is given by Taketani et al [8]. Briefly, the four sites sampled are composed of two Amazonian Dark Earth (ADE) and their two correspondent adjacent soils (Haplic Acrisol, ADJ). These sites are under a ~35 year-old secondary forest (SF) or under manioc (*Manihot esculenta*) cultivation (M). Hereafter these sites are termed as ADE-SF, ADJ-SF, ADE-M and ADJ-M. Soil samples were taken in triplicate from the topsoil layer (ca. top 10 cm), and the overlaying litter was discarded. Each sample contained approximately 300 g of soil and was transported to the laboratory at 4°C to further processing (<24 h). A portion of the samples were frozen (−20°C) for total DNA extraction while the other portion was kept at 4°C for chemical measurements.

Chemical analyses were performed at Amazon Embrapa (Manaus, Brazil), according to instructions provided by the Embrapa protocol [30]. Briefly, soil samples were analysed in triplicate for pH (H_2O, 1:2.5); H+Al (calcium extractor 0.5 mol L^{-1}, pH 7.0); sum of bases (SB); soil organic matter (SOM); soil organic carbon (SOC; Walkely-Black method); extractable fraction of Al, Ca, and Mg (1 M KCl); extractable fraction of P and K (double acid solution of 0.025 M sulphuric acid and 0.05 M hydrochloric acid Mehlich 1); and effective cation exchange capacity (eCEC).

Total soil DNA extraction and PCR amplifications for T-RFLP

Total DNA was extracted using 0.25 g of soil as an initial material. Extractions were carried out in triplicate for each site, using the PowerSoil DNA isolation kit (MoBio Laboratories, Carlsbad, CA, USA), according to the manufacturer's protocol. DNA quality and quantity were measured spectrophotometrically using NanoDrop 1000 (Thermo Scientific, Waltham, EUA).

For T-RFLP analyses the bacterial 16S rRNA gene was amplified with the primer set 27F - FAM labelled (5′ AGA GTT TGA TCC TGG CTC AG 3′) and 1492r (5′ ACC TTG TTA CGA CTT 3′) [31]. The *bph* gene was amplified with the primer set BPHD F1 – FAM labelled (5′ TAY ATG GGB GAR GAY CCI GT 3′) and BPHD R0 (5′ ACC CAG TTY TCI CCR TCG TC 3′) [21]. For 16S rRNA gene amplification, PCR reactions were carried out in a volume of 25 µL containing 2.5 µL reaction buffer 10× (Invitrogen, Carslbad, CA, USA), 1.5 µL $MgCl_2$ (50 mM), 1 µL of each primer (5 pmol $µL^{-1}$), 0.2 µL (5 U) of Platinum Taq DNA polymerase (Invitrogen), 0.5 mL of deoxyribonucleotide triphosphate mixture (2.5 mM), 0.25 µL of bovine serum albumin (1 ng mL^{-1}), 1 µL of DNA template (ca. 10 ng) and 18.05 µL of sterilized ultrapure water. Amplifications were performed in the GeneAmp PCR System 9700 thermal cycler (Applied Biosystems, Foster City, CA, USA). Reaction conditions were 94°C for 3 min, followed by 35 cycles of 94°C for 30 s, 59°C for 45 s, and 72°C for 1 min with a final extension step at 72°C for 15 min. For the *bph* gene amplification, PCR reactions were carried out in a volume of 25 µL containing 2.5 µL reaction buffer 10× (Invitrogen, Carslbad, CA, USA), 1.5 µL $MgCl_2$ (50 mM), 1.25 µL of each primer (5 pmol $µL^{-1}$), 0.5 µL (5 U) of Platinum Taq DNA polymerase (Invitrogen), 0.3 µL of deoxyribonucleotide triphosphate mixture (2.5 mM), 1 µL of DNA template (ca. 10 ng) and 16.7 µL of sterilized ultrapure water. For *bph* gene, similar PCR cycling conditions were used, except the annealing temperature that was set at 60°C. Negative PCR controls (without DNA template) and positive controls (using the *Escherichia coli* ATCC 25922 DNA for the 16S rRNA gene and the DSM 6899 *Pseudomonas putida* DNA for *bph*) were run in parallel for both amplifications. After the amplifications, 5 µL of obtained products (ca. 60 ng) was digested with the endonuclease *HhaI* (Invitrogen) in 15 µL reaction for 3 h at 37°C. Obtained fragments were further purified using sodium acetate/EDTA precipitation and then mixed with 0.25 µL of the Genescan 500 ROX size standard (Applied Biosystems) and 9.75 µL of deionized formamide. Prior to fragment analysis, samples were denatured at 95°C for 5 min and chilled on ice. Analysis of terminal restriction fragment (T-RF) sizes and quantities was performed on an ABI PRISM 3100 genetic analyzer (Applied Biosystems).

T-RFLP profiles were analysed using PeakScanner v1.0 software (Applied Biosystems, Foster City, CA, USA). Terminal restriction fragments (T-RFs) of less than 25 bp were excluded prior to the analysis. The total values of T-RFs for each soil sample were pulled together to construct a Venn's diagram showing shared T-RFs among samples. The relative abundance of a single T-RF was calculated as percent fluorescence intensity relative to total fluorescence intensity of the peaks [32]. Data from individual samples were combined to soil chemical parameters and subjected to multivariate analysis using Canoco 4.5 (Biometris, Wageningen,The Netherlands) and Primer6 (PrimerE, Ivybridge, United Kingdom). All matrices were initially analysed using de-trended correspondence analysis (DCA) to evaluate the length of the gradient of the species distribution; this analysis indicated linearly distributed data (length of gradient <3), revealing that the best-fit mathematical model for the data was the redundancy analysis (RDA). Forward selection (FS) and the Monte Carlo permutation test were applied with 1,000 random permutations to verify the significance of soil chemical properties upon the microbial community. In addition to P values for the significance of each soil chemical property, RDA and Monte Carlo permutation test supplied information about the marginal effects of environmental variables, quantifying the amount of variance explained by each factor. We used ANOSIM based on relative abundance of T-RFs to test for statistical differences between samples.

454-Pyrosequencing analyses of the bacterial *bph* gene

A partial region of the *bph* gene was amplified for the 454-pyrosequencing using the primer set BPHD F3 (5′ ACT GGA ART TYG CIG CVG A 3′) and BPHD R0 (5′ ACC CAG TTY TCI CCR TCG TC 3′) [20] containing specific Roche 454-pyrosequencing adaptors and barcodes of 8 bp. The expected fragment size was ca. 520 bp. Three independent amplifications were performed for each sample. The 20 μL PCR mixture contained 1× FastStart High Fidelity Reaction Buffer (Roche Diagnostics, Basel, Switzerland), 1.25 mM of each primer, 150 ng mL^{-1} of bovine serum albumin (New England BioLabs, Ipswich, MA, USA), 0.2 mM of dNTPs, 0.5 mL (2.5 U) of FastStart High Fidelity PCR System Enzyme Blend (Roche Diagnostics) and 4 ng of template DNA. The PCR conditions were optimized using the genomic DNA of *Burkholderia xenovorans* LB400 [33], which carries one of the target dioxygenase genes. Amplifications were performed as follows: 95°C for 3 min, 30 cycles of 95°C for 45 s, 60°C for 45 s and 72°C for 40 s, with the final extension of 72°C for 4 min. Triplicate PCR products containing the expected fragment size were purified using the QIAquick Gel Extraction Kit (Qiagen, Hilden, Germany) and QIAquick PCR Purification Kit (Qiagen). DNA concentrations were determined using the NanoDrop ND-1000 spectrophotometer (NanoDrop Technologies, Wilmington, DE, USA). Purified PCR products were pooled and subjected to pyrosequencing using the FLX sequencing system (454 Life Sciences, Branford, CT, USA).

Raw data was filtered for valid sequences using the FunGene Pipeline Repository (http://fungene.cme.msu.edu/FunGenePipeline/). Quality sequences were translated in the correct frame of aminoacids using the RDP FrameBot tool. The RDP pipeline extracted a set of representative sequences from known *bph* sequences to use as subject sequences with FrameBot. The FrameBot produces an optimal alignment between the query and the subject sequences in the presence of frameshifts. Only the protein pairwise alignment with the best score was reported. Protein sequences passing FrameBot were aligned with HMMER using a model trained on the same set of representative sequences used by FrameBot. The aligned protein sequences were chopped at position 351 of the reference sequence of *Pseudomonas putida* F1 (YP_001268196). The total valid sequences were rarefied to the smallest number of sequences per sample in order to minimize effects of sampling effort upon analysis.

Distance matrices were constructed using the MOTHUR software [34]. The resulting matrices were used to estimate the number of operational protein families (OPF) (i.e. group of proteins that share a common evolutionary origin) and to estimate richness (i.e. Chao1, Jackknife, and ACE indices) and diversity (i.e. Shannon and Simpson indices). Rarefaction curves were constructed at a cutoff level of 94% of amino acid identity. MOTHUR was also used to perform ∫-Libshuff comparisons between the four studied sites. The Good's coverage estimator was used to calculate the sample coverage using the formula $C = 1-(n_i/N)$, where N is the total number of sequences analysed and n_i is the number of reads that occurs only once among the total number of reads analysed at a cutoff value of 94% of amino acid identity [35]. Unweighted UniFrac distances among communities were estimated using a tree constructed *de novo* using FastTree. One representative sequence per OPF was selected and subjected to taxonomic affiliation by the comparison tool of NCBI Tblastx (GenBank) using Blast2Go [36].

Sequence data generated by 454-pyrosequencing are available at the MG-RAST server (http://metagenomics.anl.gov) under the project 'Diversity of bhp gene in Amazon soils' (ID 8489) and accession numbers 4557319.3 (ADE_SF), 4557318.3 (ADE_M), 4557321.3 (ADJ_SF) and 4557320.3 (ADJ_M).

Quantitative PCR (qPCR) of the bacterial 16S rRNA and *bph* genes

The bacterial 16S rRNA gene was amplified with the primer set U968F (5' AAC GCG AAG AAC CTT AC 3 ') and R1387 (5' CGG TGT GTA CAA GGC CCG GGA ACG 3 ') [37], which amplify a fragment of approximately 400 bp. The 520 bp fragment of the *bph* gene was amplified with the primer set BPHD F3 (5′ ACT GGA ART TYG CIG CVG A 3′) and BPHD R0 (5′ ACC CAG TTY TCI CCR TCG TC 3′) [21]. qPCR reactions were performed in 10 μL containing 5 μL of SYBR green PCR master mix (Fermentas, Brazil), 1 μL of each primer (5 pmol μL^{-1}), 1 μL of DNA template (ca. 10 ng) and 2 μL of sterilized ultrapure water. Thermocycling conditions for the 16S rRNA gene were set as follows: 94°C for 10 min; 40 cycles of 94°C for 30 s, 56°C for 30 s and 72°C for 40 s. Amplification specificity was checked by a melting curve and the data collection was performed at every 0.7°C. qPCR reactions for the *bph* gene were performed at similar conditions, except for the annealing temperature set at 60°C. Reactions were performed in a StepOnePlus system (Applied Biosytems). The Cts values (cycle threshold) were used as standers for determining the amount of DNA template in each sample. Standard curves were produced for the 16S rRNA and *bph* genes using specific cloned fragments. Gene fragments were quantified in a spectrophotometer (190 a 840 nm - NanoDrop ND-1000) and diluted (10^7 to 10^3 genes μL^{-1} for the 16S rRNA gene and 10^6 to 10^2 genes μL^{-1} for *bph*) to generate each specific standard curves. The gene copy numbers in different soil samples were expressed as log copy numbers of the gene per gram of soil. Statistical comparisons were performed using one-way ANOVA (Tukey's test).

Results

Variation in soil chemical properties

Soil chemical properties were measured for each individual sample collected in ADE and ADJ sites (for a detailed description see Table S1). Statistical differences were observed using Tukey's test. Overall, soil chemical properties of ADE-SF were chemically similar to ADE-M. Likewise, ADJ-SF chemical properties were also very similar to ADJ-M. As expected, major differences were attributed mostly to soil type rather than the land use history.

Higher soil pH values were observed in ADE rather than ADJ sites. While ADJ soils were very acidic with a pH of 3.53 (ADJ-SF) and 3.74 (ADJ-M), ADE sites were only weakly acidic with a pH of 5.51 (ADE-SF) and 5.41 (ADE-M). Sites at ADE showed lower total and exchangeable Al (H+Al), a phenomenon that is likely directly connected to observed variations in soil pH.

Soil organic carbon (SOC), soil organic matter (SOM) and effective cation exchange capacity (eCEC) were higher in ADE sites. Different land uses did not influence these properties in ADJ soils; however, the same properties showed significantly higher values in the site under secondary forest rather than in manioc cultivation, in ADE soil locations. In detail, SOC, SOM and eCEC values were approximately 30% higher in the ADE-SF site when compared to the ADE-M.

Assessment of community structures based on the bacterial 16S rRNA and *bph* genes

T-RFLP analyses for the bacterial 16S rRNA and *bph* genes were performed for the four sites. The obtained profiles were used to determine the richness of terminal restriction fragments (T-RFs) and to perform the multivariate analyses. A total of 152, 144, 147 and 141 T-RFs were obtained for the analysis of the bacterial 16S rRNA gene in ADE-SF, ADE-M, ADJ-SF and ADJ-M sites, respectively. There were 14 T-RFs detected as dominant throughout all sites, accounting for >50% of the total fluorescence detected for the 16S rRNA gene analyses. Tukey's test ($P>0.05$) indicated no difference between sites in the richness of T-RFs for the obtained profiles of bacterial 16S rRNA gene.

For the *bph* gene analysis no dominant T-RFs were found, possibly due to the high heterogeneity of this gene. There were 90, 78, 73 and 69 T-RFs in ADE-SF, ADE-M, ADJ-SF and ADJ-M sites, respectively. Samples from ADE sites showed statistically higher richness of T-RFs (Tukey's test) than the observed at ADJ sites ($P>0.05$).

The Venn's diagram according to soil type showed that ADE and ADJ soils shared more common T-RFs for the bacterial 16S rRNA rather than for *bph* gene. Also, the number of unique T-RFs was higher in ADE sites for both assessed genes (Figure 1a). Conversely, Venn's diagram combining the four sites showed a core containing 122 T-RFs for the bacterial 16S rRNA gene, while the distribution of T-RFs for the *bph* gene was more site specific, and only 6 T-RFs comprised a common core (Figure 1b).

Clustering analysis of T-RFLP data for the *bph* gene segregated samples according to soil type and land use (Figure 2). This analysis revealed the formation of two main clusters: the first cluster (a) included samples from ADE soils (SF and M) and the second (b) samples from ADJ soils (SF and M). This analysis also revealed that ADE and ADJ sites segregated at 12% of similarity. Concerning the ADE sites, land use systems separated different land uses at 27% of similarity. Conversely, for ADJ sites, land use systems differed at 20% similarity (Figure 2).

Redundancy analysis based on T-RFLP data explained 78.1% of the variation in the first two axes, thus confirming the segregation of sites primary according to soil type, and further in relation to different land use types (Figure 3). Replicates within each soil site were very consistent, evidenced by the formation of concise clusters. We also observed that different soil types also correlated differently to measured chemical parameters. More precisely, the *bph* community structure from ADE-SF correlated mostly with pH, eCEC, SOM and SOC, while sites at ADJ-FS presented a significant correlation to H+Al.

ANOSIM analysis indicated statistical differences between the two soil types and land use systems (Table 1). R-values revealed a clear segregation of *bph* gene structures (R>0.75) in ADJ soils, while the 16S rRNA gene differed to a lower extent across sites (R<0.2).

Diversity of the bacterial *bph* gene across soil types and land uses

To access the composition of *bph* gene, samples were sequenced using 454-pyrosequencing. A total of 7,710 reads matched the barcodes, of which 6,877 reads passed the initial filtering (89.2%) and 5,965 (86.7%) were effectively translated into amino acid sequences using the FrameBot. A total of 4,690 valid amino acid sequences were further rarified to the depth of 750 sequences per sample (the minimum in a single sample) for comparative analysis.

The diversity indices for the *bph* gene (Table 2) revealed that ADE sites (SF and M) ($H' = 4.24$ and 4.05, $L' = 0.024$ and 0.028, respectively) were more diverse than ADJ sites (SF and M) ($H' = 3.19$ and 3.17, $L' = 0.098$ and 0.107, respectively). These indices also showed that sites under SF were more diverse than sites under M. Richness estimators (i.e. Chao1, ACE and Jackknife) also revealed ADE sites (SF and M) (Chao 1 = 238 and 229, ACE = 248 and 286, Jackknife = 271 and 255, respectively) to present higher values than ADJ (Chao 1 = 151 and 127, ACE = 210 and 166, Jackknife = 170 and 130, respectively), with SF sites being also higher than sites under M.

ADE sites presented a total of 159 and 129 OPFs for SF and M sites, respectively. Conversely, these values for ADJ were lower (95 and 90, for SF and M, respectively). These sites also presented a different number of singletons (number of unique reads per OPF): 66, 57, 43 and 36 for ADE-SF, ADE-M, ADJ-SF and ADJ-M, respectively.

Statistical differences among sites for the composition of the *bph* gene were confirmed by ∫-Libshuff ($P<0.001$). Venn's diagrams highlight the number of shared OPFs among samples (Figure 4). The number of shared OPFs between ADE soils under different land uses systems (SF and M) was 84 (41% of the total OPFs presented in ADE sites). Conversely, the number of shared OPFs between ADJ soils under different land uses was 55 (42% of the total OPFs in ADJ sites). Soils under SF presented higher numbers of unique OPFs for both soils (ADE and ADJ) (75 and 40 unique OPFs for SF sites and 45 and 35 unique OPFs for ADE sites, respectively).

The estimation of Good's coverage revealed higher values for ADJ sites (0.86 for ADJ-SF and 0.87 for ADJ-M) than for ADE sites (0.79 for ADE-SF and 0.80 for ADE-M), suggesting a highest number of unique sequences in ADE sites. Rarefaction curves (Figure S1) indicate that ADE (SF and M sites) presented a more diverse community than ADJ (SF and M). Sites under SF (ADE and ADJ) were also comparatively more diverse than sites under M (ADE and ADJ). For all comparative analysis the sampling effort did not covered the richness of *bph* gene. The exception was observed for samples from ADJ sites, where a trend towards a "plateau" was observed.

The Principal Coordinate Analysis (PCoA) based on Unweighted UniFrac distances revealed distinct patterns in phylogenetic community composition (Figure 5). The first axis explained 59.59% of the data variation, and this axis separated samples according to soil type. The second axis explained 22.63% of the data variation, and this axis segregated samples according to land use system.

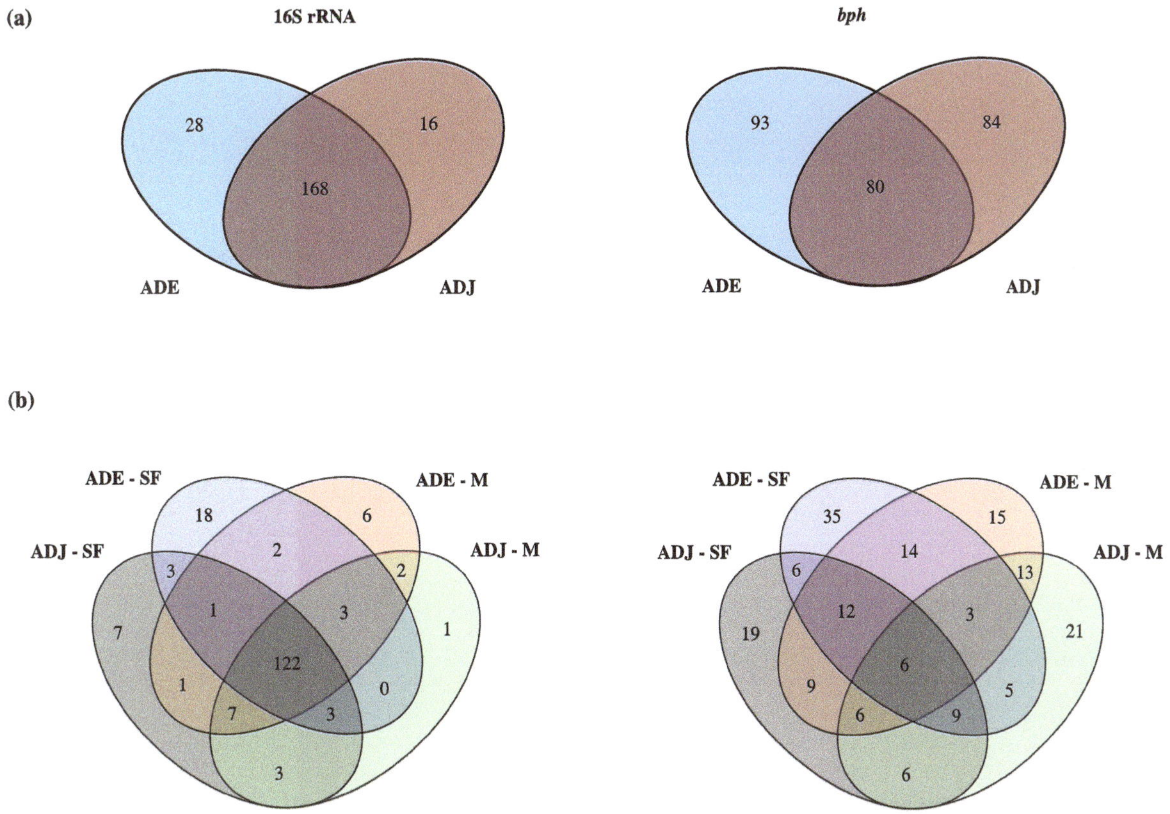

Figure 1. Venn's diagram of T-RFs for 16S rRNA and *bph* genes according to (a) soil types and (b) soil types and land uses. ADE = Amazon Dark Earth; ADJ = Adjacent soils; SF = Secondary Forest; M = Manioc cultivation.

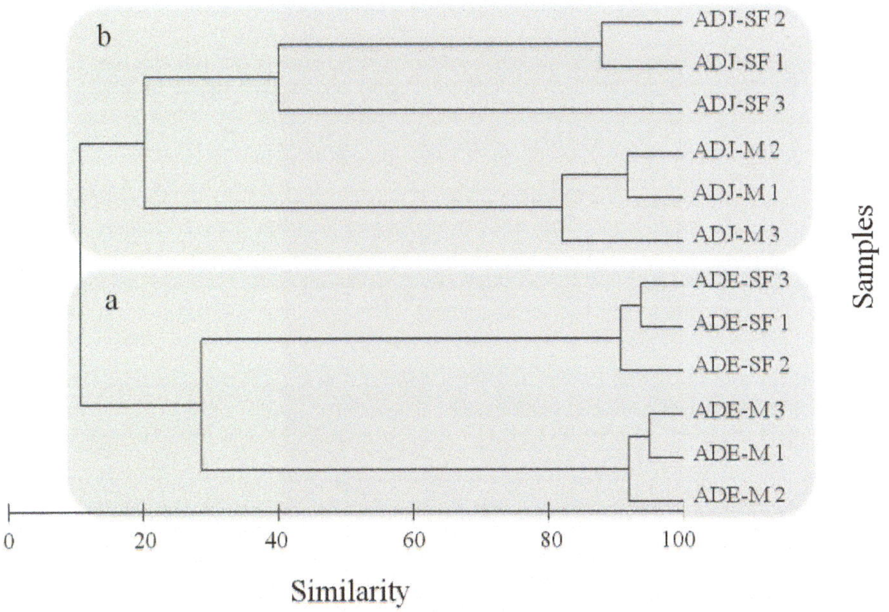

Figure 2. Clustering analysis of T-RFLP data based on Bray-Curtis similarity for the *bph* gene. 'a' and 'b' indicate the segregation patterns according to soil type. ADE = Amazon Dark Earth; ADJ = Adjacent soils; SF = Secondary Forest; M = Manioc cultivation.

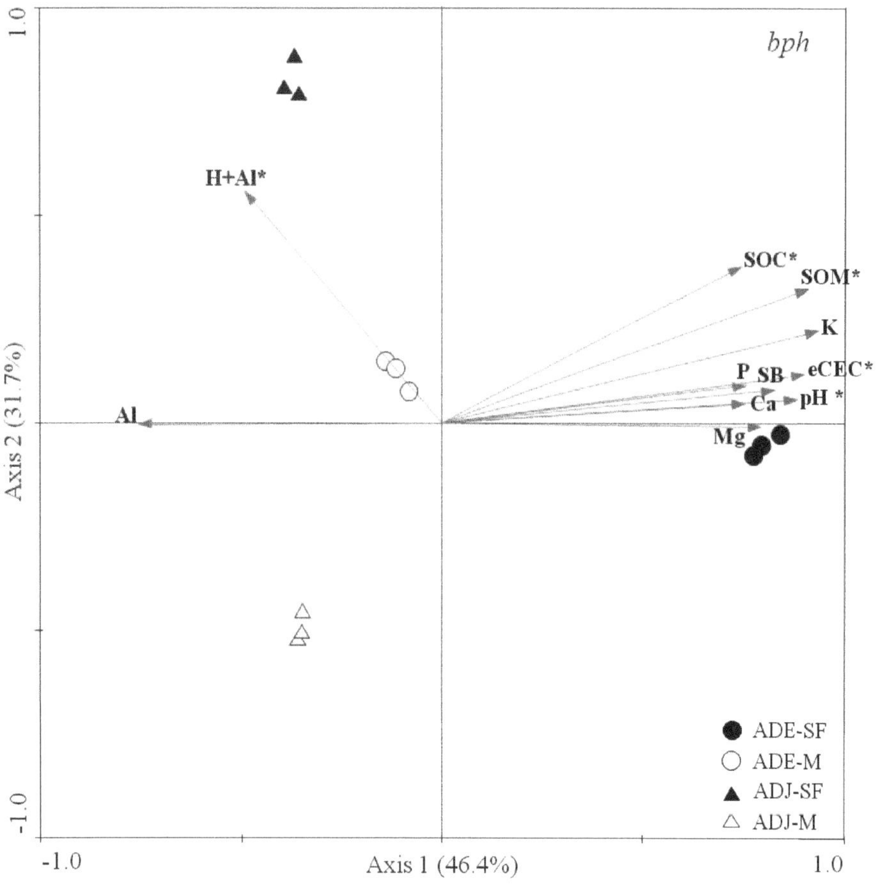

Figure 3. Redundancy analysis (RDA) based on T-RFLP data obtained for the *bph* gene, and soil properties, at the four studied sites. Arrows indicate correlation between the chemical parameters and community structure of samples. The significance of correlations was evaluated via Monte Carlo permutation test and it is indicated as follows: * $p<0.05$. ADE = Amazon Dark Earth; ADJ = Adjacent soils; SF = Secondary Forest; M = Manioc cultivation.

Taxonomic composition of the *bph* gene

Each obtained OPF were further compared to sequences from the GenBank database for taxonomic assignment. All analyzed sequences matched translated proteins described as dioxygenases or putative dioxygenases, with E-values $<10^{-3}$. The most abundant differences in dioxygenases (established as dioxygenases at least ten-fold higher in one site than another) are shown in Figure 6. Sequence matches were associated with aromatic hydrocarbon degradation genes belonging mostly to the genera *Streptomyces*, *Sphingomonas*, *Rhodococcus*, *Mycobacterium*, *Conexibacter* and *Burkholderia*, and uncultured bacterial clones. The taxonomic affiliation of the reads also revealed the predominance of the dioxygenase sequence belonging to unculturable organisms (rdh cf33) in ADE sites, and the predominance of the dioxygenase sequences similar to those previously described at Australian soil (od16) in ADJ sites.

Quantification of the bacterial 16S rRNA and *bph* genes

Variations in bacterial 16S rRNA gene abundances were clear among the evaluated sites. Bacterial 16S rRNA gene ranged from 2.7×10^7 (ADJ-M) up to 9.7×10^7 (ADE-SF) copies per gram of soil. The results indicated that, soil type and land use influenced the

Table 1. ANOSIM test for the bacterial 16S rRNA and *bph* gene based on T-RFLP data of the Amazonian Dark Earth (ADE) and Adjacent soil (ADJ) under secondary forest (SF) and under manioc cultivation (M).

	16S rRNA[a]		*bph*[a]	
	R values	p values	*R* values	p values
ADE × ADJ	0.17	<0.001	1.00	<0.001
ADE-SF × ADE-CULT	0.11	<0.001	1.00	<0.001
ADJ-SF × ADJ-CULT	0.03	<0.001	1.00	<0.001

[a]Samples were compared using T-RF peak height as a measure of abundance.

Table 2. Comparison of diversity indices and richness estimators for the *bph* gene.

Site	OPFs [a]	Richness			Diversity		ESC [b]
		Chao 1	ACE	Jackknife	Shannon (H')	Simpson (L')	
ADE-SF	159	238	248	271	4.24	0.024	0,79
ADE-M	129	229	286	255	4.05	0.028	0,80
ADJ-SF	95	151	210	170	3.19	0.098	0,86
ADJ-M	90	127	166	130	3.17	0.107	0,87

[a]The operational protein family (OPFs), richness estimators (ACE, Chao1 and Jackknife), diversity indices (Shannon and Simpson) and [b]estimated sample coverage were calculated at a cutoff value of 94% of sequence identity. ADE = Amazon Dark Earth soils; ADJ = Adjacent soils; SF = Secondary forest; M = Manioc cultivation.

abundance of bacteria in the analyzed samples, and higher values were found at ADE-SF site (Figure 7a).

The abundance of the *bph* gene ranged from 1.6×10^6 (ADJ-M) to 2.9×10^6 (ADE SF) copies per gram of soil. Conversely to the data obtained for the total bacteria (i.e. 16S rRNA gene data), the abundance of the *bph* gene was also higher at ADE-SF site (Figure 7b).

Discussion

The aim of this study was to assess bacterial hydrocarbon degrading genes in anthropogenic sites from Brazilian Amazon comparatively to their adjacent locations, under two different land use systems. Recent studies have described the high fertility of ADE soils when compared to adjacent soils in the same area (ADJ), mostly because of their increased pH, higher cation exchange capacity, nutrient content and incompletely combusted biomass [3,6,8,28,38]. Although it is well-known that the taxonomic composition of bacterial communities is strongly influenced by pH [39,40], the pH variation in our data indicate that this is also a strong predictor of the composition and diversity of the bacterial *bph* gene. Variations in soil pH, together with eCEC, SOC and SOM, collectively accounted for 78.1% of the total variation explained in the RDA plot based on T-RFLP data. Higher P values in ADE sites are likely to be an effect of pH, seeing that low acidity soils are known to increase the P solubility [28]. Also, the historical formation of ADE sites (constantly amended with bones and vegetation burning activities) is an intrinsic characteristic of this system, which could possibly explain the high content of phosphorus. There is also a direct relationship between soil P content and eCEC values in ADE sites mainly because the P adsorption decreases due to the formation of complex compounds between P and the organic matter present in the upper layer of the soil[41]. The values of eCEC in our samples were mostly correlated to the organic matter concentration, which was 2-fold higher in ADE than ADJ sites. These findings are likely explained by the higher amount of biochar found in these anthropogenic sites. Biochar is known for retaining soil nutrients due to its specific surface and negative charge density per unit of surface area [20,42,43]. In short, the collective chemical properties intrinsic of ADE sites may play an important role in their high levels of nutrient availability and, ultimately, their fertility. In this context, we hypothesize that such fertility has caused the ADE soils to harbor a higher microbial diversity and functionality than their adjacent soils. Agricultural practices can also alter soil properties, mostly by interfering on the biochemistry of the organic matter available in the system [44]. Our data revealed such an influence in the measurements of SOC, SOM and eCEC, which were significantly higher in ADE sites under secondary forest than under manioc cultivation system. Other chemical properties such as pH, SB and amount of P, Mg and H+Al, did not statistically differ between different land uses in ADE and ADJ sites.

Soil chemical data was also used in regression analyses to understand patterns in the microbial community structure of the *bph* gene. Microbial community structure can be defined by patterns of species abundance and population within a given community [45], which are mostly regulated by the ability of the microorganisms to interact among them and with local conditions [46]. In this study, T-RFLP results did not reveal significant differences among the richness of T-RFs between ADE and ADJ samples for the total bacterial community (16S rRNA). However, ANOSIM revealed significant differences in total bacterial community structure between ADJ and ADE soils types ($R = 0.17$) and between land uses ($R = 0.11$ for ADE-SF versus

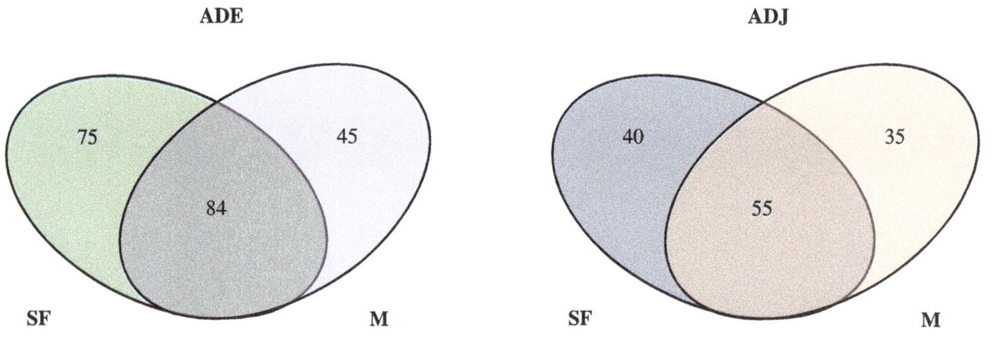

Figure 4. Venn's diagram of *bph* data belonging to operational protein families (OPFs) for different soil types under different land uses. ADE = Amazon Dark Earth; ADJ = Adjacent soils; SF = Secondary Forest; M = Manioc cultivation. Sequences were grouped into OPFs based on sequence identity of 94%.

ADE-M; and R = 0.03 for ADJ-SF versus ADJ-M; Table 1). Lima [47], who similarly assessed the variation of ADE-SF, ADE-M, ADJ-SF and ADJ-M sites, and Cannavan [48], who investigated changes across different ADE sites, also found differences in 16S rRNA community structure between ADE and ADJ sites. Collectively, these studies support the idea that archaeological sites with a long history of anthropogenic activity influence the inhabiting microbiota of their respective soils.

Analysis of the community patterns for the bacterial *bph* gene revealed higher richness of T-RFs in ADE than in ADJ sites. Our results show that *bph* community changes are clear among sites (Figure 2), which can be observed by changes in the abundance of specific T-RFs across sites. In the same way, ANOSIM revealed significant differences in *bph* gene community structure between ADJ and ADE soils types (R = 1.00) and between land uses (R = 1.00 for ADE-SF versus ADE-M; and R = 1.00 for ADJ-SF versus ADJ-M; Table 1). We suspect, despite all variables present in our system, that the high presence of biochar, which chemically is formed by an aromatic polycyclic structure, could be a factor influencing the observed differences. Táncsics et al [49], by analyzing the structure community of catechol 2,3-dioxygenase genes in aromatic hydrocarbon contaminated environments by T-RFLP, observed that T-RFLP chromatograms obtained from contaminated samples had entirely different T-RFs compared to the control non-contaminated sample.

Redundancy analysis revealed *bph* gene structure to be correlated with different soil parameters (Figure 4). Briefly, ADE-SF correlated with pH, eCEC, SOM and SOC. ADJ-SF showed a significant correlation with H+Al. Several studies have suggested that variations in soil pH and properties related to soil acidity (e.g., K, Al, base saturation) are stronger predictors of the richness and diversity of inhabiting microbial communities [6,18,50,51,52,53]. We extend this concept by advocating that other factors also might play a role in structuring the *bph* gene communities. For instance, the quantity and/or quality of soil organic matter (SOM) and its fractions are likely to regulate

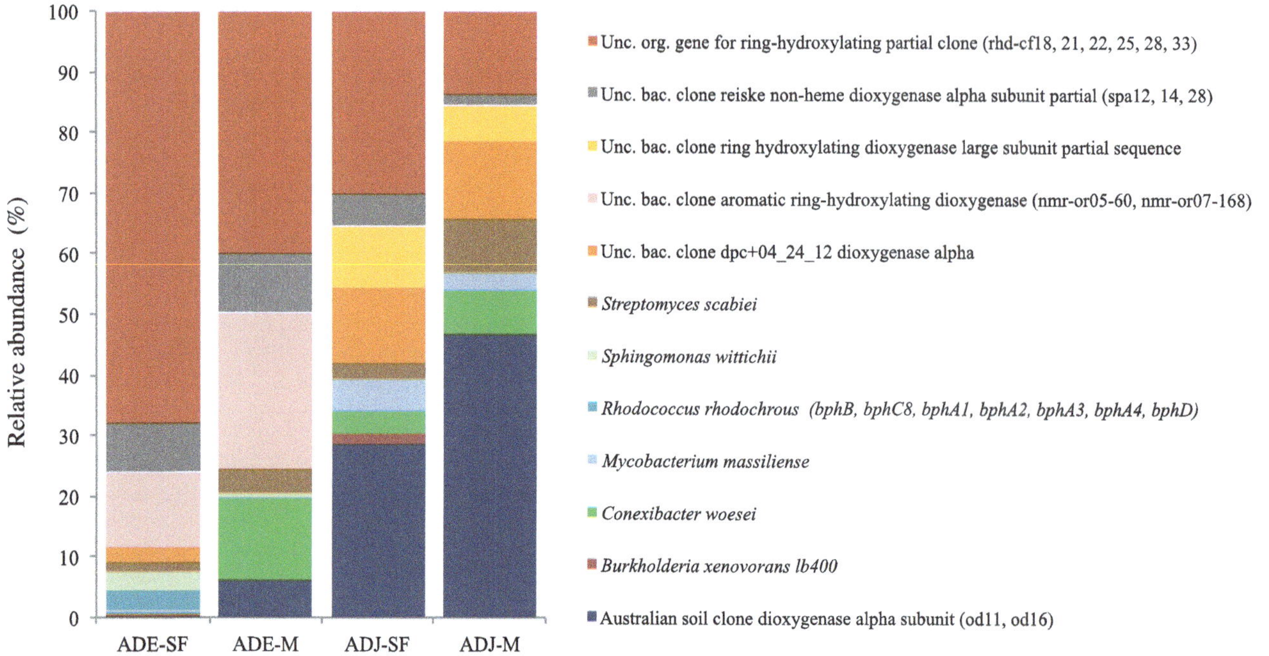

Figure 5. Bar charts representing the taxonomic affiliation of *bph* gene sequences. Sequences were affiliated using the TBlastX tool available in the GenBank database. ADE = Amazon Dark Earth; ADJ = Adjacent soils; SF = Secondary Forest; M = Manioc cultivation.

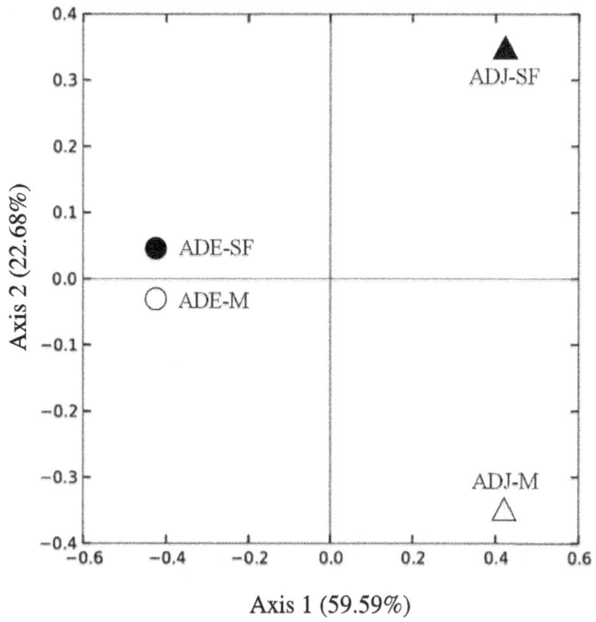

Figure 6. Principal Coordinate Analysis (PCoA) performed with the sequences obtained for the *bph* gene (based on Unweighted Unifrac distance). ADE = Amazon Dark Earth; ADJ = Adjacent soils; SF = Secondary Forest; M = Manioc cultivation.

microbial community composition and associate function [54]. Since hydrocarbon degradation is one step into the carbon cycle, *bph* degraders might have an advantage in ADE sites by harnessing energy from elevated levels of SOC and SOM.

Quantification of total bacterial (16S rRNA) and *bph* gene abundances (Figure 7) revealed higher copy numbers of these genes in ADE sites, peaking at ADE-SF. These results indicate that soil characteristics of ADE, as well as their land uses has an influence on the abundance of both analyzed genes. Ding et al [55] analyzed the abundance of total bacteria using the16S rRNA gene and the abundance of polycyclic aromatic hydrocarbons ring-hydroxylating dioxygenase (PAH-RHD$_{\alpha}$) genes by qPCR in two different phenanthrene-contaminated soils (i.e. Luvisol and Cambisol). Their results revealed a significantly higher 16S rRNA gene copy number per gram of soil in both phenanthrene-contaminated soils compared to their controls. PAH-RHD genes were detected only in contaminated soil samples, and the values ranged from 2.0×10^7 (Luvisol phenanthrene-contaminated soil) to 1.7×10^6 (Cambisol phenanthrene-contaminated soil) copies per gram of soil, showing similar values to those found in our study. Similarly, a study about the diversity of naphthalene dioxygenase genes (*nahAc*) in soil environments from the Maritime Antarctic revealed that the quantities detected in bulk and rhizospheric soils from PAH-affected sites ranged from 6.4×10^4 to 1.7×10^6 *nahAc* gene copies per gram of soil and presented significantly higher abundance compared with the corresponding counterparts of bulk and rhizospheric soils from non-polluted sites [56].

We used 454-pyrosequencing to investigate the community composition of the *bph* gene. The use of a high-throughput culture-independent approach enabled an overview of the taxonomic groups occurring in the sampled sites. Taxonomic analyses of 3,000 partial sequences of the *bph* gene revealed that this gene was most represented by bacteria belonging to the genera *Streptomyces, Sphingomonas, Rhodococcus, Mycobacterium, Conexibacter* and *Burkholderia,* in addition to sequences matching uncultured bacteria (Figure 5). Overall, these groups encompass mixed sequences of toluene/biphenyl dioxygenases, such as *bph*B, *bph*C8, *bph*A1, *bph*A2, *bph*A3, *bph*A4 and *bph*D, and enzymes encoding the alpha subunit of dioxygenases for the degradation of Polycyclic Aromatic Hydrocarbons (PAHs), which are described as environmental widely distributed dioxygenases. Other studies targeting dioxy-

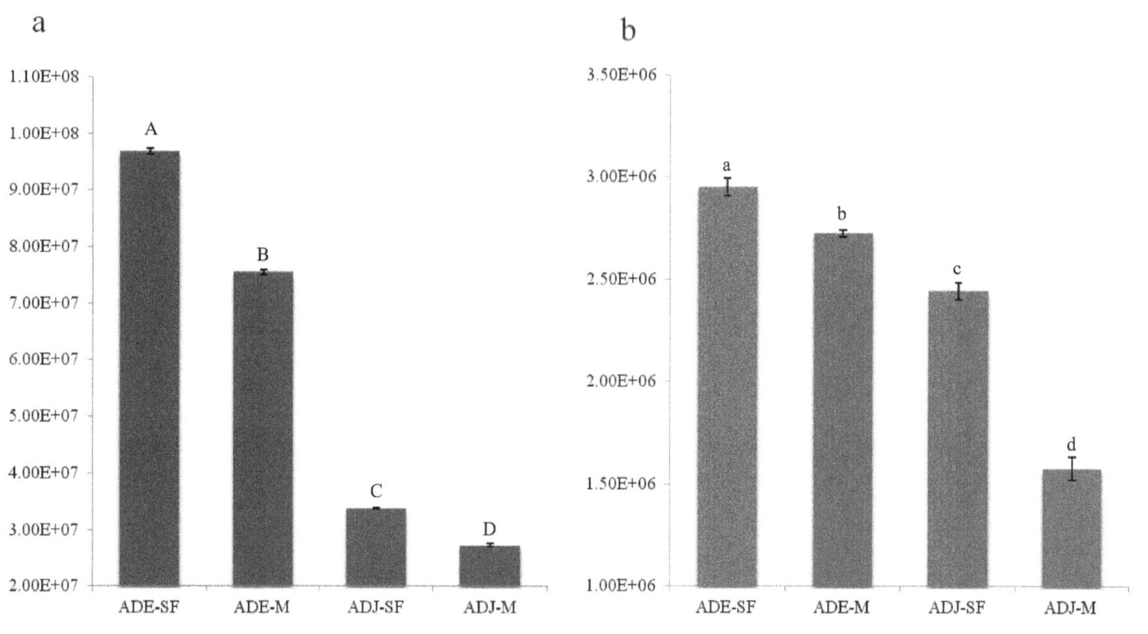

Figure 7. Log gene copy number (*y* axis) of (a) 16S rRNA and (b) *bph* genes across the studied sites. Sites are indicated in the *x* axis. Error bars represent the standard deviation of three independent replicates. ADE = Amazon Dark Earth; ADJ = Adjacent soils; SF = Secondary Forest; M = Manioc cultivation. Different upper case letters refer to differences for the 16S rRNA across sites; while different lower case letters refer to differences for the *bph* gene ($P<0.01$, Tukey test).

genases have also reported sequences belonging to *Sphingomonas*, *Rhodococcus*, *Mycobacterium*, *Conexibacter* and *Burkholderia* as major taxonomic groups [21,28,57]. Zhou et al [58] were also able to isolate aromatic hydrocarbon degrading bacteria from mangrove sediments, which were classified as the genera *Mycobacterium* and *Sphingomonas*. Members of these genera have commonly been isolated from diverse sediments and soils [59,60], and they play an important role in hydrocarbon biodegradation [61,62,63]. For instance, Ding et al [55] who studied the diversity of dioxygenases using clone libraries for aromatic ring hydroxylating dioxygenases (ARHD) genes identified gene sequences corresponding to the *phnAc* gene belonging to the *Burkholderia* sp. strain Eh1-1 and PAH-RHD genes of the *Mycobacterium* sp. strain JLS.

Confirming the results of T-RFLP analysis, differences among *bph* communities for ADE and ADJ sites and between SF and M land uses were also observed by differences in their Shannon and Simpson diversity indices (Table 2). Similarly, Germano et al [28] reported a higher diversity of ARHD genes in ADE sites under secondary forest rather than under agricultural cultivation, suggesting that deforestation in these sites has an influence on the diversity of these catabolic genes. We observed that in both soil types, the number of OPF (Table 2) was higher in SF samples, also indicating an influence of the land use system on the richness of this gene. According to Jesus et al [51], the land conversion of tropical forest to agricultural use modifies the size, activity and composition of soil microbial communities. This deeply influences specific bacterial functions, including those acting on organic matter decomposition and nutrient cycling in soils.

The intensive land use by agricultural practices and the conversion of Amazon soil into agricultural areas has been reported to cause significant shifts in the chemical properties of the soil, such as variations in SOC, SOM and eCEC, leading towards a homogenization of the inhabiting bacterial community [64]. However, we hypothesized that, despite an effect of land use, greater differences in community composition would be observed according to soil type. We also expected communities in these different soils to respond differently to agricultural practices. In this context, the literature describes the resilience phenomenon as the ability of the soil to cope with external disturbances and to retain its functional capacity upon the imposition of a stress [65,66,67]. We observed that higher differences in the diversity of *bph* occurred between the different land uses in ADJ sites rather than in ADE sites. Thus, land use type appeared to have a stronger effect on the *bph* community in ADJ sites, maybe due to the higher resilience of the ADE soil against agricultural practices. Principal Coordinate Analysis (PCoA) performed for the *bph* gene also supports these results (Figure 6), revealing that ADE sites (SF and M) clustered closer to each other than observed for ADJ sites (SF and M).

The taxonomic analyses of sequences also revealed that ADE sites harbored distinct *bph* phylogenetic structure from ADJ. These results suggest that the heterogeneity of bacterial *bph* communities could be related to their ability to respond to differences of land use type and soil chemical properties. In a previous study Germano et al. [28] compared the phylogenetic structure of *bph* sequences from ADE sites and revealed that most of the protein clusters from these sites group apart from the main well-known dioxygenase groups previously proposed by Kweon et al [27].

In conclusion, we have taken a distinct and highly diverse soil to elucidate the ecological properties and taxonomic affiliation of bacterial communities characterized by the presence of the *bph* gene, which is crucial to the biodegradation of aromatic compounds. These results enable us to understand differences in the structure, abundance and composition of the main active organisms in ADE soils when compared to their adjacent locations. Further studies focusing on the catabolic activities of these communities are needed to enable a collective view of the formation, dynamics and maintenance of functional properties in ADE soils.

Supporting Information

Figure S1 Rarefaction curves of bacterial *bph* gene sequences were grouped into OPF based on distance sequence of 0.06. ADE = Amazonian Dark Earth soils; ADJ = Adjacent soils; SF = Secondary forest; M = Manioc Cultivation.

Table S1 Soil chemical properties of the studied sites. Amazonian Dark Earth (ADE) and Adjacent soil (ADJ). Sites were located under two different land uses: Secondary forest (SF) and under Manioc (*Manihot esculenta*) cultivation (M). Significant differences between sites are followed by different letters ($P<0.05$, Tukey test).

Acknowledgments

We thank Wenceslau Geraldes Teixeira and Western Amazon Embrapa for technical support. We are grateful to ACG Souza, AK Silveira, RS Macedo and TT Souza for helping with sample collection. We thank J Quensen and the Michigan State University Genomics Technology Support Facility team for the 454-pyrosequencing support; and José Elias Gomes and Fábio Duarte for their assistance in molecular analysis. We also thank Cyrus A. Mallon for English revision and helpful comments on the manuscript.

Author Contributions

Conceived and designed the experiments: MJLB MGG ABL SMT. Performed the experiments: MJLB MGG ABL. Analyzed the data: MJLB LWM. Contributed reagents/materials/analysis tools: SMT. Wrote the paper: MJLB SMT.

References

1. Lehmann J, da Silva JP Jr, Steiner C, Nehls T, Zech W, et al. (2003) Nutrient availability and leaching in an archaeological Anthrosol and a Ferralsol of the Central Amazon basin: fertilizer, manure and charcoal amendments. Plant Soil 249: 343–357.
2. Teixeira WG, Martins GC (2003) Soil physical characterization. In: Lehmann J, Kern DC, Glaser B, Woods WI, editors.Amazonian Dark Earths: Origin, properties, management.Dordrecht: Kluwer Academic. pp. 271–286.
3. Glaser B (2007) Prehistorically modified soils of central Amazonia: a model for sustainable agriculture in the twenty-first century. Philos T Roy Soc B 362: 187–196.
4. Kim JS, Sparovek G, Longo RM, De Melo WJ, Crowley D (2007) Bacterial diversity of terra preta and pristine forest soil from the Western Amazon. Soil Biol Biochem 39: 684–690.
5. O'neill B, Grossman J, Tsai SM, Gomes JE, Lehmann J, et al. (2009) Bacterial community composition in Brazilian anthrosols and adjacent soils characterized using culturing and molecular identification. Microbial Ecol 58: 23–35.
6. Taketani RG, Tsai SM (2010) The influence of different land uses on the structure of archaeal communities in Amazonian anthrosols based on 16S rRNA and *amo*A genes. Microbial Ecol 59: 734–743.
7. Navarrete AA, Cannavan FS, Taketani RG, Tsai SM (2010) A molecular survey of the diversity of microbial communities in different Amazonian agricultural model systems. Diversity 2: 787–809.
8. Taketani RG, Lima AB, Jesus EC, Teixeira WG, Tiedje JM, et al. (2013) Bacterial community composition of anthropogenic biochar and Amazonian anthrosols assessed by 16S rRNA gene 454 pyrosequencing. A van Leeuw J Microb 104: 233–242.

9. Thies J, Suzuki K (2003) Amazonian dark earths: biological measurements. In: Lehmann J, Kern D, Glaser B, Woods W, editors.Amazonian Dark Earths: Origin, Properties, Management.Dordrecht: Kluwer Academic. pp 287–332.
10. Fitter AH, Gilligan CA, Hollingworth K, Kleczkowski A, Twyman RM, et al. (2005) Biodiversity and ecosystem function in soil. Funct Ecol 19: 369–377.
11. Horner-Devine M, Carney K, Bohannan B (2004) An ecological perspective on bacterial biodiversity. Proc R Soc Lond B Biol Sci 271: 113–122.
12. Wawrik B, Kerkhof L, Kukor J, Zylstra G (2005) Effect of different carbon sources on community composition of bacterial enrichments from soil.Appl Environ Microbiol 71: 6776– 6783.
13. Zhou JZ, Xia BC, Treves DS, Wu LY, Marsh TL, et al. (2002) Spatial and resource factors influencing high microbial diversity in soil. Appl Environ Microbiol 68: 326–334.
14. Vogel TM, Simonet P, Jansson JK, Hirsch PR, Tiedje JM, et al. (2009) TerraGenome: a consortium for the sequencing of a soil metagenome. Nat Rev Microbiol 7: 252.
15. Bissett A, Brown MV, Siciliano SD, Thrall PH (2013) Microbial community responses to anthropogenically induced environmental change: towards a systems approach. Ecol Lett 16: 128–139.
16. Buckley DH, Schmidt TM (2001) The structure of microbial communities in soil and the lasting impact of cultivation. Microbial Ecol 42: 11–21.
17. Buckley DH, Schmidt TM (2003) Diversity and dynamics of microbial communities in soils from agro-ecosystems. Environ Microbiol 5: 441–452.
18. Fierer N, Jackson RB (2006) The diversity and biogeography of soil bacterial communities. Proc Natl Acad Sci USA 103: 626– 631.
19. Anderson CR, Condron LM, Clough TJ, Fiers M, Stewart A, et al. (2011) Biochar induced soil microbial community change: Implications for biogeochemical cycling of carbon, nitrogen and phosphorus. Pedobiologia 54: 309–220.
20. Liang B, Lehmann J, Sohi SP, Thies JE, O'neill B, et al. (2010) Black carbon affects the cycling of non-black carbon in soil. Org Geochem 41: 206–213.
21. Iwai S, Chai B, Sul WJ, Cole JR, Hashsham SA, et al. (2010) Gene-targeted-metagenomics reveals extensive diversity of aromatic dioxygenase genes in the environment. ISME J 4: 279–285.
22. Wackett LP (2004) Evolution of enzymes for the metabolism of new chemical inputs into the environment. J Biol Chem 279: 41259–41262.
23. Mason JR, Cammack R (1992) The electron-transport proteins of hydroxylating bacterial dioxygenases. Annu Rev Microbiol 46: 277–305.
24. Butler CS, Mason JR (1997) Structure-function analysis of the bacterial aromatic ring-hydroxylating dioxygenases. Adv Microb Physiol 38: 47–84.
25. Hayaishi O (1962) History and scope. In Hayaishi O, editor. Oxygenases. New York: Academic Press.pp. 1–29.
26. Luz AP, Pellizari VH, Whyte LG, Greer C (2004) A survey of indigenous microbial hydrocarbon degradation genes in soils from Antarctica and Brazil. Can J Microbiol 50: 323–333.
27. Kweon O, Kim SJ, Baek S, Chae JC, Adjei MD, et al. (2008) A new classification system for bacterial Rieske non-heme iron aromatic ring-hydroxylating oxygenases. BMC Biochem 9: 11.
28. Germano MG, Cannavan FS, Mendes LW, Lima AB, Teixeira WG, et al. (2012) Functional diversity of bacterial genes associated with aromatic hydrocarbon degradation in anthropogenic dark earth of Amazonia. Pesq Agropec Bras 47: 654–664.
29. Gibson DT, Parales RE (2000) Aromatic hydrocarbon dioxygenases in environmental biotechnology. Curr Opin Biotechnol 11: 236–243.
30. Claessen MEC, Barreto WDO, Paula JL, Duarte MN (1997) Análises químicas para avaliação da fertilidade do solo. Rio de Janeiro: Brazilian Agricultural Research Corporation Press. 212p
31. Amann RI, Ludwig W, Schleifer KH (1995) Phylogenetic identification and *in situ* detection of individual microbial cells without cultivation. Microbiol Rev 59: 143–169.
32. Culman SW, Gauch HG, Blackwood CB, Thies JE (2008) Analysis of T-RFLP data using analysis of variance and ordination methods: a comparative study. J Microbiol Methods 75: 55–63.
33. Goris J, De Vos P, Caballero-Mellado J, Park JH, Falsen E, et al. (2004) Classification of the PCB and biphenyl-degrading strain LB400 and relatives as *Burkholderia xenovorans* sp. nov. Int J Syst Evol Microbiol 54: 1677–1681.
34. Schloss PD, Westcott SL, Ryabin T, Hall JR, Hartmann M, et al. (2009) Introducing mothur: open-source, platform-independent, community supported software for describing and comparing microbial communities. Appl Environ Microbiol 75: 7537–7541.
35. Good IJ (1953) The population frequencies of species and the estimation of the population parameters. Biometrika 40: 237–264.
36. Conesa A, Götz S, Garcia-Gomez JM, Terol J, Talon M, et al. (2005) Blast2GO: a universal tool for annotation, visualization and analysis in functional genomics research. Bioinformatics 21: 3674–3676.
37. Heuer H, Krsek M, Baker P, Smalla K, Wellington EMH (1997) Analysis of actinomycete communities by specific amplification of genes encoding 16S rRNA and gelelectrophoresis separation in denaturing gradients. Appl Environ Microbiol 63: 3233–3241.
38. Grossman JM, O'neill BE, Tsai SM, Liang B, Neves E, et al. (2010) Amazonian anthrosols support similar microbial communities that differ distinctly from those extant in adjacent, unmodified soils of the same mineralogy. Microb Ecolol 60: 192–205.
39. Fierer N, Bradford MA, Jackson RB (2007) Toward an ecological classification of soil bacteria. Ecology 88: 1354–1364.
40. Lauber CL, Hamady M, Knight R, Fierer N (2009) Pyrosequencing-based assessment of soil pH as a predictor of soil bacterial community structure at the continental scale. Appl Environ Microbiol 75: 5111–5120.
41. Falcão N, Moreira A, Comenford NB (2009) A fertilidade dos solos de Terra Preta de Índio da Amazônia Central. In: Teixeira WG, Kern DC, Madari BE, Lima HN, Woods W, editors.As Terras Pretas de Índio da Amazônia: sua caracterização e uso deste conhecimento na criação de novas áreas.Manaus: Embrapa Amazônia Ocidental. pp. 189–200.
42. Glaser B, Haumaier L, Guggenberger G, Zech W (2001) The 'Terra Preta' phenomenon: a model for sustainable agriculture in the humid tropics. Naturwissenschaften 88: 37–41.
43. Cunha TJF, Novotny EH, Madari BE, Benites VM, Martin-Neto L, et al. (2009) O Carbono Pirogênico. In: Teixeira WG, Kern DC, Madari BE, Lima HN, Woods W, editors.As Terras Pretas de Índio da Amazônia: sua caracterização e uso deste conhecimento na criação de novas áreas.Manaus: Embrapa Amazônia Ocidental. pp. 264–285.
44. Bünemann EK, Marschner P, Smernik RJ, Conyers M, Mcneill AM (2008) Soil organic phosphorus and microbial community composition as affected by 26 years of different management strategies. Biol Fert Soils 44: 717–726.
45. Ricklefs RE, Miller G (1999) Ecology. New York: W H Freeman. 822p.
46. Bernhard AE, Colbert D, McManus J, Field KG (2005) Microbial community dynamics based on 16S rRNA gene profiles in a Pacific Northwest estuary and its tributaries. FEMS Microbiol Ecol 52: 115–128.
47. Lima AB (2012) Influência da cobertura vegetal nas comunidades de bactérias em Terra Preta de Índio na Amazônia Central brasileira. Piracicaba: University of São Paulo. 116 p.
48. Cannavan FS (2012) A estrutura e composição de comunidades microbianas (Bacteria e Archaea) em fragmentos de carvão pirogênico de Terra Preta de Índio da Amazônia Central. Piracicaba: University of São Paulo. 116 p.
49. Táncsics A, Szabóc I, Bakab E, Szoboszlayc S, Kukolyad J, et al. (2010) Investigation of catechol 2,3-dioxygenase and 16S rRNA gene diversity in hypoxic, petroleum hydrocarbon contaminated groundwater. Syst Appl Microbiol 33: 398–406.
50. Nicol GW, Tscherko D, Chang L, Hammesfahr U, Prosser JI (2006) Crenarchaeal community assembly and microdiversity in developing soils at two sites associated with deglaciation. Environ Microbiol 8: 1382–1393.
51. Jesus EC, Marsh TL, Tiedje JM, Moreira FMS (2009) Changes in land use alter the structure of bacterial communities in Western Amazon soils. ISME J 3: 1004–1011.
52. Nielsen UN, Osler GHR, Campbell CD, Burslem DFRP, van der Wal R (2010) The influence of vegetation type, soil properties and precipitation on the composition of soil mite and microbial communities at the landscape scale. J Biogeogr 37: 1317–1328
53. Wessen E, Hallin S, Philippot L (2010) Differential responses of bacterial and archaeal groups at high taxonomical ranks to soil management. Soil Biol Biochem 42: 1759–1765.
54. Murphy DV, Cookson WR, Braimbridge M, Marschner P, Jones DL, et al. (2011) Relationships between soil organic matter and the soil microbial biomass (size, functional diversity, and community structure) in crop and pasture systems in a semi-arid environment. Soil Res 49: 582–594.
55. Ding GC, Heuer H, Zuhlke S, Spiteller M, Pronk JG, et al. (2010) Soil Type-Dependent Responses to Phenanthrene as Revealed by Determining the Diversity and Abundance of Polycyclic Aromatic Hydrocarbon Ring-Hydroxylating Dioxygenase Genes by Using a Novel PCR Detection System. Appl Environ Microbiol 76: 4765–4771.
56. Flocco CG, Gomes NC, Mac CW, Smalla K (2009) Occurrence and diversity of naphthalene dioxygenase genes in soil microbial communities from the maritime Antarctic. Environ Microbiol 11: 700–714.
57. Leigh MB, Pellizari VH, Uhlik O, Sutka R, Rodrigues J, et al. (2007) Biphenyl-utilizing bacteria and their functional genes in a pine root zone contaminated with polychlorinated biphenyls (PCBs). ISME J 1: 134–148.
58. Zhou HW, Guo CL, Wong YS, Tam NFY (2006) Genetic diversityof dioxygenase genes in polycyclic aromatic hydrocarbon-degrading bacteria isolated from mangrove sediments. FEMS Microbiol Lett 262: 148–157.
59. Leys NMEJ, Ryngaert A, Bastiaens L, Verstraete W, Top EM, et al. (2004) Occurrence and phylogenetic diversity of Sphingomonas strains in soils contaminated with polycyclic aromatic hydrocarbons. Appl Environ Microbiol 70: 1944–1955.
60. Miller CD, Hall K, Liang YN, Nieman K, Sorensen D, et al. (2004) Isolation and characterization of polycyclic aromatic hydrocarbon-degrading Mycobacterium isolates from soil. Microb Ecol 48: 230–238.
61. Khan AA, Wang RF, Cao WW, Doerge DR, Wennerstrom D, et al. (2001) Molecular cloning, nucleotide sequence, and expression of genes encoding a polcyclic aromatic ring dioxygenase fromMycobacterium sp strain PYR-1. Appl Environ Microbiol 67: 3577–3585.
62. Krivobok S, Kuony S, Meyer C, Louwagie M, Willison JC, et al. (2003) Identification of pyrene-induced proteins in Mycobacterium sp strain 6PY1: evidence for two ringhydroxylating dioxygenases. J Bacteriol 185: 3828–3841.
63. Demaneche S, Meyer C, Micoud J, Louwagie M, Willison JC, et al. (2004) Identification and functional analysis of two aromatic-ring-hydroxylating dioxygenases from a Sphingomonas strain that degrades various polycyclic aromatic hydrocarbons. Appl Environ Microbiol 70: 6714–6725.

64. Rodrigues JLM, Pellizari VH, Mueller R, Baek K, Jesus EC, et al. (2013) Conversion of the Amazon rainforest to agriculture results in biotic homogenization of soil bacterial communities. Proc Natl Acad Sci USA 110: 988–993.
65. Arthur E, Schjønning P, Moldrup P, de Jonge LW (2012) Soil resistance and resilience to mechanical stresses for three differently managed sandy loam soils. Geoderma 173–174: 50–60.
66. Gregory AS, Watts CW, Whalley WR, Kuan HL, Griffiths S, et al. (2007) Physical resilience of soil to field compaction and the interactions with plant growth and microbial community structure. Eur J Soil Sci 58: 1221–1232.
67. Schjønning P, Elmholt S, Christensen BT (2004) Soil quality management: concepts and terms. In: Schjønning P, Elmholt S, Christensen BT, editors.Challenges in Modern Agriculture.Wallingford: CABI Publishing. pp. 1–15.

3

Much beyond Mantel: Bringing Procrustes Association Metric to the Plant and Soil Ecologist's Toolbox

Francy Junio Gonçalves Lisboa[1,4]*, Pedro R. Peres-Neto[2], Guilherme Montandon Chaer[3], Ederson da Conceição Jesus[3], Ruth Joy Mitchell[4], Stephen James Chapman[4], Ricardo Luis Louro Berbara[1]

1 Soil Science Department, Agronomy Institute, Federal Rural University of Rio de Janeiro, Seropédica-RJ, Brazil, **2** Canada Research Chair in Spatial Modelling and Biodiversity; Université du Québec à Montréal, Département des sciences biologiques, Québec, Canada, **3** Embrapa Agrobiologia, Seropédica-RJ, Brazil, **4** The James Hutton Institute, Craigiebuckler, Aberdeen, United Kingdom

Abstract

The correlation of multivariate data is a common task in investigations of soil biology and in ecology in general. Procrustes analysis and the Mantel test are two approaches that often meet this objective and are considered analogous in many situations especially when used as a statistical test to assess the statistical significance between multivariate data tables. Here we call the attention of ecologists to the advantages of a less familiar application of the Procrustean framework, namely the Procrustean association metric (a vector of Procrustean residuals). These residuals represent differences in fit between multivariate data tables regarding homologous observations (e.g., sampling sites) that can be used to estimate local levels of association (e.g., some groups of sites are more similar in their association between biotic and environmental features than other groups of sites). Given that in the Mantel framework, multivariate information is translated into a pairwise distance matrix, we lose the ability to contrast homologous data points across dimensions and data matrices after their fit. In this paper, we attempt to familiarize ecologists with the benefits of using these Procrustean residual differences to further gain insights about the processes underlying the association among multivariate data tables using real and hypothetical examples.

Editor: Andrew R. Dalby, University of Westminster, United Kingdom

Funding: This work was carried out with the aid of a grant from Conselho Nacional de Desenvolvimento Científico e Tecnológico (CNPq - 563304/2010-3 and 562955/2010-0), Fundação de Amparo à Pesquisa de Minas Gerais (FAPEMIG CRA - APQ-00001-11), and the Inter-American Institute for Global Change Research (IAI-CRN II-021). Francy Lisboa greatly acknowledges a research scholarship from CAPES/EMBRAPA (Carbioma) and Dr. Beata Madari. The funders had no role in study design, data collection and analysis, decision to publish, or preparation of the manuscript.

Competing Interests: The authors have declared that no competing interests exist.

* Email: agrolisboa@gmail.com

Introduction

In multidimensional data analysis, ecologists often encounter situations where they need to choose between two or more numerical approaches that are able to tackle the same question of interest. The preference between approaches is based, among other factors, on the familiarity of the user with the method, which in turn depends on the time a particular method has been available in statistical packages and the ease in implementing and interpreting its results. Another relevant factor to consider is "literature-induced use" in which renowned research groups involved in the development, improvement and generation of statistical ecological approaches have a strong influence on the types of statistical approaches other ecologists use.

Determining the strength of the relationships between multivariate datasets is a routine analysis when trying to understand the environmental factors driving the composition and structure of ecological communities. Two approaches, the Mantel test [1] and Procrustes analysis [2], though considered analogous by the literature in the questions they can tackle [3], have not been used to the same extent. Despite the advantages of Procrustes analysis over the Mantel test [3] regarding greater statistical power in detecting significant relationships (i.e., lower type II errors) and the possibility of analyzing further the patterns of association between multivariate matrices (visually and by further statistical analyses), the Procrustean approach remains relatively unused in tackling questions regarding the relationships between data matrices involving plant and soil information or between soil matrices (Fig. 1).

The Mantel test and the Procrustes approach can be both used in many similar situations where the aim is to assess how multivariate data matrices are associated (correlated), though for unknown reasons they have been used in quite different ways in the ecological literature. For example, while the Mantel test has often been applied when testing the relationship between above and below ground data matrices [4], [5], [6], [7], [8], [9], [10], [11], [12], Procrustes analysis has predominantly been used to contrast the results of different ecological ordinations on the same data [13], [14], [15], [16], to compare fingerprinting tools for assessing microbial communities [17], [18], [19] and for deciding between methodological choices [20], [21]. Indeed the Procrustean framework has been rarely used to make inferences about plant and soil relationships [22], [23], [24], [25], [26] and other types of ecological associations between data sets. However there are instances in which the Procrustean and Mantel tests cannot be used interchangeably. Unlike Mantel, the Procrustean approach can be used to compare multiple data matrices. However, when ecologists are interested in correlating distance (or similarity)

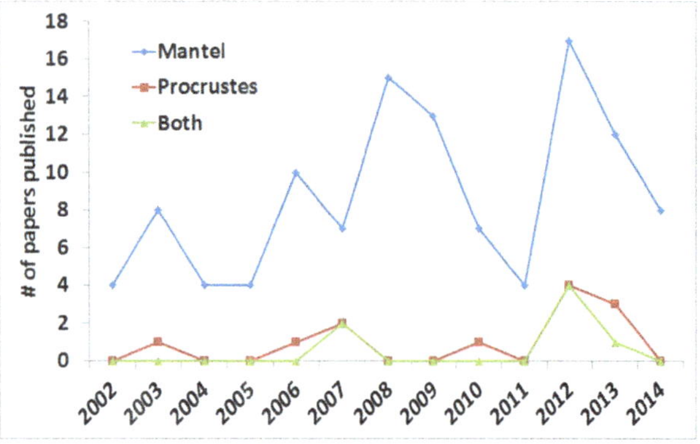

Figure 1. Papers published using Mantel and Procrustes for relating data matrices from soil or plant studies in the ten years since [3] stated the advantages of Procrustes over the Mantel approach. Data obtained using Thompson Reuters database (May, 12, 2014). We searched for papers using uniquely the Mantel approach, uniquely the Procrustes approach and papers using both approaches. The search was based on Procrust* (Procrustean or Procrustes) and PROTEST.

matrices, rather than testing the association among data matrices in their raw form (i.e., not transformed by the property of distance measures), Mantel, rather than Procrustes, is more appropriate. One particular case is the distance-decay of similarity in ecological communities [27] in which one is interested in testing the hypothesis that the similarity in community composition decreases in relation to linear (or log transformed) geographic distance between communities. The differences between raw-based and distance-based approaches have been discussed extensively elsewhere [28], [29].

Despite the relative merits of the Procrustean framework over the Mantel test shown by the relatively well-cited paper by Peres-Neto and Jackson [3], its potential has not yet been tapped. Perhaps the reason for Procrustes analysis not being as popular as the Mantel test among ecologists is the lack of a paper showing that in many situations traditionally investigated by Mantel, the Procrustean analysis can be equally well used. Here, we attempt to familiarize ecologists with the use of Procrustes analysis by using real and hypothetical examples where the Mantel test tends to be preferred. Most importantly, we highlight little explored limits of Procrustes by using its residual vector of association between data tables, hereafter referred as to PAM, in three common statistical approaches: multivariate ordination, variation partitioning and ANOVA.

Procrustes analysis: a foundation for soil and plant ecologists

In ancient Greek mythology there was a character named Procrustes who was a resident of Eleusis Mountain, a known travelers' route. As a "good" host, Procrustes always invited travelers to spend the night at his home; more specifically, he invited them to lie down on his iron bed, which was tailored to fit Procrustes' own body. The guests who did not fit the dimensions of his bed either had their limbs cut off or were stretched until their dimensions approached those of Procrustes's bed. Ironically, none of the guests ever fitted the iron bed because Procrustes secretly had two beds of different sizes [30]. One can easily make a parallel here with ecological data in which data from different sources will almost never easily compare or fit to one another.

Procrustes analysis is based on the search for the best fit between two data tables, hereafter referred to as matrices, where one is kept fixed ("Procrustes' bed" or target matrix), while the other ("Procrustes' guest" or rotated matrix) undergoes a series of transformations (translation, mirror reflection and rotation; [2]) to fit the fixed matrix. Although in this paper we concentrate on fitting two matrices, the extension of Procrustes analysis to multiple matrices is straightforward [3] in which the reference matrix can be either one of the original matrices or their averages (or medians). Hereafter, the target matrix (target) will be referred as to $\mathbf{X}$, and the data matrix to be fitted as $\mathbf{Y}$. $\mathbf{X}$ and $\mathbf{Y}$ are both $n \times p$ matrices, where n is the number of rows and p is the number of columns. The goal of the transformations in $\mathbf{Y}$ is to minimize the residual sum of squared differences between the corresponding n dimensions between $\mathbf{X}$ and $\mathbf{Y}$; the sum of the squares of these residual differences is termed m^2 (Gower's statistic), representing the optimal fit between the two data matrices, such that the higher the value of m^2, the weaker the relationship between the two data tables is. The significance of m^2 can be estimated through a permutation test (termed PROTEST after [31]; see [3] for further details).

Procrustean association metric (PAM)

The least squares superimposition between the corresponding n observations of $\mathbf{X}$ and $\mathbf{Y}$ is one of the main advantages (in addition to the increased statistical power) of the Procrustean framework in contrast to the Mantel test. The Procrustes superimposition generates a ($n \times p$) matrix of residuals that can be further used to contrast the differences between homologous observations (rows) across matrices in the form of a vector (PAM). Given that within the Mantel approach differences between observations across all dimensions are packed down into a single distance, it cannot be used to assess differences across observations across dimensions. Consistent small and large differences across homologous observations across matrices in regard to other factors of interest can further assist in understanding how $\mathbf{X}$ and $\mathbf{Y}$ are related. For example, we could use PAM to assess the degree of observation matching between a plant function trait matrix and a composition matrix and assess whether smaller or greater residual values are a function of the time elapsed since some disturbance event.

PAM is simply a vector of residual differences between the corresponding n observations. For example, assuming that an ecologist wants to correlate two matrices of data $\mathbf{X}$ and $\mathbf{Y}$, both of which are formed by four rows (i.e. sites, plots, observational units), Procrustes analysis will generate four residual differences between the $\mathbf{X}$ and $\mathbf{Y}$ configurations. The compilation of these residual differences between homologous rows (observations) across dimensions in the form of a vector – PAM – represents a useful way to represent information on the relationship between two matrices and make it available for further statistical analysis, both parametric and non-parametric; this feature is not offered by the Mantel approach.

The use of the residual vector from Procrustes (PAM) has been quite restricted in the plant and soil ecological literature. To our knowledge the first study was by [32] who assessed the plant-pollinator interaction during three consecutive summers in the southeastern portion of California, USA. These authors employed the PAM to identify which pollinating species exhibited the greatest deviation between two consecutive years. Singh et al. [22] used the PAM in a study on soil microbiology to verify the effect of soil pH on the relationship between arbuscular mycorrhizal fungi (AMF) and plant assemblages. These authors employed the following strategy: 1) Procrustes analysis was applied between the matrices representing the AMF community and that representing the plant community; 2) after detecting a significant relationship ($m_{12} = 0.28$; $P<0.001$), these authors extracted the PAM and used it as a response in a simple regression analysis with the soil pH. No effect of pH on the association between the AMF and plant communities was detected, suggesting that neither the pH nor the identity of the plant species that composed the community affected the AMF community. Other applications can be certainly found (e.g., [24], [25], [26][33]) but its flexibility and general usage remains largely unexplored.

Constructing a practical roadmap for applying PAM

There are few studies in the ecological literature that have used PAM for analyzing relationships between plant and soil datasets. The lack of examples partially explains the low popularity of Procrustes analysis among plant and soil ecologists and ecologists in general as an alternative tool to the more traditional Mantel test. In order to make the possible uses of Procrustean residuals more familiar, we will introduce a number of examples in the form of schematic roadmaps for applying PAM in association with three common statistical approaches: ordination, regression analysis and ANOVA.

Plant and soil ecologists must keep in mind that Procrustes analysis requires that the **X** and **Y** have the same number of rows and columns, though the last dimension is less restricting (see below). Given that the data for both matrices usually originate from the same sites, it is most common in ecology that only the number of columns (descriptors or variables) varies between the two matrices. Therefore, the question arises of how to make the number of columns equal across the two matrices, i.e., how to reduce them to the same dimensionality. Although Procrustes analysis can be performed between matrices having different number of dimensions (i.e., the fit is based on a singular value decomposition (svd) of X^TY, where X and Y are scaled prior to svd and T stands for matrix transpose), traditionally the matrix with the fewer number of columns ("missing columns") is made equal in dimension to the larger matrix by adding columns of zeros in order to keep (Fig. 2a; [2]). Although there are some criticisms related to this practice and alternatives have been suggested [34], the addition of zero columns does not affect the distances between columns among observations and is a convenient device rather than a hurdle [35].

Another convenient way to make **X** and **Y** have the same number of columns is to represent most of the variation in their raw data by matrices formed by the same number of orthogonal axes (Fig. 2b; [3],[35, [36]), i.e., matrices formed by axes derived through ordination methods such as Principal Components Analysis (PCA), Non-metric dimensional scaling, Correspondence Analysis (CA), Principal Coordinate Analysis (PCoA), the choice being dependent on the nature of the data (continuous, presence-absence data, abundance data). Moreover, raw data matrices can be transformed prior to ordination (see [37] for different transformations and their characteristics) or alternatively have pairwise distance matrices calculated from the data matrices that are then orthogonolized via PCoA to extract ordination axes based on the chosen distance measure (e.g., Bray-Curtis, Jaccard, Sorensen, Gower).

Here, for simplicity, we use a PCA in all applications. In cases, where species data (presence/absence or abundance) was used, the data was Hellinger-transformed and PCAs were extracted on species correlation matrix calculated from the transformed data. The Hellinger transformation alleviates the issue of double-zeros in species data matrix transformed into correlation or Euclidean-distance pairwise matrices prior to PCA in which sites sharing no species in common can be found to be more similar than sites sharing a reduced number of species in common (e.g., the horse shoe effect in ordination plots).

The general strategy is as follows:

1) Subject the raw data matrices to an ordination method (here PCA but see above for other strategies);
2) After ordinating **X** and **Y**, use the same number of ordination axes for both matrices (Fig. 2b).

Given that the higher the number of ordination axes used, the higher is the amount of variation explained in **X** and **Y**, it would be interesting run the Procrustean analysis sequentially using matrices made up of an increasing number of ordination axes. It could help ecologists check the consistency of the relationship between **X** and **Y** based on different numbers of ordination axes, which will give more reliability to the results.

The use of PAM in ecological ordination

The first form of PAM shown here is based on ordination methods. Ordination is the graphical representation of the variation of objects (sites), descriptors (species/environmental parameters) or both, in a reduced space formed by orthogonal axes [38].

To illustrate the use of Procrustes analysis associated with ordination we use data derived from Mitchell et al. [39]. This study aimed to compare the plant communities and soil chemistry in their ability to predict changes in the structure of the soil microbial community in three moorland areas established in Northern Scotland called Craggan, Kerrow and Tulchan. The plant community matrices from each area were based on the percent cover. Three matrices for the soil microbial community were obtained for each site: one based on the fatty acid profile of the soil (PLFA analysis), and the other two on the T-RFLP analysis of the communities of fungi and bacteria, respectively. The matrix representing the soil chemistry was based on the concentrations of Na, K, Ca, Mg, Fe, Al, P, total C, total N in addition to pH, loss on ignition and moisture.

There is some consensus that the variation in vegetation can act as a proxy for changes in the soil microbial community, either directly in the case of symbionts, for example, or indirectly via changes in soil chemistry itself. We use Procrustes analysis associated with ordination techniques to verify potential drivers of the soil microbial community and to determine if plant community and soil chemistry are equally related to the microbiological variation. The sequence of analysis was as follows:

1) Ordination analysis: All data matrices (community plant, soil chemistry and soil microbial communities) containing the three chronosequences were subjected to separate PCAs based on correlation matrix. The community plant was Hellinger-transformed prior to PCA. Then, the first six PCA ordination axes from each matrix were retained in order to assemble four PCA matrices representing the variation summarized in the first 3, 4, 5 and 6 PCA axes. Thus, four PCA matrices were obtained from each dataset: plant community, soil chemistry and soil microbial community (PLFA, bacterial and fungal T-RFLP) (Fig. 3a).
2) Procrustes analysis: The PCA matrices of plant community and soil chemistry were used to run Procrustean analyses with the PCA matrices of soil microbial community based on PLFA, and fungal and bacterial T-RFLP datasets.

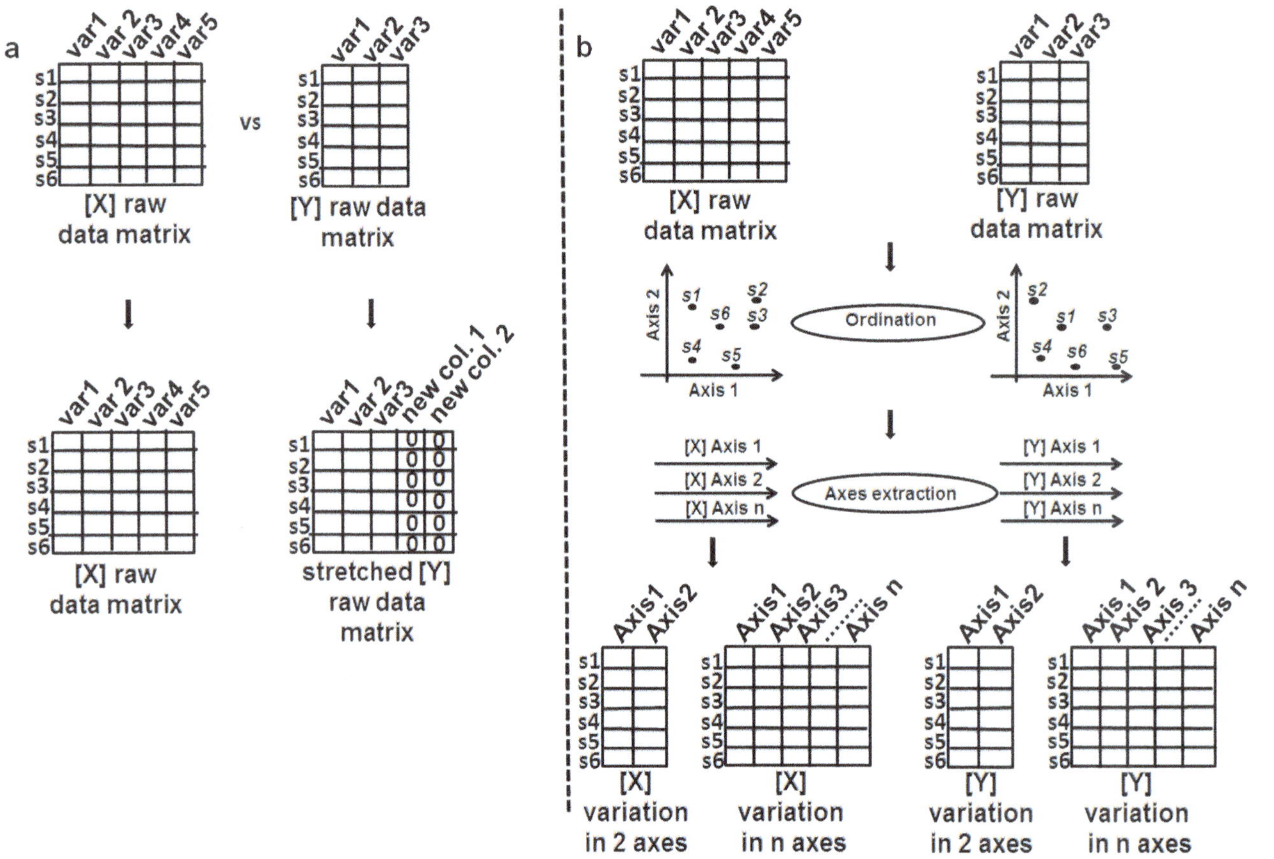

Figure 2. Roadmap for two alternative ways to reach the same dimensionality between matrices, and so relating it by Procrustes analysis. a) Addition of columns containing zeros to the **Y** raw data matrix for matching the **X** raw data matrix dimension; **b)** Application of ordination to raw data matrices to make matrices have equal dimensionality prior to Procrustes analysis.

3) PAM extraction: Since all Procrustean relationships based on PCA matrices with *n* axes were significant, for simplicity, only the PAM obtained from relationships of PCA matrices with 6 axes were used for subsequent analyses. Six PAMs were generated: PAM1 (soil chemistry on PLFA), PAM2 (soil chemistry on bacteria), PAM3 (soil chemistry on fungi), PAM4 (plant on PLFA), PAM5 (plant on bacteria), and PAM6 (plant on fungi) (Fig. 3c).
4) PAM ordination: The PAMs were assembled in a single matrix ("effect matrix") with one PAM per row (Fig. 3c). Therefore, the effect matrix compiled the effects of plant community and soil chemistry on soil microbial community structure derived from the three methods. This effect matrix was submitted to PCA ordination to verify whether the plant community effect on soil microbial community structure differed from the effect of soil chemistry (Fig. 3c).

The results showed that for all chronosequences the plant effect on microbial structure was divergent in relation to the soil chemistry effect, as suggested by the separation along the axis of greatest variation (Fig. 4). Although we cannot apply a proper statistical significance test in one-table based ordination methods (PCA, NMDS, PCoA, etc), visual inferences can be made. For example the Craggan area exhibited a clear distortion between plant community and soil chemistry variation in terms of their effects on the soil microbial community structure depicted by PLFA, bacterial T-RFLP and fungal T-RFLP (Fig. 4a). Also in this area, the response of the microbial community based on PLFA was distant from the response based on molecular data (T-RFLP) (Fig. 4a),

We expected that the effects of soil chemistry on microbial structure were closer to the effect of plant community once the plant community is considered to be a direct and indirect driver for the biotic component of soil [39]. However, these results suggest that plant communities and soil chemistry are acting differently on the soil microbial community structure [24], [40]. They also suggest that the effects of soil chemical properties on the microbial communities may be weakly mediated by above ground alterations [24]. This example shows the usefulness of Procrustes analysis to raise additional evidence in plant and soil ecology studies. (See Text S1 containing the R code used for this example).

The PAM and regression analysis

In regression analysis, 'response' and 'predictor' are common terms. In ecology, predictors can have different natures. Space, time, organic matter and moisture, among other factors, are some examples of predictors. On the other hand the microbial communities are often used as a response variable because they are considered better indicators of a given ecosystem.

Some authors familiar with soil microbial ecology have been using the Mantel test to assess the individual contribution of deterministic and stochastic processes on the soil microbial structure variation [41], [42]. As an example of the utility of the

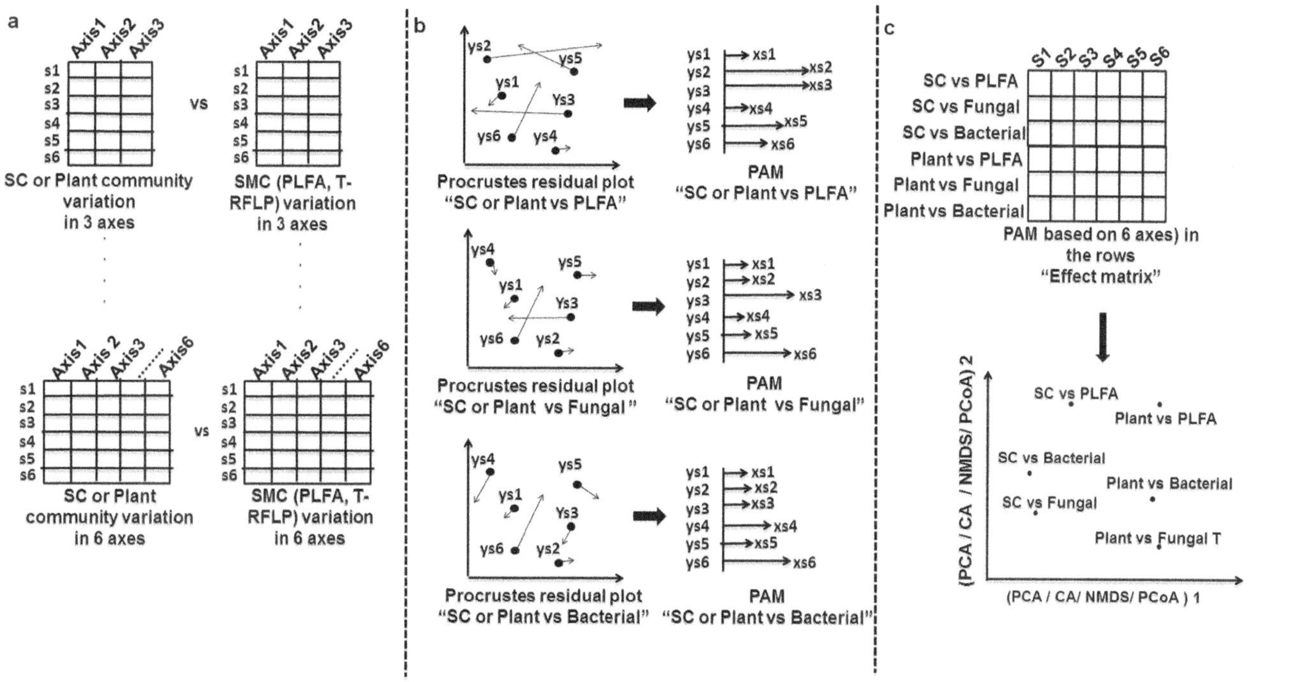

Figure 3. Roadmap for applying the Procrustes association metric (PAM) in the multivariate ordination context using data of [39]. **a)** Assembling matrices with different ordination axes, through Procrustes analysis, soil chemistry (SC) and plant community with soil microbial community (PLFA, and bacterial and fungal T-RFLP); **b)** Extraction of PAM from Procrustean relationships based on matrices with 6 ordination axes; **c)** Assembling of PAM based PCA matrices with 6 axes as rows in a single matrix ("effect matrix"), and using it in an ordination technique (e.g., PCA, PCoA, NMDS) to verify if the different effects diverge.

Procrustes analysis in the context of variation partitioning we can take a hypothetical scenario with four datasets from a given area, corresponding to soil microbial community structure (PLFA), soil microbial functioning (enzyme activities), soil properties and spatial variation. Spatial variation can be represented, for example, by 100 sampling points generated from a 10 m×10 m transects. The matrix of geographical coordinates of the sampling points can be submitted to PCNM (principal coordinates neighbour matrix) analysis generating a matrix of spatial eigenfunctions termed PCNMs [34]. In this scenario, we can assume that the ecologist aims to assess the relative contributions of individual soil properties (deterministic processes) and spatial variation (stochastic event) on the relationship between microbial community structure and soil microbial functioning rather than on these components individually. To use the Procrustean association metric (PAM) in this context, one can use the following steps:

1) Ordinate the two matrices (i.e., the soil microbial community and soil microbial functioning) via PCA (the soil microbial community matrix was Hellinger-transformed) and select a similar number of ordination axes. The multivariate scores of the two matrices across the selected number of axes are subjected to a Procrustes analysis and a PAM was then calculated.
2) Use individual PAMs (based on 2, 3 or more PCA axes) as response variable and soil properties and spatial variation as independent (predictor) variables in a multiple regression framework (Fig. 5b).

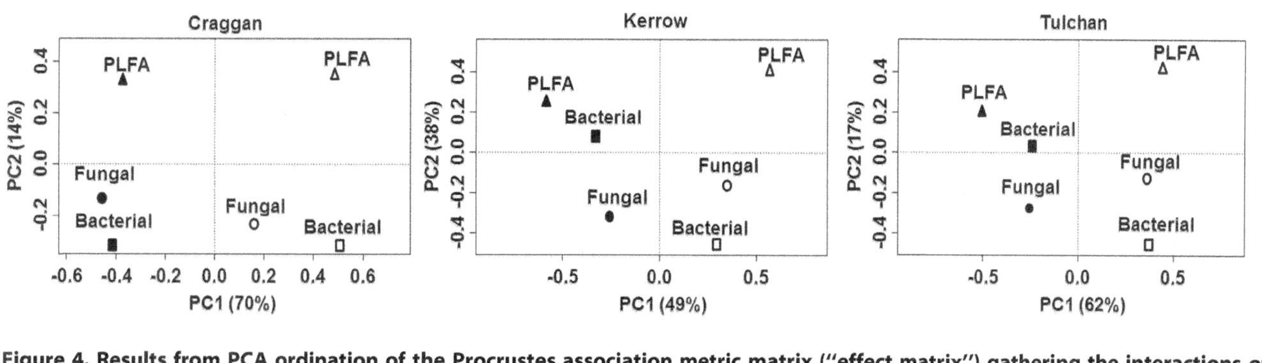

Figure 4. Results from PCA ordination of the Procrustes association metric matrix ("effect matrix") gathering the interactions of soil chemistry and plant community with soil microbial matrices (PLFA, and bacterial and fungal T-RFLP). The filled symbols are the Procrustes relationships between soil chemistry and soil microbial matrices, and the open symbols between plant community and soil microbial matrices. Data from three chronosequences (Craggan, Kerrow and Tulchan) obtained by [39].

3) Finally, the independent contributions of soil properties (independent of space) and unmeasured spatial process and/or factors (spatial variation independent of soil properties) to the microbial structure can be estimated via variation partitioning [43] and represented by a Venn diagram (Fig. 5c). (See Text S2 containing the R code for this example).

The PAM and Analysis of Variance

Although regression and analysis of variance are ultimately the same analysis in which the response is either continuous (regression) or ascribed to factors (ANOVA), we provide examples for each of them in different sections given that often they are seen as distinct forms of analyses. Evaluation of the effects of land use on soil microbial communities has been a common case-study issue in soil ecology. Some of these studies have been carried out using the Mantel approach to assess how land use type effects soil microbial structure and functioning [44], [45]. However, Mantel does not yield a vector of structure – functioning relationship, that is, a continuous variable, able to be partitioned by categorical variables like land use types. In the following example we show how to use PAM to evaluate the effect of land use type on the relationship between microbial community structure and microbial function in the form of PAM.

In a hypothetical scenario, a researcher is interested in studying whether four different land use types within the Amazon biome are affecting the relationship between microbial structure and microbial functioning. In each of the land uses (original forest fragment, silvipastoral system, improved pasture and unimproved pasture) six plots (10 m×10 m) were established and one composite soil sample (0–10 cm) collected per plot (Fig. 6a). The X dataset (soil microbial structure) was represented by PLFA data, and the Y dataset (microbial functioning) by the abundance of genes associated with microorganisms involved in greenhouse gas emission processes, such as nitrifiers, denitrifiers and methanotrophic organisms. The researcher's hypothesis is that in the original forest (non-altered environment) there is a better matching between microbial structure and microbial function. Thus, in anthropogenically disturbed environments (silvipastoral system, improved pasture, and unimproved pasture) the change in microbial structure relative to the original (forest) is not followed by a change in the microbial functioning to the same magnitude. This hypothesis can be tested using an integration of Procrustes analysis and ANOVA through the following steps:

1) Reduce the datasets **X** (soil microbial structure) and **Y** (soil microbial functioning) to similar dimensions using PCA. Then, run the Procrustean analysis between the PCA matrix of the soil microbial community structure and the PCA matrix of soil microbial functioning and extract the PAM (Fig. 6a).
2) Run an ANOVA with land use type as fixed factor and the PAM as the response variable (Fig. 6b).
3) If the F value of ANOVA is significant, a means test can be performed to compare the mean PAMs of the land use types (Fig. 6c). (See Text S3 containing the R code for this example).

Discussion

In this paper we have attempted to show the advantages of the Procrustean analysis over the Mantel test, in which the former can be used for gaining further information on underlying drivers of data table associations. In particular we have shown how the Procrustean association metric (PAM) constructed of the residuals

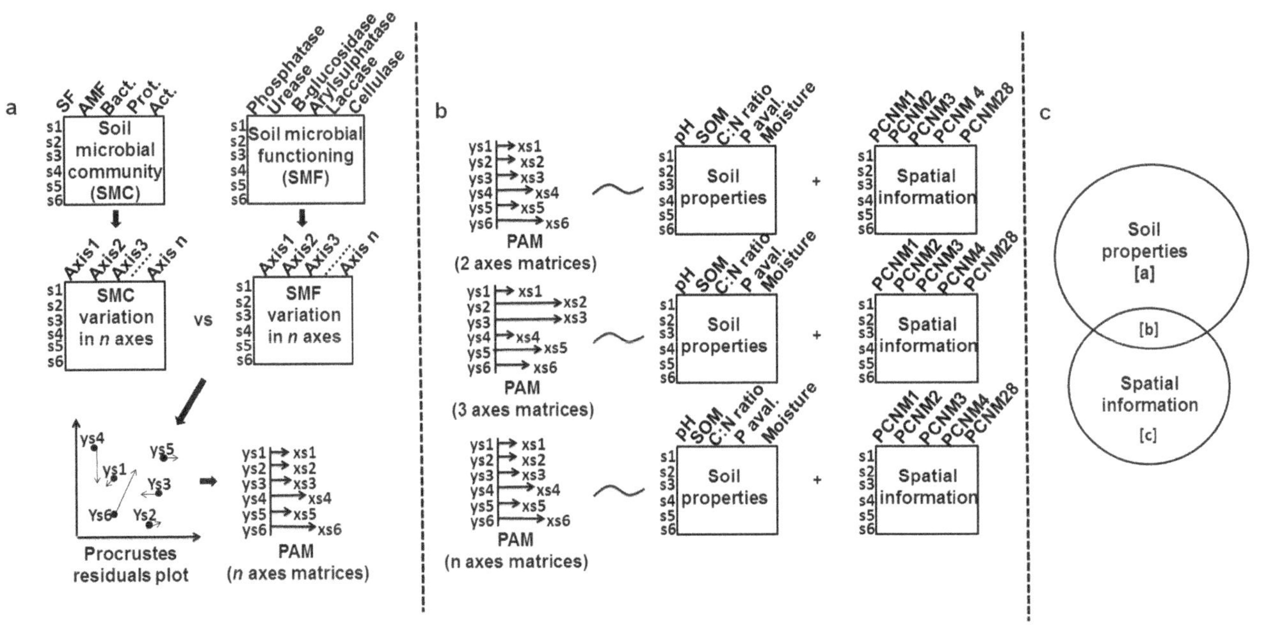

Figure 5. Roadmap for using Procrustes Association Metric (PAM) in a multiple regression analysis framework (variation partitioning). a) Soil microbial community (SMC) and soil microbial functioning (SMF) matrices are submitted to an ordination to reach the same dimensionality, and SMC and SMF matrices formed by 2, 3 and *n* axes related through Procrustes analysis in order to generate PAMs; **b)** PAMs generated were used as response variables in a variation partitioning to verify the individual contribution of soil properties and spatial information (PCNM eigenfunctions) on the SMC-SMF relationship; **c)** Venn diagram depicting the relative contribution of soil properties (niche processes **[a]**) and unmeasured spatial factors (neutral processes **[c]**).

of the vectors after the Procrustes analysis. We concentrated on showing how patterns of concordance between data matrices can be displayed and individual observations contrasted separately using the Procrustean framework, allowing further examination of the common and different association patterns among multiple data matrices. Given that in the Mantel framework, multivariate information is translated into a pairwise distance matrix, we lose the ability to contrast homologous data points across dimensions and data matrices. It is important to notice that it was not our goal to show the statistical advantages of Procrustes over Mantel as done by previous work [3]. Instead, we concentrated on generating different analytical schemes, especially for plant and soil ecologists, to incorporate Procrustes into their statistical toolbox.

What is unique about Procrustean framework? There are at least four characteristics of the approach not shared by others. First, because the approach is correlative rather than regressive, the number of observations (e.g., sites) in the matrices does not have to be greater than the number of columns as in common regression approaches such as RDA and CCA. Second, we can fit as many matrices as we have available; this latter issue is particularly restrictive under a regression approach given the limitation of number of rows versus number of columns. Moreover, all matrices are treated in equal footing as no matrix is treated as response or predictor. Third, the relationships within (only across) matrix columns do not affect the analysis. Fourth, residual values across observations and dimensions can be calculated and explored as shown here. These characteristics should not be necessarily seen as advantages per se over other methods but rather features that are unique and may be useful in many situations. There are certainly other tools that can be used to look at the associations between data sets. RDA and CCA are well-established tools in ecology and are based on regression (asymmetric) methods. Traditionally these approaches may have been thought to be more appropriate for analysis of the examples given in this paper, since they establish relations of cause and effect. However, because these analyses include a regression step, they are limited to situations where the number of rows (sites) in the environmental matrix **X** is higher than the number of columns (variables) [36], [46]. This is not a limitation in Procrustes analysis and moreover, it is not clear how residual variation among homologous observations across dimensions should be explored in the case of RDA and CCA.

At least two other symmetric approaches are similar to the Procrustean approach, namely Co-inertia analysis [47] and symmetric Co-correspondence analysis [48], a form of Co-inertia analysis in which a correspondence analysis is applied to two species matrices prior to the analysis. The main difference resides in the fact that fit is influenced by all variables pairs in Co-inertia analysis (within and between matrices), whereas fit is influenced only by variation between matrices in Procrustean. Co-inertia is always based on ordination within data matrices, whereas in Procrustes either the raw data or their ordination axes can be used. Co-inertia can also take into account row (e.g., sites) and column (e.g., species) weights in the analysis, though the standardization and fit processes in Procrustean analysis could also take these into account [36]. Co-inertia and Procrustean analysis are certainly related in the sense that they both treat matrices as symmetrical during the fitting process, though more studies are necessary to assess in which conditions (e.g., correlation within and across matrices, differences in dimensionality between matrices, outliers within and across matrices) they differ. Finally Dray et al. [36] showed the advantages of merging Co-inertia and Procrustean analysis, where the latter is used as a precursor of the former. In reality, future studies are required to contrast Co-inertia and Procrustean analysis, but in either form of analyses we can produce residual vectors (PAM) that can be further analyzed.

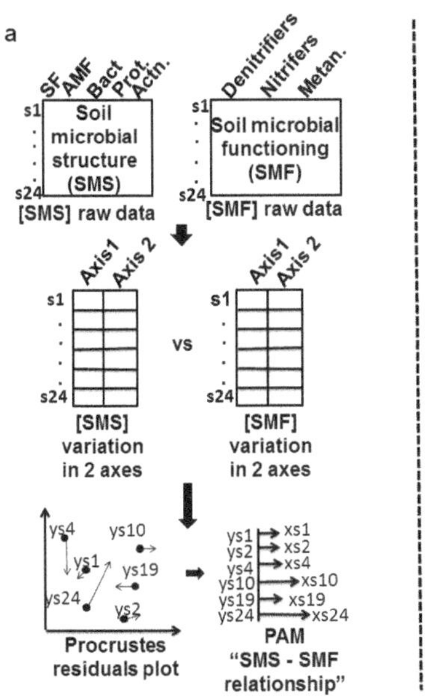

b

Table. One-way ANOVA of the effect of different land use types on the Procrustean relationship between soil microbial structure (SMS) and soil microbial functioning (SMF).

Land use type (fixed factor)	PAM (SMS-SMF relationship)	
	F	*P*
Land use	7.047	0.003**
Residuals	1.345	>0.05

c

Table. Mean effect of different land use types on the Procrustean relationship between soil microbial structure (SMS) and soil microbial functioning (SMF).

Land use types	SMS-SMF relation (PAM in 2 axes)	
	Mean	SE
Original forest	0.021 a	0.003
Silvopastoral system	0.052 b	0.001
Improved pasture	0.074 c	0.003
Unimproved pasture	0.082 c	0.002

Figure 6. Roadmap for using Procrustes association metric (PAM) in an ANOVA context. a) PCA ordination of each SMC and SMF raw data matrices, and then Procrustes correlation from 2 axes-based PCA matrices in order to generate the PAM depicting the SMC-SMF relationship. **b)** Table showing results of a one-way ANOVA for using PAM as response and land use type as fixed factor. **c)** Multiple comparisons test (Tukey, 95%) for means of the Procrustean relationship between soil microbial structure and functioning (PAM in 2 axes) across land use types.

Procrustes can be perhaps best justified when the number of predictors is greater than the number of observations or when **X** and **Y** matrices are equally applicable as explanatory and response variables. In plant-soil ecology, for example, above- and below-ground data matrices can be interchanged as explanatory and response variables. Plant community variation has been shown to be related to variation in below-ground compartments [24]. In addition, soil components such as fertility and the microbial community have been proven to influence aspects of vegetation [49]. Thus, with the literature showing that both types of datasets under analysis can structure each other, the use of Procrustes analysis, as a symmetric canonical analysis method, should be encouraged among plant and soil ecologists and ecologists in general. We hope that this paper has provided enough examples of the potential for using the Procrustes framework as a precursor to further explore ecological data.

Supporting Information

Text S1 R code showing how to use PAM associated to ordination methods (Fig. 4 in the main text). For this example we used data from Mitchell et al. [39].

Text S2 R code showing how to use the PAM in a Regression framework (Fig. 5 in the main text).

Text S3 R code showing how to use the PAM in an ANOVA framework (Fig. 6 in the main text).

Author Contributions

Conceived and designed the experiments: FJGL. Analyzed the data: FJGL. Wrote the paper: FJGL PRPN GMC ECJ SJC RJM RLLB. Contributed with real datasets: SJC RJM.

References

1. Mantel N (1967) The detection of disease clustering and a generalized regression approach. Cancer Res 27: 209–220.
2. Gower JC (1971) Statistical methods of comparing different multivariate analyses on the same data. In: Hodson FR, Kendall DG, Tautu P, editors. Mathematics in the archeological and historical sciences. Edinburgh University Press, Edinburgh. 138–149.
3. Peres-Neto PR, Jackson DA (2001) How well do multivariate data sets match? The advantages of a Procrustean superimposition approach over the Mantel test. Oecologia 129: 169–178.
4. Tuomisto H, Poulsen AD, Ruokolainen K, Moran RC, Quintana C, et al. (2003a) Linking floristic patterns with soil heterogeneity and satellite imagery in Ecuadorian Amazonia. Ecol App 2: 352–371.
5. Tuomisto H, Ruokolainen K, Yli-Halla M (2003b) Dispersal, environment, and the floristic variation of western Amazonian forests. Science 299: 241–244.
6. Tuomisto H, Ruokolainen K, Aguilar M, Sarmiento A (2003c) Floristic patterns along 43-km long transect in an Amazonian rain forest. J Ecol 91: 743–756.
7. Kang S, Mills AL (2004) Soil bacterial community structure changes following disturbance of the overlying plant community. Soil Sci 169: 55–65.
8. Poulsen AD, Tuomisto H, Balsev H (2006) Edaphic and florist variation within a 1-ha plot of lowland Amazonian rain forest. Biotropica 38: 468–478.
9. Fitzsimons MS, Miller RM, Jastrow JD (2008) Scale-dependent niche axes of arbuscular mycorrhizal fungi. Oecologia 158: 117–127.
10. Powers JS, Becknell JM, Irving J, Pères-Aviles D (2009) Diversity and structure of regenerating tropical dry forests in Costa Rica: Geographic patterns and environmental drivers. Forest Ecol Manag 258: 959–970.
11. Castilho-Monroy AP, Bowker MA, Maestre FT, Rodriguez-Echeverria S, Martinez I, et al. (2011) Relationships between biological soil crusts, bacterial diversity and abundance, and ecosystem functioning: insights from a semi - arid Mediterranean environment. J. Veg. Sci 22: 165–174.
12. Pomara LY, Ruokolainen K, Tuomisto H, Young K (2012) Avian composition co-varies with floristic composition and soil nutrient concentration in Amazonian upland forests. Biotropica 44: 545–553.
13. Artz RRE, Chapman SJ, Campbell CD (2006) Substrate utilization profiles of microbial communities in peat are depth dependent and correlate with whole soil FTIR profiles. Soil Biol Biochem, 38: 2958–2962.
14. Trivedi MR, Marecroft MD, Berry PM, Dowson TP (2008) Potential effects of climate change on plant communities in three montane nature reserves in Scotland, UK. Bio Conserv 141: 1665–1675.
15. Jesus EC, Marsh TL, Tiedje JM, Moreira FMS (2009) Changes in land use alter the structure of bacterial communities in Western Amazon soils. ISME J 3: 1004–1011.
16. Merilä P, Lämsa- MM, Stark S, Spetz P, Vierikko K, et al. (2010) Soil organic matter quality as a link between microbial community structure and vegetation composition along a successional gradient in a boreal forest. App Soil Ecol 46: 259–267.
17. Grayston SJ, Campbell CD, Bardgett RD, Mawdsley JL, Clegg CD, et al. (2004) Assessing shifts in microbial community structure across a range of grasslands of differing management intensity using CLPP, PLFA, and community DNA techniques. Appl Soil Ecol 25: 63–84.
18. Singh BK, Munro S, Reid E, Ord B, Potts M, et al. (2006) Investigating microbial community structure in soils by physiological, biochemical and molecular fingerprinting methods. Eur J Soil Sci 57: 72–82.
19. Vinten AJA, Artz RRE, Thomas N, Potts JM, Avery L, et al. (2011) Comparison of microbial community assays for the assessment of stream biofilm ecology. J Microbiol Methods 85: 190–198.
20. Hirst CN, Jackson DA (2007) Reconstructing community relationships: the impact of sampling error, ordination approach, and gradient length. Divers. Distrib 13: 361–371.
21. Poos MS, Jackson DA (2012) Addressing the removal rare species in multivariate bioassessments: The impact of methodological choices. Ecol Indic 18: 82–90.
22. Singh BK, Nunan N, Ridgway KP, Mcnicol J, Young JPW, et al. (2008) Relationship between assemblages of mycorrhizal fungi and bacteria on grass roots. Environ Microbiol 10: 534–542.
23. Burke L, Irwin R (2009) The importance of interannual variation and bottom up nitrogen enrichment for plant – pollinator networks. Oikos 118: 1816–1829.
24. Lisboa FJG, Chaer GM, Jesus EC, Gonçalves FS, Santos FM, et al. (2012) The influence of litter quality on the relationship between vegetation and below-ground compartments: a Procrustean approach. Plant Soil 367 551–562.
25. Landeiro VL, Bini LM, Costa FRC, Franklin E, Nogueira A, et al. (2012) How far can we go in simplifying biomonitoring assessments? An integrated analysis of taxonomy surrogacy, taxonomic sufficiency and numerical resolution in a mega diverse region. Ecol Indic 23: 366–373.
26. Siqueira T, Bini LM, Roque FO, Cottiene K (2012) A metacommunity framework for enhancing the effectiveness of biological monitoring strategies. PLoS One 7: e43626.
27. Nekola JC, White PS (1999) The distance decay of similarity in biogeography and ecology. J Bio 26: 867–878.
28. Legendre P, Borcard D, Peres-Neto PR (2005) Analysing beta diversity: partitioning the spatial variation of community composition data. Ecol Mon 75: 435–450.
29. Tuomisto H, Ruokolainen K (2006) Analysing or explaining beta diversity? Understanding the targets of different methods of analysis. Ecology 87: 2697–2708.
30. Kuehnelt-Leddihn ER (2007) The menace of the herd or Procrustes at large. The Bruce Publishing Company Milwaukee, Auburn. 385 p.
31. Jackson DA (1995) PROTEST: a Procrustean randomization test of community environment concordance. Ecoscience 2: 297–303.
32. Alárcon R, Waser NM, Orlleton J (2008) Year-to-year variation in the topology of a plant - pollinator interaction network. Oikos 117: 1796–1807.
33. Burke L, Irwin R (2009) The importance of interannual variation and bottom up nitrogen enrichment for plant – pollinator networks. Oikos 118: 1816–1829.
34. ten Berge JMF, Kiers HAL, Commandeur JJF (1993) Orthogonal Procrustes rotation for matrices with missing values. B J Math Stat Psyc 46: 119–134.
35. Dijksterhuis GB, Gower JC (1992) The interpretation of Generalized Procrustes Analysis and allied methods. Food Qual Prefer 3: 67–87.
36. Dray S, Chessel D, Thioulouse J (2003) Procrustean Co-inertia analysis for the linking of multivariate datasets. Ecoscience 10: 110–119.
37. Legendre P, Gallagher ED (2001) Ecologically meaningful transformations for ordination of species data. Oecologia 129: 271–280.
38. Legendre P, Legendre L (2012) *Numerical Ecology*, 3rd English edn. Elsevier Science BV, 516 Amsterdam.
39. Mitchell RJ, Hester AJ, Campbell CD, Chapman SJ, Cameron CM (2010) Is vegetation composition or soil chemistry the best predictor of soil microbial community? Plant Soil 333: 417–430.
40. Mitchell RJ, Hester AJ, Campbell CD, Chapman SJ, Cameron CM, et al. (2012) Explaining the variation in the soil microbial community: do vegetation composition and soil chemistry explain the same or different parts of the microbial variation? Plant Soil 351: 355–362.
41. Dumbrell AJ, Nelson M, Helgason T, Dytham C, Fitter AH (2009) Relative roles of niche and neutral processes in structuring a soil microbial community. ISME J doi:10.1038/ismej.2009.122.

42. Zheng YM, Cao P, Fu B, Hughes JM, He JZ (2013) Ecological Drivers of Biogeographic Patterns of Soil Archaeal Community. PloS One 8: e63375.
43. Peres-Neto PR, Legendre P, Dray S, Borcard D (2006) Variation partitioning of species data matrices: estimation and comparison of fractions. Ecology 87: 2603–2613.
44. Chaer GM, Fernandes MF, Myrold DD, Bottomley PJ (2009) Shifts in microbial community composition and physiological profiles across a gradient of induced soil degradation. Soil Sci Soc Am J 73: 1327–1334.
45. Peixoto RS, Chaer GM, Franco N, Reis Junior FB, Mendes IC, et al. (2010) A decade of land use contributes to changes in the chemistry, biochemistry and bacterial community structures of soils in the Cerrado. A Van Leeuw 98: 403–413.
46. Thioulouse J, Simier M, Chessel D (2004) Simultaneous analysis of a sequence of paired ecological tables. Ecology 85: 272–283.
47. Dolédec S, Chessel D (1994) Co-inertia analysis: an alternative method for studying species–environment relationships. Freshwater Biol 31: 277–294.
48. Ter Braak, CJF, Schaffers AP (2004) Co-correspondence analysis: a new ordination method to relate two community compositions. Ecology 85: 834–846.
49. van der Heijden MGA, Klironomos J, Ursic M, Moutoglis P, Streitwolf-Engel R, et al. (1998) Mycorrhizal fungi determines plant biodiversity, ecosystem variability and productivity. Nature 396: 69–72.

4

Long Term Effect of Land Reclamation from Lake on Chemical Composition of Soil Organic Matter and Its Mineralization

Dongmei He, Honghua Ruan*

Faculty of Forest Resources and Environmental Science, and Key Laboratory of Forestry and Ecological Engineering of Jiangsu Province, Nanjing Forestry University, Nanjing, Jiangsu, China

Abstract

Since the late 1950s, land reclamation from lakes has been a common human disturbance to ecosystems in China. It has greatly diminished the lake area, and altered natural ecological succession. However, little is known about its impact on the carbon (C) cycle. We conducted an experiment to examine the variations of chemical properties of dissolved organic matter (DOM) and C mineralization under four land uses, i.e. coniferous forest (CF), evergreen broadleaf forest (EBF), bamboo forest (BF) and cropland (CL) in a reclaimed land area from Taihu Lake. Soils and lake sediments (LS) were incubated for 360 days in the laboratory and the CO_2 evolution from each soil during the incubation was fit to a double exponential model. The DOM was analyzed at the beginning and end of the incubation using UV and fluorescence spectroscopy to understand the relationships between DOM chemistry and C mineralization. The C mineralization in our study was influenced by the land use with different vegetation and management. The greatest cumulative CO_2-C emission was observed in BF soil at 0–10 cm depth. The active C pool in EBF at 10–25 cm had longer (62 days) mean residence time (MRT). LS showed the highest cumulative CO_2-C and shortest MRT comparing with the terrestrial soils. The carbohydrates in DOM were positively correlated with CO_2-C evolution and negatively correlated to phenols in the forest soils. Cropland was consistently an outlier in relationships between DOM chemistry and CO_2-evolution, highlighting the unique effects that this land use on soil C cycling, which may be attributed the tillage practices. Our results suggest that C mineralization is closely related to the chemical composition of DOM and sensitive to its variation. Conversion of an aquatic ecosystem into a terrestrial ecosystem may alter the chemical structure of DOM, and then influences soil C mineralization.

Editor: Han Y.H. Chen, Lakehead University, Canada

Funding: This study was supported by the MOST of China (973 Program No. 2012CB416904), the National Science Foundation of China (No. 31170417) and partially supported by PAPD and Collaborative Innovation plan of Jiangsu higher education, Doctorate Fellowship Foundation of Nanjing Forestry University. The funders had no role in study design, data collection and analysis, decision to publish, or preparation of the manuscript.

Competing Interests: The authors have declared that no competing interests exist.

* E-mail: hhruan@njfu.edu.cn

Introduction

Microbial respiration of soil organic matter (SOM) causes a large flux of CO_2 to the atmosphere, and understanding the controls on this flux is a critical component of society's effort to cope with increasing atmospheric CO_2 and climate change [1–2]. Many abiotic and biotic factors affect the global carbon (C) cycle, especially, human activities, such as the fossil fuel combustion and land use change [3]. Of these factors, land use changes, such as deforestation and afforestation, draining of wetlands, converting grassland to arable cropping and reclaiming lakes, directly affect microbial respiration [4–5]. Many studies on the influence of land use change have shown it results in drastic changes to soil C cycling [6–8]. Land use determines the vegetation type grown on the soil, and therefore the type of organic matter input to the ecosystem [5,9]. Land use change can also increase microbial C mineralization, which may exacerbate the trend of global warming and other effects related to climate change [7].

In the late 1950s, reclaiming land from lakes became a common type of land use in China. China is a traditional agricultural country with a growing population, and it was deemed necessary to expand cultivated land to develop agriculture as well as feed an increasing population. Thus, after several decades more than 1.4×10^4 km^2 of land was reclaimed from lakes [3,10]. Taihu Lake as one of the five largest freshwater lakes in China also experienced land reclamation. Taihu Lake is situated in the south of the Yangtze Delta among 30°55′–31°33′N and 119°55′–120°36′E and with a land area of 2.3×10^3 km^2. From the 1950s to 1980s, more than 160 lakes around the Taihu Lake basin were disappeared and the area of water surface had reduced by 13.6% [11]. In recent years, many studies have focused on the effects of reclamation on ecosystem. They observed that many negative consequences caused by reclaiming land from lakes, such as species decline for habitat loss, eliminating the natural buffers and flood control capacity, water pollution and eutrophication caused by agriculture measures and so on [12]. However, there are few studies about the effect of lake reclamation on the properties of soil organic matter in the terrestrial ecosystem and little is known about its impact on soil C cycling.

Solid soil organic matter must pass through the dissolved phase to be decomposed by microbes [13–14]. Although dissolved organic C (DOC) makes up only a small portion of total soil organic carbon (mostly less than 1%), it represents the active or labile C pool [15–17] and turn over more than 4000 times annually [18]. Thus, the DOM pools is closely correlated to the labile fraction of soil organic matter and is a sensitive indicator of its overall dynamics [16,19]. Also, many studies observed the close relationships between biodegradability of DOC and C mineralization [14,16,20].

Land use is considered as the factor with the greatest influence on soil DOM because it determines the quantity and quality of organic matter input into the soil [5]. However, the mechanism of how land use influences dynamics of DOM is still not clear. Several authors have found that biodegradability of DOM largely depends on its chemical composition [21–23]. It is generally assumed that the easily degradable DOM consists mainly of simple carbohydrate monomers (i.e., glucose, fructose), low molecular organic acids (i.e., citric, oxalic, succinic acid), amino acids, amino sugars and low molecular weight proteins [13,18]. The stable DOM fraction generally contains polyphenolic, aromatic structures and other complex macromolecules [24]. So, the chemical structure of DOM and the complexity of its molecules were considered to correlate with C mineralization. The chemical structure of DOM often described with UV and fluorescence spectroscopy [25–26]. Coupling and understanding of the chemical composition of DOM and C mineralization could be a useful tool for evaluating the influence of land use change on soil C dynamics.

In a previous study, Wang et al. [3] reported the labile SOC concentrations were different from each other under four types of land use from land reclaimed from Taihu Lake, China. Here we endeavored to link the C mineralization potential to the chemical makeup of DOM in soils of land that had been converted from lake sediments to forests and cropland. We hypothesized that this conversion of aquatic ecosystems to terrestrial ecosystems resulted in increased recalcitrance in chemical structure of DOM and ultimately affected potential soil C mineralization, because terrestrial organic matter is usually rich in aromatic and phenolic structures while DOM from aquatic environments has a lower content of aromatic C and more aliphatic C [27–28]. Our objectives were to (1) test for differences in C mineralization in lake sediments and terrestrial soils under different land use on reclaimed land; (2) compare the chemical composition of DOM in lake sediments and soils under different land use; (3) relate C mineralization and chemical properties of DOM in order to explore the role of DOM in C cycling in reclaimed soils.

Materials and Methods

Site description

The study was conducted at the Xiaodian Lake Forest (31°10′N, 120°48′E), located in the northeast of Wujiang City, Jiangsu province, southeast of China (fig. 1) (The Department of Agriculture & Forestry of Jiangsu Province is the authority responsible for it. There were no specific permissions required for the study area. Our field studies did not involve endangered or protected species). The climate in this region is humid north subtropical monsoon with a mean annual temperature of 16°C, annual rainfall of approximately 1100 mm (mainly during the summer months), annual average relative humidity of 78%, and an annual non-frost period of up to 240 days [3]. This area used to be a part of the Taihu Lake, and the lake area was converted to farmland in the early 1960s. The soil properties of the reclaimed land were unsuitable for food crops and so, afforestation projects were carried out in 1969 in much of the region. After more than 40 years of forest management, it has developed into a forest park with 1.33 km^2 forest-covered area. The dominant forest types in the park are dominated by coniferous forest (*Metasequoia glyptostroboides*), evergreen broadleaf forest (*Cinnamomum camphora*), bamboo forest (*Phyllostachys heterocycla*) and all of them are the planted tree species monoculture. Additionally, there is approximately 7.0 ha of cropland with rotations of rice and canola in parcels around the park and with the common tillage practices of ploughing, irrigation, mineral and organic fertilization. Soils in this region are all derived from lake sediment and were similar following reclamation. The general characteristics of each site were described by Wang et al. [3].

Soil sampling

We chose four sites within coniferous forest (CF), evergreen broadleaf forest (EBF), bamboo forest (BF) and cropland (CL), respectively. Four 10 m×10 m plots were randomly established within each site. Soil samples were collected randomly using a 3 cm diameter soil sampler at 0–10 cm and 10–25 cm depths from each plot in October 2011. After removal of the litter layer, ten soil cores were taken inside each plot and pooled to form a composite soil sample. Visible roots, residues and stones were immediately removed after sampling and then field-moist soil samples were sieved through a 2 mm mesh. Soil samples were divided into two parts: one stored at 4°C for analysis of soil microbial biomass, DOM and C mineralization incubation experiments. The other part was air dried and analyzed for pH, soil total carbon (TC) and total nitrogen (TN). Lake sediments (LS) were sampled with plexiglass tube (11 cm I.D., 50 cm in length) by cylindrical sampler (Rigo Co. Φ110 mm×500 mm) at four points in the water area of the lake. Sediments in the tube were push-back and sliced at 10 cm for 0–10 cm depth and then 15 cm for 10–25 cm depth using a stainless steel spatula [29–30]. A total of 40 sediment samples were collected and kept in room temperature of 4°C.

Soil physical and chemical characteristics

Soil bulk density was measured with the core method in each layer by using a 5 cm×5 cm metal cylinder [31]. Soil moisture was determined gravimetrically by oven drying at 105°C for 24 h. Soil pH was measured on air-dried soil in a 1:2.5 (w/w) soil-water suspension using a glass electrode. Soil TC and TN were determined via dry combustion and thermal conductivity detection using an elemental analyzer (Vario EL III, Elementar, Germany). Microbial biomass carbon (MBC) was determined by chloroform fumigation as described by Vance et al. [32].

Incubation experiment and analytical methods

Carbon mineralization. C mineralization was measured in each soil sample during incubation in the dark at 25°C for 360 days [33]. From each incubation sample, 100 g fresh moist soil was incubated in a 1000 ml sealed Mason jar. A 50 ml beaker containing 20 ml 1 M NaOH solution was placed in each jar to trap the evolved CO_2. Three additional jars containing 20 ml 1 M NaOH solution but no soil were used as a control. During incubation, the evolved CO_2 trapped in NaOH was determined by titration with 1 M HCl after precipitating the carbonate with 1 M $BaCl_2$ solution. After the CO_2 traps were taken out, the jars were left open for 2 h to maintain aerobic conditions in each jar, and resealed for further incubation. Water loss in the jars was monitored by weighing the jars and replenished by adding ultrapure water after opening [34]. The cumulative C mineralization was expressed as g CO_2-C kg^{-1} soil.

DOM extraction and characterization. DOM was extracted from each soil sample before and after soil incubation. 10 mM

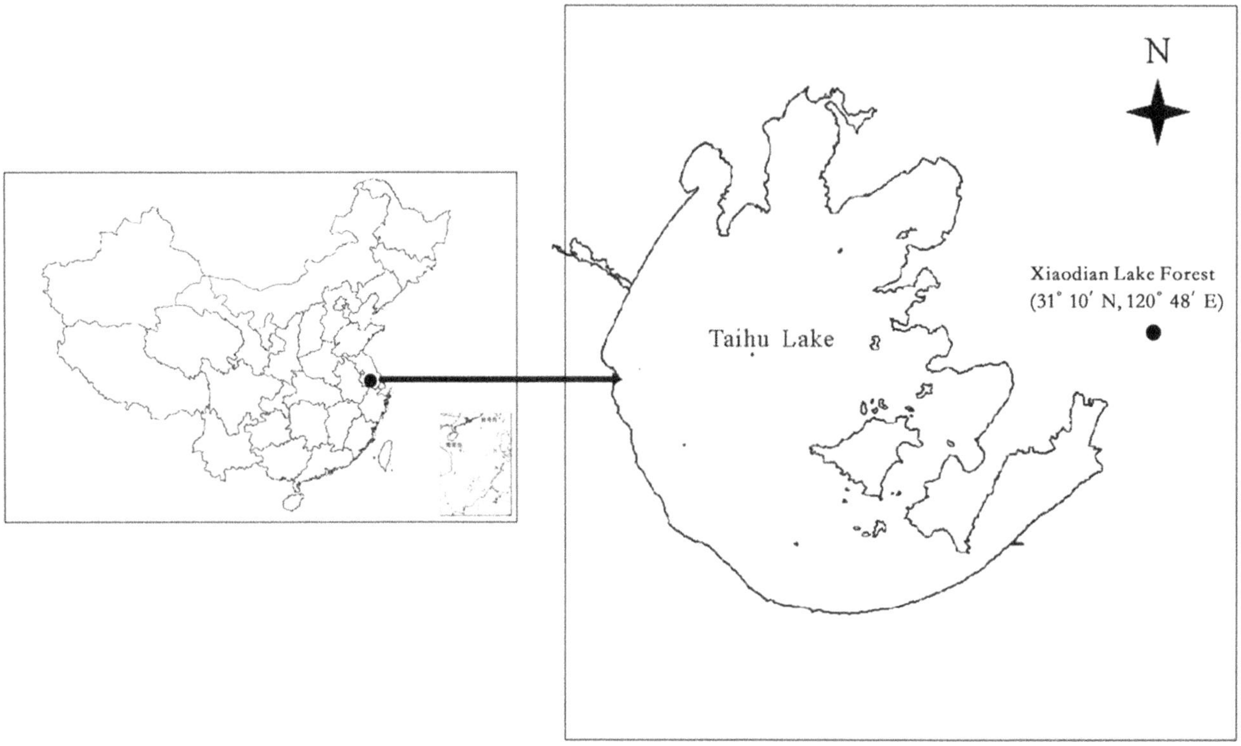

Figure 1. The map of study site in Xiaodian Lake area of the Taihu Lake basin China.

$CaCl_2$ solution was added to an aliquot of soil samples at a soil-solution ratio of 1:2 (w/v) and shaken for 30 min in a horizontal shaker at room temperature. The suspension was spun in a centrifuge for 20 min at 4000 rpm and filtered through a 0.45 μm pore-size cellulose acetate membranes (Schleicher & Schuell, OE 67). The filtered solutions were stored frozen (−20°C) for additional analyses.

DOC concentration was determined by Shimadzu TOC-5050 total organic carbon analyzer. Hydrophilic and hydrophobic fractions in DOM were separated by Amberlite XAD-8 resin (Rohm and Haas, Philadelphia, PA) [35]. Briefly, DOM solutions were acidified to pH 2 with HCl, and then eluted through glass columns filled with Amberlite XAD-8 resin. The column effluent representing the hydrophilic fraction of DOM (Hi) was analyzed for organic C and the amount of organic C in hydrophobic fraction was calculated by the difference between organic C in DOM solution and in hydrophilic fraction. Total carbohydrates in DOM (CH) were measured by phenol-sulfuric acid method [36], using glucose as a standard. Total soluble phenolic compounds (Phe) were analyzed by Folin-Ciocalteau method [37], using tannic acid as a standard.

For spectroscopic measurements, all samples were diluted to 10 mg C L^{-1} to avoid concentration effects and were brought to a constant pH of 7.7 [21] by adding NaOH. UV absorption at 254 nm of DOM was measured using a UV/Vis spectrophotometer (Shimadzu UV-2550, Japan). The specific UV absorbance values ($SUVA_{254}$) were determined as the ratio of the UV absorbance at 254 nm to the DOM concentration and multiplying the value of 100, i.e. $(UV_{254}/DOC) \times 100$. $SUVA_{254}$ can serve as an indicator of the aromaticity of DOM.

Fluorescence emission spectra were obtained with a Varian Cary Eclipse Fluorescence Spectrophotometer (λ_{ex} 254 nm, slit 10 nm, λ_{em} 300–480 nm, slit 10 nm, and scan speed 1200 nm min^{-1}) using 1 cm cuvettes. Fluorescence efficiency index (FE) proposed by Ewald et al. [38] was considered to be proportional to the quantum efficiency and was defined as the ratio of maximum fluorescence intensity (F_{max}) divided by UV absorption at the excitation wavelength of 254 nm (F_{max}/Abs). The emission humification index (HIX_{em}, dimensionless) was determined as the ratio between the area in the upper quarter (Σ435–480 nm) of the usable fluorescence emission spectrum and the area in the lower usable quarter (Σ300–345 nm) [39–41]. The HIX_{em} can show the degree of complexity and condensation of the DOM.

Statistical analysis

The double exponential model [20,21,42], separating the mineralized organic C into an active C pool and slow C pool can be presented as:

$$C_m(\%) = a \times [1 - \exp(k_1 t)] + (100 - a) \times [1 - \exp(-k_2 t)]$$

where, t is incubation days; C_m is cumulative value of mineralized C at t time presented as percent of initial C in the soil; a is the portion of organic C that is readily decomposed (in % of initial C in the soil = labile C); (100 − a) is the part of organic C that is slowly decomposed (stable C); k_1 and k_2 are mineralization rate constants for active and stable C pools (day^{-1}). Mean residence time (MRT) for each pool is calculated as the reciprocal of the decomposition rate constant in the double exponential model.

This double decomposition equation for two distinct C pools with different mineralization rate constants was fitted with the nonlinear regression that was used in the Marquardt algorithm and an iterative process to find the parameter values that could minimize the residual sum of squares. The model that gave the least squared error was chosen to be the best. For a model to be chosen, all the parameters had to generate plausible results in realistic ranges. For example, the rate constants could not be negative, and the sum of active and stable C pools should not exceed initial C in the soil.

All results were reported as the mean ± SE of the four field replicates. Differences in soil properties, mineralization parameters and DOM chemical properties among land use types for each site were tested using ANOVA followed by least significant difference (LSD). T-test was used to assess the differences between soil depths. Statistical significance was determined at $p<0.05$ level. Linear regression analysis was used to determine relationships between C mineralization and chemical properties of DOM. We calculated Pearson correlation coefficients among properties of soil, C mineralization parameters and the chemical characteristics of DOM to discuss the influence of selected parameters on C mineralization. All statistical analyses were performed with SPSS 13.0.

Results

Soil properties

We observed variation in several soil characteristics under different land uses and soil depth (Table 1). Soil bulk density was lower at 0–10 cm soil depth than 10–25 cm depth and BF was significantly lower than all other land uses ($p<0.05$). Soil pH in all soils and sediments were ranging from 4.13 to 6.30 with the highest value in LS. TC and TN concentrations from the terrestrial soils were significantly higher at 0–10 cm soil depth (TC 14.95–35.35 g C/kg soil, TN 1.77–3.41 g N/kg soil), but both of these two factors for LS were lower at upper layer. BF soils were observed remarkably higher TC and TN than other land uses. There was significant difference in MBC and DOM between the land use types and soil depth (Table 1). MBC and DOM were significantly higher at 0–10 cm soil depth in terrestrial land uses. Under the four land uses, MBC showed highest value in CL and lowest value in CF at both soil depths, while DOM was significantly greater in CF at 0–10 cm depth and significantly lower in EBF at 10–25 cm depth. MBC and DOM in LS showed an inverse trend compared with the other land uses, which was greater at 10–25 cm depth.

Chemical characteristics of DOM

The chemical composition of DOM was compared for each sample before and after incubation (Table 2). Before incubation, the fraction of CH at 0–10 cm soil depth ranged from 11.25 to 28.39% of total DOM, with the highest value in BF. The Phe at 0–10 cm soil depth ranging from 1.06 to 3.74% of total DOM was significantly larger than at 10–25 cm depth, with the highest value in CL. The Hi fraction at 0–10 cm depth was varied in the land uses with the highest in CL (61.75%) and lowest in BF (21.22%). At 10–25 cm soil depth, the fraction of CH in BF and LS was significantly larger than the other soils. The largest Phe concentration was observed in LS (3.49%). The proportion of hydrophilic C at 10–25 cm soil depth was no significant difference between each site, but it is significantly higher than at 0–10 cm depth except for CL soil.

From the initial samples before incubation, the spectral indicators were found to be significantly different among the land uses and between soil depths (Table 2). At 0–10 cm depth, $SUVA_{254}$ was significantly greater in CF while FE was significantly greater in CL. At 10–25 cm soil depth, $SUVA_{254}$ and HIX_{em} were both lower in CL. Difference in FE at 10–25 cm was statistically significant between each site. There were no significant differences between soil depths in $SUVA_{254}$ for EBF and LS, but FE for all the

Table 1. Soil total C concentration (TC), total N concentration (TN), dissolved organic M (DOM), microbial biomass C (MBC), pH and the bulk density of different land uses soils collected on reclaimed land from Taihu Lake, China.

Site		TC	TN	DOM	MBC	pH	Bulk density
Depth	Land use	(g C/kg Soil)	(g N/kg Soil)	(mg C/kg Soil)	(mg C/kg Soil)		(g cm^{-3})
0–10 cm	CF	19.12±1.64b*	2.24±0.15b*	126.63±2.75a*	214.50±5.77d*	4.45±0.07d*	1.22±0.03a*
	EBF	16.05±0.69bc*	1.77±0.04b*	93.50±2.56d*	370.40±9.64c*	4.78±0.08c	1.18±0.03a
	BF	35.35±1.79a*	3.41±0.22a*	110.50±2.40b*	423.01±5.13b*	4.13±0.06e*	0.89±0.06b
	CL	14.95±0.96c*	1.84±0.10b*	105.80±2.81bc*	521.98±7.72a*	5.67±0.08b*	1.16±0.03a*
	LS	16.14±1.38bc	2.26±0.23b	99.95±2.11cd*	350.03±7.64c*	6.19±0.02a	Nd
10–25 cm	CF	8.34±0.53c	1.18±0.07b	97.72±4.50b	112.77±4.43a	4.89±0.15bc	1.31±0.01a
	EBF	8.81±0.77c	1.16±0.08b	78.39±3.06c	309.94±3.90c	4.98±0.07b	1.25±0.02a
	BF	23.76±0.94a	2.59±0.25a	95.88±2.46b	232.28±4.65d	4.30±0.02c	0.99±0.06b
	CL	9.30±1.16c	1.23±0.13b	91.40±3.13b	455.99±5.20b	6.18±0.16a	1.31±0.03a
	LS	17.51±1.03b	2.34±0.13c	115.48±4.21a	558.82±10.28e	6.30±0.07a	Nd

Values are mean ± SE (n = 4). Means within a column of the corresponding depth followed by different letters are significantly different and * indicates the significant difference between the soil depth. (Significance at $p<0.05$). CF: coniferous forest; EBF: evergreen broadleaf forest; BF: bamboo forest; CL: cropland; LS: lake sediment; Nd: Not determined.

Table 2. Chemical properties of soil DOM at two soil depths under different land uses reclaimed from Taihu Lake, China.

Site			$SUVA_{254}$	FE	HIX_{em}	CH	Phe	Hi
Depth	Land use	DOM	($l\ mg\ C^{-1}\ m^{-1}$)			(% of total DOM)		
0–10 cm	CF	Initial	2.30±0.06a*	321.28±2.28b*	1.76±0.11ab	23.86±1.21b	2.50±0.04b*	32.71±1.38c*
		Δ	2.65±0.05a*	29.62±1.95c	4.17±0.11a*	32.87±2.02a*	4.68±0.22a*	12.15±2.20b
	EBF	Initial	1.00±0.09d	301.18±3.40b*	1.64±0.03b	21.83±0.49b*	2.08±0.04c	37.47±1.35b*
		Δ	2.38±0.12b*	95.03±5.61b*	4.50±0.69a*	25.40±0.51b*	2.81±0.17b*	8.72±1.91b
	BF	Initial	2.07±0.07b*	252.31±3.26d*	1.94±0.07a	28.39±1.13a	2.14±0.06c*	29.22±1.14c*
		Δ	1.05±0.06c	216.67±6.46a*	5.59±0.70a*	4.70±0.37c*	1.37±0.10c*	19.90±2.17a
	CL	Initial	1.20±0.04c*	430.40±2.29a*	1.11±0.08c*	18.98±1.02c*	3.74±0.12a*	61.75±0.70a
		Δ	0.55±0.06d*	−59.55±6.02e*	2.78±0.43b	6.09±0.37c*	−2.61±0.11e*	−3.61±1.35c*
	LS	Initial	1.15±0.06c	289.30±17.80c*	1.68±0.11b	11.25±0.72d*	1.06±0.07d*	59.58±1.19a*
		Δ	0.47±0.07d*	−4.65±1.88d	0.79±0.01c	−4.00±0.31d*	0.25±0.07d*	10.87±3.92b*
10–25 cm	CF	Initial	0.98±0.05b	253.32±1.30d	1.69±0.02a	21.37±0.46b	1.33±0.13c	47.77±1.12a
		Δ	0.60±0.07b	24.40±2.49bc	1.84±0.23c	6.74±0.17a	1.72±0.06a	7.31±1.48c
	EBF	Initial	1.20±0.06a	408.93±4.16a	1.07±0.02b	15.35±0.79c	2.18±0.09b	52.48±0.70a
		Δ	0.18±0.06c	11.25±2.46c	2.24±0.15b	2.78±0.38b	0.60±0.03a	6.98±2.83c
	BF	Initial	1.30±0.06a	170.44±2.17e	1.88±0.02a	26.14±0.81a	1.51±0.02c	46.18±0.49a
		Δ	1.10±0.04a	87.96±2.75a	2.85±0.45a	−2.14±0.85c	0.42±0.05a	17.15±0.69b
	CL	Initial	0.78±0.05c	316.91±2.46c	0.48±0.02c	11.28±0.55d	0.92±0.04d	56.06±1.65a
		Δ	0.15±0.05c	25.32±6.10b	3.52±0.16a	3.61±0.50b	−0.08±0.02c	18.33±3.58b
	LS	Initial	1.23±0.05a	350.29±11.95b	1.60±0.27a	24.81±0.37a	3.49±0.12a	40.76±1.08a
		Δ	0.20±0.04c	−48.88±11.18d	0.40±0.23d	−13.39±0.80d	0.38±0.99b	43.26±3.23a

Results shown are characteristics of initial DOM and the variation of each characteristic at the end of 360 days incubation. Values are mean ± SE (n = 4). Means within a column of the corresponding depth followed by different letters are significantly different and * indicates the significant difference between the soil depth. (Significance at $p<0.05$).

CF: coniferous forest; EBF: evergreen broadleaf forest; BF: bamboo forest; CL: cropland; LS: lake sediment; $SUVA_{254}$: specific UV absorbance at 254 nm; FE: fluorescence efficiency (F_{max}/A254); HIX_{em}: humification index using emission fluorescence spectra (ratio of areas: 435–480 nm/300–345 nm); CH: carbohydrate C; Phe: phenol C; Hi: hydrophilic C; initial: properties of initial samples (soils before incubation); Δ: variation of DOM chemical properties between the initial and final value during incubation.

sites were observed significantly different between soil depths. Differences in HIX_{em} between soil depths were only found in CL.

After 360 days incubation, there were significant amounts of variation in DOM chemical properties among the land uses and lake sediments and most of the variation between soil depths was also significant (Table 2). The proportion of CH at 0–10 cm soil depth increased 17%–138% in the land uses with significantly greater value in CF. The proportion of Phe and Hi respectively increased 23%–188% and 18%–68% for CF, EBF, BF, and LS at 0–10 cm depth, but the values of these two factors for CL showed decreased trend after incubation. The increased values of Phe were significantly different from each site with the greatest variation in CF, while increased amount of Hi was lager in BF, at 0–10 cm depth. The proportion of CH fraction had the largest increase in CF, but it was observed reducing in BF and LS, at 10–25 cm soil depth. At this layer, both the proportion of Phe and Hi increased after incubation, and the largest increase was found in LS. There was no significant difference in the variation of Phe among the three forest sites. $SUVA_{254}$ and HIX_{em} from the samples at the end of incubation were increased at both 0–10 cm and 10–25 cm soil depth (Table 2). CF had the greatest variation in $SUVA_{254}$ and HIX_{em} at 0–10 cm depth and the variation of FE in BF was larger the other sites. There was also no significant difference in the variation of HIX_{em} among the three forest sites. The variation of FE values was significantly different between each other site and the FE values in CL and LS decreased after incubation. At 10–25 cm depth, the variation of $SUVA_{254}$ and FE in BF was significantly larger than other sites.

Carbon mineralization

The cumulative CO_2 production under different land uses and sediments ranged from 0.88 to 7.72 g CO_2-C kg^{-1} soil and there were significant difference between soil depths (Fig. 2a). At 0–10 cm soil depth, cumulative C mineralization in BF soils (2.87 g CO_2-C kg^{-1} soil) was significantly greater than other land uses and CL had the lowest amounts of cumulative CO_2-C (1.62 g CO_2-C kg^{-1} soil). The cumulative CO_2-C for the three forest sites decreased with the increasing soil depth. At 10–25 cm soil depth, C mineralization was found to be significantly different in the following order of CL>BF>CF>EBF (Table 3). Representing the precursor of the four terrestrial soils, LS had different C mineralization patterns than the four terrestrial soils, and its evolution of cumulative CO_2-C (4.59 and 7.72 g CO_2-C kg^{-1} soil, respectively) was significantly larger than other soils, at 0–10 cm and 10–25 cm soil depth. After 360 days of incubation, the percentage of cumulative mineralized C (as % of TC) ranged from 5.11% to 47.19% with the highest C mineralization in LS at 10–25 cm depth (Fig. 2b). At 0–10 cm soil depth, the percentage of mineralized C was significantly different between each other site with largest value in EBF (15.23%) and lowest value in BF (8.11%). At 10–25 cm soil depth, CL had the highest percentage of cumulative mineralized CO_2-C (21.07%) among the four land uses, approximately four times larger than BF, which had the lowest percentage of mineralized C (5.11%).

The rate of CO_2-C evolution from all land use types were highest at the beginning and then decreased progressively with the advancement of time (Fig. 3). The CO_2-C evolution patterns in all land uses and lake sediments were best described by the double exponential model, with the model fits resulting in r^2 values between 0.96 and 0.99 (Table 3). The size of the labile C pool in different soils was comprised of 0.2%–7.09% with the highest value in LS at 10–25 cm depth. The k_1 values (mineralization rate constant) of labile C at 0–10 cm depth ranged from 0.02 to 0.19 with the mean residence time (MRT_1) of 21 days (average of all soils), varying from 0.02 to 0.12 at 10–25 cm depth with an average MRT_1 of 33 days. Determination of the size and turnover of the stable C pool showed values at 0–10 cm soil depth were larger than those at 10–25 cm soil depth (Table 3). The mineralization rate constants of the stable C pool (k_2) were 2–3 orders of magnitude lower than k_1, ranging from 0.00022 to 0.00098 in upper soil layer with an average of MRT_2 of 9 years and ranging from 0.00011 to 0.00129 in lower soil layer with an average of MRT_2 of 12 years. At 0–10 cm soil depth, the mean residence time of active and stable C pools in CL sites was larger than the others, whereas at 10–25 cm soil depth, the largest mean residence time of stable C pools was found in BF. The mean residence time of active and stable C pools in LS was lower than all other land use types.

C mineralization in relation to DOM characteristics

C mineralization was significantly correlated with DOM characteristics and its change (the variation of DOM properties between beginning and ending of incubation) under the four land uses (Table 4, Fig. 4). The CO_2-C evolution at the 0–10 cm soil depth had a positive correlation with the proportion of CH, variation of the Hi fraction (ΔHi), fluorescence efficiency variation (ΔFE) and HIX_{em} variation (ΔHIX_{em}), while had a negative correlation with initial Phe, Hi, FE and variation of CH (ΔCH), Phe (ΔPhe) and $SUVA_{254}$ ($\Delta SUVA_{254}$) (Fig. 4). The high proportion of labile C pool (*a* value of the double exponential model) from upper soil was related to the high proportion of Phe, Hi fraction and high values of FE. In contrast, it was negatively related to most of the parameters including CH fraction, ΔPhe, ΔHi, HIX_{em}, $SUVA_{254}$, ΔFE, ΔHIX_{em} and $\Delta SUVA_{254}$ (Table 4). The k_1 values of upper soils were correlated very well with proportion of Phe and Hi, the values of FE, HIX_{em}, ΔFE and ΔHIX_{em}, whereas the k_2 values were only correlated with Phe, ΔCH, ΔPhe and $\Delta SUVA_{254}$ (Table 4).

In 10–25 cm soil layer, cumulative CO_2-C positively correlated with proportion of CH and the values of HIX_{em}, ΔHi, ΔFE, ΔHIX_{em}, $\Delta SUVA_{254}$, while negatively correlated with proportion of Phe and Hi, as well as the values of FE and $SUVA_{254}$ (Table 4). The proportion of labile C pool showed positive correlations with Phe and FE, and showed the negative relationships with ΔHi, HIX_{em} and ΔFE. The k_1 values highly related to proportion of CH, Phe and Hi, as well as values of $SUVA_{254}$ and ΔHIX_{em}. The relationships between k_2 values and the properties of DOM were similar with the k_1 values (Table 4).

Discussion

Chemical characteristics of DOM

The decomposition rate of organic matter is in part controlled by its chemical fractions and structures [13,21]. The highest portion of CH fraction in BF was due to a lot of litter and root exudate input. For the four land uses, the percentage of CH was observed lower in CL, which may be explained by two reasons: (i) soil microbial community may deplete CH fraction to a greater degree because fertilization can increase the microorganism activity [9] and (ii) the type of crop residue input may influence the chemical composition of DOM [9]. Different from the land uses, the CH for LS was greater at 10–25 cm, probably due to the deeper sediments in anoxic condition would limit the microbial activity [43]. The lowest proportion of phenol C of the DOM from the un-incubated soils was found in LS at 0–10 cm soil depth, possibly because most of the DOM is derived from algae, bacteria and macrophytes, which contain lower phenolic groups [44]. The highest percentage of Hi fraction in CL agreed with Chantigny [5]

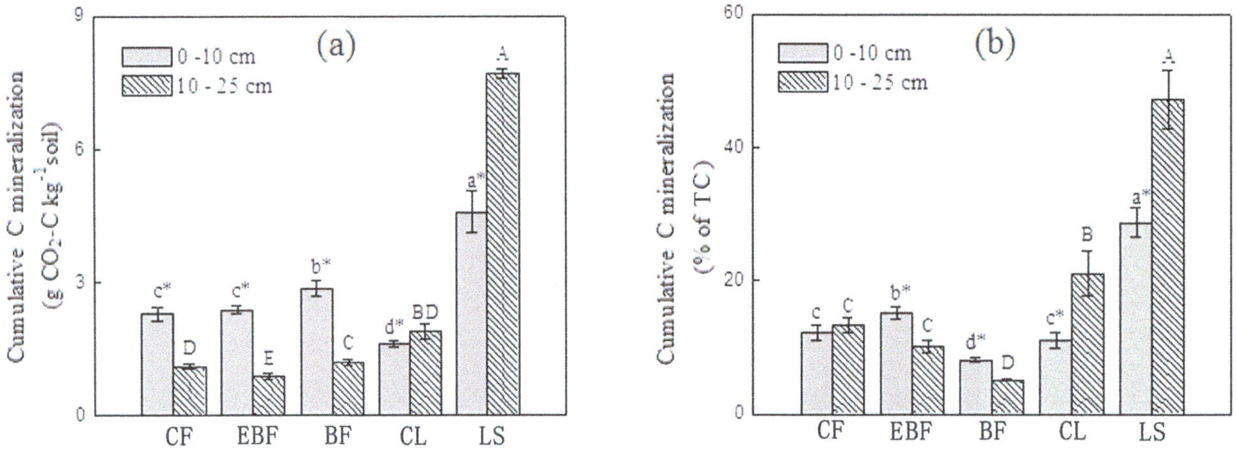

Figure 2. Carbon mineralized at the end of the 360 days incubation period for different land uses and lake sediments. Bars represent standard errors (n = 4). Different letters above bars indicate significant differences ($p<0.05$) of mean values of different sites. * indicates the significant difference ($p<0.05$) between the soil depth.

who reported that DOM from crop residue contained more hydrophilic fractions. During the incubation period, the quality and quantity of organic matter in soils changed, which should lead to variation in the chemical structure of DOM. The proportion of Hi fraction, CH and Phe in DOM under most of the sites were observed to increase at the end of incubation, indicating that some insoluble components were converted to soluble fractions, possibly due to desorption of adsorbed organic matter or other biotic regulatory mechanisms [13,45]. Kalbitz et al. [24] reported that relative increase in polysaccharides after incubation is likely caused by microbial formation and that many bacteria and fungi can release diverse polysaccharides. The relative enrichment in phenols was possibly due to oxidation and fragmentation of lignin in soils [46]. Although both of the percentage of labile and stable fractions in DOM increased at the end of the incubation due to complement of soil organic matter [20], there was a greater increase in the proportion of Phe than CH and Hi fractions. The larger increase in the proportion of Phe at the end of incubation suggested that phenols were the most stable fractions, even though these compounds can experience partial degradation [47]. Our results suggested that CH and Hi fractions were preferentially utilized by microorganisms while phenols were resistant to degradation during incubation. Unlike the other sites, both the proportion of Phe and Hi fractions in CL DOM decreased slightly, probably due to the influence of management practices such as fertilization which changed the soil environment and accelerated both of them degradation [9,48].

The specific UV absorbance at 254 nm ($SUVA_{254}$), humification indices (HIX_{em}) and fluorescence efficiency indices (FE) have been used to characterize the content of aromatic structures and complexity degree in DOM [49–51]. In our study, $SUVA_{254}$ and HIX_{em} decreased with soil depth, consistent with Corvasce et al. [52] and Bu et al. [53], suggesting that the aromatic fraction of partially degraded lignin-derived compounds might be gradually adsorbed by upper mineral soil and protected from microbial degradation [54]. CF and BF soils had the highest $SUVA_{254}$ and HIX_{em} values, indicating that there were more complex and condensed polyaromatic structures in the DOM due to the presence of ligninolitic compounds [55]. Our results agree with Khomutova et al. [56] who also found values of specific UV absorbance at 260 nm were larger in coniferous forest soils compared to deciduous forest and pasture soils, suggesting that DOM under conifer soils is enriched in hydrophobic aromatic compounds [57]. Just like $SUVA_{254}$ and HIX_{em}, fluorescence efficiency (FE) was also used to express the degree of condensation of DOM. The low FE in BF soils may be partially attributed to its highly substituted aromatic structural features and its inter- or intra-molecular bonding which could result in self-quenching within macromolecules [26]. Ewald et al. [38] drew a similar conclusion that fluorescence efficiency had an inverse linear correlation with molecular weight of the fulvic acid fractions, which was due to internal quenching in macromolecules caused by energy transfers. The smaller values of $SUVA_{254}$ and HIX_{em} as well as larger value of FE were in CL consistent with Chantigny [5] who reported there were more lignin and other recalcitrant compounds in forest litter compared to crop residues. The increase of $SUVA_{254}$ and HIX_{em} under different sites suggested aromatic structures were accumulating with the decomposition of the labile components of DOM during incubation [21,50]. Hur et al. [27,44] also found the increase of SUVA values for all the DOM samples after microbial incubation and attributed this result to preferential microbial utilization of non-aromatic fractions and/or to microbial transformation of labile compounds into aromatic C structures. At 0–10 cm soil depth, the increase of $SUVA_{254}$ and HIX_{em} in CL were significantly smaller than the forest soils, indicating the same conclusion that DOM in CL contains smaller complex compounds. Although the changes in spectra parameters during incubation exhibited increasing trends under most land use types, the increased DOM for LS was smaller than other land uses. This result suggests that the accumulation of condensed aromatic structures was relatively small during incubation, supporting the previous statement that accumulation of phenols was small compared to hydrophilic fractions for LS DOM. For the soil DOM samples, a pronounced increase in FE value was inconsistent with the changes in HIX_{em} or $SUVA_{254}$, which couldn't be explained by fluorescent quenching. This inconsistency may be attributed to the characterization method of FE influenced by the changes in DOM properties.

Soil carbon mineralization in relation to DOM characteristics

The C mineralization was significantly higher at 0–10 cm soil depth regardless of vegetation types for all land uses except CL.

Table 3. C mineralization kinetics of soil after 360 days incubation at 25°C: cumulative CO_2-C, sizes of the labile and stable C pools, mineralization rate constants and mean residence times of the labile and the stable C pools.

Site		C_m	Labile C	Stable C	k_1	k_2	MRT_1	MRT_2	r^2
Depth	Land use	(%)	(%)	(%)	(day^{-1})	(day^{-1})	(days)	(years)	
0–10 cm	CF	12.24	0.75	99.25	0.03	0.00036	30.48	7.68	0.993
	EBF	15.23	0.72	99.28	0.13	0.00048	7.82	5.80	0.979
	BF	8.11	0.20	99.80	0.10	0.00023	10.17	12.13	0.998
	CL	10.95	2.62	97.38	0.02	0.00022	53.13	12.60	0.999
	LS	28.71	1.13	98.87	0.19	0.00098	5.15	2.85	0.983
10–25 cm	CF	13.45	3.07	96.93	0.03	0.00032	33.27	8.67	0.995
	EBF	10.11	4.81	95.19	0.02	0.00016	61.71	17.80	0.999
	BF	5.11	1.42	98.58	0.02	0.00011	48.62	26.11	0.999
	CL	21.07	1.70	98.30	0.07	0.00065	13.53	4.26	0.993
	LS	47.19	7.09	92.91	0.12	0.00128	8.11	2.16	0.957

CF: coniferous forest; EBF: evergreen broadleaf forest; BF: bamboo forest; CL: cropland; LS: lake sediment; C_m: cumulative mineralized C as percentage of initial C in the soil. Labile C: rapidly mineralizable C (calculated using a double exponential model); Stable C: slowly mineralizable C (calculated using a double exponential model); k_1: mineralization rate constant of the labile C pool (double exponential model); k_2: Mineralization rate constant of the stable C pool (double exponential model); MRT_1: Mean residence times of the labile C pool ($MRT_1 = 1/k_1$); MRT_2: Mean residence times of the stable C pool ($MRT_1 = 1/k_2$); r^2: coefficient of determination of the double exponential model.
Values are mean (n = 4).

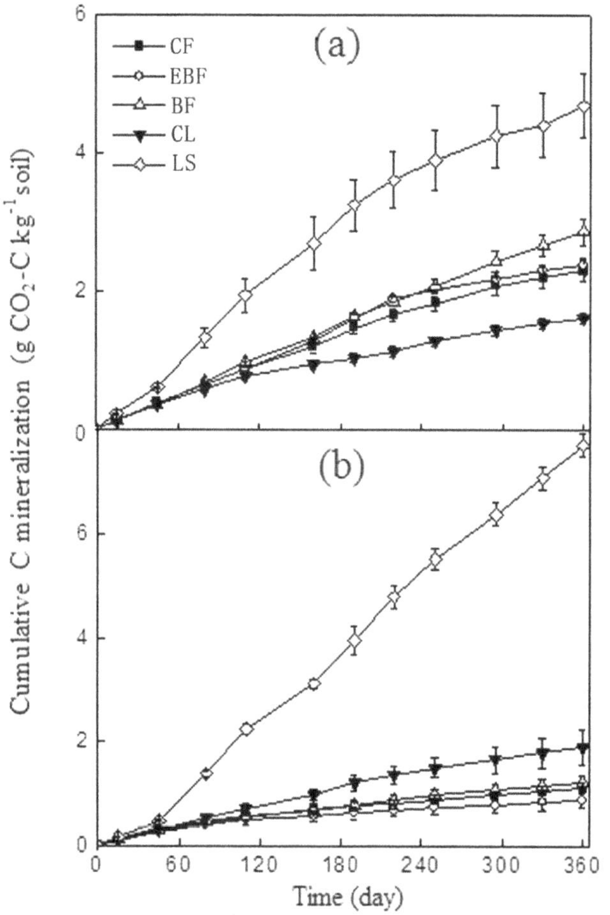

Figure 3. Dynamics of C mineralization of soils under different land use types reclaimed from Taihu Lake, China. (a): C mineralization at 0–10 cm depth; (b): C mineralization at 10–25 cm depth. Bars indicate the standard errors (n = 4).

The higher C mineralization observed at 10–25 cm soil depth in CL was likely due to the tillage practice of ploughing which stimulated the C mineralization in deeper layers [58]. At 0–10 cm soil depth, cumulative C mineralization in BF soils was greater than other land uses, also possibly due to plenty of SOC deriving from degradation of bamboo litter and rhizomatous. While cumulative C emission for CL was significantly lower than the forest soils at 0–10 cm and Zhao et al. [20] observed the similar result that the average cumulative mineralized CO_2-C in forest sites and arable sites was 382 and 279 mg C/kg soil, respectively. The amount of cumulative mineralized CO_2-C during 360 days incubation was significantly higher in lake sediments compared to the terrestrial soils. This result was consistent with our primary hypothesis. Mora et al. [33] observed that during 10 days incubation 0.5% of the sediment C, as well as 0.25% of soil C was mineralized.

The analysis of double exponential model showed that emission of CO_2-C increased as the labile C pool (*a*) decreased, which was inconsistent with the results of Kalbitz et al. [21] and Bu et al. [59]. During the incubation period, lake sediments had a higher percentage of C mineralized (as % of TC), a larger pool of labile C, and shorter residence times for labile and stable C. This suggested that lake sediments were enriched in less complex, more easily mineralized forms of C [60–62]. The smaller fraction of mineralized C and lower mineralization rate constants for the stable C pool found in BF soils suggested that organic C in BF soils was composed mainly of polyphenols and aromatic structures deriving from the decomposition of bamboo rhizomes and recalcitrant to biodegradation [63–65]. Compared to forest soils, both labile and stable C pools in CL soils had lower mineralization rate constants, which was different from the result of Kalbitz et al. [21]. This discrepancy was likely explained by DOM representing a small fraction of labile C in CL soils, and mineralization of soil C can't be explained only by consumption of DOM. Zhao et al. [20] also found no correlations between the decrease in quantity of DOM and soil C mineralization. At 0–10 cm depth, mineralization rate constants for both the labile and stable C pool of CF soils were significantly smaller than EBF soils, consistent with previous studies which found C mineralization rate in a conifer forest to be lower than in an evergreen broadleaf forest [66].

In our study, we found the chemical compositions and spectra characteristics of DOM were correlated with C mineralization, but the relationship varied across land uses. The cumulative CO_2-C emission showed a significant positive correlation with proportion of CH at the beginning of the incubation and a significant negative relationship with the proportion of initial Phe. The results suggest that CH and Phe respectively represent the labile and refractory fractions in DOM, and they may regulate the mineralization potential of SOM. Soil carbohydrates are readily available for biodegradation and often have short half-lives due to rapid uptake and assimilation by soil microbial communities [22]. At the end of incubation, the increase in proportion of CH at 0–10 cm soil depth had a negative relationship with cumulative CO_2-C emission, similar to Tian et al. [34], and emphasizing that carbohydrates are utilized preferentially by microorganisms [13,21]. With the consumption of labile fractions, the increasing proportion of Phe was closely related to the less CO_2-C emission, explained by inhibiting the activity of various enzymes [34,46]. The correlation between ΔCH (0–10 cm), ΔPhe (0–10 cm), CH (10–25 cm) and cumulative CO_2-C emission in CL statistically deviated from the regression line (fig. 4a, 4b and 4g), suggesting that the contribution rate to C mineralization made by Phe or CH in CL was different from other land uses. Hi fractions of DOM separated by XAD-8 resin can be characterized as labile soluble organic moieties, particularly carbohydrates, amino sugars and low-molecular-weight organic acids and show a higher degree of biodegradation [13,23,35,57]. In our study, the increase of Hi fraction was strongly tied to increasing C mineralization, which was inconsistent with the general results of previous studies [21,44,67]. This result attributed to the fact that part of the hydrophilic fractions such as hydrophilic neutral (HiN) is composed mainly of carbohydrates that are bound to aromatic compounds and are converted into typical hydrophobic fractions during long term incubation [21]. However, the correlation of Hi and C mineralization at 10–25 cm depth in CL also deviated from the regression line.

Recently, fluorescence and UV spectroscopy has been used successfully to describe the chemical properties of DOM [21,53,68–69]. In many studies, the extent of C mineralization was inversely related to the specific UV absorbance of DOM [13,53,70–71]. This was confirmed by our results observed in 10–25 cm soil depth, indicating that UV-inactive substances were degraded preferentially. The relative increase in $SUVA_{254}$ at the end of incubation was positively related to cumulative CO_2-C emission, which suggests that more aromatic structures were cumulated during the C mineralization. Fluorescence spectroscopy can provide additional information relating to structure, functional groups, conformation, and heterogeneity, as well as dynamic

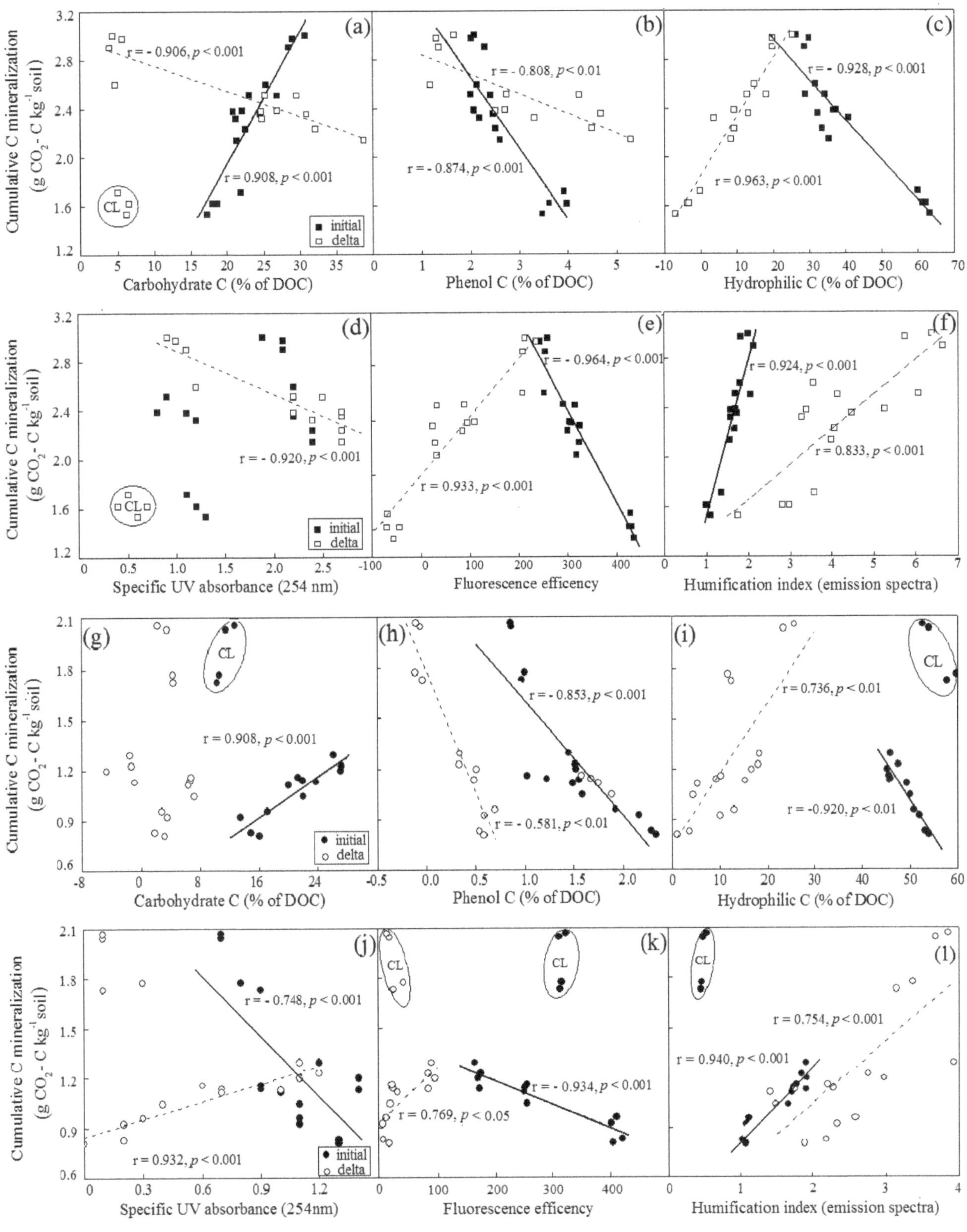

Figure 4. Relationships between cumulative C mineralization and DOM chemical properties of the four land use types. 0–10 cm soil depth: a–f; 10–25 cm soil depth: g–l. (a and g) between carbohydrate carbon percentages and cumulative CO_2-C, (b and h) between phenol carbon percentages and cumulative CO_2-C, (c and i) between hydrophilic carbon and cumulative CO_2-C, (d and j) between $SUVA_{254}$ values and cumulative CO_2-C, (e and k) between the fluorescence efficiency and cumulative CO_2-C, (f and l) between the humification index (emission fluorescence spectra) and cumulative CO_2-C. The curves for the figures are generated with the data of the CL samples excluded.

Table 4. Pearson correlation coefficients (r-value) between C mineralization and soil properties and chemical characteristics of DOM under four terrestrial land use types reclaimed from Taihu Lake, China.

		DOC	CH	ΔCH	Hi	ΔHi	Phe	ΔPhe	$SUVA_{254}$	$ΔSUVA_{254}$	FE	ΔFE	HIX_{em}	$ΔHIX_{em}$
0–10 cm	*a*	−0.140	−0.762**	−0.315	0.976**	−0.893**	0.951**	−0.782**	−0.504*	−0.538*	0.982**	−0.871**	−0.902**	−0.702**
	k_1	−0.586*	0.373	0.048	−0.555*	0.499*	−0.792**	0.339	−0.263	0.275	−0.737**	0.752**	0.510*	0.532*
	k_2	−0.285	−0.148	0.803**	−0.346	0.090	−0.548*	0.666**	−0.289	0.849**	−0.283	0.043	0.210	0.106
	MRT_1	0.270	−0.600*	−0.203	0.822**	−0.742**	0.948**	−0.604*	−0.101	−0.447	0.928**	−0.876**	−0.755**	−0.665**
	MRT_2	0.130	0.099	−0.886**	0.414	−0.145	0.562*	−0.762**	0.142	−0.921**	0.305	−0.031	−0.268	−0.117
	CO_2-C	0.085	0.908**	0.044	−0.928**	0.963**	−0.873**	0.573*	0.432	0.279	−0.964**	0.934**	0.925**	0.833**
10–25 cm	*a*	−0.600*	−0.263	0.390	0.135	−0.696**	0.755**	0.368	0.200	−0.499	0.776	−0.688**	−0.954**	−0.534*
	k_1	0.170	−0.653**	0.276	0.642**	0.466	−0.796**	−0.483	−0.819**	−0.456	0.090	−0.230	0.575	0.586*
	k_2	−0.420	0.408	−0.383	−0.394	−0.410	0.916**	0.189	0.825**	0.235	0.172	0.125	0.079	0.459
	MRT_1	0.142	−0.710**	0.469	0.662**	0.318	−0.758**	−0.322	−0.886**	−0.560*	0.199	−0.404	−0.700*	−0.436
	MRT_2	−0.088	0.694**	−0.794**	−0.578*	0.028	0.581*	−0.074	0.885**	0.661**	−0.360	0.680**	0.191	−0.122
	CO_2-C	0.310	−0.446	0.014	0.423	0.736**	−0.853**	−0.581*	−0.748**	−0.223	−0.122	0.012	0.940*	0.754**

n = 16; abbreviations see Table 2 and Table 3;
*Significance at $p<0.05$;
**Significance at $p<0.01$.

properties of DOM [25–26,49]. In our study, the FE and HIX_{em} deduced from fluorescence emission spectra exhibited strong correlation with CO_2 emission. The FE and HIX_{em} for initial DOM respectively showed negative and positive correlation with C mineralization, suggesting that in the long-term incubation not only labile C but also stable C contributed to the C mineralization. The ΔFE and ΔHIX_{em} during incubation were positively correlated with C mineralization, supporting expectation that with the oxidation or degradation of labile fractions, more aromatic compounds and other complex stable molecules are accumulated [68]. The relationship between part of spectroscopy parameters and cumulative CO_2 emission in CL also deviated from the regression line (fig. 4d, 4k and 4l) possibly due to the effluence of land use and management.

Conclusions

Soils from varied land uses under reclaimed area from Taihu Lake had significant differences in soil DOM chemical properties and C dynamics. The C mineralization of lake sediments was much larger than that in terrestrial ecosystems, confirming our hypothesis. For the terrestrial soils, C mineralization in the upper soil layer was higher in BF and lower in CL. We have found that C mineralization in our study was fitting the double exponential model and the kinetics parameters were closely related to the chemical properties of DOM and sensitive to its variation. Cumulative CO_2-C was positively correlated with the carbohydrates while negatively correlated with phenols and the aromaticity of DOM, indicating labile compounds are preferentially utilized by microbes. Moreover, the variation of Phe (ΔPhe) and humificaton (ΔHIX_{em}) at the end of incubation showed reverse correlation with cumulative CO_2, suggesting in the long-term incubation stable compounds also make contribution to C mineralization.

Overall, our study suggests that in conversion from aquatic ecosystem to terrestrial ecosystem, different land use types with different vegetation cover and varied management practices lead to significant changes in the chemical structure of DOM, which can influence the C dynamics. Inconsistent trend observed in CL soils may be attributed to the tillage practices (e.g. fertilization, crop rotation).

Acknowledgments

We gratefully acknowledge Yiling Luan, Juhua Yu, Jiaojiao Guo, Zilong Ma, Yupeng Zhao and zhiqin Hua for help with field and lab work. Special thanks go to Dr. Jason Vogel for his helpful comments.

Author Contributions

Conceived and designed the experiments: HR DH. Performed the experiments: DH. Analyzed the data: DH. Contributed reagents/materials/analysis tools: DH. Wrote the paper: DH.

References

1. Leinweber P, Jandl G, Baum C, Eckhardt K, Kandeler E (2008) Stability and composition of soil organic matter control respiration and soil enzyme activities. Soil Biology & Biochemistry 40: 1496–1505.
2. Hansson K, Kleja DB, Kalbitz K, Larsson H (2010) Amounts of carbon mineralized and leached as DOC during decomposition of Norway spruce needles and fine roots. Soil Biology & Biochemistry 42: 178–185.
3. Wang Y, Ruan HH, Huang LL, Feng YQ, Zhou JZ, et al. (2010) Soil labile organic carbon with different land uses in reclaimed land area from Taihu lake. Soil Science 175: 624–630.
4. Lal R (2002) Soil carbon dynamics in cropland and rangeland. Environmental Pollution 116: 353–362.
5. Chantigny MH (2003) Dissolved and water-extractable organic matter in soils: a review on the influence of land use and management practices. Geoderma 113: 357–380.
6. Caravaca F, Masciandaro G, Ceccanti B (2002) Land use in relation to soil chemical and biochemical properties in a semiarid Mediterranean environment. Soil & Tillage Research 68: 23–30.
7. Martin D, Lal T, Sachdey CB, Sharma JP (2010) Soil organic carbon storage changes with climate change, landform and land use conditions in Garhwal hills of the Indian Himalayan mountains. Agriculture, Ecosystems and Environment 138: 64–73.
8. Tobiašová E (2011) The effect of organic matter on the structure of soils of different land uses. Soil & Tillage Research 114: 183–192.
9. Kalbitz K, Solinger S, Park JH, Michalzik B, Matzner E (2000) Controls on the dynamics of dissolved organic matter in soils: a review. Soil Science 165: 277–304.
10. Ge QS, Zhao MC, Zheng JY (2000) Land use change of China during the 20th century. Acta Geographica Sinica 55: 698–706.
11. Qin BQ, Xu PZ, Wu QL, Luo LC, Zhang YL (2005) Environmental issues of Lake Taihu, China. Hydrobiologia 581: 3–13.
12. Søndergaard M, Jeppesen E (2007) Anthropogenic impacts on lake and stream ecosystems, and approaches to restoration. Journal of Applied Ecology 44: 1089–1094.
13. Marschner B, Kalbitz K (2003) Controls of bioavailability and biodegradability of dissolved organic matter in soils. Geoderma 113: 211–235.
14. Li ZP, Han CW, Han FX (2010) Organic C and N mineralization as affected by dissolved organic matter in paddy soils of subtropical China. Geoderma 157: 206–213.
15. Cookson WR, Abaye DA, Marschner P, Murphy DV, Stockdale EA, et al. (2005) The contribution of soil organic matter fractions to carbon and nitrogen mineralization and microbial community size and structure. Soil Biology & Biochemistry 37: 1726–1737.
16. Marinari S, Liburdi K, Fliessbach A, Kalbitz K (2010) Effects of organic management on water-extractable organic matter and C mineralization in European arable soils. Soil & Tillage Research 106: 211–217.
17. Kowalczuk P, Durako MJ, Young H, Kahn AE, Cooper WJ, et al. (2009) Characterization of dissolved organic matter fluorescence in the South Atlantic Bight with use of PARAFAC model: Interannual variability. Marine Chemistry 113: 182–196.
18. Boddy E, Hill PW, Farrar J, Jones DL (2007) Fast turnover of low molecular weight components of the dissolved organic carbon pool of temperate grassland field soils. Soil Biology & Biochemistry 39: 827–835.
19. Marschner B, Bredow A (2002) Temperature effects on release and ecologically relevant properties of dissolved organic carbon in sterilised and biologically active soils. Soil Biology & Biochemistry 34: 459–466.
20. Zhao MX, Zhou JB, Kalbitz K (2008) Carbon mineralization and properties of water-extractable organic carbon in soils of the south Loess Plateau in China. European Journal of Soil Biology 44: 158–165.
21. Kalbitz K, Schmerwitz J, Schwesig D, Matzner E (2003) Biodegradation of soil-derived dissolved organic matter as related to its properties. Geoderma 113: 273–291.
22. Don A, Kalbitz K (2005) Amounts and degradability of dissolved organic carbon from foliar litter at different decomposition stages. Soil Biology & Biochemistry 37: 2171–2179.
23. Said-Pullicino D, Kaiser K, Guggenberger G, Gigliotti G (2007) Changes in the chemical composition of water-extractable organic matter during composting: Distribution between stable and labile organic matter pools. Chemosphere 66: 2166–2176.
24. Kalbitz K, Schwesig D, Schmerwitz J, Kaiser K, Haumaier L, et al. (2003) Changes in properties of soil-derived dissolved organic matter induced by biodegradation. Soil Biology & Biochemistry 35: 1129–1142.
25. Chen J, Gu BH, LeBoeuf EJ, Pan HJ, Dai S (2002) Spectroscopic characterization of the structural and functional properties of natural organic matter fractions. Chemsophere 48: 59–68.
26. Chen J, LeBoeuf EJ, Dai S, Gu BH (2003) Fluorescence spectroscopic studies of natural organic matter fractions. Chemosphere 50: 639–647.
27. Hur J, Kim G (2009) Comparison of the heterogeneity within bulk sediment humic substances from a stream and reservoir via selected operational descriptors. Chemosphere 75: 483–490.
28. Teixeira MC, Azevedo JCR, Pagioro TA (2011) Spatial and seasonal distribution of chromophoric dissolved organic matter in the Upper Paraná River floodplain environments (Brazil). Acta Limnologica Brasiliensia 23: 333–343.
29. You BS, Zhong JC, Fan CX, Wang TC, Zhang L, et al. (2007) Effects of hydrodynamics processes on phosphorus fluxes from sediment in large, shallow Taihu Lake. Journal of Environmental Sciences 19: 1055–1060.
30. Yin HB, Fan CX, Ding SM, Zhang L, Zhong JC (2008) Geochemistry of iron, sulfur and related heavy metals in metal-Polluted Taihu Lake sediments. Pedosphere 18: 564–573.
31. Blake GR, Hartge KH (1986) Bulk density. In: Klute A ed. Methods of soil analysis. Part I. Physical and mineralogical methods. Madison, WI: American Society of Agronomy and Soil Science Society of America 363–375.

32. Vance ED, Brookes PC, Jenkinson DS (1987) An extraction method for measuring soil microbial biomass carbon. Soil Biology & Biochemistry 19: 703–707.
33. Mora JL, Guerra JA, Armas CM, Rodríguez-Rodríguez A, Arbelo CD, et al. (2007) Mineralization rate of eroded organic C in Andosols of the Canary Islands. Science of the Total Enviroment 378: 143–146.
34. Tian L, Dell E, Shi W (2010) Chemical composition of dissolved organic matter in agroecosystems: Correlations with soil enzyme activity and carbon and nitrogen mineralization. Applied Soil Ecology 46: 426–435.
35. Simonsson M, Kaiser K, Danielsson R, Andreux F, Ranger J (2005) Estimating nitrate, dissolved organic carbon and DOC fractions in forest floor leachates using ultraviolet absorbance spectra and multivariate analysis. Geoderma 124: 157–168.
36. Chantigny MH, Angers DA, Kaiser K, Kalbitz K (2007) Extraction and characterization of dissolved organic matter. In: Carter MR, Gregorich EG eds. Soil sampling and methods of analysis, chap 48. CRC Press. 617–635.
37. Kalbitz K, Meyer A, Yang R, Gerstberger P (2007) Response of dissolved organic matter in the forest floor to long-term manipulation of litter and throughfall inputs. Biogeochemistry 86: 301–318.
38. Ewald M, Berger P, Visser SA (1988) UV-visible absorption and fluorescence properties of fulvic acids of microbial origin as function of their molecular weights. Geoderma 43: 11–20.
39. Zsolnay Á, Baigar E, Jimenez M, Steinweg B, Saccomandi F (1999) Differentiating with fluorescence spectroscopy the sources of dissolved organic matter in soil subjected to drying. Chemosphere 38: 45–50.
40. Kalbitz K, Geyer W (2001) Humification indices of water-soluble fulvic acids derived from synchronous fluorescence spectra-effects of spectrometer type and concentration. Journal Plant Nutrition and Soil Science 164: 259–265.
41. Kalbitz K, Geyer W, Geyer S (1999) Spectroscopic properties of dissolved humic substances-areflection of land use history in a fen area. Biogeochemistry 47: 219–238.
42. Yang LX, Pan JJ, Yuan SF (2006) Predicting dynamics of soil organic carbon mineralization with a double exponential model in different forest belts of China. Journal of Forestry Research 17: 39–43.
43. Wang XW, Li XZ, Hu YM, Lv JJ, Sun J, et al. (2010) Effect of temperature and moisture on soil organic carbon mineralization of predominantly permafrost peatland in the Great Hing'an Mountains, Northeastern China. Journal of Environmental Sciences 22: 1057–1066.
44. Hur J, Lee B, Shin H (2011) Microbial degradation of dissolved organic matter (DOM) and its influence on phenanthrene-DOM interactions. Chemosphere 85: 1360–1367.
45. Kemmitt SJ, Lanyon CV, Waite IS, Wen Q, Addiscott TM, et al. (2008) Mineralization of native soil organic matter is not regulated by the size, activity or composition of the soil microbial biomass-a new perspective. Soil Biology & Biochemistry 40: 61–73.
46. Rovira P, Vallejo VR (2007) Labile, recalcitrant, and inert organic matter in Mediterranean forest soils. Soil Biology & Biochemistry 39: 202–215.
47. Wieder WR, Cleveland CC, Townsend AR (2008) Tropical tree species composition affects the oxidation of dissolved organic matter from litter. Biogeochemistry 88: 127–38.
48. Majcher EH, Chorover J, Bollag JM, Huang PM (2000) Evolution of CO_2 during birnessite-induced oxidation of ^{14}C-labeled catechol. Soil Science Society of America Journal 64: 157–163.
49. Zsolnay Á (2003) Dissolved organic matter: artefacts, definitions, and functions. Geoderma 113:187–209.
50. Akagi J, Zsolnay Á, Bastida F (2007) Quantity and spectroscopic properties of soil dissolved organic matter (DOM) as a function of soil sample treatments: Air-drying and pre-incubation. Chemosphere 69: 1040–1046.
51. Wang LY, Wu FC, Zhang RY, Li W, Liao HQ (2009) Characterization of dissolved organic matter fractions from Lake Hongfeng, Southwestern China Plateau. Journal of Environmental Sciences 21: 581–588.
52. Corvasce M, Zsolnay A, D'Orazio V, Lopez R, Miano TM (2006) Characterization of water extractable organic matter in a deep soil profile. Chemosphere 62: 1583–1590.
53. Bu XL, Wang LM, Ma WB, Yu XN, McDowell WH, et al. (2010) Spectroscopic characterization of hot-water extractable organic matter from soils under four different vegetation types along an elevation gradient in the Wuyi Mountains. Geoderma 159: 139–146.
54. Traversa A, D'Orazio V, Senesi N (2008) Properties of dissolved organic matter in forest soils: influence of different plant covering. Forest Ecology and Management 256: 2018–2028.
55. Fuentes M, González-Gaitano G., García-Mina JM (2006) The usefulness of uv-visible and fluorescence sepectroscopies to study the chemical nature of humic substances from soils and composts. Organic Geochemistry 37: 1949–1959.
56. Khomutova TE, Shirshova LT, Tinz S, Rolland W, Richter J (2000) Mobilization of DOC from sandy loamy soils under different land use (Lower Saxony, Germany). Plant and Soil 219: 13–19.
57. Kiikkilä O, Kitunen V, Smolander A (2006) Dissolved soil organic matter form surface organic horizons under birch and conifers: Degradation in relation to chemical characteristics. Soil Biology & Biochemistry 38: 737–746.
58. Cookson WR, Murphy DV, Roper MM (2008) Characterizing the relationships between soil organic matter components and microbial function and composition. Soil Biology & Biochemistry 40: 763–777.
59. Bu XL, Ding JM, Wang LM, Yu XN, Huang W, et al. (2011) Biodegradation and chemical characteristics of hot-water extractable organic matter from soils under four different vegetation types in the Wuyi Mountains, southeastern China. European Journal of Soil Biology 47: 102–107.
60. Jacinthe PA, Lal R, Owens LB, Hothem DL (2004) Transport of labile carbon in runoff as affected by land use and rainfall characteristics. Soil and Tillage Research 77: 111–123.
61. Rodríguez-Rodríguez A, Guerra A, Arbelo C, Mora JL, Gorrín SP, et al. (2004) Forms of eroded soil organic carbon in andosols of the Canary Islands. Geoderma 121: 205–219.
62. Juarez S, Rumpel C, Mchnu C, Vincent C (2011) Carbon mineralization and lignin content of eroded sediments from a grazed watershed of South-Africa. Geoderma 167–168: 247–253.
63. Faikd HA (2000) Primary studies on lactone in bamboo leaf. MS degree Thesis. Zhejiang University, China.
64. Zhang LY (2005) The application and Study of Flavone in Bamboo Leaf. Modern Food Science and Technology 22: 247–249 (in Chinese).
65. Zhou Y (2009) Soil organic carbon pools and the characteristics of mineralization along an elevation gradient in Wuyi Mountain, China. Ph.D. Dissertation. Nanjing Forestry University, China.
66. Rey A, Jarvis P (2006) Modelling the effect of temperature on carbon mineralization rates across a network of European forest sites (FORCAST). Global Change Biology 12: 1894–1908.
67. Nguyen HV, Hur J (2011) Tracing the sources of refractory dissolved organic matter in a large artificial lake using multiple analytical tools. Chemosphere 85: 782–789.
68. Glatzel S, Kalbitz K, Dalva M, Moore T (2003) Dissolved organic matter properties and their relationship to carbon dioxide efflux from restored peat bogs. Geoderma 113: 397–411.
69. Matilainen A, Gjessing ET, Lahtinen T, Hed L, Bhatnagar A, et al. (2011) An overview of the methods used in the characterisation of natural organic matter (NOM) in relation to drinking water treatment. Chemosphere 83: 1431–1442.
70. Embacher A, Zsolnay A, Gattinger A, Munch JC (2007) The dynamics of water extractable organic matter (WEOM) in common arable topsoils: I. Quantity, quality and function over a three year period. Geoderma 139: 11–22.
71. Embacher A, Zsolnay A, Gattinger A, Munch JC (2008) The dynamics of water extractable organic matter (WEOM) in common arable topsoils: II. Influence of mineral and combined mineral and manure fertilization in a Haplic Chernozem. Geoderma 148: 63–69.

5

Bacterial Communities in Malagasy Soils with Differing Levels of Disturbance Affecting Botanical Diversity

Leah C. Blasiak[1], Alex W. Schmidt[2], Honoré Andriamiarinoro[3], Temesgen Mulaw[1], Rado Rasolomampianina[4], Wendy L. Applequist[3], Chris Birkinshaw[3], Félicitée Rejo-Fienena[4], Porter P. Lowry, II[3,5], Thomas M. Schmidt[2], Russell T. Hill[1]*

1 Institute of Marine and Environmental Technology, University of Maryland Center for Environmental Science, Baltimore, Maryland, United States of America, **2** The Center for Microbial Sciences, University of Michigan, Ann Arbor, Michigan, United States of America, **3** Missouri Botanical Garden, St. Louis, Missouri, United States of America, **4** Laboratoire de Microbiologie de Environnement, Centre National de Recherches sur l'Environnement, Antananarivo, Madagascar, **5** Département Systématique et Évolution (UMR 7205), Muséum National d'Histoire Naturelle, Paris, France

Abstract

Madagascar is well-known for the exceptional biodiversity of its macro-flora and fauna, but the biodiversity of Malagasy microbial communities remains relatively unexplored. Understanding patterns of bacterial diversity in soil and their correlations with above-ground botanical diversity could influence conservation planning as well as sampling strategies to maximize access to bacterially derived natural products. We present the first detailed description of Malagasy soil bacterial communities from a targeted 16S rRNA gene survey of greater than 290,000 sequences generated using 454 pyrosequencing. Two sampling plots in each of three forest conservation areas were established to represent different levels of disturbance resulting from human impact through agriculture and selective exploitation of trees, as well as from natural impacts of cyclones. In parallel, we performed an in-depth characterization of the total vascular plant morphospecies richness within each plot. The plots representing different levels of disturbance within each forest did not differ significantly in bacterial diversity or richness. Changes in bacterial community composition were largest between forests rather than between different levels of impact within a forest. The largest difference in bacterial community composition with disturbance was observed at the Vohibe forest conservation area, and this difference was correlated with changes in both vascular plant richness and soil pH. These results provide the first survey of Malagasy soil bacterial diversity and establish a baseline of botanical diversity within important conservation areas.

Editor: Christina A. Kellogg, U.S. Geological Survey, United States of America

Funding: This study was supported by a grant from the National Science Foundation (NSF) Directorate for Biological Sciences Integrative Organismal Systems (IOS-0919728), Fogarty International Center, National Cancer Institute, National Institute of Allergy and Infectious Diseases, National Institute of Mental Health, National Institute on Drug Abuse, National Heart Lung and Blood Institute, National Center for Complementary and Alternative Medicine, Office of Dietary Supplements, National Institute of General Medical Sciences, Biological Sciences Directorate of the NSF, and the Office of Biological and Environmental Research of the U.S. Department of Energy under Cooperative Agreement U01 TW00313 with the International Cooperative Biodiversity Groups. This project was also supported by the National Research Initiative of the Cooperative State Research, Education and Extension Service, USDA, Grant #2008-35621-04732. The funders had no role in study design, data collection and analysis, decision to publish, or preparation of the manuscript.

Competing Interests: The authors have declared that no competing interests exist.

* E-mail: hill@umces.edu

Introduction

While the effects of land-use on the biodiversity of plants and animals are well-studied, much less is known about how disturbance shapes microbial communities. Ecological studies on the effects of disturbance on plant communities have shown that smaller disturbances may increase floral species richness, while larger ones can lead to decreased richness, loss of botanical biodiversity, or even extinction of locally endemic species (reviewed in [1]). Large disturbances such as fires or conversion of native habitats to agriculture may take centuries to completely reverse, and the re-establishment of a primary or secondary plant community often follows a predictable pattern of community succession. Advances in sequencing technology, using deep pyrosequencing of 16S rRNA gene amplicons, now permit interrogation of soil bacterial community composition in sufficient detail to begin to investigate the effects of land-use disturbance on below-ground microbial diversity. Additionally, it is now possible to investigate whether impacts to above-ground plant diversity are paralleled by changes in soil bacterial diversity. Understanding the effects of land-use on soil bacterial diversity may add a new dimension to conservation goals because soil bacteria provide critical ecosystem services and are an important source of biodiversity for drug discovery, and because bacterial communities may be early indicators of ecosystem disturbance [2,3,4].

Recent studies have used multiplexed pyrosequencing of 16S rRNA genes and other non-culture-based methods to identify parameters that correlate with soil bacterial community diversity or composition, but their causal relationships in driving soil bacterial community structure remain unclear. Many studies have now documented the important role of soil pH, which is strongly correlated to soil bacterial diversity (richness and evenness) at local and global scales [5,6,7,8]. Other soil edaphic parameters, including soil nitrogen (N) amendment [9,10] and total N or carbon (C) [6,11] have been found to be significantly associated with shifts in bacterial community composition. The taxonomic composition of soil bacterial communities can also be differentiated by land-use. For example, vegetation type [7], conversion of native vegetation to agriculture [12,13,14], and agriculture management practices [14,15,16] all correlate with changes in taxonomic composition, although these changes often co-vary with changes in soil chemistry. Vegetation type correlates with bacterial community composition locally within a given soil type or as a secondary factor after soil edaphic properties [7,17,18,19]. These correlations with plant diversity may represent indirect interactions (covariance of biogeographic patterns in plant and microbial communities) or direct interactions through influence of plant root exudates, organic matter deposition, and/or specific interactions with plant rhizospheres. For example, bacterial isolation from roots of endemic Malagasy rosewood trees has yielded a wide diversity of root-nodulating bacteria [20].

We hypothesized that loss of plant species richness in Malagasy forests through both human and non-human impacts is correlated with a loss of bacterial diversity and a change in bacterial community composition. Several studies have documented decreases in bacterial diversity or functional diversity with different types of disturbance, including impacts from wildfires, camping, and agriculture [9,21,22,23,24]. However, grassland soils have generally been found to harbor higher bacterial diversity than forest soils, and so disturbance of forests and conversion to grasslands may actually increase soil bacterial diversity and richness [6,13]. The most detailed study thus far on the impact of forest disturbance on microbial communities demonstrated a significant effect of timber harvesting, organic matter removal, and soil compaction on soil bacterial community composition in coniferous forests even 10–15 years after the disturbance and despite replanting with tree seedlings [3].

In this study, we aimed to quantify the changes in soil bacterial communities with differing levels of disturbance to forest plant communities in Madagascar. Madagascar is a biodiversity hotspot with significant conservation value, and the country contains an astonishing botanical diversity including an estimated 13,000–14,000 species of vascular plants of which nearly 90% are endemic to this island nation [25,26,27]. The study sites included three Malagasy forests that have been prioritized for conservation efforts based on work by the Missouri Botanical Garden as part of an International Biodiversity Cooperative Group grant. Within each forest, we established two 0.1 ha plots that were representative of different disturbance levels present at the site as assessed by their plant community structure. Vascular plant species richness and plant community structure, including the number and size of stems, were quantified within each plot, and the soil bacterial communities were investigated using barcoded 454 pyrosequencing of the 16S rRNA gene. This study represents the most detailed characterization to date of plant community structure and soil bacterial diversity in Malagasy forests. The resulting data provide an important initial reference for monitoring the effects of future conservation efforts within the selected forests. The observed patterns of bacterial community composition may also influence the design of sampling strategies to maximize access to bacterial taxonomic and genetic diversity for drug discovery efforts.

Materials and Methods

Site description

This study was conducted within three humid forests (as defined by [28]) located in eastern Madagascar: the forests of Analalava, Pointe à Larrée, and Vohibe (Table 1). All three sites are located in a per-humid climate zone [29], but while Analalava and Vohibe are located on basement rock, Pointe à Larrée is situated on old sand dunes [30]. The loose sand geology at Pointe à Larrée and the proximity of this forest to the sea result in a distinct subclass of humid forest known as littoral forest [28]. The Vohibe Forest is at a higher elevation than the other two sites. Each of the three forests is owned by the Malagasy government, and they are currently the focus of activities that will lead to their being classified and managed as new protected areas within Madagascar's protected area network. Permits for access to sampling sites and for collection of botanical and soil samples were obtained from the Malagasy Ministère de l'Environnement et des Forêts.

A total of six plots of 50 m×20 m were established, with two plots at each of the three sites (Table 1). The two plots within each forest were selected subjectively to maximize the difference between the levels of disturbance, both human and natural, on the vegetation. All sites have been subject to historical and/or recent disturbance from cyclones. At Analalava and Pointe à Larrée, no areas of the forest have been completely protected from human impact; sites were thus selected that represent moderate and high levels of selective timber exploitation. In the Vohibe forest, a site that has been well protected from human disturbance was contrasted with one that was subject to a single cycle of shifting cultivation prior to 2002 and is now largely a monoculture of a tall rhizomatous herb, Madagascar Cardamom [*Aframomum angustifolium* (Sonn.) K. Schum.].

Botanical diversity

The diversity of vascular plants within each 0.1 ha plot was characterized by botanists from the Missouri Botanical Garden (Table 2). Plants were identified by means of nested plots. Within the entire 0.1 ha plot all plant stems with diameter at breast height (DBH) ≥5 cm were identified. Additionally, within five 16 m^2 plots, arranged at regular intervals along the longest transect of the plot, all plants with DBH between 1 cm to 5 cm and all herbs were identified. We did not attempt to identify tree seedlings or non-vascular plants. When possible, plants were identified to a formally named species, but in many cases such precise identification could not be made because the plants were infertile (i.e. neither flowering nor fruiting). In such cases, the plants were classified only to morphospecies: distinct identities defined on the basis of habit and leaf morphology. Plants could rarely be identified with confidence in the field, so voucher herbarium specimens were made and identified later in the laboratory using the botanical literature, mounted reference herbarium specimens available in the herbarium at the Parc Botanique et Zoologique de Tsimbazaza, Antananarivo, and the electronic database of Malagasy flora (http://www.tropicos.org/project/mada). Fertile specimens were preserved at the herbaria of the Missouri Botanical Garden and the Parc Botanique et Zoologique de Tsimbazaza (MO and TAN respectively). The measurements of stem diameters were also used to describe the structure of the vegetation in the plots in terms of stem basal area and stem abundance in various DBH classes.

Soil sampling and chemical analysis

A total of 30 soil cores of 4 cm diameter and 10 cm depth were sampled from the six plots (3 forests ×2 impact levels ×5 soil cores). Five soil cores were taken in 10 m intervals along the central length of each 50×20 m plot. Soil samples were collected from Analalava and Pointe à Larrée in September 2011 and from Vohibe in October 2011, and were transported on ice to the Centre National de Recherches sur l'Environnement (CNRE) laboratory in Antananarivo, Madagascar, and stored at −20°C

Table 1. Description of the 0.1 ha plots.

Site	Forest	Subjective evaluation of disturbance	Georeference (and elevation)	Known disturbance history
A1	Analalava	High	S 17° 42′ 42,3″/E 049° 27′ 18,7″ (49 m)	Subjected to selective exploitation of medium and large trees and the collection of fence posts until 2005 when the forest was protected. Impacted by Cyclone Ivan in 2008 but sustained little damage.
A2	Analalava	Moderate	S 17° 42′ 18,9″/E 049° 27′ 22,9″ (56 m)	As at A1, but with selective exploitation of mainly large trees.
P1	Pointe à Larrée	High	S 16° 47′ 14,5″/E 049° 44′ 27,3″ (15 m)	In past decades and until present this forest is subjected to the selective exploitation of large trees. The forest was impacted by Cyclone Ivan in 2008 with consequent major damage.
P2	Pointe à Larrée	Moderate	S 16° 47′ 45,5″/E 049° 44′ 38,0″ (15 m)	As at P1, but with selective exploitation of large trees now rare.
V1	Vohibe	Very high	S 19° 09′ 45,7″/E 048° 35′ 02,3″ (598 m)	Subjected to one cycle of shifting cultivation (trees cut, left to dry and burnt, and then land cultivated until soil exhausted). The last crops were cultivated on this land in about 2002.
V2	Vohibe	Very low	S 19° 09′ 32,2″/E 048° 34′ 39,1″ (646 m)	In the past possibly subjected to very rare selective exploitation of precious woods and timber.

until processing. Soil cores were homogenized and sieved through a 2 mm mesh. Soil pH, C, N, and C:N ratio (CN) were determined at FOFIFA, the Centre National de la Recherche Appliquée au Développement Rural in Madagascar (Table S1). Briefly, soil pH was measured as a water/soil suspension, total organic carbon (C) was determined according to the ANNE method [31], and the total N by the Kjeldahl method [32].

DNA extraction and pyrosequencing

DNA was extracted from 0.25 g of soil from homogenized soil cores using the PowerSoil DNA isolation kit (Mo Bio, Carlsbad, CA) according to the manufacturer's instructions, except that the DNA yield was increased by pooling the DNA from two DNA extraction kits for each sample. A total of 60 DNA samples were prepared (two technical replicates per soil core ×5 cores per plot ×6 plots). DNA samples were dried at 50°C immediately after extraction at the CNRE laboratory and shipped dry at room temperature to Michigan for sequencing. The V4–V6 region of the 16S rRNA gene was PCR amplified in triplicate using the following barcoded primers: forward primer 518F = CCAGCAGCYGCGGTAAN, forward primer adapter = CCTATCCCCTGTGTGCCTTGGCAGTCTCAG, reverse primer 1046R = CGACRRCCATGCANCACCT, reverse primer adapter = CCATCTCATCCCTGCGTGTCTCCGACTCAG (http://www.hmpdacc.org/doc/16S_Sequencing_SOP_4.2.2.pdf) [33]. The PCR products were then purified using Agencourt® AMPure® XP beads (Beckman Coulter, Brea, CA) and the Life Technologies Dynamag-2 Magnet. DNA concentrations were quantified and samples were combined into an equimolar pool and sequenced using a Roche GS Junior 454 Sequencer. Sequences and associated metadata were deposited in the MG-RAST database and are publically accessible (http://metagenomics.anl.gov/linkin.cgi?project=6283) with accession numbers 4538937.3–4538996.3.

Sequence processing

The V4–V6 region of the 16S rRNA gene was sequenced from 60 soil samples (2 technical replicates ×5 cores per plot ×6 plots). Sequence processing was performed in mothur v1.28.0 [34] and generally followed the recommendations described by Schloss *et al.* [35]. Raw sequences (598,730 reads) were denoised with PyroNoise [36] and then sequences with mismatches to forward primer or barcode sequences, ambiguous bases, and homopolymer stretches greater than 8 bp were removed. Barcodes and primer sequences were removed, short sequences were discarded, and the remaining sequences were trimmed to exactly 250 bp. The denoised and quality filtered sequences were then aligned to the Silva reference alignment (v102) [37]. Preclustering of aligned sequences was employed using an abundance weighted single-linkage preclustering step at ~2% difference (maximum 2 bp difference over 250 bp) [38]. Chimeras were identified and removed using the UChime algorithm [39] and sequences were classified with an 80% bootstrap cutoff to the Ribosomal Database Project (RDP) 16S rRNA reference database Release 9 [40] using the RDP Classifier [41]. Sequences classified as *Archaea*, *Eukaryota*, chloroplasts, mitochondria, and unclassified sequences were removed (<0.01% of sequences), leaving only sequences classified as *Bacteria*. Pairwise distances were calculated on aligned sequences and then sequences were clustered into operational taxonomic units (OTUs) at 3% difference using the average neighbor method. The taxonomy of each OTU was assigned in mothur as the consensus taxonomy for at least 51% of the sequences within the OTU.

Sequence processing was performed with technical replicates separated (60 samples) and then with technical replicates pooled (30 samples). Sequences were subsampled to either 5,800 sequences (technical replicates separated) or 9,800 sequences (technical replicates pooled), to allow comparisons between samples with different sequencing depth. Beta diversity analysis using NMDS ordination of Bray-Curtis distances demonstrated that technical replicates clustered together and thus had similar bacterial communities (Figure S1). The mean Bray-Curtis dissimilarity between pairs of technical replicates was 0.21±0.06 (standard deviation) while the mean dissimilarity between all pairs of samples within each site was 0.39±0.05. This result indicates that the soil core homogenization was sufficient, because replicate soil samples from the same core gave reproducible bacterial communities. Thus, each sequenced soil sample is representative of the entire soil core. Therefore, technical replicates were pooled for all further data analysis. The final quality-filtered and

Table 2. Description of plant community structure by plot.

Site	Subjective evaluation of disturbance	# of morpho-species of vascular plants	Stem basal area (m^2/0.1 ha) by diameter size class			Number woody stems per 0.1 ha by stem diameter size class (cm)						
			≥10 cm	≥5 to <10 cm	≤5 cm	≥1-<5	≥5-<10	≥10-<20	≥20-<30	≥30-<40	≥40-<50	≥50
A1	High	214	1.26	1.01	2.30	1875	272	53	8	2	0	0
A2	Moderate	225	1.99	1.45	3.44	2338	391	96	12	2	0	0
P1	High	103	1.86	0.64	2.50	1300	180	47	14	5	1	0
P2	Moderate	89	2.44	0.76	3.20	1250	213	61	20	5	0	1
V1	Very high	32	0.74	0.17	0.91	163	40	17	4	0	2	0
V2	Very low	215	4.10	0.66	4.75	1225	179	73	23	9	2	3

subsampled dataset contained 294,000 sequences with 30 samples and 9,800 sequences per sample.

Statistical analyses

Comparisons of soil parameters across plots were made using one-way analysis-of-variance (ANOVA) in R [42]. Bacterial diversity measures including the Inverse Simpson's Index and Bray-Curtis dissimilarity matrices were calculated on 3% OTUs in mothur [34]. Relationships between bacterial diversity and soil chemical data were investigated by plotting and with the function cor.test in R. Nonmetric multidimensional scaling (NMDS) ordinations on Bray-Curtis dissimilarities, permuted analysis of variance (PERMANOVA), and permuted analysis of multivariate dispersions were performed using the functions metaMDS, adonis, and betadisper from the vegan package in R [43,44]. Because of the large differences between forests, including elevation, climate, and sampling date, the forests were treated as three separate experiments for pairwise comparisons of disturbance levels by PERMANOVA. Canonical analysis of principal coordinates (CAP) implemented through the capscale function in vegan was used to perform constrained ordination to test for linear relationships between bacterial community composition (Bray-Curtis dissimilarities) and soil chemical parameters within each forest. All permutation tests were performed with 999 permutations. Figures were prepared with R and the package ggplot2 [45].

Results and Discussion

Botanical diversity

The botanical diversity of the six plots is summarized in Table 2. At Analalava and Vohibe, the plots in less disturbed forest had more vascular plant morphospecies than those in more disturbed forest. The reverse was true for the plots at Pointe à Larrée, with the more disturbed plot having higher plant species richness. The moderately disturbed plot at Pointe à Larrée included much lower diversity of vascular plants than the moderately disturbed plot at Analalava or the little disturbed plot at Vohibe. The total stem DBH and stem abundance in various size classes showed a striking structural difference between the little disturbed plot at Vohibe and its very disturbed counterpart. The plots at Analava and Pointe à Larrée also had a generally consistent decrease in stem DBH and stem abundance with disturbance, but the differences between the disturbance levels much less pronounced. While the little disturbed forest plot at Vohibe included more large trees and a much higher total stem basal area than the two plots at Analalava, both of the latter included a morphospecies richness of vascular plants similar to that of the little disturbed plot at Vohibe.

Bacterial community composition

In order to compare the relative abundance of bacterial taxonomic groups observed in this study, 3% OTUs were classified according to the RDP taxonomy and grouped at the phylum level, except for proteobacterial groups, which were further divided by class (Figure 1). The most abundant bacterial group in the Pointe à Larrée and Vohibe plots was *Betaproteobacteria*, representing an average of about 20–30% of sequences within each plot. In contrast, the Analalava plots contained mostly *Acidobacteria* and *Verrucomicrobia*. Other abundant groups with an average of >5% of sequences across all plots were *Actinobacteria*, *Alphaproteobacteria*, *Gammaproteobacteria*, and *Firmicutes*. The bacterial community composition was further analyzed by grouping OTUs at the genus level (Figure 2). Abundant genera across all soil samples included *Burkholderia*, the *Acidobacteria* Group 1, *Rhodococcus*, and *Bacillus*. The functional and ecological roles of the *Acidobacteria* remain poorly characterized, but there is evidence for strong correlations between abundance of acidobacterial groups and pH [8,46], as well as indications that many *Acidobacteria* may be adapted to an oligotrophic lifestyle [47]. Naether *et al.* observed that the Group 1 *Acidobacteria* were dominant in temperate forest soils, while the Group 6 Acidobacteria were most prevalent in grassland soils [46]. The abundance of the betaproteobacterial genus *Burkholderia* observed here is consistent with the recent identification of this genus as an acid tolerant group that may outcompete other taxa in acidic soils such as those in our study [48]. A BLAST analysis of a representative sequence from the dominant *Burkholderia* OTU in our dataset showed that the closest relative was *Burkholderia sabiae* strain STM7319, which clusters within the plant-beneficial-environmental (PBE) group of *Burkholderia* [49]. The next two most abundant *Burkholderia* OTUs in our dataset also clustered with this group rather than the typically pathogenic *Burkholderia cepacia* complex. Members of the PBE clade are often associated with plants and may be capable of nitrogen fixation or otherwise promote plant growth. Amplicon-based community profiling has many well-documented limitations, including DNA extraction and primer bias, which constrain our ability to draw conclusions about absolute abundance of bacterial groups. However, despite any bias, sequencing results are reproducible, so comparisons can be made concerning relative abundance of taxa in samples with consistent DNA extraction and sequencing methods [50].

Bacterial diversity and disturbance

The Inverse Simpson diversity index is a measure of alpha diversity that accounts for the richness and evenness of OTUs observed within a single sample. We observed a trend of lower bacterial diversity in the less disturbed plots (Table 3, Figure S2), although this difference was not significant. We observed no significant difference in the Inverse Simpson index between each of the plots (treating each of the six plots separately) or between each of the forests (combining the two plots in each forest). Comparisons made using richness estimators (number of OTUs, Chao1) and a phylogenetic diversity measure (Faith's PD) gave similar results (data not shown). Comparisons using Smith and Wilson's metric of evenness demonstrated a significant difference at the level of plots ($p = 0.002$) and forests ($p = 0.008$). A Tukey's honestly significant difference (HSD) test showed that the high impact plot at Pointe à Larrée had significantly lower evenness than the low impact plot at Vohibe ($p = 0.003$) and the moderate and high impact plots at Analalava ($p = 0.002$ and 0.04 respectively). However, there was no significant difference in evenness between any of the pairs of plots within each forest ($p > 0.05$).

These results indicate that disturbance from selective timber harvesting or historical cultivation did not significantly affect the soil bacterial diversity, although there was a slight trend towards higher bacterial diversity with disturbance in all forests. We did not detect a significant difference in bacterial diversity or richness even between the Vohibe plots where we observed dramatic differences in plant species richness and vegetation structure.

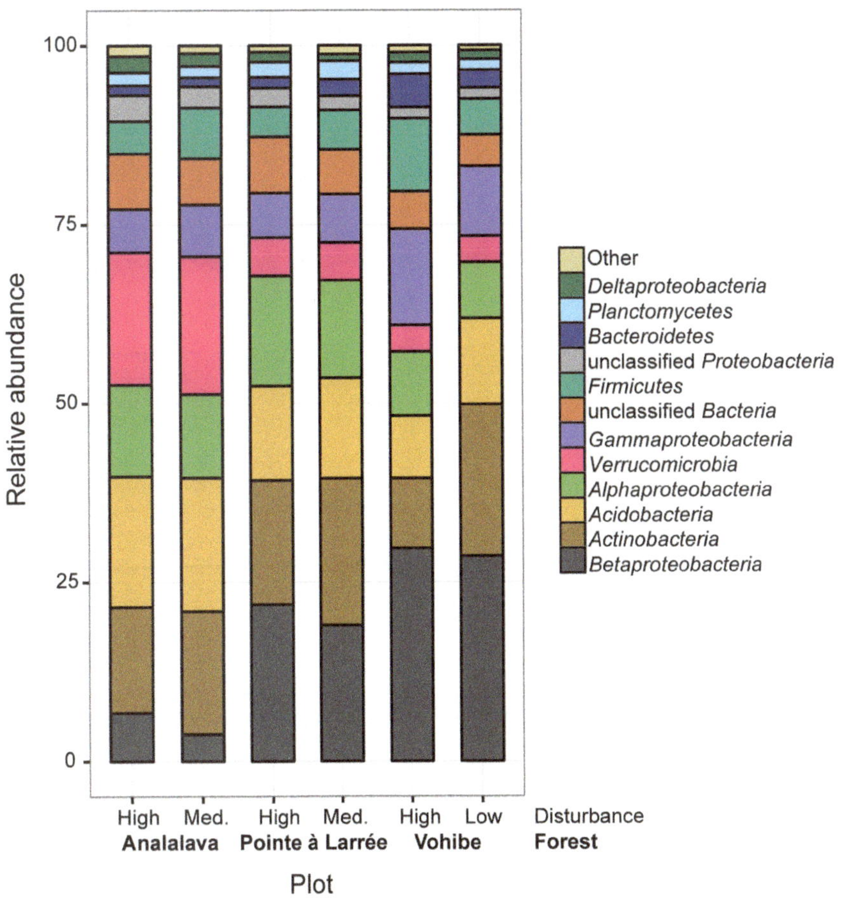

Figure 1. Average relative abundance of dominant bacterial taxa by plot. Sequences in 3% OTUs were classified and grouped at the phylum level, with *Proteobacteria* further divided by class. Taxa representing <1% of the total sequences were grouped as Other.

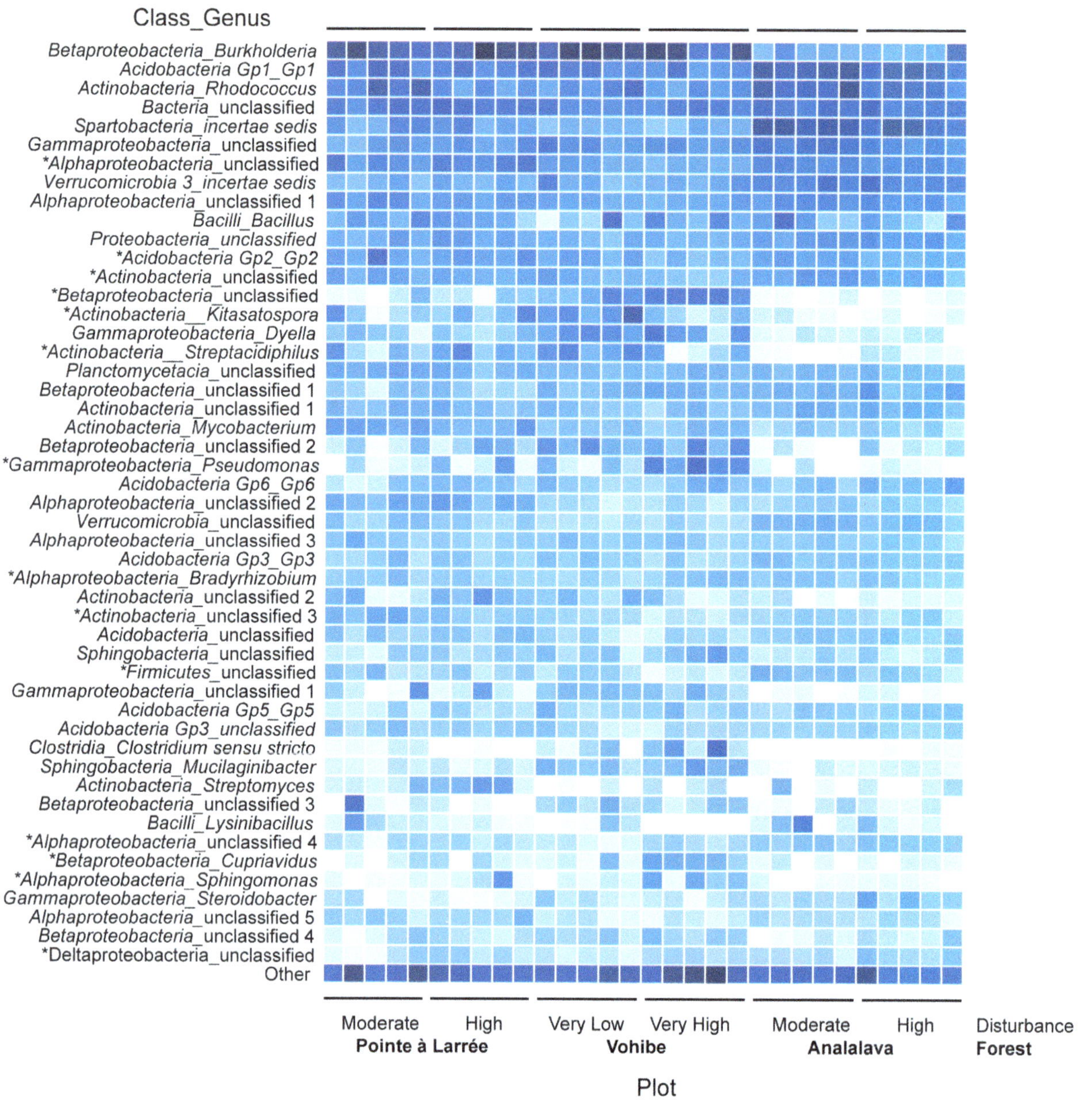

Figure 2. Log transformed abundance of dominant bacterial genera by sample. Sequences in 3% OTUs were classified and grouped at the genus level. Taxa representing <0.3% of the total sequences were grouped as Other. Darker color indicates higher relative abundance. Asterisks indicate taxa with significant shifts in abundance between disturbance levels at Vohibe (see Table 5).

Table 3. Bacterial diversity statistics[a] calculated at the OTU (3%) level.

Plot	Good's coverage %	# OTUs	Inverse Simpson	Shannon	Smith and Wilson's Evenness
A1	95.8 (0.9)	970 (178)	51.0 (23.3)	5.08 (0.36)	0.34 (0.02)
A2	96.5 (0.5)	819 (78)	37.1 (10.0)	4.77 (0.17)	0.37 (0.01)
P1	94.0 (0.9)	1212 (174)	40.0 (25.1)	5.12 (0.47)	0.29 (0.03)
P2	95.7 (0.9)	972 (199)	34.0 (13.6)	4.88 (0.42)	0.33 (0.04)
V1	94.9 (0.8)	1031 (186)	43.1 (14.7)	4.96 (0.37)	0.33 (0.03)
V2	96.0 (0.8)	838 (120)	29.1 (14.6)	4.64 (0.32)	0.37 (0.02)

[a]Shown are the mean and (sd) of each statistic calculated for 5 sequencing samples per plot.

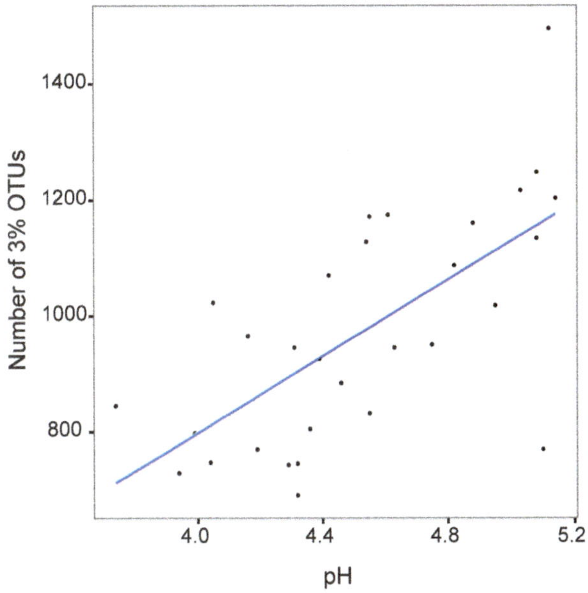

Figure 3. Richness and soil pH. Bacterial community richness is shown as the number of 3% OTUs. Pearson's $r = 0.66$, $p < 0.0001$.

Other studies have found correlations in bacterial diversity with land-use, with agricultural sites having higher diversity than forested sites in tropical soils, but these sites also differed in soil environmental parameters [6,13]. The results of our study more closely resemble those of Suleiman *et al.* in the Brazilian Pampa, where no significant difference in alpha diversity was found between eight-year-old de-forested pasture as compared to surrounding undisturbed forest [12]. While we did not find significant effects of land-use on soil bacterial richness at the resolution afforded by the 16S rRNA gene, there could be finer level changes in the diversity of functional genes or natural product biosynthetic genes [51]. It remains an open question whether bacterial taxonomic diversity (richness) correlates with ecosystem function, and if loss of microbial diversity affects community resilience or function [21,52,53]. Moreover, a recent study found large fluctuations in soil bacterial diversity over time, with temporal variability exceeding that attributed to land-use [53]. More sampling replicates obtained from different seasons might be required to reveal differences in bacterial diversity with disturbance level or botanical diversity.

Bacterial diversity and soil properties

Several studies have documented a strong relationship between bacterial richness and soil pH at both global and local scales [5,6,8]. Although we did not observe site-dependent changes in diversity with pH, we hypothesized that changes in soil properties irrespective of site might be related to bacterial community richness. In fact, when looking at all samples collectively, the number of OTUs per sample was positively correlated with soil pH (Pearson's $r = 0.66$, $p < 0.0001$), which varied from 3.7–5.1 in our study (Figure 3). Measures of diversity, including the Inverse Simpson's Index, were also positively correlated with soil pH across all samples. This relationship was significant but weaker than that observed by Rousk *et al.* at a single site, a gradient of pH 4–8 at the Hoosfield acid strip long-term agricultural research site in Harpenden, UK [8]. We found no significant correlations between richness and the other measured soil parameters including organic C, total N, or C:N ratio.

Bacterial community composition varies with forest more than with disturbance level

An NMDS ordination of Bray-Curtis dissimilarities showed strong clustering of bacterial communities by forest as compared to levels of disturbance (Figure 4). The significance of this clustering was tested using a nested PERMANOVA design (sites nested within forests) with Bray-Curtis dissimilarities calculated on 3% OTUs. We observed significant differences between bacterial community compositions at the three forest sites (pseudo-F = 10.47, $p = 0.001$) and a significant interaction effect between forests and sites (pseudo-F 1.87, $p = 0.009$). Pairwise comparisons

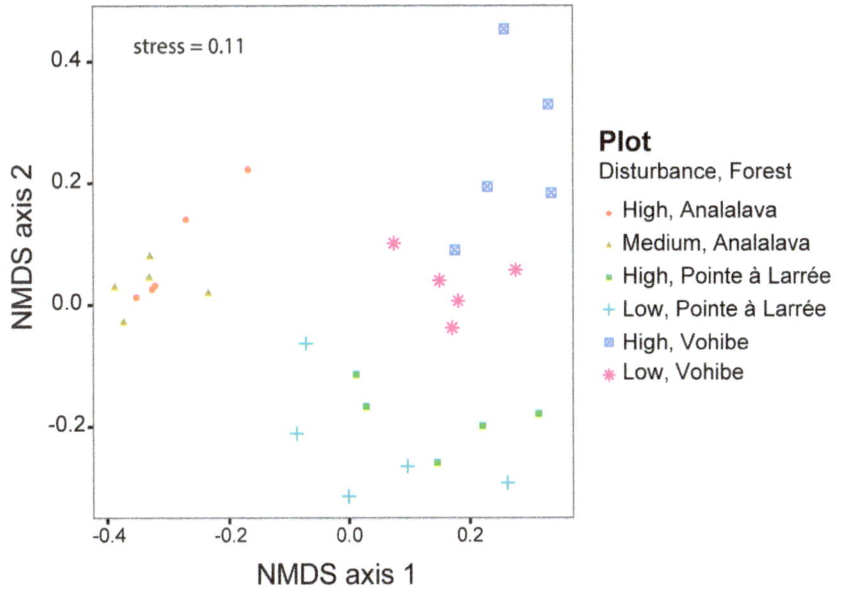

Figure 4. Nonmetric multidimensional scaling (NMDS) plot of pairwise Bray-Curtis dissimilarities between bacterial communities in soil calculated with 3% OTUs.

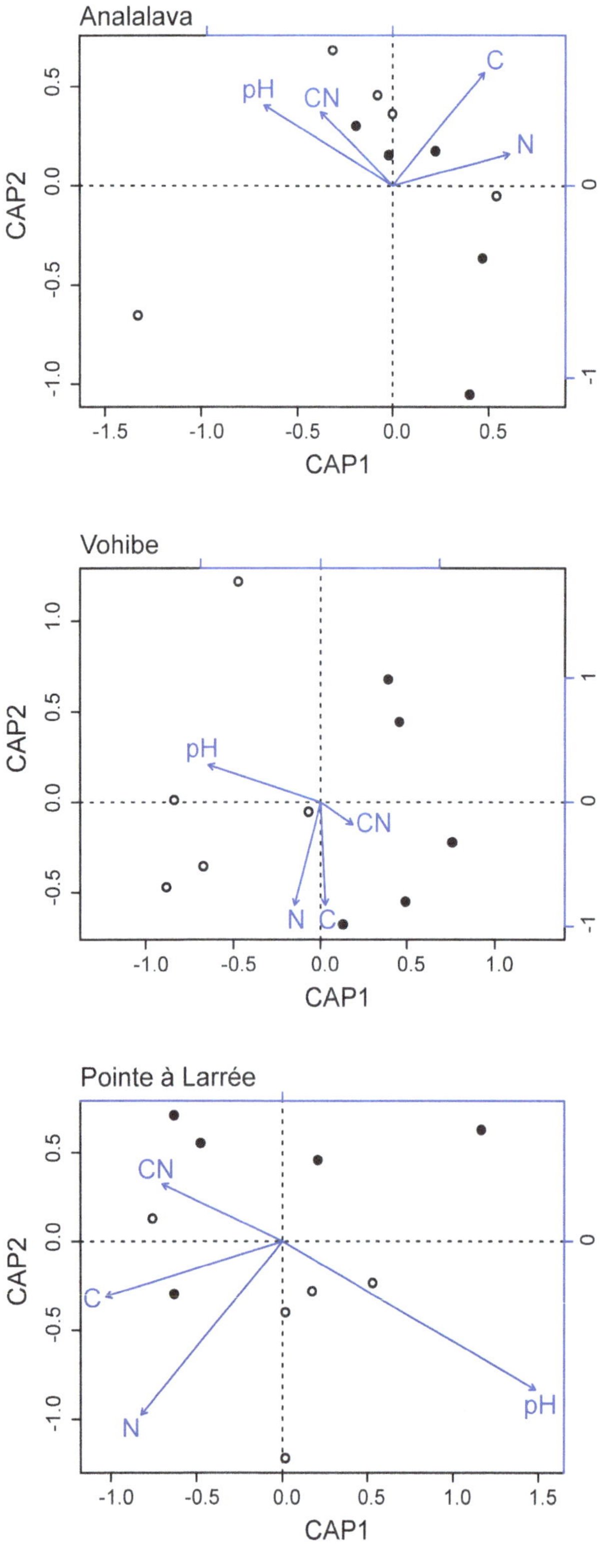

Figure 5. CAP ordination biplots of the bacterial community composition data and soil chemical properties (arrows) within each forest. Closed circles indicate sites from the lower disturbance plot and open circles indicate sites from the higher disturbance plot.

between forests also showed strong differences in the bacterial community structures (A-V: $t=12.95$; A-P: $t=11.20$; V-P: $t=5.44$, by PERMANOVA, Bonferroni corrected $p=0.003$ for these pairwise comparisons). The samples from Analalava had significantly lower spread than those from Pointe à Larrée, but no other significant differences were detected in the dispersion of samples within forests (pairwise comparisons with betadisper, A-V: $t=3.80$, $p=0.20$; A-P: $t=7.03$, $p=0.02$; V-P: $t=0.065$, $p=2.41$, with Bonferroni correction).

Pairwise PERMANOVAs were then performed between the impact levels within each forest in order to test whether soil bacterial community composition varied with the level of disturbance. Only the bacterial communities at Vohibe varied significantly with disturbance level ($t=3.0$, $p=0.007$). The NMDS ordination and smaller PERMANOVA pseudo-F/t statistic indicate that the difference between the bacterial communities of the two sites at Vohibe is less than that observed between forests. Pairwise tests for homogeneity of group variance between the impact levels within each forest were not significant ($p>0.05$ for all pairwise comparisons). The two Vohibe sites were the most different in terms of impact level (very high vs. very low impact), as indicated by the measured differences in plant morphospecies richness and tree cover (Table 2). It is therefore perhaps not surprising that we were able to observe a difference in bacterial structure between this pair of plots. By contrast, no areas representing a low level of disturbance were available at Analalava and Pointe à Larrée, and so the selected pairs of sites represent moderate and high levels of impact from selective harvesting of trees. To test for differences in soil parameters between plots, we used one-way ANOVAs followed by Tukey's HSD test, and the only significant differences between paired plots within each forest were in pH at Vohibe and in total N at Pointe à Larrée. The pH levels of the soil samples from the high disturbance Vohibe forest were higher compared to those of the low disturbance plot, perhaps resulting from the history of burning followed by shifting cultivation in the disturbed plot. Thus, the observed change in bacterial community structure between sites with different impact levels at Vohibe is correlated with differences in botanical diversity and with a change in soil pH.

Beta diversity and soil properties

We expected that soil samples with more similar soil chemistry would also host more similar bacterial communities. To test this hypothesis, CAP ordinations of the community composition data (relative abundance of 3% OTUs) constrained by the soil chemical data were performed within each forest (Figure 5). Within all forests, pH was significantly correlated with bacterial community structure ($p<0.05$ for all tests). The total N was only significant within the Pointe à Larrée forest. Total C and CN were not significant within any forest and C tended to covary with N. As expected given the previous unconstrained ordination, pH was significantly aligned with CAP axis 1 and the two disturbance sites at Vohibe also separate along this axis. Past studies have demonstrated associations of bacterial community composition with pH, C, N, and/or C:N ratio, depending on the sample type or treatment [6,11,24]. Fierer *et al.* [9] also observed a correlation between bacterial community phylogenetic distances and high soil N amendment levels in temperate agricultural soils. Strong shifts in soil bacterial community composition have been reported with nitrogen amendment to soil samples and year-long incubation under controlled, laboratory conditions [54]. While pH is strongly correlated with bacterial community composition in the majority of soil bacterial community studies, the importance of other soil edaphic properties may vary depending on the soil type or treatment.

Table 4. Bacterial phylotypes at the Class level with differing abundance between Vohibe high impact (V1) and low impact (V2) sites (Metastats $p<0.05$).

RDP Classification	V1 rel. abundance (%)	V2 rel. abundance (%)	Abundance at higher impact plot
Phylum_Class	mean (sd)	mean (sd)	
*Firmicutes*_unclassified	0.14 (0.04)	0.46 (0.10)	lower
Acidobacteria_Gp2	1.1 (0.2)	2.3 (0.3)	lower
Verrucomicrobia_Opitutae	0.09 (0.02)	0.03 (0.02)	higher
Actinobacteria_Actinobacteria	9.8 (1.6)	21.2 (4.5)	lower

Differentially abundant groups between plots at Vohibe

The program Metastats [55] was used to determine which bacterial groups were differentially abundant between the Vohibe plots and therefore contributed to the significant difference in bacterial community composition between these plots. At the OTU level, this analysis detected greater than 200 bacterial groups with significant changes in abundance (data not shown). In order to relate these changes to classified bacterial taxa, the Metastats analysis was performed with OTUs grouped at the class and genus levels of the RDP taxonomy (Table 4 and Table 5). Differences in abundance of bacterial taxa were considered

Table 5. Bacterial phylotypes at the Genus level with differing abundance between Vohibe high impact (V1) and low impact (V2) sites (Metastats $p<0.05$).

RDP Classification	V1 rel. abundance (%)	V2 rel. abundance (%)	Abundance at higher impact plot
Class_Genus	**mean (sd)**	**mean (sd)**	
Actinobacteria_Solirubrobacter	0.00008 (0.00002)	0.00029 (0.00004)	lower
*Deltaproteobacteria*_unclassified	0.0037 (0.0004)	0.0016 (0.0004)	higher
Gammaproteobacteria_Pseudomonas	0.05 (0.01)	0.006 (0.003)	higher
*Actinobacteria*_unclassified	0.0023 (0.0005)	0.0045 (0.0003)	lower
Alphaproteobacteria_Bradyrhizobium	0.010 (0.001)	0.0044 (0.0006)	higher
*Betaproteobacteria*_unclassified	0.061 (0.006)	0.026 (0.009)	higher
*Firmicutes*_unclassified	0.001 (0.003)	0.0045 (0.0009)	lower
*Alphaproteobacteria*_unclassified	0.0039 (0.0006)	0.0014 (0.0004)	higher
Betaproteobacteria_Cupriavidus	0.014 (0.003)	0.003 (0.002)	higher
Acidobacteria Gp2_Gp2	0.011 (0.002)	0.023 (0.003)	lower
*Alphaproteobacteria*_unclassified	0.013 (0.001)	0.021 (0.003)	lower
Actinobacteria_Streptacidiphilus	0.010 (0.005)	0.036 (0.008)	lower
Planctomycetacia_Gemmata	0.00008 (0.00004)	0.0006 (0.0002)	lower
Gammaproteobacteria_Coxiella	0.00014 (0.00005)	0	higher
*Gammaproteobacteria*_unclassified	0.0007 (0.0002)	0.00002 (0.00002)	higher
Actinobacteria_Nocardioides	0.0018 (0.0007)	0.00002 (0.00002)	higher
Caldilineae_Caldilinea	0.00006 (0.00002)	0	higher
Betaproteobacteria_Georgfuchsia	0.00006 (0.00002)	0	higher
Alphaproteobacteria_Sphingomonas	0.012 (0.004)	0.0008 (0.0002)	higher
Sphingobacteria_Ferruginibacter	0.0010 (0.0004)	0.0024 (0.0005)	lower
*Anaerolineae*_unclassified	0.00010 (0.00005)	0	higher
Actinobacteria_Kitasatospora	0.007 (0.003)	0.05 (0.02)	lower
Acidobacteria Gp22_Gp22	0.00008 (0.00004)	0	higher
Alphaproteobacteria_Hyphomicrobium	0.00008 (0.00004)	0	higher
Gammaproteobacteria_Salmonella	0.00008 (0.00004)	0	higher
Planctomycetacia_Planctomyces	0.00016 (0.00005)	0.00004 (0.00002)	higher
*Actinobacteria*_unclassified	0.010 (0.002)	0.018 (0.004)	lower

significant with $p<0.05$. At the class level, the relative abundance of bacterial groups *Acidobacteria* Group 2, unclassified *Firmicutes*, and *Actinobacteria* were all lower in the high impact Vohibe plot as compared to the low impact plot (Table 4). Conversely, the abundance of class *Opitutae* within the phylum *Verrucomicrobia* was higher in the high impact plot compared to the low impact plot samples from Vohibe. Changes in acidobacterial groups with pH have been well documented. For example, Rousk *et al.* [8] also observed a decrease in the relative abundance *Acidobacteria* Gp 2 with increasing pH in temperate agricultural soil. In the same study, no consistent trend in the abundance of *Actinobacteria* was observed from pH 4–8, but within the relevant pH range of our samples (~4–5), Rousk *et al.* did observe a decrease in actinobacterial abundance. At a larger geographic scale, Fierer *et al.* [9] found that the abundance of the phylum *Acidobacteria* decreased with increasing pH, while the relative abundance of *Actinobacteria* increased [56]. However, the increasing abundance of *Actinobacteria* was observed mainly between pH 5–7. *Verrucomicrobia* were not dominant groups in these studies, perhaps because of primer bias, and so differential abundances were not reported.

When the shifts in the bacterial community composition at Vohibe were examined in greater detail at the genus level, we also observed higher abundance of the genus *Bradyrhizobium* and of an unclassified group within the *Oxalobacteraceae* in the high impact plot; both of these clades contain nitrogen-fixing members (Table 5). Interestingly, several bacterial genera known for their roles as human and animal pathogens had significantly higher abundance with disturbance, including *Pseudomonas*, *Coxiella*, and *Salmonella*. This shift might reflect the history of shifting cultivation in the higher impact site at Vohibe. Corroborating the results at the class level, the abundance of several actinobacterial genera, including *Solirubrobacter*, *Streptacidiphilus*, and *Kitasatospora* as well as unclassified *Actinobacteria*, were lower in the high impact plot (Table 5 and Figure S3). Only one actinobacterial genus, the *Nocardioides*, was observed in significantly higher abundance in the higher impact plot.

Members of the phylum *Actinobacteria* are well known for their role in degrading complex organic matter in soils and for their potential as producers of bioactive natural products. The significant decrease in the relative abundance of *Actinobacteria* amplicons in the disturbed plot at Vohibe indicates that sampling across disturbance levels could improve the sampled biodiversity for this natural product discovery target phylum. Similarily, Zabinski & Gannon [22] showed that disturbance from recreational camping decreased the percentage of culturable bacteria with *Actinomycete*-like morphology in subalpine soils. However, the sequencing survey performed in this study measures relative abundance of amplicons and cannot distinguish between true shifts in the numbers of *Actinobacteria* and changes in abundance due to shifts in other groups. The results from our work suggest that disturbance from shifting cultivation at Vohibe could decrease the proportion of *Actinobacteria,* and we plan future experiments to test this hypothesis directly using comparisons of culturable bacterial isolates, direct sequencing, and characterization of the diversity of natural product biosynthetic genes.

Conclusions

Contrary to our initial hypothesis, we found no significant differences in soil bacterial diversity with different disturbance levels, although there was a non-significant trend towards higher richness with disturbance. Comparison of technical replicates from each bulk soil core demonstrated that the observed bacterial community composition within samples from the same homogenized core were reproducible. We detected a significant change in soil bacterial community composition between the low and high disturbance plots in Vohibe forest, but not at Analalava or Pointe à Larrée where there was less difference between disturbance levels in the plots sampled. The Vohibe sites exhibited a large decrease in woody plant morphospecies richness and a decrease in tree cover in the high disturbance plot compared to relatively intact, native forest in the low disturbance plot. The change in bacterial community composition with disturbance at Vohibe was also correlated with an increase in soil pH, which may have resulted from historical burning during shifting cultivation. The study design did not allow us to differentiate between co-varying parameters and so future experiments will be required to distinguish direct and indirect effects. Future work might also look for shifts in microbial functional genes that could be masked by this OTU level taxonomic analysis. Our study focused on the soil bacterial diversity, but changes in botanical diversity and land use would also be expected to influence archaeal and fungal taxa that are known to serve important roles within soil microbial communities. The relative changes in bacterial community composition between sites were much greater than between disturbance levels, suggesting that widespread geographic sampling across forests would maximize access to bacterial diversity rather than collecting multiple samples from different disturbance levels within a given forest.

Supporting Information

Figure S1 Nonmetric multidimensional scaling (NMDS) plot of pairwise Bray-Curtis distances between soil samples calculated with 3% OTUs. Technical replicates (two DNA samples from the same homogenized bulk soil core) are labeled with the same number.

Figure S2 Mean bacterial alpha diversity in each plot as measured by the Inverse Simpson's index of diversity. Errors bars represent the standard error of the mean for five soil samples.

Figure S3 Average relative abundance of actinobacterial genera by plot. Sequences in 3% OTUs were classified and grouped at the genus level. Taxa representing <0.1% of the total sequences were grouped as Other.

Table S1 Chemical data for homogenized soil cores.

Acknowledgments

We thank David Kingston for his leadership as Principal Investigator of the International Cooperative Biodiversity Group "Biodiversity Conservation & Drug Discovery in Madagascar". We would also like to thank the site-based Conservation Facilitators at the three study sites for their help during the fieldwork: Anselme Tilahimena at Analalava, Adolphe Lehavana at Pointe à Larrée, and Fortunat Rakotoarivony at Vohibe. This is IMET contribution No. 13-113 and UMCES contribution No. 4827.

Author Contributions

Conceived and designed the experiments: LCB TM RR FRF WLA CB PPL TMS RTH. Performed the experiments: HA TM RR CB AWS. Analyzed the data: LCB AWS CB WLA. Contributed reagents/materials/analysis tools: AWS TMS. Wrote the paper: LCB AWS CB PPL TMS RTH.

References

1. Keddy P (2005) Putting the plants back into plant ecology: Six pragmatic models for understanding and conserving plant diversity. Annals of Botany 96: 177–189.
2. Griffiths BS, Philippot L (2013) Insights into the resistance and resilience of the soil microbial community. Fems Microbiology Reviews 37: 112–129.
3. Hartmann M, Howes CG, VanInsberghe D, Yu H, Bachar D, et al. (2012) Significant and persistent impact of timber harvesting on soil microbial communities in Northern coniferous forests. ISME J 6: 2199–2218.
4. Daniel R (2004) The soil metagenome—a rich resource for the discovery of novel natural products. Curr Opin Biotechnol 15: 199–204.
5. Lauber CL, Hamady M, Knight R, Fierer N (2009) Pyrosequencing-based assessment of soil pH as a predictor of soil bacterial community structure at the continental scale. Appl Environ Microbiol 75: 5111–5120.
6. Tripathi BM, Kim M, Singh D, Lee-Cruz L, Lai-Hoe A, et al. (2012) Tropical soil bacterial communities in Malaysia: pH dominates in the equatorial tropics too. Microb Ecol 64: 474–484.
7. Griffiths RI, Thomson BC, James P, Bell T, Bailey M, et al. (2011) The bacterial biogeography of British soils. Environ Microbiol 13: 1642–1654.
8. Rousk J, Baath E, Brookes PC, Lauber CL, Lozupone C, et al. (2010) Soil bacterial and fungal communities across a pH gradient in an arable soil. ISME J 4: 1340–1351.
9. Fierer N, Lauber CL, Ramirez KS, Zaneveld J, Bradford MA, et al. (2012) Comparative metagenomic, phylogenetic and physiological analyses of soil microbial communities across nitrogen gradients. ISME J 6: 1007–1017.
10. Ramirez KS, Lauber CL, Knight R, Bradford MA, Fierer N (2010) Consistent effects of nitrogen fertilization on soil bacterial communities in contrasting systems. Ecology 91: 3463–3470; discussion 3503–3414.
11. Nemergut DR, Cleveland CC, Wieder WR, Washenberger CL, Townsend AR (2010) Plot-scale manipulations of organic matter inputs to soils correlate with shifts in microbial community composition in a lowland tropical rain forest. Soil Biology & Biochemistry 42: 2153–2160.
12. Suleiman AK, Manoeli L, Boldo JT, Pereira MG, Roesch LF (2013) Shifts in soil bacterial community after eight years of land-use change. Syst Appl Microbiol 36: 137–144.
13. Rodrigues JL, Pellizari VH, Mueller R, Baek K, Jesus Eda C, et al. (2013) Conversion of the Amazon rainforest to agriculture results in biotic homogenization of soil bacterial communities. Proc Natl Acad Sci U S A 110: 988–993.
14. Buckley DH, Schmidt TM (2003) Diversity and dynamics of microbial communities in soils from agro-ecosystems. Environ Microbiol 5: 441–452.
15. Berthrong ST, Buckley DH, Drinkwater LE (2013) Agricultural management and labile carbon additions affect soil microbial community structure and interact with carbon and nitrogen cycling. Microb Ecol 66: 158–170.
16. Wu T, Chellemi DO, Graham JH, Martin KJ, Rosskopf EN (2008) Comparison of soil bacterial communities under diverse agricultural land management and crop production practices. Microb Ecol 55: 293–310.
17. Mitchell RJ, Hester AJ, Campbell CD, Chapman SJ, Cameron CM, et al. (2010) Is vegetation composition or soil chemistry the best predictor of the soil microbial community? Plant and Soil 333: 417–430.
18. Andrew DR, Fitak RR, Munguia-Vega A, Racolta A, Martinson VG, et al. (2012) Abiotic factors shape microbial diversity in Sonoran Desert soils. Appl Environ Microbiol 78: 7527–7537.
19. Liliensiek AK, Thakuria D, Clipson N (2012) Influences of plant species composition, fertilisation and *Lolium perenne* ingression on soil microbial community structure in three Irish grasslands. Microbial Ecology 63: 509–521.
20. Rasolomampianina R, Bailly X, Fetiarison R, Rabevohitra R, Bena G, et al. (2005) Nitrogen-fixing nodules from rose wood legume trees (*Dalbergia* spp.) endemic to Madagascar host seven different genera belonging to *alpha-* and *beta-Proteobacteria*. Molecular Ecology 14: 4135–4146.
21. Levine UY, Teal TK, Robertson GP, Schmidt TM (2011) Agriculture's impact on microbial diversity and associated fluxes of carbon dioxide and methane. ISME J 5: 1683–1691.
22. Zabinski CA, Gannon JE (1997) Effects of recreational impacts on soil microbial communities. Environ Manage 21: 233–238.
23. Lupwayi NZ, Rice WA, Clayton GW (1998) Soil microbial diversity and community structure under wheat as influenced by tillage and crop rotation. Soil Biology & Biochemistry 30: 1733–1741.
24. Ferrenberg S, O'Neill SP, Knelman JE, Todd B, Duggan S, et al. (2013) Changes in assembly processes in soil bacterial communities following a wildfire disturbance. ISME J 7: 1102–1111.
25. Myers N, Mittermeier RA, Mittermeier CG, da Fonseca GA, Kent J (2000) Biodiversity hotspots for conservation priorities. Nature 403: 853–858.
26. Callmander MW, Phillipson PB, Schatz GE, Andriambololonera S, Rabarimanarivo M, et al. (2011) The endemic and non-endemic vascular flora of Madagascar updated. Plant Ecology and Evolution 144: 121–125.
27. Phillipson PB, Schatz GE, Lowry PP II, Labat JN (2006) A catalogue of the vascular plants of Madagascar. In: Ghanzanfar SA, Beentje HJ, editors. Taxonomy and ecology of african plants: their conservation and sustainable use. Addis Ababa, Ethiopia. 613–627.
28. Moat P, Smith P (2007) Atlas of the vegetation of Madagascar. Kew: Royal Botanic Gardens.
29. Cornet A (1974) Essai de cartographie bioclimatique à Madagascar. Notice Explicative 55. Paris: ORSTOM. 1–28.
30. Du Puy DJ, Moat J, editors (1996) A refined classification of the primary vegetation of Madagascar based on underlying geology: using GIS to map its distribution and assess its status. Paris: ORSTOM. 205–218.
31. Aubert G (1978) Methode d'analyse des sols. Marseille.
32. Bremner JM, Mulvaney CS (1982) Regular Kjeldahl Method. In: Page AL, Miller RH, Keeney DR, editors. Methods of Soil Analysis Part 2: Chemical and Microbiological Properties. 2nd ed. Madison: American Society of Agronomy, Inc. 610–614.
33. Mitreva M, and the Human Microbiome Project Consortium (2012) Structure, function and diversity of the healthy human microbiome. Nature 486: 207–214.
34. Schloss PD, Westcott SL, Ryabin T, Hall JR, Hartmann M, et al. (2009) Introducing mothur: open-source, platform-independent, community-supported software for describing and comparing microbial communities. Appl Environ Microbiol 75: 7537–7541.
35. Schloss PD, Gevers D, Westcott SL (2011) Reducing the effects of PCR amplification and sequencing artifacts on 16S rRNA-based studies. PLoS One 6: e27310.
36. Quince C, Lanzen A, Davenport RJ, Turnbaugh PJ (2011) Removing noise from pyrosequenced amplicons. BMC Bioinformatics 12: 38.
37. Quast C, Pruesse E, Yilmaz P, Gerken J, Schweer T, et al. (2013) The SILVA ribosomal RNA gene database project: improved data processing and web-based tools. Nucleic Acids Res 41: D590–596.
38. Huse SM, Welch DM, Morrison HG, Sogin ML (2010) Ironing out the wrinkles in the rare biosphere through improved OTU clustering. Environ Microbiol 12: 1889–1898.
39. Edgar RC, Haas BJ, Clemente JC, Quince C, Knight R (2011) UCHIME improves sensitivity and speed of chimera detection. Bioinformatics 27: 2194–2200.
40. Cole JR, Wang Q, Cardenas E, Fish J, Chai B, et al. (2009) The Ribosomal Database Project: improved alignments and new tools for rRNA analysis. Nucleic Acids Res 37: D141–145.
41. Wang Q, Garrity GM, Tiedje JM, Cole JR (2007) Naive Bayesian classifier for rapid assignment of rRNA sequences into the new bacterial taxonomy. Appl Environ Microbiol 73: 5261–5267.
42. Team RC (2012) R: A language and environment for statistical computing. Vienna, Austria: R Foundation for Statistical Computing.
43. Warton DI, Wright ST, Wang Y (2012) Distance-based multivariate analyses confound location and dispersion effects. Methods in Ecology and Evolution 3: 89–101.
44. Oksanen J, Blanchet FG, Kindt R, Legendre P, Minchin PR, et al. (2013) vegan: Community Ecology Package. R package version 2.0-7 ed.
45. Wickham H (2009) ggplot2: elegant graphics for data analysis: Springer New York.
46. Naether A, Foesel BU, Naegele V, Wust PK, Weinert J, et al. (2012) Environmental factors affect Acidobacterial communities below the subgroup level in grassland and forest soils. Appl Environ Microbiol 78: 7398–7406.
47. Fierer N, Bradford MA, Jackson RB (2007) Toward an ecological classification of soil bacteria. Ecology 88: 1354–1364.
48. Stopnisek N, Bodenhausen N, Frey B, Fierer N, Eberl L, et al. (2013) Genus-wide acid tolerance accounts for the biogeographical distribution of soil *Burkholderia* populations. Environ Microbiol. doi: 10.1111/1462-2920.12211.
49. Suarez-Moreno ZR, Caballero-Mellado J, Coutinho BG, Mendonca-Previato L, James EK, et al. (2012) Common features of environmental and potentially beneficial plant-associated *Burkholderia*. Microb Ecol 63: 249–266.
50. Pilloni G, Granitsiotis MS, Engel M, Lueders T (2012) Testing the limits of 454 pyrotag sequencing: reproducibility, quantitative assessment and comparison to T-RFLP fingerprinting of aquifer microbes. PLoS One 7: e40467.
51. Reddy BV, Kallifidas D, Kim JH, Charlop-Powers Z, Feng Z, et al. (2012) Natural product biosynthetic gene diversity in geographically distinct soil microbiomes. Appl Environ Microbiol 78: 3744–3752.
52. Philippot L, Spor A, Henault C, Bru D, Bizouard F, et al. (2013) Loss in microbial diversity affects nitrogen cycling in soil. ISME J 7: 1609–1619.
53. Lauber CL, Ramirez KS, Aanderud Z, Lennon J, Fierer N (2013) Temporal variability in soil microbial communities across land-use types. ISME J 7: 1641–1650.
54. Ramirez KS, Craine JM, Fierer N (2012) Consistent effects of nitrogen amendments on soil microbial communities and processes across biomes. Global Change Biology 18: 1918–1927.
55. White JR, Nagarajan N, Pop M (2009) Statistical methods for detecting differentially abundant features in clinical metagenomic samples. PLoS Comput Biol 5: e1000352.
56. Fierer N, Jackson RB (2006) The diversity and biogeography of soil bacterial communities. Proc Natl Acad Sci U S A 103: 626–631.

6

Changes in Soil Carbon and Nitrogen following Land Abandonment of Farmland on the Loess Plateau, China

Lei Deng[1], Zhou-Ping Shangguan[1]*, Sandra Sweeney[2]

1 State Key Laboratory of Soil Erosion and Dryland Farming on the Loess Plateau, Northwest A&F University, Yangling, Shaanxi, China, 2 Institute of Environmental Sciences, University of the Bosphorus, Istanbul, Turkey

Abstract

The revegetation of abandoned farmland significantly influences soil organic C (SOC) and total N (TN). However, the dynamics of both soil OC and N storage following the abandonment of farmland are not well understood. To learn more about soil C and N storages dynamics 30 years after the conversion of farmland to grassland, we measured SOC and TN content in paired grassland and farmland sites in the Zhifanggou watershed on the Loess Plateau, China. The grassland sites were established on farmland abandoned for 1, 7, 13, 20, and 30 years. Top soil OC and TN were higher in older grassland, especially in the 0–5 cm soil depths; deeper soil OC and TN was lower in younger grasslands (<20 yr), and higher in older grasslands (30 yr). Soil OC and N storage (0–100 cm) was significantly lower in the younger grasslands (<20 yr), had increased in the older grasslands (30 yr), and at 30 years SOC had increased to pre-abandonment levels. For a thirty year period following abandonment the soil C/N value remained at 10. Our results indicate that soil C and TN were significantly and positively correlated, indicating that studies on the storage of soil OC and TN needs to focus on deeper soil and not be restricted to the uppermost (0–30 cm) soil levels.

Editor: Ben Bond-Lamberty, DOE Pacific Northwest National Laboratory, United States of America

Funding: The study was funded by the Strategic Priority Research Program of the Chinese Academy of Sciences (XDA05060403) and the Scholarship Award for Excellent Doctoral Student granted by Ministry of Education. The funders had no role in study design, data collection and analysis, decision to publish, or preparation of the manuscript.

Competing Interests: The authors have declared that no competing interests exist.

* E-mail: shangguan@ms.iswc.ac.cn

Introduction

Changes in land use have important effects on regional ecological processes and global climate change [1–2]. During the past two decades, many studies have focused on the effects of land-use change on soil organic carbon (SOC) and total nitrogen (TN) in terrestrial ecosystems [3–5]. The large differences in climatic conditions [6], soil properties [7], and type of land use change [8] are three factors whose effects are not yet well understood.

The revegetation of degraded land is one of the principal strategies for the control of both soil erosion and ecosystem recovery in fragile regions where either anthropogenic activities or severe environmental conditions can lead to disturbance [9]. Revegetation also greatly influences soil quality, C and N cycling, land management, as well as regional socioeconomic development [10–11]. The development of managed grassland and forest accelerates ecosystem restoration [12], affecting C and N cycles and C and N pools stored in soils [10,13]. Altered C and N cycles and C and N pools influence the production of biomass and ecosystem function [14]. Furthermore, C–N interactions are very important in determining whether the C sink in land ecosystems can be sustained over the long term [15–16]. Luo et al. [15] have proposed that N dynamics are a key factor in the regulation of long-term terrestrial C sequestration. N may become progressively more limiting as C accumulates in ecosystems under elevated CO_2 if the total N content in an ecosystem does not change [15]. Therefore, studying the dynamics of organic carbon (OC) and N in soils along a restoration succession gradient and analyzing the relationships between C and N storage dynamics following restoration may be of importance in improving our knowledge of the sustainable management of land resources and predictions of future global C and N cycling.

Large-scale monocropping and over-grazing [17] have affected the semi-arid northern Loess Plateau in China. In addition, over the last century, the expanding human population, combined with a changing lifestyle, has accelerated ecosystem fragmentation and degradation [9]. To stabilize the fragile natural ecosystems characteristic of the Loess Plateau and to alleviate the degradation of land, the Chinese government has launched a series of nationwide conservation projects. One such project converts degraded farmland to either grassland or forest [18–19] to control soil erosion, increase storage of SOC and N, and prevent the occurrence of soil desiccation on the Loess Plateau [11]. Restoration succession may affect SOC and N decomposition. With regard to these ecosystem functions, the retention of C and N in soil is crucial [20]. This is of particular concern at sites with substantial N saturation, which are becoming increasingly widespread due to elevated atmospheric N deposition.

Previous studies report that the revegetation of degraded land can increase OC and N storages in soil [9,13]. Wang et al. [21] reported that both SOC and TN increase as a linear function of years abandoned. However, many previous studies have focused primarily on the topsoil (0–30 cm) [21–24]. Little is known about long-term changes to SOC and N in the deeper soil layers of the restoration succession on the Loess Plateau. This information can be useful in estimating the temporal distribution of storage of SOC and N and for evaluating OC and N dynamics throughout the conversion from managed to natural communities in semi-arid regions.

The objectives of this work were to investigate changes in SOC and TN concentration, soil OC and N storage and the relationship between SOC and TN with time since abandonment of farmland with depth in the soil profile.

Materials and Methods

Site description

This study was conducted in the Zhifanggou watershed in Ansai County, Shaanxi Province, NW China (36°46′28″–36°46′42″N, 109°13′03″–109°16′46″E; 1,010–1,431 m a.s.l., 8.27 km^2) (Figure 1). The study area is characterized by a semi-arid climate and a deeply incised hilly-gully Loess landscape. Slopes vary between 0°–65°. The mean annual temperature range is 9.1±0.1°C from 1970 to 2010, and the annual mean temperature has increased over time (Figure 2a), in summer, the highest temperature is 35.3±2.3°C and in winter, the lowest temperature is −20.3±3.7°C from 1970 to 2010; the average frost-free period is 157 days. Mean annual precipitation is 503±15 mm (from 1970 to 2010) (Figure 2b), of which 70% falls between July and September. The loess-derived soils are fertile but extremely susceptible to erosion. The sand, silt and clay contents are 65%, 24% and 11%, respectively [21]. The main grassland species are *Stipa bungeana*, *Bothriochloa ischaemum*, *Artemisia sacrorum*, *Potentilla acaulis*, *Stipa grandis*, *Androsace erecta*, *Heteropappus altaicus*, *Lespedeza bicolor*, *Artemisia capillaries*, and *Artemisia frigid*, of which, *S. bungeana* is the most widely distributed. In addition, shrubland species such as *Rosa xanthina*, *Spiraea pubescens*, and *Hippophae rhamnoides* can be found in gullies. The primary planted trees in the study area are *Robinia pseudoacacia*, *Populus simonii*, *Caragana microphylla*, and *Platycladus orientalis*.

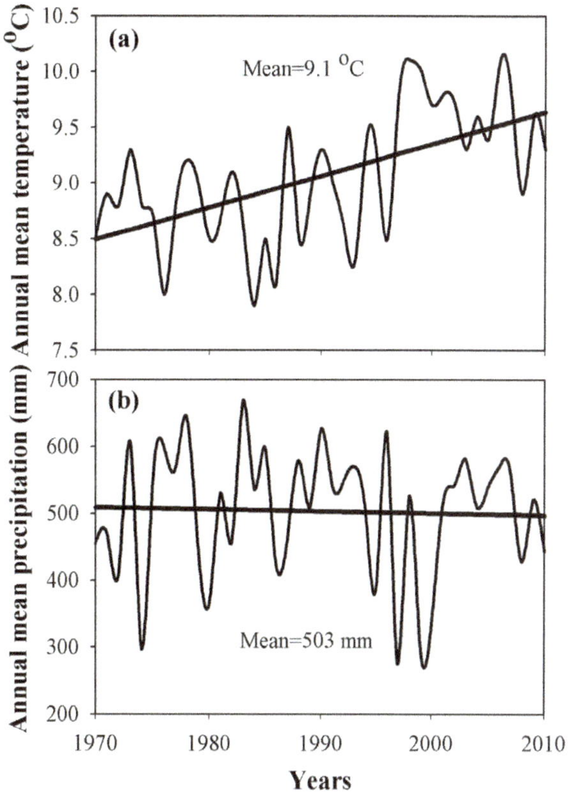

Figure 2. Annual mean temperature (a) and Annual mean precipitation (b) at the study site 1970–2010. Note: the solid lines in the figures a, b are the dynamic fitting curve for annual mean temperature, annual mean precipitation changes with time.

Experimental design and sampling

A common method used to study vegetation restoration is to monitor plants and soils under similar climatic conditions following the sequence of vegetation development [25]. This chronosequence method is widely adopted in applied ecosystem research [26–27] and is considered a "retrospective" research method because it compares existing conditions with original conditions and treatments [28]. The substitution of "space" for "time" is an effective way of studying change over time [28–29]. It makes the critical assumption that each site in the sequence differs only in age and that each site has traced the same history in both its abiotic and biotic components [27]. Before revegetation soil conditions are largely driven by geomorphological processes. Thus, sites stabilized through revegetation for different periods of time offer an ideal opportunity to understand vegetation succession processes in extreme environments.

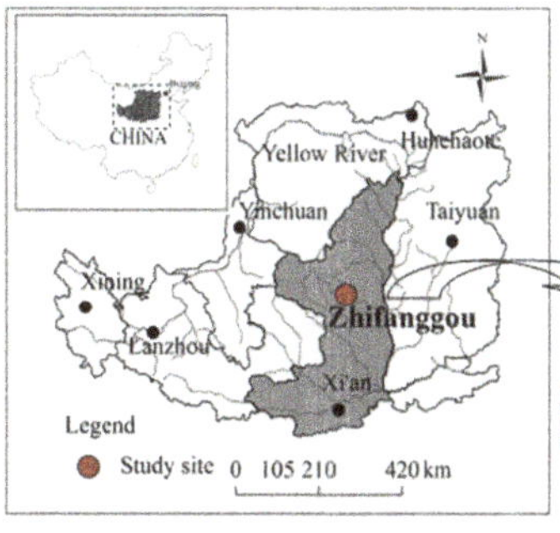

Figure 1. Location and DEM model of the Zhifanggou watershed (adapted from Wang et al., 2011).

An abandoned farmland chronosequence in the watershed was selected for study after the history of the sites was determined through interviews with local farmers (Mr. Haibin Zhang, Soil and Water Conservation Experiment Station, Northwest A&F University, Ansai County, Shaanxi, NW China). Five age classes, 1, 7, 13, 20, and 30 years were selected. In August 2011, when the grassland community biomass peaked, five sites were established in each of the age classes, and the sites were separated by 0.5–1.5 km apart. At each site we set up a plot of 20 m×20 m. In each plot, five quadrats (1 m×1 m) were separately chosen in each of the four corners and center of the plot. In total, we surveyed five plots with twenty-five quadrats in each age class, and twenty-five plots with one hundred and twenty-five quadrats for our study. In each quadrat, the coverage and height, above- and belowground biomass, litter biomass, and soil samples in 0–100 cm soil cores were observed. The morphological traits of the herbage in each age group are listed in Table 1. The plots were all located near the top of loess mounds and there was little difference among the sites in altitude (1185–1341 m), aspect (half south-facing and half north-facing), gradient (18°–37°), or previous farming practices. In the study area, the soils were loess-derived. In addition, five sites on maize farmland (CK) were selected for comparison. Before the farmland was abandoned maize (*Zea mays*) had been widely seeded. The average amount of fertilizer applied was 225–300 kg ha^{-1} of sheep manure as the base fertilizer in April and 300–450 kg ha^{-1} of urea which was applied in June as topdressing. We state clearly

Table 1. Dominant species and their biomass, total cover, height at different number of years abandoned.

Years abandoned (yr)	Above- ground biomass (g m^{-2})	Below- ground biomass (g m^{-2})	Accumulative litter biomass (g m^{-2})	Coverage (%)	Height (cm)	Dominant species
1	169.34±49.67a	78.92±13.57c	28.90±4.26c	8.8±0.58c	66±6.96ab	*Artemisia scoparia*
7	73.73±7.96a	138.73±6.72c	118.51±14.93bc	19.4±2.86bc	41.6±7.88b	*Lespedeza bicolor+ Setaira viridis*
13	100.26±19.01a	218.71±36.29c	130.78±33.00bc	24.4±5.61abc	68.2±6.01a	*Agropyron cristatum+ Heteropappus altaicus*
20	210.42±56.73a	575.21±129.16b	209.79±35.99ab	41±10.77ab	69±5.79a	*Artemisia sacrorum+ Bothriochloa ischaemum*
30	159.08±12.03a	1080.34±111.18a	281.86±25.77a	48±6.04a	52±2.07ab	*Bothriochloa ischaemum+ Stipa bungeana*

Different letters indicate significant differences at $P<0.05$ among years abandoned.
Values are mean ±SE of 5 sites.

that no specific permissions were required for the location. We confirm that the location is not privately-owned or protected in any way. We confirm that the field studies do not involve endangered or protected species.

In each quadrat, all the aboveground parts of the green plants were cut, collected from the ground and put into envelops and tagged, as was all litter. Because the biomass samples were large, they were weighed fresh and then a part of each sample was dried and weighed. The aboveground biomass of the samples was calculated by multiplying the ratio of the dry weight/fresh weight ratio by the fresh weight.

Soil samples were taken at five points in the quadrats of each plot. These were the four corners and center of the biomass sampling sites as described above. Litter horizons were removed before soil sampling. Soil sampling, using a soil drilling sampler (9 cm inner diameter), was done in seven soil layers, 0–5, 5–10, 10–20, 20–30, 30–50, 50–70, and 70–100 cm. We then mixed the same layers together to make one sample. All samples were sieved through a 2 mm screen, and roots and other debris were removed. Each sample was air-dried and stored at room temperature for the determination of soil physical and chemical properties. The soil bulk density (g cm^{-3}) of the different soil layers was measured using a soil bulk sampler with a 5 cm diameter and 5 cm high stainless steel cutting ring (5 replicates) at points adjacent to the soil sampling quadrats. The original volume of each soil core and its dry mass after oven-drying at 105°C were measured.

To measure belowground biomass, soil sampling was done three times in seven soil layers, 0–5, 5–10, 10–20, 20–30, 30–50, 50–70, and 70–100 cm at a depth of 0–100 cm in each quadrat using a 9 cm diameter root auger. The majority of the roots were found in the soil samples thus obtained and then isolated using a 2 mm sieve. The remaining fine roots taken from the soil samples were isolated by spreading the samples in shallow trays, overfilling the trays with water and allowing the outflow from the trays to pass

Table 2. One-way analysis of variance (ANOVA) for soil properties in the seven soil layers among abandoned farmland in different years.

Soil layer (cm)	Df	SOC		TN	
		F	sig. (*P*)	F	sig. (*P*)
0–5	5	10.367	0.0001**	14.557	0.0001**
5–10	5	2.604	0.0668	3.651	0.0217*
10–20	5	0.88	0.4936	0.397	0.8081
20–30	5	5.322	0.0044**	0.132	0.9688
30–50	5	6.892	0.0012**	3.348	0.0298*
50–70	5	5.787	0.0029**	6.963	0.0011**
70–100	5	2.567	0.0697	1.825	0.1636
Soil depth (cm)	**Df**	**C storages**		**N storages**	
		F	sig. (*P*)	F	sig. (*P*)
0–5	5	5.71	0.0031**	5.809	0.0029**
0–10	5	4.053	0.0145*	4.527	0.0091**
0–20	5	0.783	0.5496	0.732	0.5811
0–30	5	1.404	0.2688	0.139	0.966
0–50	5	4.483	0.0095**	1.014	0.4237
0–70	5	7.042	0.0010**	3.089	0.0392*
0–100	5	7.905	0.0005**	3.783	0.0190*

*indicates significant at $P<0.05$ and **indicates significant at $P<0.01$.

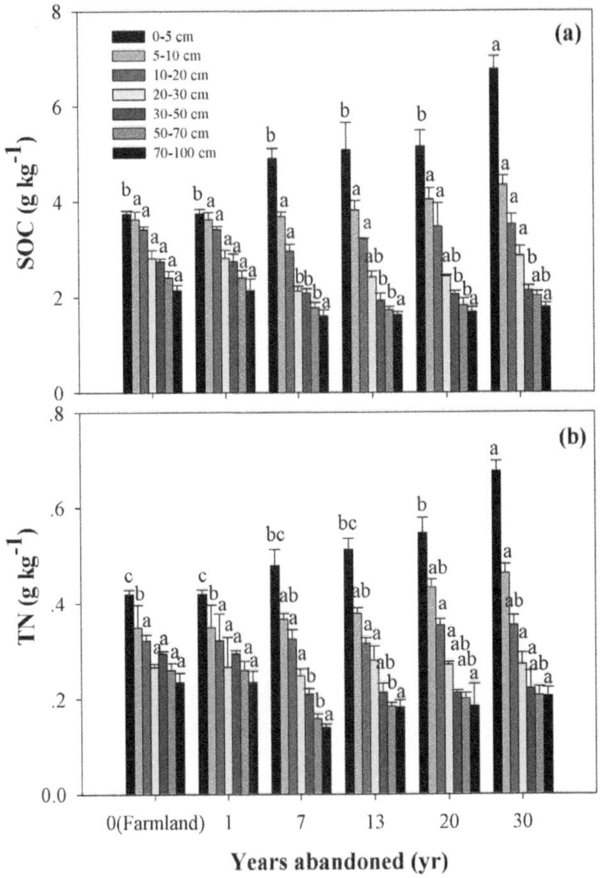

Figure 3. SOC and TN dynamics for seven soil layers of 0–100 cm following the conversion of farmland to grassland. Values are in the form of Mean ± SE and the sample size n = 5. Different lower-case letters above the bars mean significant differences in the same soil layers among abandoned land in different years ($P<0.05$).

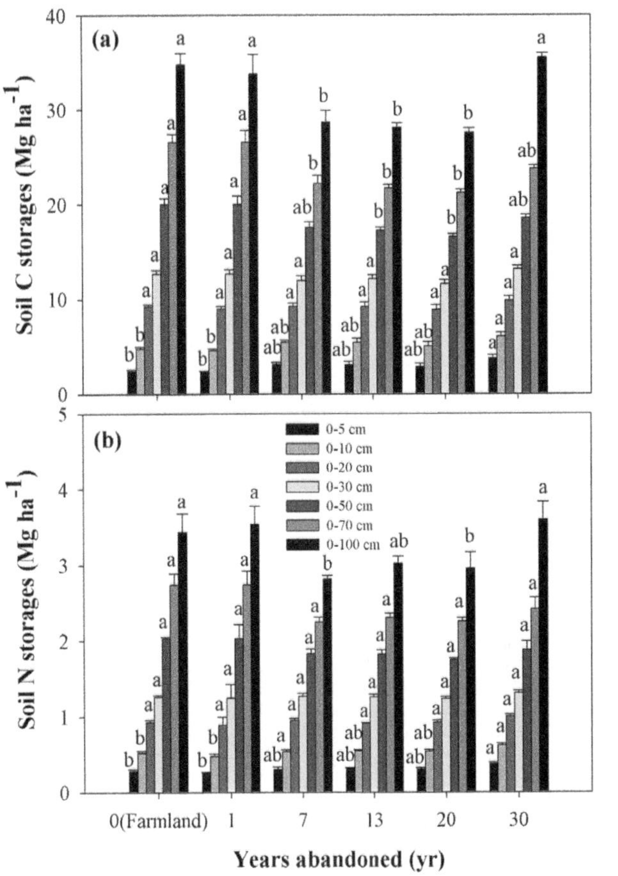

Figure 4. Dynamics of soil OC and N storages in 0–100 cm soil depth following the conversion of farmland to grassland. Values are in the form of Mean ± SE and the sample size n = 5. Different lower-case letters above the bars mean significant differences in the same soil layers among abandoned land in different years ($P<0.05$).

through a 0.5 mm mesh sieve. No attempts were made to distinguish between living and dead roots. All the roots thus isolated were oven-dried at 65°C and weighed to within 0.01 g.

Physical and chemical analysis

Soil bulk density (BD) was calculated depending on the inner diameter of the core sampler, sampling depth and the oven dried weight of the composite soil samples [30]. Soil OC content was assayed by dichromate oxidation [31] and soil TN content was assayed using the Kjeldahl method [32].

Calculation of soil C and N storages

Our sample soils did not have any coarse fraction (>2 mm). Therefore, the study used the following equation to calculate soil organic carbon storage (Cs) [22]:

$$Cs = BD \times SOC \times D/10 \tag{1}$$

in which, Cs is SOC storages (Mg ha^{-1}); BD is soil bulk density (g cm^{-3}); SOC is soil organic carbon concentration (g kg^{-1}); and D is soil thickness (cm).

The following equation was used to calculate soil N storage (Ns) [33]:

$$Ns = BD \times TN \times D/10 \tag{2}$$

in which, Ns is soil N storage (Mg ha^{-1}); BD is soil bulk density (g cm^{-3}); TN is soil TN concentration (g kg^{-1}); and D is soil thickness (cm).

Statistical analysis

One-way ANOVA was used to analyze the means of the same soil layers among the different abandoned years. Differences were evaluated at the 0.05 significance level. When significance was observed at the $P<0.05$ level, Tukey's post hoc test was used to carry out the multiple comparisons. All statistical analyses were performed using the software program SPSS, ver. 17.0 (SPSS Inc., Chicago, IL, USA).

Results

Dynamics of SOC and TN

SOC in 0–5 cm soil was higher in older grassland following abandonment (Table 2, Figure 3a). In the first 20 years, 0–5 cm SOC showed no significant changes; after 20 years, it had significantly increased (Figure 3a). The 5–20 cm SOC had not significantly increased (Table 2). The 20–70 cm SOC was significantly lower in the younger grasslands (<20 yr) and higher

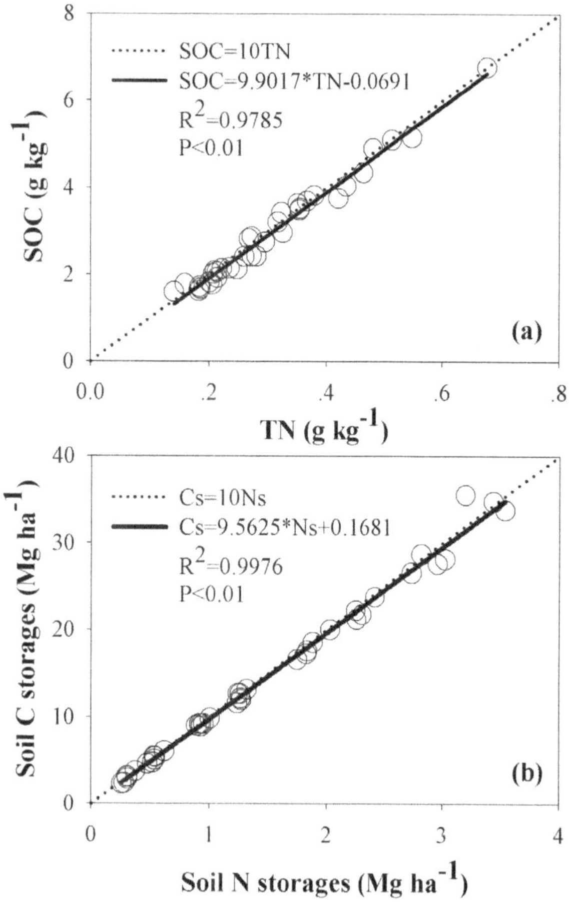

Figure 5. Relationship between SOC and TN, soil OC storages and soil N storages, and accumulative soil C storages and soil N storage.

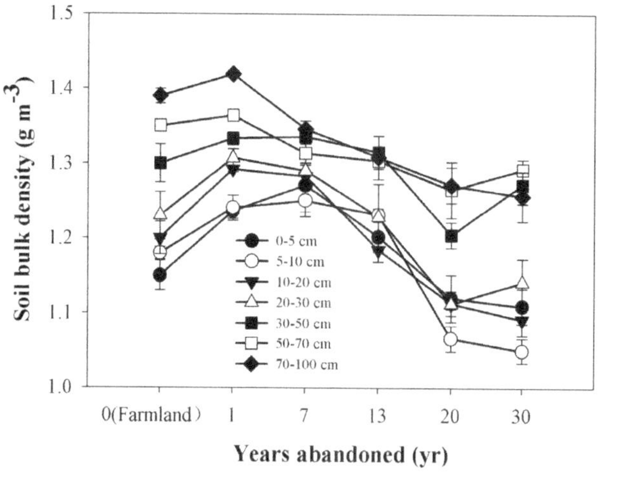

Figure 6. Dynamics of soil bulk density in 0–100 cm soil depth following the conversion of farmland to grassland. Values are in the form of Mean ± SE and the sample size n = 5.

in the older grasslands (30 yr) (Figure 3a). When abandoned for 30 years, 20–70 cm SOC had increased to the level prior to abandonment (maize) (Figure 3a). The 70–100 cm SOC showed no significant changes since having been abandoned, but also showed a tendency to be low in younger grasslands and higher in the older grasslands (Table 2, Figure 3a).

Soil TN in 0–5 cm soil was also higher in older grassland following abandonment (Table 2, Figure 3b). In the first 13 years after having been abandoned, TN in the 0–5 cm soil layer showed no significant changes; after 13 years, it had significantly increased (Figure 3b). The TN in the 5–10 cm soil layer showed no significant changes in the first 13 years, whereas after 13 years, it had significantly increased (Figure 3b). The 10–30 cm soil TN did not increase significantly (Table 2, Figure 3b). The 30–70 cm soil TN first decreased significantly and then increased to the level prior to abandonment (maize) (Figure 3b). Similar to the SOC, 70–100 cm soil TN had showed no significant changes since abandonment, but also showed a tendency to be lower in younger grasslands and higher in older grasslands (Table 2, Figure 3b).

C and N storages dynamics

Soil OC storage in the 0–10 cm soil was also higher in older grassland since abandonment (Table 2, Figure 4a). After having been abandoned for 30 years, the 0–10 cm soil OC storage had increased significantly. 0–30 cm soil OC storage had not increased significantly (Table 2, Figure 4a). 0–100 cm soil OC storage was significantly lower in the younger grasslands (<20 yr) and higher in the older grasslands (30 yr); at 30 years it had increased to pre-abandonment levels (Figure 4a). Soil C storage changed mainly in the top 10 cm of soil following the conversion of farmland to grassland for a period of 30 years.

Similar to the accumulative soil OC storage, soil N storage had changed mainly in the top 10 cm soil layer 30 years after having been abandoned. The conversion of farmland into grassland significantly increased the soil N storage in the 0–10 cm soil (Table 2, Figure 4b). After having been abandoned for 30 years, 0–10 cm soil N storage had significantly increased. 0–50 cm soil N storage had not significantly increased (Table 2, Figure 4b). 0–100 cm soil N storage was also significantly lower in younger grasslands (<20 yr) and higher in older grasslands (30 yr), until after thirty years it had increased to pre-abandonment levels (maize) (Figure 4b).

Relationship between Soil C and N

Soil C and N showed significant positive correlations (Figure 5). The relationship between SOC and TN, soil C storage and soil N storage were significant ($P<0.01$). In the process of the revegetation, soil OC and TN, soil C storage and N storage approximately represents SOC = 10TN (Figure 5a) and Cs = 10Ns (Figure 5b).

Discussion

The soil C and N results supported the hypothesis that soil organic C and N conditions in both the top soil and in the deeper soil layer are significantly affected by land use change on the northern Loess Plateau. In our study, top soil OC and TN were higher in older post-abandonment grassland, especially in the 0–5 cm soil depth (figure 3), indicating the accumulation of soil OC and TN by revegetation [23–24]. These results agree with those of Wang et al. [21], who studied changes to the physico-chemical properties of top soil during natural succession on abandoned farmland in the Zhifanggou watershed. The evident increase may be partly attributed to a lower fraction of non-soluble materials in more readily decomposed plant residues. In the farmland, cultivation breaks up soil aggregates, decreases total soil porosity, and accelerates composition and mineralization of soil organic

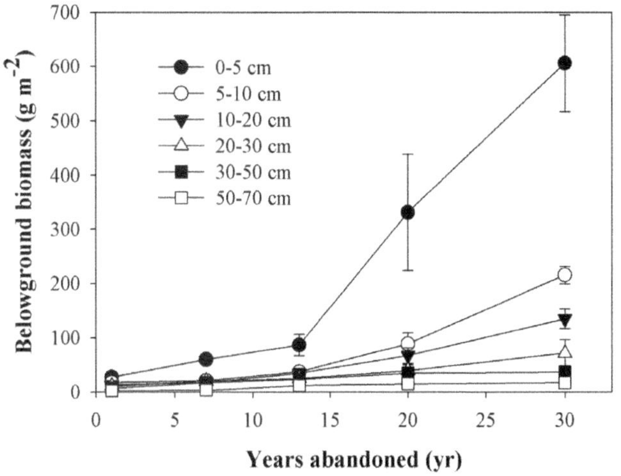

Figure 7. Dynamics of belowground biomass in 0–100 cm soil depth following the conversion of farmland to grassland. Values are in the form of Mean ± SE and the sample size n = 5. It didn't have roots in the 70–100 cm soil layers.

matter (SOM) due to exposure of previously accessible SOM to microbial attack [34]. This results in a reduction in the amounts of intra-aggregate light fraction organic carbon (LFOC) and some organomineral SOC [35]. In addition, the reduction of crop residue return to soil may also be a factor as farmers take away straw with grain harvesting each year. This speculation can be supported by the results of Wu et al. [36] in this region where plant residue in the top soil layer was reduced considerably after native grasslands were cultivated, contributing to the decrease in LFOC and SOC. Conversely, the conversion of farmland into grassland increases SOC and its fractions [37], and increases total soil porosity, thus resulting in a reduction in soil BD (Figure 6, [21]).

Aboveground vegetation plays an important role in regulating the biogeochemistry of ecosystems by fixing C and nutrients and preventing the loss of nutrients under disturbance, such as plant decline, acid-rain and climate change [38]. It is also clear from these results that revegetated grassland has a great impact on the storage of soil OC and N. In our study, both soil OC and N storage in the 0–10 cm soil depth was higher in older grassland that had been abandoned for thirty years (Figure 4). This indicates that during the 30 year period soil OC and N storage mainly had changed in the top 10 cm of soil since abandoned. There may be a range of potential mechanisms through which soil OC and N in the top soil increased with revegetation. A prime candidate is the return of both C and N from increased aboveground biomass and litter. As soil OC and N input are mainly derived from the decomposition of litter [21], primary productivity is the main driver of soil carbon sequestration [39], which resulted primarily in soil OC and N storage increasing in the top soil. Secondly, belowground biomass (dead roots, mycorrhizae, and exudates) is an important element of soil carbon sequestration [40]. Belowground biomass increased in the time after the farmland had been abandoned (Table 1, Figure 7). Thirdly, changes in vegetation composition and the dominant plant functional group could affect the sequestration of C and N in the soil [39,41]. Plant functional composition strongly influences the chemical and physical composition of litter inputs, and thereby their decomposability, carbon loss through soil respiration and leaching, and carbon immobilization in hummified plant residues [39], and the increase of C3 plants can increase soil C and N accumulation in meadow soils [41]. Revegetation had a direct effect on the dominant vegetation species, vegetation cover, height, above- and belowground biomass (Table 1).

SOC and TN at deeper soil levels (>30 cm) were lower in younger grasslands and higher in older grasslands (Figure 3), this is probably due to long-term natural organic fertilizer and inorganic fertilizer inputted into the soil resulting in higher SOC and N in the farmland stage. So, this initial loss of soil OC and N following the conversion of farmland has been commonly attributed to the net effect of decreased organic matter inputs and losses through decomposition [42]. During the course of grasslands development, we observed an increase in the soil SOC and N storage, after 30 years of revegetation, where deeper soil C and N had returned to pre-abandonment levels (Figure 3); the return of belowground biomass to the deeper soil layer is another reason (Figure 7). Previous studies of soil C dynamics have emphasized the role of physical protection from different particle fractions (sand, silt and clay) [43], microbes and enzymes within aggregates [44], microaggregates and macroaggregates [45], and bacterial and fungal [46], therefore, the mechanism in the soil C and N dynamics in the deeper soil during the process of conversion from farmland into grassland probably relate to those factors. However, our lack of a full understanding of this process calls for more attention to be paid to soil C and N dynamics in the deeper soil layers. Wang et al. [21] found that SOC and TN had a negative relationship with soil bulk density, and Singh et al. [23–24] reported that SOC and TN were significantly greater while BD was significantly lower during the restoration of degraded lands, our results agreed with them (Figures 3 and 6). So, we can infer the trend of SOC and TN according to the trend of soil bulk density.

We observed that soil C and N were significantly and positively correlated (Figure 5). In the restoration process, soil OC and TN, soil C storage and N storage approximately represents SOC = 10TN (Figure 5a) and Cs = 10Ns (Figure 5b). So, we can conclude that soil C/N value was 10 in the process of 30 years of conversion from farmland to grassland in the Loess Plateau, a value greater that reported by Liu et al. [35] (C/N = 7.62). Ammonium-N and the sum of NO_3^- -N and NO_2^- -N are the readily available forms of soil N for root uptake. Liu et al. [35] reported that ammonium-N and the sum of NO_3^- -N and NO_2^- -N were lower in grassland than in cropland with the addition of chemical fertilizer N, because the conversion of farmland into grassland increased the C/N ratio and reduced soil mineral N by enhancing soil N immobilization.

Grasslands play an important role in the global C and N cycles [3–4,10]. Grassland in good condition should be in balance in terms of C and N input and output or in the state where C and N input is greater than their output [41]. At least one study has shown that carbon input is greater than carbon output in enclosed grassland [47]. In our study, 0–100 cm soil OC and N storage was significantly lower in the younger grasslands (<20 yr) and higher in the older grasslands (30 yr), at 30 years they had increased to pre-abandonment levels, and that the values in farmland were higher compared to grasslands abandoned for 20 years. Our results indicate that the study of the storage of soil OC and N need to include the deeper soil layer and not focus solely on the top soil. Li et al. [48] also found that SOC density in the deeper soil layer was significantly higher than that of the farmland in the Zhifanggou watershed, Loess Plateau. Li et al. [49] found that soil C and N storage in the deeper layers (mineral layer) show significant difference in top soil layers (organic layer) with time after afforestation at the global scale. Therefore, estimating soil OC and N input or output requires that research consider not only soil depth but also time since abandonment.

Conclusions

The results of this study indicate that plant succession after land has been abandoned resulted in a significant improvement in the physico-chemical properties of soil. Thirty years following abandonment, soil OC and N storage had increased primarily in the top 10 cm of the soil depth. After 30 years of restoration, deeper soil C and N storage had increased to pre-abandonment levels. This finding indicates that deeper soil has a higher potential to fix both C and N in the future (>30 yr). Thus, in the semi-arid environment of the Loess Plateau, vegetation recovery following abandonment is slow and the improvement of soil properties is likely to require a considerably long period of time (>30 yr). Therefore, the findings are important for assessing the resilience of these degraded ecosystems and developing a more effective strategy of vegetation restoration for the management of degraded grassland from a long-term perspective. More research, for example, on soil physico-chemical properties, soil enzyme activities, soil microbial, animal and plant function and composition, is required to better understand the mechanism behind how the soil fixes C and N in the deeper soil profile of sub-arid regions, for example, the Loess Plateau, China.

Author Contributions

Conceived and designed the experiments: LD ZS. Analyzed the data: LD. Contributed reagents/materials/analysis tools: LD. Wrote the paper: LD ZS SS.

References

1. Kalnay E, Cai M (2003) Impact of urbanization and land-use change on climate. Nature 423: 528–531.
2. Ficetola GF, Maiorano L, Falcucci A, Dendoncker N, Boitani L, et al. (2010) Knowing the past to predict the future: land-use change and the distribution of invasive bullfrogs. Global Change Biol 16: 528–537.
3. Russell AE, Laird DA, Parkin TB, Mallarino AP (2005) Impact of nitrogen fertilization and cropping system on carbon sequestration in Midwestern Mollisols. Soil Sci Soc Am J 69: 413–422.
4. Brown J, Angerer J, Salley SW, Blaisdell R, Stuth JW (2010) Improving estimates of rangeland carbon sequestration potential in the US Southwest. Rangeland Ecol Manag 63: 147–154.
5. Qiu SJ, Ju XT, Ingwersen J, Qin ZC, Li L, et al. (2010) Changes in soil carbon and nitrogen pools after shifting from conventional cereal to greenhouse vegetable production. Soil Till Res 107: 80–87.
6. Jobbágy EG, Jackson RB (2000) The Vertical Distribution of Soil Organic Carbon and Its Relation to Climate and Vegetation. Ecol Appl 10: 423–436.
7. Piao SL, Fang JY, Ciais P, Peylin P, Huang Y, et al. (2009) The Carbon balance of terrestrial ecosystems in China. Nature 458: 1009–1013.
8. Arora VK, Boer GJ (2010) Uncertainties in the 20th century carbon budget associated with land use change. Global Change Biol 16: 3327–3348.
9. Jia XX, Wei XR, Shao MA, Li XZ (2012) Distribution of soil carbon and nitrogen along a revegetational succession on the Loess Plateau of China. Catena 95: 160–168.
10. Eaton JM, McGoff NM, Byme KA, Leahy P, Kiely G (2008) Land cover change and soil organic carbon stocks in the Republic of Ireland 1851–2000. Climate Change 91: 317–334.
11. Fu XL, Shao MA, Wei XR, Horton R (2010) Soil organic carbon and total nitrogen as affected by vegetation types in Northern Loess Plateau of China. Geoderma 155: 31–35.
12. She DL, Shao MA, Timm LC, Reichardt K (2009) Temporal changes of an alfalfa succession and related soil physical properties on the Loess Plateau, China. Pesquisa Agropecuária Brasileira 44: 189–196.
13. Wei XR, Shao MA, Fu XL, Horton R (2010) Changes in soil organic carbon and total nitrogen after 28 years of grassland afforestation: effects of tree species, slope position, and soil order. Plant Soil 331: 165–179.
14. Foster D, Swanson F, Aber J, Burke I, Brokaw N (2003) The importance of land-use legacies to ecology and conservation. BioScience 53: 77–88.
15. Luo YQ, Field CB, Jackson RB (2006) Does nitrogen constrain carbon cycling, or does carbon input stimulate nitrogen cycling? Ecology 87: 3–4.
16. Luo YQ, Su B, Currie WS, Dukes JS, Finzi A, et al. (2004) Progressive nitrogen limitation of ecosystem responses to rising atmospheric carbon dioxide. BioScience 54: 731–739.
17. Fu BJ, Chen LD, Ma KM, Zhou HF, Wang J (2000) The relationships between land use and soil conditions in the hilly area of the Loess Plateau in northern Shaanxi, China. Catena 39: 69–78.
18. Deng L, Shangguan ZP (2011) Food security and farmers' income: impacts of the Grain for Green Programme on rural households in China. J Food Agric Environ 9: 826–831.
19. Deng L, Shangguan ZP, Li R (2012) Effects of the grain-for-green program on soil erosion in China. Int J Sediment Res 27: 120–127.
20. Prietzel J, Bachmaan S (2012) Changes in soil organic C and N stocks after forest transformation from Norway spruce and Scots pine into Douglas fir, Douglas fir/spruce, or European beech stands at different sites in Southern Germany. Forest Ecol Manag 269: 134–148.
21. Wang B, Liu GB, Xue S, Zhu BB (2011) Changes in soil physico-chemical and microbiological properties during natural succession on abandoned farmland in the Loess Plateau. Environ Earth Sci 62: 915–925.
22. Guo LB, Gifford RM (2002) Soil carbon storage and land use change: a meta analysis. Global Change Biol 8: 345–360.
23. Singh K, Pandey VC, Singh B, Singh RR (2012) Ecological restoration of degraded sodic lands through afforestation and cropping. Ecol Eng 43: 70–80.
24. Singh K, Singh B, Singh RR (2012) Changes in physico-chemical, microbial and enzymatic activities during restoration of degraded sodic land: Ecological suitability of mixed forest over monoculture plantation. Catena 96: 57–67.
25. Bhojvaid PP, Timmer VR (1998) Soil dynamics in an age sequence of Prosopis Juliflora planted for sodic soil restoration in India. Forest Ecol Manag 106: 81–193.
26. Fang W, Peng SL (1997) Development of species diversity in the restoration process of establishing a tropical man-made forest ecosystem in China. Forest Ecol Manag 99: 185–196.
27. Johnson EA, Miyanishi K (2008) Testing the assumptions of chronosequences in succession. Ecol Let 11: 419–431.
28. Li XR, Kong DS, Tan HJ, Wang XP (2007) Changes in soil and vegetation following stabilization of dunes in the southeastern fringe of the Tengger Desert, China. Plant Soil 300: 221–231.
29. Sparling GP, Schipper LA, Bettjeman W, Hill R (2003) Soil quality monitoring in New Zealand: practical lessons from a 6-year trial. Agr Ecosyst Environ 104: 523–534.
30. Jia GM, Cao J, Wang CY, Wang G (2005) Microbial biomass and nutrients in soil at the different stages of secondary forest succession in Ziwulin, northwest China. Forest Ecol Manag 217: 117–125.
31. Kalembasa SJ, Jenkinson DS (1973) A comparative study of titrimetric and gravimetric methods for the determination of organic carbon in soil. J Sci Food Agr 24: 1085–1090.
32. Bremner JM (1996) Nitrogen-total. In: Sparks, D.L. (Ed.), Methods of Soil Analysis, Part 3. America Society of Agronomy, Madison, pp. 1085–1121 (SSSA Book Series: 5).
33. Rytter RM (2012) Stone and gravel contents of arable soils influence estimates of C and N stocks. Catena 95: 153–159.
34. Shepherd TG, Saggar S, Newman RH, Ross CW, Dando JL (2001) Tillage-induced changes to soil structure and organic carbon fractions in New Zealand soils. Aust J Soil Res 39: 465–489.
35. Liu X, Li FM, Liu DQ, Sun GJ (2010) Soil Organic Carbon, Carbon Fractions and Nutrients as affected by Land Use in Semi-Arid Region of Loess Plateau of China. Pedosphere 20: 146–152.
36. Wu T, Schoenau JJ, Li F, Qian P, Malhi SS, et al. (2004) Influence of cultivation and fertilization on total organic carbon and carbon fractions in soils from the Loess Plateau of China. Soil Till Res 77: 59–68.
37. Zeng ZX, Liu XL, Jia Y, Li FM (2007) The effect of conversion of cropland to forage legumes on soil quality in a semiarid agroecosystem. J Sustain Agr 32: 335–353.
38. Bormann BT, Sidle RC (1990) Changes in productivity and distribution of nutrients in a chronosequence at Glacier Bay National Park, Alaska. Journal of Ecology 78: 561–578.
39. De Deyn GB, Cornelissen JHC, Bardgett RD (2008) Plant functional traits and soil carbon sequestration in contrasting biomes. Ecol Lett 11: 516–531.
40. Langley JA, Hungate BA (2003) Mycorrhizal controls on belowground litter quality. Ecology 84: 2302–2312.
41. Wu GL, Liu ZH, Zhang L, Chen JM, Hu TM (2010) Long-term fencing improved soil properties and soil organic carbon storage in an alpine swamp meadow of western China. Plant Soil 332: 331–337.
42. Zhao WZ, Xiao HL, Liu ZM, Li J (2005) Soil degradation and restoration as affected by land use change in the semiarid Bashang area, Northern China. Catena 59: 173–186.
43. He NP, Wu L, Wang YS, Han XG (2009) Changes in carbon and nitrogen in soil particle-size fractions along a grassland restoration chronosequence in northern China. Geoderma 150: 302–308.
44. Udawatta RP, Kremer RJ, Adamson BW, Anderson SH (2008) Variation in soil aggregate stability and enzyme activities in a temperate agroforestry practice. Appl Soil Ecol 39: 153–160.
45. Chen FS, Zeng DH, Fahey TJ, Liao PF (2010) Organic carbon in soil physical fractions under different-aged plantations of Mongolian pine in semi-arid region of Northeast China. Appl Soil Ecol 44: 42–48.

46. Six J, Frey SD, Thiet RK, Batten KM (2006) Bacterial and fungal contributions to carbon sequestration in agroecosystems. Soil Sci Soc Am J 70: 555–569.
47. Li YQ, Zhao HL, Zhao XY, Zhang TH, Chen YP (2006) Soil respiration, carbon balance and carbon storage of sandy grassland under post-grazing natural restoration. Acta Prataculturae Sinica 15: 25–31. (in Chinese with English abstract).
48. Li MM, Zhang XC, Pang GW, Han FP (2013) The estimation of soil organic carbon distribution and storage in a small catchment area of the Loess Plateau. Catena 101: 11–16.
49. Li DJ, Niu SL, Luo YQ (2012) Global patterns of the dynamics of soil carbon and nitrogen stocks following afforestation: a meta-analysis. New Phytol 195: 172–181.

7

Quantifying Spatial Variability of Selected Soil Trace Elements and Their Scaling Relationships Using Multifractal Techniques

Fasheng Zhang[1,2]*, Guanghua Yin[1], Zhenying Wang[1], Neil McLaughlin[2], Xiaoyuan Geng[2], Zuoxin Liu[1]*

1 Institute of Applied Ecology, Chinese Academy of Sciences, Shenyang, China, **2** Eastern Cereal and Oilseed Research Centre, Agriculture and Agri-Food Canada, Ottawa, Canada

Abstract

Multifractal techniques were utilized to quantify the spatial variability of selected soil trace elements and their scaling relationships in a 10.24-ha agricultural field in northeast China. 1024 soil samples were collected from the field and available Fe, Mn, Cu and Zn were measured in each sample. Descriptive results showed that Mn deficiencies were widespread throughout the field while Fe and Zn deficiencies tended to occur in patches. By estimating single multifractal spectra, we found that available Fe, Cu and Zn in the study soils exhibited high spatial variability and the existence of anomalies ($[\alpha(q)_{max} - \alpha(q)_{min}] \geq 0.54$), whereas available Mn had a relatively uniform distribution ($[\alpha(q)_{max} - \alpha(q)_{min}] \approx 0.10$). The joint multifractal spectra revealed that the strong positive relationships ($r \geq 0.86$, $P < 0.001$) among available Fe, Cu and Zn were all valid across a wider range of scales and over the full range of data values, whereas available Mn was weakly related to available Fe and Zn ($r \geq 0.18$, $P < 0.01$) but not related to available Cu ($r = -0.03$, $P = 0.40$). These results show that the variability and singularities of selected soil trace elements as well as their scaling relationships can be characterized by single and joint multifractal parameters. The findings presented in this study could be extended to predict selected soil trace elements at larger regional scales with the aid of geographic information systems.

Editor: Helmut Ahammer, Medical University of Graz, Austria

Funding: This work was funded by the National Natural Science Foundation of China (41201212), the National Food Production Project of China (2011BAD16B12) and the National Major Technology Program of China (2012BAD09B02). The funders had no role in study design, data collection and analysis, decision to publish, or preparation of the manuscript.

Competing Interests: The authors have declared that no competing interests exist.

* E-mail: fasheng.zhang@yahoo.com (FSZ); liuzuoxin@iae.ac.cn (ZXL)

Introduction

Although trace elements are needed by plants in only very small quantities [1,2], their deficiencies in soils may result in severe plant malnutrition as well as animal and human health problems [3,4]. Quantifying the spatial variability of soil trace elements enables prediction of trace elements at intermediate locations where measurements have not been made [5]. The thematic map produced by the prediction is critical to understanding, assessing and managing trace element-deficient soils [6].

Plant availability of trace elements in soils have been shown to have a high degree of spatial variability [7,8]. This arises mainly from the existence of anomalies in their distributions, which is a consequence of local extreme soil and weather conditions [3] such as alkalinity, soil infertility, erosion, drought, etc. From a practical standpoint, anomalies may carry valuable information for identifying specific trace element deficiencies. Including anomalies is essential when developing models that characterize the spatial variability of soil trace elements.

Under an assumption of intrinsic stationarity, the semivariogram has been the most common method for quantifying the spatial variability of soil trace elements [7–11]. In modern geostatistics, anomalies in a trace element dataset are usually transformed or eliminated in advance to avoid erratic behavior of a semivariogram [12]. As a second-order moment statistic, semivariogram cannot differentiate subtle variations in a trace element dataset. These facts reveal that current semivariogram models may not be able to fully characterize the spatial variability of soil trace elements, especially the singularities of anomalies.

Over a given area, plant availability of soil trace elements is commonly associated with similar soil and environmental conditions. An understanding of the relationships among soil trace elements can help in the analysis of the underlying processes controlling their spatial variability [7,8]. Thus, there is a need to examine the spatial variability of soil trace elements in a way that takes account of the relationships among them.

Factors and processes affecting soil properties including plant availability of trace elements operate at different spatial scales and influence these properties in different ways [9]. Relationships among soil trace elements may be heterogeneous across nested spatial scales. Although correlation analysis can quantify the relationship among soil trace elements at the measured scale, this relationship may not be valid at other spatial scales. As such, relationships among soil trace elements across a range of spatial scales, i.e., scaling relationships, are needed.

Over the past three decades, single multifractal spectrum has been applied in soil science to quantify the spatial variability of soil properties [13–18] including pH, organic matter content, cation exchange capacity, macro- and moderate-nutrient content, texture, structure, hydraulic properties, etc. The relationships among these properties and their associations with other environmental variables across spatial scales have also been investigated by researchers using a joint multifractal spectrum [18−20]. To date, however, studies examining spatial variability and singularity of soil trace elements and their scaling relationships using multifractal techniques are still rare [21]. Usually, the spatial distributions of pH, organic matter content, macro-nutrient content, texture, and other basic soil properties are relatively uniform. Geostatistical methods are able to characterize their spatial variability well and predict them with high accuracy. However, the distribution of soil trace elements is quite different from that of other soil properties, and is often characterized by high spatial variability and anomalies that are accompanied by singularities. As previously mentioned, semivariogram analysis and correlation analysis may not be able to completely characterize the singularity of soil trace elements and their scaling relationships. Panahi and Cheng pointed out that an exploratory dataset with anomalies may be more likely to follow a multifractal rather than a normal or lognormal distribution [22]. Meanwhile, the multifractal techniques have long been known to be able to characterize singularities and scaling relationships. Thus, the application of multifractal techniques in the analysis of soil spatial variability and their scaling relationships may be more appropriate for spatial analysis of soil trace elements rather than for other soil properties.

The objective of this study was therefore to examine whether the spatial distribution of selected soil trace elements in an agricultural field in northeast China exhibits a multifractal property, and if so, to quantify their spatial variability and scaling relationships using multifractal techniques. We also discuss the practical significance of the findings for prediction of selected soil trace elements.

Theory of Multifractal Analysis

Single Multifractal Spectrum

For a spatial dataset, there are several methods for estimating a single multifractal spectrum (e.g., the moment method, Rényi method). Here we used the moment method discussed by Agterberg [23], which is much easier to apply yet still provides accurate results.

Using this method, a normalized variable, $P_i(\varepsilon)$, is defined first according to:

$$P_i(\varepsilon) = \frac{\varphi_i(\varepsilon)}{\sum_{i=1}^{n} \varphi_i(\varepsilon)}$$

where $\varphi_i(\varepsilon)$ is the measured value of variable φ in the ith segment of scale ε in a plot, and n is the number of segments of scale ε over the whole plot. A partition function, $\chi_q(\varepsilon)$, which varies with the moment order q and scale ε, is then calculated as:

$$\chi_q(\varepsilon) = \sum_{i=1}^{n} [P_i(\varepsilon)]^q$$

For a spatial dataset, the power-law relationship between the partition function and scale is used to determine whether or not it exhibits a multifractal property. This power-law relationship can be confirmed if the logarithm of the partition function varies linearly against a logarithm of scale with a slope of the mass exponent, $\tau(q)$:

$$\log \chi_q(\varepsilon) \propto \tau(q) \cdot \log(\varepsilon)$$

By plotting $\log\chi_q(\varepsilon)$ against $\log\varepsilon$ at many moment orders q, we compute the generalized fractal dimension, $D(q)$:

$$D(q) = \frac{\tau(q)}{q-1} = \frac{1}{q-1} \times \lim_{\varepsilon \to 0} \frac{\log \chi_q(\varepsilon)}{\log \varepsilon}, \qquad q \neq 1$$

$$D(q) \approx \lim_{\varepsilon \to 0} \frac{\sum_{i=1}^{n} [P_i(\varepsilon) \cdot \log P_i(\varepsilon)]}{\log \varepsilon}, \qquad q = 1$$

Then the Hölder exponent (i.e., singularity index), $\alpha(q)$, with respect to variable φ, is computed by Legendre transformation:

$$\alpha(q) = \frac{d\tau(q)}{dq}$$

The fractal dimension, $f(q)$, with respect to variable φ, is given as:

$$f(q) = q\alpha(q) - \tau(q)$$

Plotting $f(q)$ against $\alpha(q)$ produces the so-called single multifractal spectrum. $\alpha(q)$ can be interpreted as an index of crowding of variable φ at the moment order q when ε approaches zero. $f(q)$ is the fractal dimension of the subsets of variable φ having specific $\alpha(q)$. Thus, the abundance of $\alpha(q)$ reflects the extent of spatial variability in variable φ. More $\alpha(q)$ (i.e., wider spectrum) represent greater spatial variability in distribution of variable φ and vice versa. In addition, with the amplification of high positive or negative q values, the spatial variability controlled by the higher or lower values of variable φ is separated from each other and represented by the $\alpha(q)$ in the left and right parts of the single multifractal spectrum, respectively. In particular, the $\alpha(q)_{\max}$ and $\alpha(q)_{\min}$ correspond to the singularity strength of the lowest and highest values of variable φ, respectively. The single multifractal spectrum therefore can detect a spatial pattern and separate anomalous variation from the background variation [13,23].

Chhabra and Jensen [24] proposed a method to avoid the difficult task of estimating $\alpha(q)$ by Legendre transformation. In the Chhabra and Jensen method, we define another normalized variable, $\delta_i(q,\varepsilon)$:

$$\delta_i(q,\varepsilon) = \frac{[P_i(\varepsilon)]^q}{\sum_{i=1}^{n} [P_i(\varepsilon)]^q}$$

Then the singularity index and fractal dimension are computed respectively as:

$$\alpha(q) = \lim_{\varepsilon \to 0} \frac{\sum_{i=1}^{n} [\delta_i(q,\varepsilon) \cdot \log P_i(\varepsilon)]}{\log \varepsilon}$$

$$f(q) = \lim_{\varepsilon \to 0} \frac{\sum_{i=1}^{n} [\delta_i(q,\varepsilon) \cdot \log \delta_i(q,\varepsilon)]}{\log \varepsilon}$$

Joint Multifractal Spectrum

For coexisting variables, we need to extend the single multifractal spectrum of each variable to a joint multifractal spectrum for two variables [25]. Two normalized variables, $P_i(\varepsilon)$ and $Q_i(\varepsilon)$, are first defined respectively for variables φ and μ. Then the joint normalized variable for variables φ and μ, $\delta_i(q,t,\varepsilon)$, is calculated as:

$$\delta_i(q,t,\varepsilon) = \frac{[P_i(\varepsilon)]^q [Q_i(\varepsilon)]^t}{\sum_{i=1}^{n} [P_i(\varepsilon)]^q [Q_i(\varepsilon)]^t}$$

where q and t are both moment orders and ε is the scale. The singularity indices, $\alpha(q,t)$ and $\beta(q,t)$, with respect to variables φ and μ, are computed respectively as:

$$\alpha(q,t) = \lim_{\varepsilon \to 0} \frac{\sum_{i=1}^{n} [\delta_i(q,t,\varepsilon) \cdot \log P_i(\varepsilon)]}{\log \varepsilon}$$

$$\beta(q,t) = \lim_{\varepsilon \to 0} \frac{\sum_{i=1}^{n} [\delta_i(q,t,\varepsilon) \cdot \log Q_i(\varepsilon)]}{\log \varepsilon}$$

Then the joint fractal dimension, $f(q,t)$, with respect to variables φ and μ, is given as:

$$f(q,t) = \lim_{\varepsilon \to 0} \frac{\sum_{i=1}^{n} [\delta_i(q,t,\varepsilon) \cdot \log \delta_i(q,t,\varepsilon)]}{\log \varepsilon}$$

Plotting $f(q,t)$ against $\alpha(q,t)$ and $\beta(q,t)$ produces the so-called joint multifractal spectrum, contours of which describe the scaling relationships between variables φ and μ. For positive scaling relationships, the joint distribution of higher $\alpha(q,t)$ and $\beta(q,t)$ values at the upper right part of the joint multifractal spectrum indicates the positive correlation between lower values of variables φ and μ, while the joint distribution of lower $\alpha(q,t)$ and $\beta(q,t)$ values at the lower left part indicates the positive correlation between higher values of variables φ and μ. For negative scaling relationships, higher $\alpha(q,t)$ values correspond to lower $\beta(q,t)$ values and vice versa at the diagonal parts of the joint multifractal spectrum, which indicates that lower and higher values of variables φ and μ are distributed together. If variables φ and μ are not correlated, there will be no correlation between $\alpha(q,t)$ and $\beta(q,t)$ [25,26].

Materials and Methods

Site and Soil

The study site is located in Fuxin County in northeast China (42°08′14″ N, 121°44′21″ E). The climate is typical of a warm temperate zone with continental monsoons, hot summers and cold winters. The mean annual temperature is 7.2 degrees Celsius with an average of 2868 hrs of annual sunshine. This region is a typical dryland farming area with an average annual precipitation of 480 mm, over 60% of which occurs between June and August (SWCP I soil moisture sensor & WITU-WS weather station, WITU Technologies, Inc., China).

The main agricultural soil in the region is cinnamon soil which develops through a combination of calcium carbonate leaching, illuviation and humification. It is characterized by a thin humus layer and medium or thick solum. Soil pH (H_2O) is 7.5−8.5 and soil organic matter content is 10.2 g·kg^{-1}. The cation exchange capacity is 19.2 meq/100 g, and mean concentrations of available Fe, Mn, Cu, and Zn are 9.9 mg·kg^{-1}, 36 mg·kg^{-1}, 1.32 mg·kg^{-1}, and 1 mg·kg^{-1}, respectively (The Second National Soil Survey of China, 1985).

Data Collection and Analysis

A total of 1024 soil samples were collected from a flat agricultural field of 10.24-ha (320 m by 320 m) in Matiyingzi Village, 10 km from the center of Fuxin County, in April 2008 (before annual planting). The soils were sampled over the 20-cm depth within a 10-m grid. The samples were taken with a stainless steel sampler and brought to the laboratory. After removing visible plant debris and air drying, soil samples were ground to pass through a 2-mm plastic sieve. Available Fe, Mn, Cu and Zn in each sample were measured by a DTPA (diethylene triamine pentaacetic acid) soil test [27]. The extractant consisted of 0.005 M DTPA, 0.1 M TEA (triethanolamine), and 0.01 M $CaCl_2$, with a pH of 7.3. Hooda and Alloway reported that DTPA method could provide accurate information about plant availability of trace elements in soils [28]. All trace element concentrations were determined by inductively coupled plasma mass spectrometry (Perkin Elmer Optima 3000). All necessary permits were obtained for the described field studies. The land user and owner, Mingfu Zhang, approved the field-work activities at each sampling location. The field examined in this study is not protected in any way, and the study did not involve any endangered or protected species.

A series of five scales were superimposed on the sample plot: 10-m pixel, 20-m pixel, 40-m pixel, 80-m pixel, and 160-m pixel with 1,024, 256, 64, 16 and 4 samples, respectively. The dataset for the minimum scale was the measured trace element data in the initial 10-m pixel grid, while the dataset for the other scales was the average of the measured trace element data of all initial 10-m pixel grids included within the larger pixel in the respective grid.

The single and joint multifractal spectra parameters for selected soil trace element data were calculated using the equations in the part of theory of multifractal analysis by an Office VBA project and were presented in graphs by SigmaPlot 10.0. The correlation analysis between $\alpha(q,t)$ and $\beta(q,t)$ was conducted using SPSS 16.0.

Results

Descriptive Statistics

A summary of descriptive statistics for selected trace elements in the study soils is presented in Table 1. Available Fe, Cu and Zn varied to a high degree with coefficient of variation (CV) of 82.7%, 41.7% and 89.5%, respectively. The CV is very sensitive to the anomalies in a dataset and high CV's for available Fe, Cu and Zn data suggest that there must be some significant anomalous values in their distributions. Available Mn had a relatively uniform distribution with a CV of only 9.5%. Based on the thresholds recommended by Liu [29], we found that Mn deficiency was widely present throughout the field. Although Fe and Zn deficiencies were not widespread, 25.5% of soil samples were deficient in available Fe, and 13.3% were deficient in available Zn. This indicated that Fe and Zn deficiencies tended to occur in

Table 1. Descriptive statistics for selected soil trace elements in the study agricultural field of northeast China.

Trace elements	Min† ($mg \cdot kg^{-1}$)	Max ($mg \cdot kg^{-1}$)	Mean ($mg \cdot kg^{-1}$)	CV	Deficiency‡
Fe	0.30	73.73	18.93	82.7%	25.5%
Mn	25.53	91.11	69.27	9.5%	100%
Cu	0.18	5.27	1.34	41.7%	1.4%
Zn	0.02	14.81	2.48	89.5%	13.3%

†Min, Max and CV are abbreviations for minimum, maximum and coefficient of variation.
‡Critical levels for available Fe, Mn, Cu and Zn recommended by Liu [29] are 4.5 $mg \cdot kg^{-1}$, 100 $mg \cdot kg^{-1}$, 0.3 $mg \cdot kg^{-1}$ and 1.0 $mg \cdot kg^{-1}$ (DTPA extracted). The concentrations of available Fe, Mn, Cu and Zn under the critical levels are recognized as deficiencies.

patches rather than throughout the field. Only a very limited percentage of soil samples were deficient in available Cu.

Empirical Tests of Multifractality

Scatter plots and their fitted relationships between the logarithm of partition function and the logarithm of superimposed scale for selected soil trace elements are shown in Figure 1. The coefficients of determination R^2 for all linear regression lines were larger than 0.99 (F-test, $P<0.001$) which confirms the assumption of the existence of the power-law relationship between partition function and scale. This finding revealed that the spatial distribution of available Fe, Mn, Cu and Zn in the study soils exhibited a multifractal property. Hence, multifractal techniques can be used to characterize the spatial variability of selected soil trace elements and their scaling relationships.

Variables showing a multifractal property usually have a $D(q)$ that demonstrates a monotonic and nonlinear trend with q values. This pattern can be observed when examining the trend of $D(q)$ for available Fe, Cu and Zn in Figure 2. The $D(q)$ values for available Fe, Cu and Zn decreased markedly as q values increased. This is a typical characteristic of a multifractal system, and also indicated that the multifractal property observed here was strong. In contrast, the $D(q)$ values for available Mn decreased very slowly against q values and showed small deviations from the capacity dimension of the sample plot (i.e., 2). This revealed that the distribution of available Mn can be represented by a monofractal or a weak multifractal. Note that a monofractal is a special case of a multifractal.

Variability and Singularity of Soil Trace Elements

To examine the spatial variability of selected soil trace elements, their single multifractal spectra were constructed and shown in Figure 3. The single multifractal spectra for available Fe, Cu and Zn all had large widths ($[\alpha(q)_{max}-\alpha(q)_{min}]\geq 0.54$). This indicated that available Fe, Cu and Zn had quite high spatial variability in studied soils. To characterize such high spatial variability of available Fe, Cu and Zn, a large number of singularity indices ranging from 1.69 to 2.83, 1.81 to 2.35, and 1.60 to 2.38

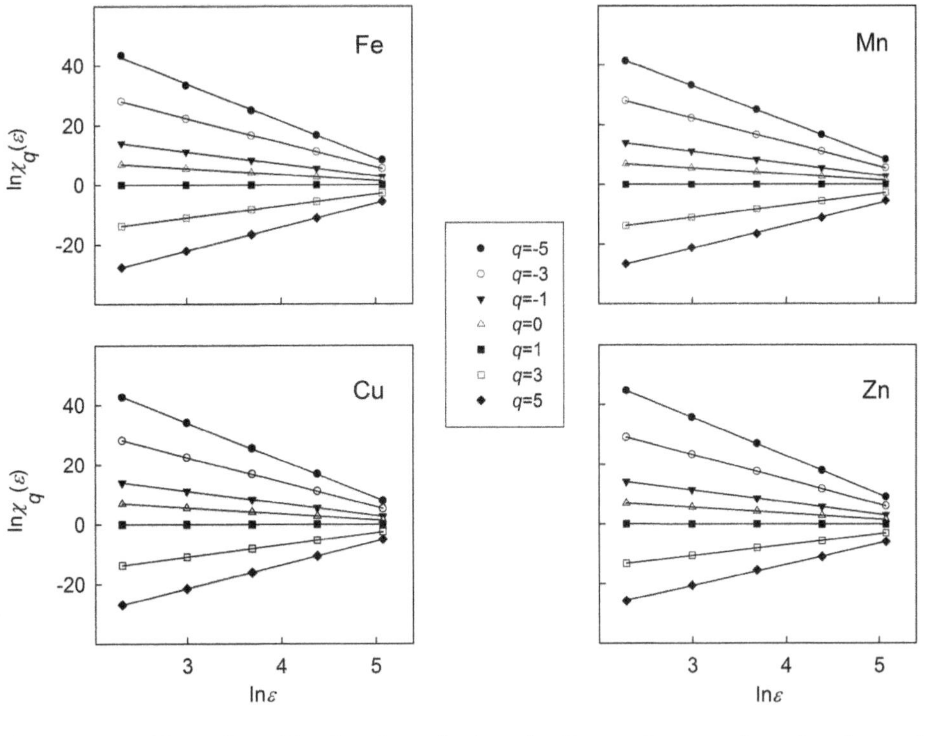

Figure 1. Logarithms of partition functions of selected soil trace elements plotted against scales. The R^2 values of the linear fit for all q considered were always greater than 0.99.

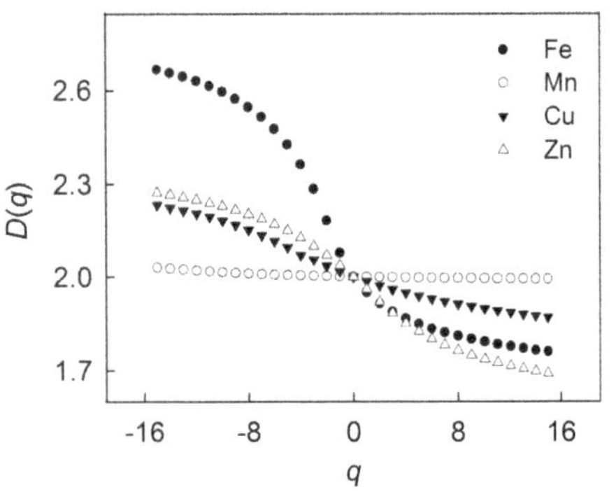

Figure 2. General fractal dimension $D(q)$ for selected soil trace elements at −15 to 15 range of q evaluated at 1 increment.

respectively were needed. In contrast, the single multifractal spectrum of available Mn had a very narrow width ($[\alpha(q)_{max}-\alpha(q)_{min}]\approx0.10$). This indicated that the spatial variability of available Mn was relatively low. Such low spatial variability can be characterized by a very small set of singularity indices, ranging from 1.99 to 2.09.

The single multifractal spectra of available Fe, Mn, Cu and Zn all had asymmetric tails on either the right or left part (Figure 3), which is another typical characteristic of a multifractal property. The left and right parts of a single multifractal spectrum respectively represent the higher and lower values of a dataset, emphasized by high positive and negative q values. The long left and right tails of the single multifractal spectra for available Fe, Cu and Zn were therefore the result of both extremely high and low values in their distributions. The singularities of these anomalies were characterized by the corresponding singularity indices in the single multifractal spectra of available Fe, Cu and Zn. The single multifractal spectrum of available Mn had a relatively long tail only on the right part. This indicated that there were no anomalies but just some relatively low values in the distribution of available Mn in the study soils.

Scaling Relationships among Soil Trace Elements

The relationships among selected soil trace elements across the spatial scales superimposed on the sample plot were analyzed using the joint multifractal spectra (Figure 4). In general, there were strong positive scaling relationships among available Fe, Cu and Zn, which were evident from the compact joint distribution of their singularity indices at the upper right and lower left parts of

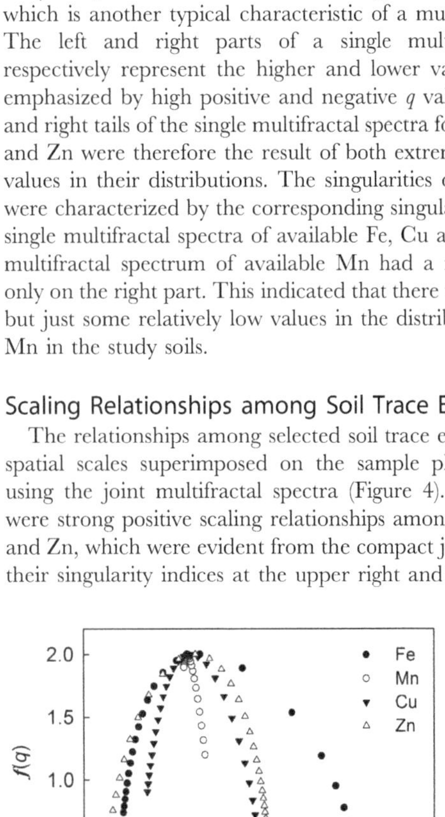

Figure 3. Single multifractal spectrum for selected soil trace elements at −15 to 15 range of q evaluated at 1 increment.

their joint multifractal spectra (Figures 4A, 4B and 4C). The contours in Figure 4B were tightly concentrated in one bundle. This indicated that there were strong positive relationships between available Fe and Zn for both higher and lower values across the superimposed spatial scales. With a small diagonal stretch at the upper right, Figure 4A and Figure 4C had a similar pattern of joint distribution of singularity indices as Figure 4B. This indicated that strong positive relationships also existed between available Cu and available Fe and Zn over a wider spatial scale and for the entire range of data values. In Figures 4D, 4E and 4F, the wide stretch of diagonal contours showed that the singularity indices of available Mn was not correlated tightly with that of available Fe, Cu and Zn. This suggested that available Mn may not be associated, or may be weakly associated, with available Fe, Cu and Zn over the range of measured concentrations and superimposed spatial scales.

To further confirm the scaling relationships among selected soil trace elements, the Pearson correlation coefficients (r) between their singularity indices were analyzed (Table 2). The singularity indices between available Fe and Zn had the highest positive correlation coefficient ($r=0.98$, $P<0.001$). The positive correlation coefficients between the singularity indices of available Fe versus available Cu and available Zn versus available Cu were relatively high as well ($r=0.87$ and 0.86, $P<0.001$). These results indicated that the positive correlation relationships among available Fe, Cu and Zn were all very strong across superimposed spatial scales. Although the associations between singularity indices of available Fe versus available Mn and available Zn versus available Mn were also significant, their correlation coefficients ($r=0.19$ and 0.18, $P<0.01$) were relatively low compared to those among available Fe, Cu and Zn. This suggested the positive correlations between available Mn and available Fe and Zn were weak. The correlation coefficient between singularity indices of available Cu and Mn was not significant ($r=-0.03$, $P=0.40$), which implied that available Cu and Mn were not correlated across superimposed spatial scales. The strong scaling relationships among available Fe, Cu and Zn were likely due to a common controlling process, which was quite different from that for available Mn.

Discussion

The influences of factors in the CLORPT function or the SCORPAN model on the variations of selected soil trace elements should be relatively consistent over a flat 10.24-ha level field [30,31]. In most soils, available Fe, Mn, Cu and Zn are not directly related to their total contents, but are more related to other soil characteristics such as pH, redox potential, organic matter content, cation exchange capacity, and texture [32]. In agricultural soils, long-term intensive agricultural practices also add another layer of complexity on the distribution of available Fe, Mn, Cu and Zn. Kravchenko et al. [13], Perrier and Bird [14], Caniego et al. [15], Guan et al. [16] and Perrier et al. [33] have reported that pH, redox potential, organic matter content, cation exchange capacity, texture and structure interact with each other and are prone to lead to a nonlinear soil system in which multifractal properties are expressed. The multifractal nature of available Fe, Mn, Cu and Zn observed in this study can be attributed to the complexity of this system.

The spatial variability of selected soil trace elements characterized by single multifractal spectrum was partly consistent with the result of CV analysis. However, the results based on the single multifractal spectrum were obtained over a wider range of spatial scales and moment orders and better reflected the true spatial variability of selected soil trace elements. The single multifractal spectrum also captured the detailed singularity strength of anomalies in available Fe, Mn, Cu and Zn data while the joint multifractal spectrum characterized the scaling relationships among them. These two points demonstrate that multifractal techniques present unique advantages over CV and semivariogram analyses.

Knowledge of soil spatial variability is a prerequisite needed to predict soil properties at unsampled locations. In this study, the spatial variability and singularity of selected soil trace elements quantified by the parameters of single multifractal spectra would have practical significance for prediction of these trace elements. Appropriate prediction procedures for variables with high spatial variability and a large amount of singularity indices are different than that for those with low spatial variability and limited singularity indices. This implication has been summarized in the study by Cheng [34] and justified by Cheng [35] and Xie et al. [36]. In our study, available Mn had a narrow multifractal spectrum with a limited number of singularity indices. Its spatial variability was very low and there were no anomalies within its spatial distribution. For a variable with those distribution characteritics and spatial autocorrelation, a simple prediction procedure such as an ordinary Kriging interpolation may be appropriate. In contrast, available Fe, Cu and Zn in this study's soils had wide multifractal spectra with many singularity indices both at $\alpha(q)>2$ and $\alpha(q)<2$. Their spatial variability was high with both extremely high and extremely low values in their spatial distributions. Under these circumstances, it is possible, but difficult, to apply a prediction procedure. For example, $\alpha(q)>2$ implied that there were some specific locations with extremely low concentrations of available Fe, Cu and Zn. An ordinary interpolation will smooth out the extremely low values of available Fe, Cu and Zn, resulting in the loss of valuable information for identifying Fe, Cu and Zn deficiencies. In this case, incorporating singularity indices $\alpha(q)>2$ into the prediction procedure can help preserve the extremely low values of available Fe, Cu and Zn.

Table 2. Pearson correlation coefficients (r) between singularity indices of joint multifractal spectra for selected soil trace elements.

	Fe	Mn	Cu
Mn	0.19 ($P<0.01$)		
Cu	0.87 ($P<0.001$)	−0.03 ($P=0.40$)	
Zn	0.98 ($P<0.001$)	0.18 ($P<0.01$)	0.86 ($P<0.001$)

Similarly, $\alpha(q)<2$ is also needed when predicting available Fe, Cu and Zn with extremely high values at specific locations.

Recent studies have delineated the spatial distribution of selected trace elements in soils using their relationships with environmental data, but those relationships were observed at only a single scale and may not be valid at other scales. Across a regional soil landscape, factors and processes impacting the plant availability of soil trace elements are expected to be much more complicated and heterogeneous than that of the 10.24-ha agricultural field observed in this study. The major controlling processes for different soil trace elements may differ at specific spatial scales, which would further impact the spatial variability of soil trace elements. The results of joint multifractal analysis in this study had demonstrated the ability of joint multifractal spectrum in characterizing the scaling relationships among selected soil trace elements at the field scale. It is well-known that fractals and multifractals are ubiquitous in complex ecosystems [37]. Further study is needed to evaluate the technique in modeling the scaling relationships of soil trace elements at larger regional scales where their variability in relation to the local environments would be expected. Using scale-specific relationships future researchers should be able to improve models when predicting soil trace elements with auxiliary environmental data.

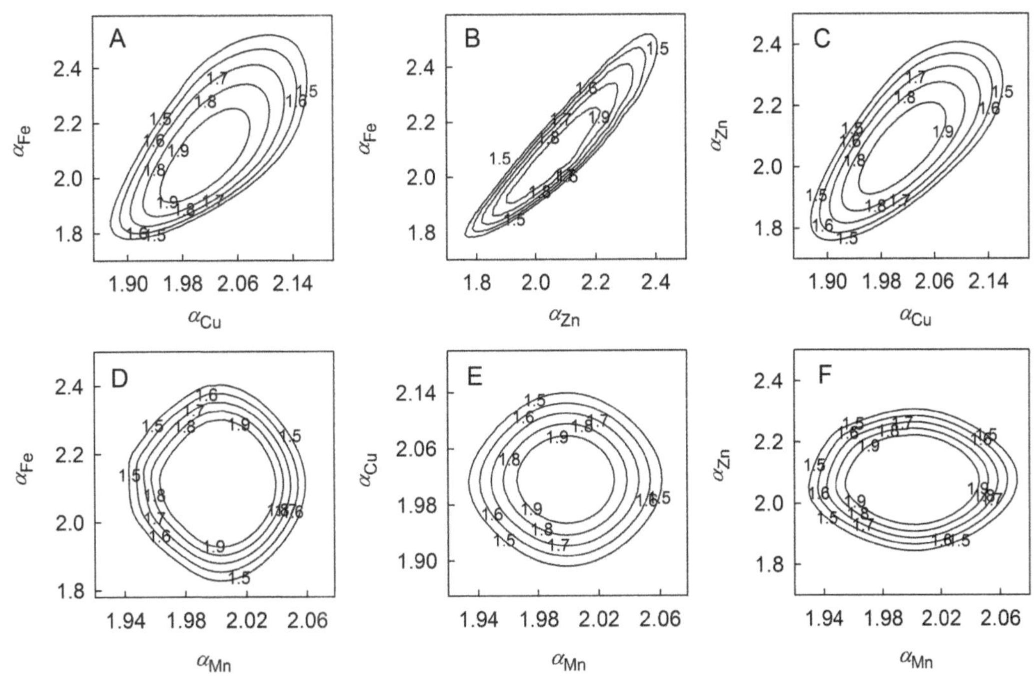

Figure 4. Joint multifractal spectrum for selected soil trace elements at −15 to 15 range of q evaluated at 1 increment.

In recent years, geographic information systems (GIS) have facilitated the process of spatial analysis and computation. If further research demonstrates that the multifractal phenomenon exists over larger regional scales, the findings in this study could be incorporated into GIS to improve the prediction of selected trace elements over a range of scales for optimizing soil use and management.

Conclusions

Spatial variability of available Fe, Mn, Cu and Zn and their scaling relationships were examined using multifractal techniques for a 10.24-ha agricultural field in northeast China. The distribution of selected soil trace elements exhibited a multifractal property. The variability and singularities of available Fe, Cu and Zn in the study soils needed to be quantified by multifractal spectra, whereas that of available Mn can be characterized by a monofractal or a weak multifractal spectrum. The joint multifractal spectra and the Pearson correlation analysis of their singularity indices captured the relationships among selected soil trace elements across a wider range of spatial scale and over the full range of data values.

The use of single and joint multifractal parameters has potential to enhance the prediction of selected soil trace elements in future research. We are currently examining the multifractality of soil trace elements at the county scale and are building a model that incorporates the scaling relationships between selected soil trace elements and environments into a prediction procedure. Such studies are essential to validate the applicability and extension of the findings presented here.

Acknowledgments

We thank Guixian Li for providing assistance in carrying out the field experiment.

Author Contributions

Conceived and designed the experiments: FSZ ZXL. Performed the experiments: FSZ GHY ZYW. Analyzed the data: FSZ ZYW. Contributed reagents/materials/analysis tools: FSZ. Wrote the paper: FSZ NM XYG ZXL.

References

1. Havlin JL, Beaton JD, Tisdale SL, Nelson WL (2005) Soil fertility and fertilizers: An introduction to nutrient management. 7th Edition. Upper Saddle River, NJ:Prentice Hall. 515 p.
2. Hood PS (2010) Trace elements in soils. Chichester, UK:Wiley-Blackwell. 616 p.
3. Alloway BJ (2008) Micronutrient deficiencies in global crop production. Dordrecht, The Netherlands:Springer. 353 p.
4. Gupta UC, Wu KN, Liang SY (2008) Micronutrients in soils, crops, and livestock. Earth Science Frontiers 15: 110–125.
5. Heuvelink GBM, Webster R (2001) Modelling soil variation: Past, present and future. Geoderma 100: 269–301.
6. White JG, Zasoski RJ (1999) Mapping soil micronutrients. Field Crops Research 60: 11–26.
7. Wu CF, Luo YM, Zhang LM (2010) Variability of copper availability in paddy fields in relation to selected soil properties in southeast China. Geoderma 156: 200–206.
8. Behera SK, Singh MV, Singh KN, Todwal S (2011) Distribution variability of total and extractable zinc in cultivated acid soils of India and their relationship with some selected soil properties. Geoderma 162: 242–250.
9. Goovaerts P (1999) Geostatistics in soil science: State-of-the-art and perspectives. Geoderma 89: 1–45.
10. Shi J, Xu J, Huang P (2008) Spatial variability and evaluation of status of micronutrients in selected soils around Taihu Lake, China. Journal of Soils and Sediments 8: 415–423.
11. Liu XM, Xu JM, Zhang MH, Si BC, Zhao KL (2008) Spatial variability of soil available Zn and Cu in paddy rice fields of China. Environmental Geology 55: 1569–1576.
12. Gringarten E, Deutsch CV (2001) Teacher's aide: Variogram interpretation and modeling. Mathematical Geology 33: 507–534.
13. Kravchenko AN, Boast CW, Bullock DG (1999) Multifractal analysis of soil spatial variability. Agronomy Journal 91: 1033–1041.
14. Perrier EMA, Bird NRA (2002) Modelling soil fragmentation: the pore solid fractal approach. Soil and Tillage Research 64: 91–99.
15. Caniego FJ, Espejo R, Martín MA, José FS (2005) Multifractal scaling of soil spatial variability. Ecological Modelling 182: 291–303.
16. Guan XY, Yang PL, Ren SM, Li YK (2007) Multifractal analysis of soil structure under long-term wastewater irrigation based on digital image technology. New Zealand Journal of Agricultural Research 50: 789–796.
17. Wang ZY, Shu QS, Liu ZX, Si BC (2009) Scaling analysis of soil water retention parameters and physical properties of a Chinese agricultural soil. Australian Journal of Soil Research 47: 821–827.
18. Biswas A, Cresswell HP, Si BC (2012) Application of multifractal and joint multifractal analysis in examining soil spatial variation: A Review. In: Ouadfeul SA, editor. Fractal analysis and chaos in geosciences. InTech. pp. 109–138.
19. Zeleke TB, Si BC (2005) Scaling relationships between saturated hydraulic conductivity and soil physical properties. Soil Science Society of America Journal 69: 1691–1702.
20. Wang ZY, Shu QS, Xie LY, Liu ZX, Si BC (2011) Joint multifractal analysis of scaling relationships between soil water-retention parameters and soil texture. Pedosphere 21: 373–379.
21. Jesus AP, Mateus A, Gonçalves MA, Munhá J (2013) Multi-fractal modelling and spatial Cu-soil anomaly analysis along the southern border of the Iberian Terrane in Portugal. Journal of Geochemical Exploration 126–127: 23–44.
22. Panahi A, Cheng Q (2004) Multifractality as a measure of spatial distribution of geochemical patterns. Mathematical Geology 36: 827–846.
23. Agterberg FP (1995) Multifractal modeling of the sizes and grades of giant and supergiant deposits. International Geology Review 37: 1–8.
24. Chhabra A, Jensen RV (1989) Direct determination of the $f(\alpha)$ singularity spectrum. Physical Review Letters 62: 1327–1330.
25. Meneveau C, Sreenivasan KR, Kailasnath P, Fan MS (1990) Joint multifractal measures: Theory and applications to turbulence. Physical Review A 41: 894–913.
26. Kravchenko AN, Bullock DG, Boast CW (2000) Joint multifractal analysis of crop yield and terrain slope. Agronomy Journal 92: 1279–1290.
27. Lindsay WL, Norvell WA (1978) Development of a DTPA soil test for zinc, iron, manganese, and copper. Soil Science Society of America Journal 42: 421–428.
28. Hooda PS, Alloway BJ (1994) The plant availability and DTPA extractability of trace metals in sludge–amended soils. Science of the Total Environment 149: 39–51.
29. Liu Z (1996) Microelements in soils of China. Nanjing;Jiangsu Sci. and Technol. Publishing House. 396 p. (in Chinese).
30. Jenny H (1941) Factors of soil formation: A system of quantitative pedology. New York: McGraw-Hill. 281 p.
31. McBratney AB, Mendonça-Santos ML, Minasny B (2003) On digital soil mapping. Geoderma 117: 3–52.
32. Fageria NK, Baligar C, Clark RB (2002) Micronutrients in crop production. Advances in Agronomy 77: 185–268.
33. Perrier E, Bird N, Rieutord T (2010) Percolation properties of 3-D multiscale pore networks: how connectivity controls soil filtration processes. Biogeosciences Discussion 7: 2997–3018.
34. Cheng Q (2005) A new model for incorporating spatial association and singularity in interpolation of exploratory data. In: Leuangthong O, Deutsch CV, editors. Geostatistics Banff, 2004, Quantitative geology and geostatistics. Dordrecht, The Netherlands:Springer. pp. 1017–1025.
35. Cheng Q (2008) Modeling local scaling properties for multiscale mapping. Vadose Zone Journal 7: 525–532.
36. Xie SY, Cheng Q, Xing XT, Bao ZY, Chen ZJ (2010) Geochemical multifractal distribution patterns in sediments from ordered streams. Geoderma 160: 36–46.
37. Solé RV, Bascompte J (2006) Self-organization in complex ecosystems. Princeton:Princeton University Press. 392 p.

8

Effects of Picoxystrobin and 4-n-Nonylphenol on Soil Microbial Community Structure and Respiration Activity

Marianne Stenrød*, Sonja S. Klemsdal, Hans Ragnar Norli, Ole Martin Eklo

Norwegian Institute for Agricultural and Environmental Research (Bioforsk), Ås, Norway

Abstract

There is widespread use of chemical amendments to meet the demands for increased productivity in agriculture. Potentially toxic compounds, single or in mixtures, are added to the soil medium on a regular basis, while the ecotoxicological risk assessment procedures mainly follow a chemical by chemical approach. Picoxystrobin is a fungicide that has caused concern due to studies showing potentially detrimental effects to soil fauna (earthworms), while negative effects on soil microbial activities (nitrification, respiration) are shown to be transient. Potential mixture situations with nonylphenol, a chemical frequently occurring as a contaminant in sewage sludge used for land application, infer a need to explore whether these chemicals in mixture could alter the potential effects of picoxystrobin on the soil microflora. The main objective of this study was to assess the effects of picoxystrobin and nonylphenol, as single chemicals and mixtures, on soil microbial community structure and respiration activity in an agricultural sandy loam. Effects of the chemicals were assessed through measurements of soil microbial respiration activity and soil bacterial and fungal community structure fingerprints, together with a degradation study of the chemicals, through a 70 d incubation period. Picoxystrobin caused a decrease in the respiration activity, while 4-n-nonylphenol caused an increase in respiration activity concurring with a rapid degradation of the substance. Community structure fingerprints were also affected, but these results could not be directly interpreted in terms of positive or negative effects, and were indicated to be transient. Treatment with the chemicals in mixture caused less evident changes and indicated antagonistic effects between the chemicals in soil. In conclusion, the results imply that the application of the fungicide picoxystrobin and nonylphenol from sewage sludge application to agricultural soil in environmentally relevant concentrations, as single chemicals or in mixture, will not cause irreversible effects on soil microbial respiration and community structure.

Editor: Hauke Smidt, Wageningen University, The Netherlands

Funding: This work was performed as part of the strategic institute programme 'Bioavailability and biological effects of chemicals - Novel tools in risk assessment of mixtures in agricultural and contaminated soil' funded by the Norwegian Research Council, project number 186901/i30. The funders had no role in study design, data collection and analysis, decision to publish, or preparation of the manuscript.

Competing Interests: The authors have declared that no competing interests exist.

* E-mail: marianne.stenrod@bioforsk.no

Introduction

Various practices have been promoted to meet the demands for increased productivity of agricultural areas, including the use of mineral fertilizers and pesticides as well as the use of organic amendments. The potential risk of non-target effects of pesticides in soils is evident, and has been widely studied. Specific attention is now given to the use of organic amendments originating from sewage sludge. Sludge has been shown to contain high levels of many organic chemicals (e.g. nonylphenols, PAHs a.o.) that might exert ecotoxicological effects upon soil addition, e.g., [1,2]. Potentially toxic compounds, single or in mixtures, are added to the soil medium on a regular basis, while the ecotoxicological risk assessment procedures mainly follow a chemical by chemical approach focusing on establishing dose-response relationships for soil fauna. The use of sewage sludge on agricultural fields where pesticides are sprayed regularly as part of conventional farming practices calls for increased attention to the potential combined effects of pesticides and known contaminants in sewage sludge.

Sewage sludge, potentially used for application on agricultural soil, is known to contain considerable amounts of nonylphenol [2–4] – an industrial by-product and degradation product of nonylphenol ethoxilate plasticizers. The occurrence, fate and toxicity of nonylphenol in the environment have been reviewed by Soares and co-workers [4], pointing at knowledge gaps as well as identified challenges with nonylphenol in soil including occurrence and possible accumulation in soil following sludge application, reduced degradation rates in soil due to sorption and reduced bioavailability, and potential toxic effects on soil microorganisms. Nonylphenol is also included on the list of priority substances of the Water Framework Directive [5] and, hence, require further attention to clarify its potential effects in soil. Picoxystrobin is a strobilurin fungicide [6] for spraying in cereals, with a maximum of one spraying per season in Norway. Picoxystrobin inhibits mitochondrial respiration by blocking electron transfer at the Qo centre of cytochrome bc1 [7]. Picoxystrobin is reported to have a high acute toxicity to earthworms (LC_{50} at 6.7 mg kg^{-1}) [8], and field assays of earthworm toxicity indicate acute toxic effects even at recommended doses for use [9], which might be caused by heavy rain shortly after spraying forcing the earthworm to migrate to the surface. Potentially negative effects on soil microbial nitrogen and carbon mineralization activity are shown to be transient (dose: 750 g ha^{-1}, duration: 28 days) [8]. Picoxystrobin require continued attention due to its demonstrated potentially negative effects to earthworms.

The importance of soil microbes and their activity in the functioning of soils, e.g., [10,11] justify their thorough investigation in risk assessments [12–14]. The development of genomic techniques over the last decades has made detailed studies of the soil microbial community possible, beyond the scope of broad-scale measures like substrate induced respiration. DNA extraction from soil followed by different molecular approaches to determine the genetic diversity and quantify the presence of single organisms or groups of related organisms in a soil sample, have been employed successfully in studies of species and functional diversity in agricultural soils [15–17]. According to OECD guidelines for the testing of chemicals carbon [18] and nitrogen [19] transformation tests (with cut-off criteria of 25% effect) are the recommended methods to assess effects concentrations of chemicals on the soil microbial community. In research, soil respiration is commonly used to assess effects of pesticides and other chemicals on soil microbes. These are measures linked to the activity level of the soil microbial community, but are crude measures that do not necessarily reveal all relevant effects. The low percentage of soil microorganisms we are able to culture ex situ stresses the need for employing molecular and genomic methods suitable for terrestrial ecotoxicological studies. T-RFLP-analysis is one much used technique showing good results when looking at effects of different environmental conditions and chemical stressors on soil microbial communities, e.g., [16,17,20]. But there are many important methodological aspects to be considered when interpreting the results, e.g., [21,22], including choice of primers and restriction enzymes, procedure for noise reduction and profile alignment, and statistical analysis using relative abundance or presence/absence data.

The main objective of this study was to assess the effects of picoxystrobin and nonylphenol, as single chemicals and mixtures, on soil microbial community structure and respiration activity in an agricultural sandy loam soil. The choice of compounds is based on their individual occurrence and effects in the environment, and expected potential for co-occurrence in the field, although the latter is scarcely documented. Further, they were chosen to represent an example of combined effects of agricultural and industrial contaminants with their expected independent effect mechanisms due to differences in mode of action. Effects of the chemicals were assessed through measurements of soil microbial respiration activity and soil bacterial and fungal community structure fingerprints, together with a degradation study of the chemicals, through a 70 d incubation period. The different measures showed corresponding results in support of a conclusion that a mixture situation with picoxystrobin and nonylphenol in soil will not increase the potential negative effects of picoxystrobin on the soil microbial community.

Materials and Methods

Soil

Bulk soil sampling in the top 10 cm of the plough-layer of an agricultural field at Norderås, Ås, South East Norway (59°41′14″ N, 10°46′22″ E), was done in middle of August 2008. After sieving (4 mm mesh) the soil was physically and chemically characterized (Table 1; all analyses performed in accordance with recognized laboratory standards at Analycen AS/Eurofins Norway) and stored moist at 4°C for about 3 weeks before use in laboratory experiments. No specific permits were required for the described field soil sampling and subsequent lab studies.

Table 1. Selected physical and chemical characteristics of the studied soil.

Parameter	Value	Unit
pH	6.3	
TOC – Organic C	1.1	%
Kjeldahl N	0.0986	%
C/N	11	
Tot C	2.0	%
Tot N	0.07	%
CEC	11.2	meq/100 g
Coarse (>2 mm)	13.7	%
Sand (2.000-0.060)	68.3	%
Silt (0.060-0.002)	20.7	%
Clay (<0.002)	11.0	%

Analysis performed by Analycen AS/Eurofins Norway.

The Water Holding Capacity (WHC) was estimated on sieved soil gently packed in plastic columns. After wetting with Milli-Q-water overnight, excess water was allowed to drip off during 1 h the next day, and the gravimetric moisture content was measured after drying overnight at 105°C.

Chemicals

Treatment solutions were prepared from analytical grade picoxystrobin (Riedel-de Häen) and 4-n-nonylphenol (Fluka) purchased from Sigma-Aldrich. The chemicals were dissolved in acetone (4 ml portions; Lab-Scan) and added to quartz sand (27 g portions; Sigma-Aldrich). After evaporation of the acetone (about 30 minutes) the quartz sand with chemical was mixed into batches of moist soil (2.7 kg dry weight equivalent (dw eq.), 60% WHC) giving 2 or 10 mg picoxystrobin kg^{-1} soil (dw eq.), or 0.5 or 10 mg 4-n-nonylphenol kg^{-1} soil (dw eq.) as single treatments as well as all mixture combinations (Table 2). A solvent control (acetone added to quartz sand, left to evaporate before addition to soil) and an untreated control (soil with no addition of chemical) were included in the experiment.

The lowest test concentration for picoxystrobin was set from reported levels for transient effects on carbon and nitrogen mineralization (750 g ha^{-1}) [8], considering that picoxystrobin has been shown to sorb to the top cm of soil (50–70% interception, sorption in top 1 cm of soil), although recommended doses in Norway are below these levels. The worst case concentration was set to 10 mg kg^{-1} surpassing reported acute toxicity levels for earthworm ($LC_{50} = 6.7$ mg kg^{-1}) [8]. The lowest test concentration for nonylphenol was set to the Predicted Initial Environmental Concentration (PIEC) estimated through a risk assessment made by the Norwegian Scientific Committee on Food Safety [2]; application of sludge containing mean expected amounts of nonylphenol (32 mg kg^{-1}) at rates in accordance with Norwegian regulations (40 tonnes ha^{-1}). From this a worst case concentration was set to 10 mg kg^{-1} considering a risk of reduced incorporation depth and possible initial effects.

Experimental Set-up

An incubation experiment was set up with all chemical treatments in five replications and the control treatments in four replications. Subsamples of about 100 g dw eq. of sieved, moist soil

Table 2. Overview of picoxystrobin and 4-n-nonylphenol treatments in the incubation experiment.

Treatment	Picoxystrobin (PI)		4-n-nonylphenol (NP)	
	2 mg kg^{-1} dry soil	10 mg kg^{-1} dry soil	0.5 mg kg^{-1} dry soil	10 mg kg^{-1} dry soil
PI low	x			
PI high		x		
NP low			x	
NP high				x
Mix low	x		x	
Mix NP high	x			x
Mix PI high		x	x	
Mix high		x		x
Control				

(60% WHC) in plastic (polypropylene) containers were placed in air-tight glass jars together with small beakers with NaOH-solution (10 mL, 1 M) to trap CO_2 evolved from the soil. All treatments were set up with five repeats to enable destructive sampling of five and four replicates, of chemical treatments and the control respectively, on each sampling occasion through the 70 d incubation period. A total of 244 samples were incubated at 20°C in the dark. Samples were taken on six time-points; before addition of chemicals (4 samples), and after 1, 7, 14, 28 and 70 days of incubation (48 samples on each occasion). Each sample was homogenized and subsamples taken for chemical residue analysis (25 g), and DNA extraction (30 g). In addition, the general microbial activity level in the soil was monitored by weekly replacement and analysis of the NaOH-traps, giving a total of 11 time-points with measurements.

Total Soil Microbial Activity

The total activity of soil microbial biomass was followed by respirometric measurements [23]. The CO_2 in the NaOH traps was determined by colourimetric continuous flow analysis (AutoAnalyzer3, Bran+Luebbe, Germany). Results are reported as mg CO_2-C kg^{-1} dry soil.

Soil Microbial Community Structure

DNA extraction procedure. Soil samples were homogenized (30 g moist soil +30 mL Milli-Q-water) in a mill (3 min forward +3 min reverse spin; Retsch PM 400) before extraction of DNA according to the Fast DNA Spin Kit for Soil (QBiogene) manual. In short, soil suspension (400 µL) was added to Lysing Matrix E Tubes before homogenization with the FastPrep® instrument (30 sec at speed 5.5; QBiogene, MP Biomedicals) and centrifugation (14000×g, 3 min). DNA was purified with guanidine thiocyanate (5.5 M, Sigma) twice (1 mL and 600 µL) after DNA binding to the Binding Matrix Suspension, before elution (100 µL sterile distilled water) from Spin™Filters. The eluate was further purified through Micro Bio-Spin Chromatography columns (Bio-Rad Laboratories Ltd.) packed with polyvinylpolypyrrolidone (PVPP) (Sigma) (4000×g, 5 min), and stored at −80°C.

PCR procedure. PCR reactions were run with fluorescently labelled PCR primer sets (Applied Biosystems), both forward and reverse primer labelled. After testing of 8 primer sets, chosen from previously published T-RFLP studies, PCR reactions were run for all samples with three selected primer sets and fluorescence labelling (Table 3).

All PCR reactions were run with 1 µL template and 1 µL of forward and reverse primer (10 pmol/µL) in TaqMan® Environmental Master Mix 2.0 (12.5 µL; Applied Biosystems) made up to a total reaction volume of 25 µL with sterile distilled water, on GeneAmp® PCR system 9700 (Applied Biosystems) or PTC-200 (MJ Research) thermal cyclers (40 cycles). An annealing temperature of 55°C was used for all primer combinations. PCR-products were purified with GenElute™PCR clean-up kit (Sigma-Aldrich) according to the instructions from the supplier, before restriction digestion and T-RFLP.

Table 3. Primer sets used in the experiments.

Primer set no.	Primer	Sequence	Reference
1	63F[1]	CAG GCC TAA CAC ATG CAA GTC	[24]
	1087R[2]	CTC GTT GCG GGA CTT ACC CC	[24]
2	EF4[1]	GGA AGG GRT GTA TTT ATT AG	[25]
	Fung5[2]	GTA AAA GTC CTG GTT CCC C	[25]
3	ITS1F[3]	CTT GGT CAT TTA GAG GAA GTA A	[26]
	ITS4[4]	TCC TCC GCT TAT TGA TAT GC	[27]

Fluorescence labeling: [1]6FAM,
[2]VIC,
[3]PET,
[4]NED.

T-RFLP procedure. After initial testing of a range of restriction enzymes (AluI, DdeI, HaeIII, HinfI, MboI, MspI, RsaI, TaqI, HhaI, MvnI, BstUI) on random samples, MspI and HaeIII, and MspI and HinfI were chosen for restriction of PCR products from bacteria and fungi, respectively. Restriction with the different enzymes were run in separate reactions and analysed separately. Analysis was performed on an ABI 3730 DNA Analyzer (Applied Biosystems, Foster City, USA) after mixing of 1 µl sample (template) with 8.75 µl Hi-Di formamid™ (Applied Biosystems) and 0.25 µl GeneScan™ 500 LIZ size standard (Applied Biosystems). The results reported for the T-RFLP analyses are based on relative comparisons of fingerprints, to show relative effects of different chemical treatments and comparison with a soil sample not treated with chemicals.

Degradation Study

Residues of picoxystrobin and 4-n-nonylphenol were extracted from 5 g subsamples of soil after mixing with 1.0 g of dehydrated $MgSO_4$ (purum, Fluka, Sigma-Aldrich GmbH) and 10 mL acetonitrile (Pestiscan, LAB-SCAN POCH SA, Gliwice, Poland) in 50 ml centrifugal tubes. One µg of 4-n-nonylphenol-2,3,5,6-d4-OD (99.4%, Chiron, Trondheim, Norway) was added as internal standard. After a short homogenization (10–15 sec. whirl mix) the samples were extracted by end-over-end shaking (1 h; Reax2, Heidolph). Extraction efficiencies derived from analysis of sterilised (autoclaved) soil at 117±2.0% and 108±0.0% for 4-n-nonylphenol and picoxystrobin, respectively. After centrifugation (1800×g, 5 min), 1.5 ml of the supernatant was transferred to GC-vials for analysis on GC-MS. Calibration standards at 0.001, 0.01, 0.05, 0.2, 1.0 and 5 µg ml^{-1} where prepared by diluting stock solutions of picoxystrobin (98%, Dr. Ehrenstorfer, Augsburg, Germany) and 4-n-nonylphenol (99.2%, Chiron, Trondheim, Norway) with acetonitrile. To balance the matrix in the samples a GC-vial was added 1 ml of a blank soil extract which was evaporated to dryness. One ml of each calibration standard was added to the GC-vial together with 0.1 µg 4-n-nonylphenol-2,3,5,6-d4-OD as internal standard. The measurements were performed on an Agilent 6890 gas chromatograph connected to an Agilent 5973 mass spectrometer using ChemStation Software version D.03.00. The gas chromatograph was equipped with a Gerstel (Mühlheim Ruhr, Germany) programmable temperature vaporizing (PTV) injector with a sintered liner. The separation was performed using a fused silica column (HP-5MSI 30 m width, 0.25 mm internal diameter, 0.25 µm film thickness, J&W Scientific) connected to a 2.5 m methyl deactivated pre column (Varian Inc., Lake Forest CA, USA) of same internal diameter as the analytical column. The temperature program was as follows; 65°C held for 1.5 min, 20°C min^{-1} to 120°C, held for 0 min, 20°C min^{-1} to 300°C, held for 0.5 min, total runtime 13.75 min. The PTV program was as follows: solvent vent temperature 60°C held for 1.40 min with a vent flow at 200 ml min^{-1}. After 1.42 min the split valve was closed and the injector temperature increased by 720°C min^{-1} to 250°C and kept for 2 min. Injection volume 10 µl. The mass spectrometer was operated in selected ion monitoring mode with target/qualifier ions as follows: 4-n-nonylphenol-D4: m/z = 111/224, 4-n-nonylphenol: m/z = 107/220 and picoxystrobin: m/z = 145/335. Transfer line temperature was set at 280°C, ion source temperature at 230°C and quadrupole temperature at 150°C.

Biological degradation of picoxystrobin and 4-n-nonylphenol was verified by spiking selected matrices (quartz sand, unsterile soil, sterilized soil) at 0.05 µg g^{-1} and left in the dark at room temperature (20±2°C) for 24 h before extraction and analysis as described above. The soil was sterilised by autoclaving (121°C, 20 min, x2) (Matachana S1000).

Data Processing and Statistical Analysis

First-order rate constants for organic carbon mineralization were estimated by linear regression (SigmaPlot 11.0, SYSTAT) using mean cumulative values for CO_2-C evolved, from the equation $\ln S = \ln S_0 + kt$, where S is the amount of CO_2-C evolved at time t (mg kg^{-1} dry soil), S_0 is the initial amount of mineralization product, k is the rate constant (d^{-1}), and t is the time (d). Due to the non-normal distribution of our CO_2-measurements, non-parametric analysis was utilized to test for differences between treatments; i.e. Kruskall Wallis test for equality of medians for all treatments, and Mann-Whitney test for equality of medians between selected treatments (Minitab 15, Minitab Inc.).

Handling of T-RFLP data with binning/identification of alleles, normalization (within project; i.e. for all samples independent of plate) and (automatic) alignment of profiles was done with GeneMapper 4.0 (Applied Biosystems) based on a 4 basepair (bp) bin width (assumed to give the more stable number of bins based on a screening of bin widths from 0.5 to 10 bp in 0.5 bp increments) and a threshold of 50 (default). Obvious pull-up peaks were removed from the dataset before further analysis. No further trimming of the data was done (i.e. no small peaks removed) to avoid exclusion of potentially important peaks for the microbial diversity analysis. Size (bp), area and height data were exported to an excel spreadsheet (Excel 2007, Microsoft corp.) for calculation of average size of each identified peak and calculation of relative peak height and area for all peaks within each individual sample. All peak data for each sample was assembled before statistical analysis, keeping the data for each primer set separate. The terminal fragments labelled from the forward primer as well as the fragments labelled from the reverse primer from each of the restriction reactions, were included in a single data row for each sample before analysis. The data were found to be not normally distributed (i.e. skewed due to rarely occurring peaks). Relative peak heights were analysed with non-metric multidimensional scaling (nmMDS) in R-software (v. 2.9.1) [28] using the metaMDS routine in the vegan package [29] with k = 2 or 3 dimensions, the isoMDS routine in the MASS package [30] with k = 2 dimensions, and with principal coordinates analyses using the PCO routine in R. All analyses resulted in similar patterns and the results presented are from metaMDS (k = 3 dimensions) as these gave the best fit to the observed data (STRESS below 0.10) [31,32]. Results are reported as two and three-dimensional plots showing the clustering of treatments at the different time-points of analysis through the incubation period. The samples taken before addition of chemical and onset of the incubation period (original soil sample) were included in the analysis of all time-points as a reference. Further, analysis of similarity was performed with the ANOSIM routine in PAST-software [33] with Bray-Curtis distance measure utilized in all tests. Results are given as ANOSIM R of a value between 0 and 1, where values close to 1 indicate large differences between treatments.

The observed T-RFLP patterns for the untreated and solvent control were not statistically significant different as analysed by nmMDS, and the results presented focus on differences between chemical treatments and the solvent control. This was to avoid overestimating any effects of the studied chemicals. A similar clustering of T-RFLP fingerprints was observed for the results obtained from using primer sets 2 and 3 (i.e. soil fungal community fingerprints; Table 3). Hence, the presented results only include primer set 3.

Soil degradation half-lives (DT50) for picoxystrobin and 4-n-nonylphenol was estimated from results from the degradation study, and calculated according to first-order kinetics from the equation $DT50 = \log_{10}2/b$, where b is the regression coefficient estimated through a linear regression procedure. All measured values were utilized in the estimation of the DT50 values except for two extreme outliers being removed before statistical analysis of the data. Due to the five replicates for each treatment this could be done without compromising the results of the data analyses, as the analyses for the affected treatments were still based on four replicate samples. Due to some differences in the observed degradation pattern for the chemicals at low and high concentrations, the soil half-lives were estimated separately for these. The estimated half-lives are based on data from both the relevant single chemical and mixture treatment, as the observed degradation pattern did not differ significantly between these. As an example

DT50 for 2 mg picoxystrobin kg^{-1} dry soil was estimated using data from both the single chemical treatment and the two mixture treatments with this concentration of picoxystrobin. The estimation was done assuming bi-phasic degradation of both chemicals, with both phases sufficiently described by first-order degradation kinetics and, hence, possible to estimate through linear regression.

A significance level of 5% was used for hypothesis testing.

Results

Total Soil Microbial Activity

No statistically significant differences between the treatments could be detected from observations of cumulative CO_2-C development after 70 d and estimated first-order rate constants for organic C-mineralization (Table 4). However, cumulative respiration curves (Fig. 1) indicated a tendency of reduced respiration activity in treatments with picoxystrobin as a single chemical, and increased respiration activity in treatments with high concentrations of nonylphenol (single chemical and mixture). The untreated control showed a lower respiration rate in the beginning of the period and a higher respiration rate at the end of the period, as compared to the other treatments. This difference did not result in any statistically significant difference in cumulative CO_2-C respiration, but indicated an effect of the solvent (acetone) used in the experiments.

The indications from the latter were supported by analysis of the separate measurements of CO_2 development from the soil samples at 11 time-points through the 70 d incubation period. Kruskall Wallis test for equality of medians between the treatments indicated statistically significant differences ($p<0.001$) when looking at all the individual CO_2 measurements in the analysis separately (i.e. not cumulative values at 70 d). The individual ranking of the treatments (data not shown) gave the solvent control a rank close to the overall rank (i.e. mean), while treatments with picoxystrobin as a single chemical (2 and 10 mg kg^{-1} dry soil) were indicated to cause a reduction in respiration activity, through a rank statistically significant lower than the overall rank. High concentrations of 4-n-nonylphenol (10 mg kg^{-1} dry soil) were indicated to cause an increase in respiration activity, through a rank statistically significant higher than the overall rank. This was also the case for the untreated control.

These indications were further supported by results from Mann-Whitney tests for equality of treatment medians, comparing the treatments two by two (test criteria: $p<0.05$). High concentration of 4-n-nonylphenol (10 mg kg^{-1} dry soil) in mixture with low concentration of picoxystrobin (2 mg kg^{-1} dry soil) gave statistically significant higher CO_2 development than treatments with picoxystrobin as a single chemical (2 and 10 mg kg^{-1} dry soil), low concentration of 4-n-nonylphenol (0.5 mg kg^{-1} dry soil), low concentration mixture (2 and 0.5 mg kg^{-1} dry soil of picoxystrobin and 4-n-nonylphenol, respectively) as well as the solvent control. Further, treatments with picoxystrobin as a single chemical (2 and 10 mg kg^{-1} dry soil) gave statistically significant reduction in CO_2 development compared to all mixture treatments, high concentration of 4-n-nonylphenol (10 mg kg^{-1} dry soil) as a single chemical, and the untreated control.

Soil Microbial Community Structure

The fungicide picoxystrobin was found to have a statistically significant effect on the soil microbial community structure, as found from nmMDS and ANOSIM of relative peak height data from T-RFLP analyses (Table S1). Looking at the effects of single chemical treatments in bacteria (Fig. S1) the largest segregation between treatments was observed after 7 d, with an overall ANOSIM R of 0.65 ($p=0.0001$). Here there was a statistically significant segregation between the low and high concentration treatment ($R=0.42$; $p=0.018$), indicating a concentration effect, and also a significant effect compared to the solvent control at low levels of picoxystrobin in the soil ($R=0.44$, $p=0.04$). Large variability was found in the data for the effects of picoxystrobin on the soil fungal community structure, with the largest spread in the data as shown by nmMDS after 7 d (Fig. 2a, Fig. S2) ($R=0.47$, $p=0.0001$). However, an effect due to different treatment concentrations of picoxystrobin was evident until 28 d ($R=0.38$, $p=0.017$) (Fig. S2). There was still a statistically significant difference between the single chemical treatments and the solvent control at 70 d for both bacteria (Fig. S1) and fungi (Fig. 2b), with an overall ANOSIM R of 0.65 and 0.4 ($p=0.0001$), respectively.

Less evident effects were observed for the single chemical treatments with 4-n-nonylphenol. Analysis of the bacterial community structure indicated an increasing change in the T-RFLP fingerprints with time, with statistically significant differences due to chemical treatment in nmMDS-analysis at 28 d (overall ANOSIM $R=0.56$, $p=0.0001$) and 70 d (overall ANOSIM $R=0.67$, $p=0.0001$) (Fig. 3, Fig. S3). A concentration effect was also evident at 28 d ($R=0.42$, $p=0.035$). The T-RFLP fingerprints of the fungal community upon treatment with 4-n-nonylphenol showed a similar pattern (Fig. S4), however, with larger variation between replicate samples giving lower ANOSIM R-values (range: 0.25–0.52).

The effects of the mixture treatments (Fig. S5 and S6) could only be separated from the solvent control after 28 and 70 d for the T-RFLP fingerprints of the soil bacterial community ($R=0.73$, $p=0.007$ and $R=0.44$, $p=0.04$, respectively) and after 70 d for the soil fungal community ($R=0.72$, $p=0.007$). When comparing the mixture treatments with the single treatments, the more prominent alteration in T-RFLP fingerprints originated from picoxystrobin as a single chemical (10 mg kg^{-1} dry soil), with the largest separation between treatment effects observed after 28 d both for the soil bacterial (Fig. 4a) and soil fungal (Fig. 4b) community structure (overall ANOSIM $R=0.61$ for both analyses). At this time-point also the single chemical treatment with 4-n-nonylphenol (10 mg kg^{-1} dry soil) could be singled out, with statistically significant changes in T-RFLP fingerprint as compared to the solvent control and other chemical treatments.

The duration of the observed changes in T-RFLP fingerprints of the soil microbial communities under the different chemical treatments was examined through nmMDS analysis of the grouping of the time-points for T-RFLP analyses. In general, these results indicated that the observed changes were transient. There was a more rapid re-establishment of the original T-RFLP fingerprints in the soil bacterial (i.e. closer resemblance between time-points 0 and 70) as compared to the soil fungal community (Fig. 5).

Degradation of Chemicals

Soil chemical residue was analysed day 0, 1, 7, 14, 28 and 70 of the incubation period. A lag phase was observed before the degradation of picoxystrobin commenced (Fig. 6a), while 4-n-nonylphenol was rapidly degraded in the soil (Fig. 6b). Calculations of soil half-lives (Table 5) were based on data from both single chemicals and mixtures as the degradation patterns were similar for these. Due to very rapid initial degradation of 4-n-nonylphenol in the low concentration treatments (0.5 mg kg^{-1} dry soil), the soil half-lives were here estimated for the 0–14 days and 14–70 days separately. For the other treatments the estimates were made for 0–28 days and 28–70 days. Soil half-life for picoxystrobin for the 28–70 day period was not calculated due to a measured

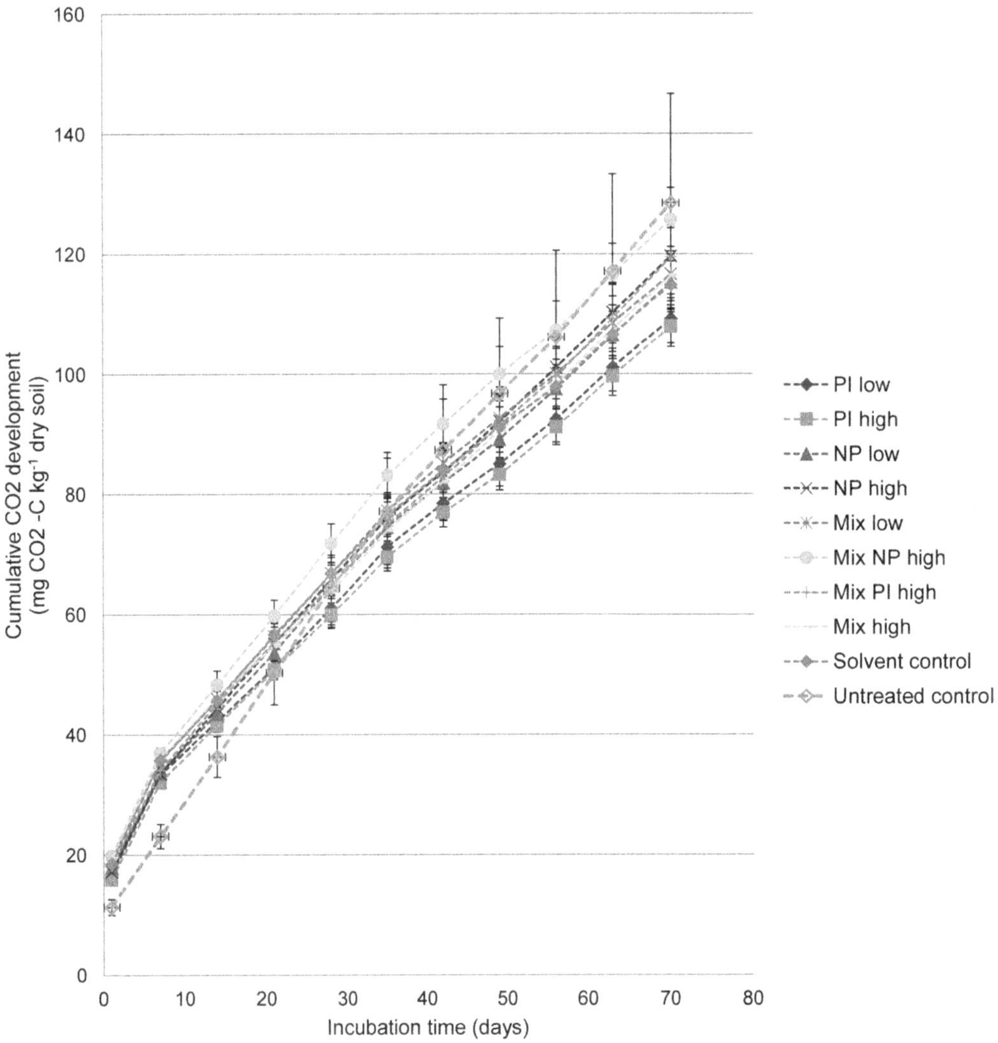

Figure 1. Effects of picoxystrobin and 4-n-nonylphenol on soil microbial respiration activity. Cumulative CO_2 development in a sandy loam soil during a 70 d incubation period after treatment with picoxystrobin and 4-n-nonylphenol as single chemicals and in mixture. Mean values ± SD. Legends: Treatments with picoxystrobin (PI) or 4-n-nonylphenol (NP) or both chemicals (Mix), in low (2 mg kg^{-1} PI, 0.5 mg kg^{-1} NP) or high (10 mg kg^{-1}) concentrations. Cf. Table 2 for full details.

increase in the soil residual concentrations at the end of the incubation period, hence, not allowing a reliable estimate of the long term degradation rate.

The rapid initial degradation of 4-n-nonylphenol was confirmed and shown to be microbially mediated (Fig. 7).

Discussion

Significance and Duration of Observed Effects

Total soil microbial activity. Our findings of a negative effect of picoxystrobin on individual measurements of CO_2 development, but not cumulative values, were in accordance with previous reports. Approval guidelines for plant protection products require documentation of effects on carbon and nitrogen mineralization, and only transient effects on nitrogen and carbon mineralization have been reported for picoxystrobin [8]. This fungicide inhibits mitochondrial respiration by blocking electron transfer at the Qo centre of cytochrome bc1 [7], and, hence, can be expected to affect the soil fungal community. We did not, however, expect an adverse effect in bacteria. The observed lag-phase before on-set of degradation of picoxystrobin (Fig. 6a) could correspond with a negative initial effect on respiration activity and microbial degradation activity. This may, however, not be concluded from the results as bacterial growth on anthropogenic substrates in soil is generally preceded by a lag-phase to allow adaptation to a new substrate. The observed slowing of the degradation rates after 28 days (Table 5) indicates limitations to microbial activity (e.g. oxygen limitation, occurrence of inhibiting factors), reduced bioavailability of the compound (e.g. time-dependent sorption) and/or a negative effect of picoxystrobin on the soil microbial community.

Our results indicated stimulating effects of high levels of 4-n-nonylphenol on soil respiration that could, in part, be explained by an increase in soil microbial activity level. As an endocrine-disrupting agent, we did not expect adverse effects of 4-n-nonylphenol on the soil bacterial community. Our findings are in accordance with others reporting low risk of adverse effects of nonylphenol in environmentally relevant concentrations on soil fungi, but with a stated need for more in-depth studies [34]. This study shows stimulation of specific fungal strains during long-term

Table 4. Organic C mineralization in soil.

Treatment	Organic C mineralization		
	Cumulative CO_2-C	Rate constant	
	(mg CO_2-C/kg dry soil 70d)	(k_{totC}, 10^{-4} day^{-1})	r^2
PI low	109.2±4.1[a]	1.7±0.17[b]	0.82
PI high	108.0±3.5	1.6±.0.17	0.83
NP low	115.3±4.6	1.7±0.18	0.83
NP high	119.6±5.1	1.8±0.17	0.85
Mix low	116.6±4.5	1.8±0.19	0.81
Mix NP high	125.7±5.3	1.9±0.20	0.83
Mix PI high	119.5±6.8	1.8±0.18	0.84
Mix high	116.4±7.9	1.8±0.19	0.81
Solvent control	114.9±3.9	1.8±0.19	0.80
Untreated control	128.5±18.2	1.7±0.19	0.96
Total		1.8±0.05	0.83

[a]SD (standard deviation of measurements),
[b]SE (standard error of estimate).

exposure, as well as appreciable sorption of nonylphenol in soil. Others have reported toxic effects of nonylphenol on soil microbes [4]. However, the general picture is that of rapid and complete mineralization in a wide range of soils [35].

Soil microbial community structure and degradation of chemicals. The observed changes in the T-RFLP patterns of the soil bacterial and fungal communities during the lag-phase before onset of picoxystrobin degradation (Fig. 6a), are in agreement with the general assumption of adaptation of the microbial community to a new substrate. The strobilurin mode of action is, however, typically very fast acting [6], and it may not be ruled out that the observed rapid effects on T-RFLP patterns are, in part, caused by negative effects of the fungicide. We observed a bi-phasic degradation, with initial degradation rate (DT50) in correspondence with previous reports [8]. However, in our studies the levels of picoxystrobin stabilized after 28 days, possibly due to sorption and reduced bioavailability [8,36]. The estimated degradation half-lives (DT50) for picoxystrobin assume that the analysed fraction is the total residual fraction available for degradation. The test for initial rapid degradation (Fig. 7, left) did not indicate a rapid sorption in moist soil, but specific sorption studies were not performed. Picoxystrobin is, however, classified as only slightly mobile [36], and non-bioavailable residues are expected to be above 20% after 100 days [8]. This would explain the lack of a concentration effect beyond 28 days. Alternatively this could be due to a lasting effect on the soil microbial community, as indicated by the lasting significant differences in T-RFLP-fingerprints in picoxystrobin treated soils compared to the solvent control. Similar percentage levels of degradation of picoxystrobin were observed for the two concentrations tested, meaning that actual concentration levels remaining in the treatments with high concentrations of picoxystrobin were much higher after 28 days than for the treatments with low concentrations. There was, however, no observable concentration effect at 70 days between the T-RFLP-profiles.

This observed levelling out of the residual soil concentration of picoxystrobin after 28 days together with the lack of a concentration effect at 70 days, indicated that the observed transience of the effects on the soil microbial community was coupled to decreasing bioavailability of the chemical. Time-dependent sorption might be the reason for reduced degradation and bioavailability of the pesticide after 70 days [37]. Observed major metabolites of picoxystrobin from degradation under aerobic conditions in soil are reported to have mean half-lives in the range of 14–29 days [8]. The chemical analyses performed in our study did not include metabolites, and we may, hence, not rule out any effects of these in our short-term laboratory experiment.

Although 4-n-nonylphenol was found to stimulate soil respiration, the effects on the microbial community structure were less evident. The observed changes in the T-RFLP fingerprints for the soil bacterial community developed more slowly after 4-n-nonylphenol treatment than for picoxystrobin. This indicated a change induced by microbial growth on the chemical, in accordance with the rapid disappearance of 4-n-nonylphenol from the soil through microbial degradation (Fig. 7, right). Residual amounts approached zero both when measured in percentage and in actual amounts. Nonylphenol originates from anthropogenic activity and accumulates and persists in environmental compartments characterized by high organic content, such as sewage sludge [3]. We observed rapid degradation in agricultural soil of comparatively low organic carbon content (quite on the average for large parts of the Norwegian agricultural area). This rapid degradation in an aerobic soil environment is in accordance with previous studies [35,38]. There are, however,

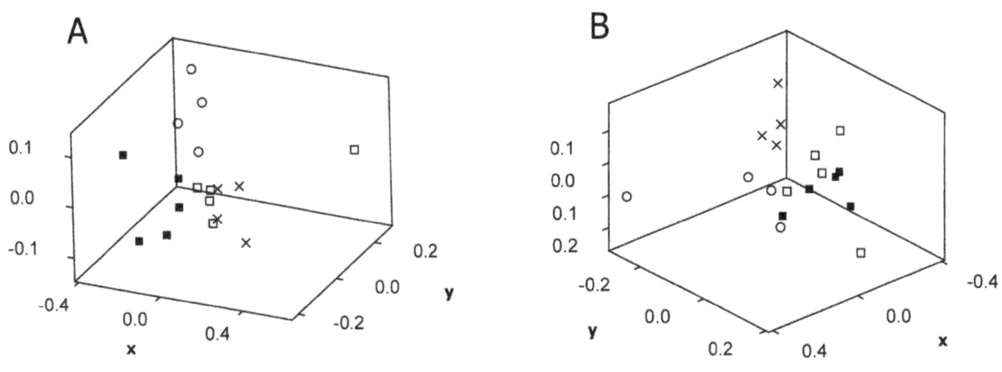

Figure 2. Effects of picoxystrobin on soil fungal community structure. Effects of picoxystrobin (PI) on soil fungal community structure, as shown from nmMDS of data from T-RFLP analyses after 7 (a) and 70 (b) days of incubation. Legends: Treatment PI low (□; 2 mg kg^{-1} dry soil) and PI high (■; 10 mg kg^{-1} dry soil), solvent control (×; no chemical added) and original soil sample (○).

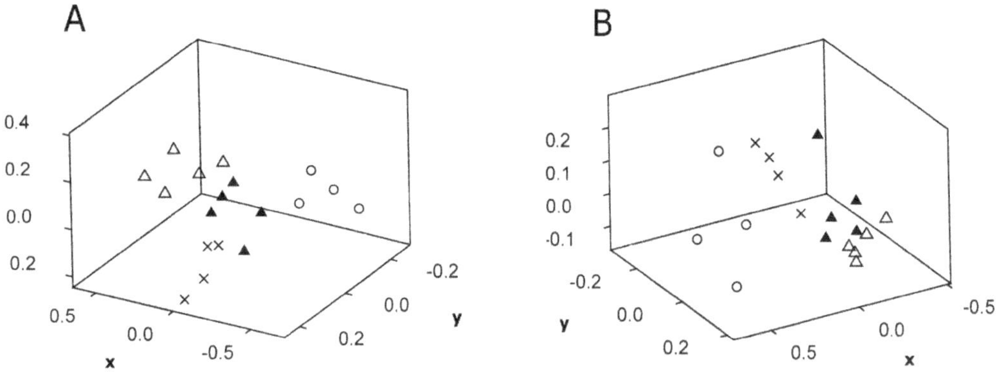

Figure 3. Effects of 4-n-nonylphenol on soil bacterial community structure. Effects of 4-n-nonylphenol (NP) on soil bacterial community structure, as shown from nmMDS of data from T-RFLP analyses after 28 (a) and 70 (b) days of incubation. Legends: Treatment NP low (Δ; 0.5 mg kg^{-1} dry soil) and NP high (▲; 10 mg kg^{-1} dry soil), solvent control (×; no chemical added) and original soil sample (○).

reports of field studies with fast initial degradation, while some residues might persist [4]. The occurrence of metabolites of 4-n-nonylphenol was not measured in this study, but the observed stimulation of respiration activity by 4-n-nonylphenol addition as compared to the solvent control, was not large enough to account for the observed disappearance of the chemical through the incubation period. This indicated that the chemical was not completely mineralized. Results from a mineralization study with a range of soils show the conversion of around 40% of applied nonylphenol to CO_2 during a 40 day incubation period [39].

Our results indicated shorter duration of the effects of the chemicals in the soil bacterial as compared to the soil fungal community. This was expected due to shorter generation times enabling a faster re-establishment of the community structure. Overall, our results were in accordance with this, showing smaller differences in T-RFLP fingerprints between start and end of the experiment for the former (Fig. 5). The more statistically significant effects of the treatments were however shown for bacteria, while the changes in the soil fungal community were less consistent and varied a lot between replicate samples.

Implications of a Mixture Situation on the Effects of Picoxystrobin on the Soil Microbial Community

Analyses of the effects of mixture treatments on the soil microbial community structure indicated that the two chemicals might have antagonistic effects. Established concepts to estimate biological effects of chemical mixtures rely on data available for single chemicals [40,41]. Single chemicals vary in how they excerpt their effect and the effects of a mixture cannot be found directly from the different constituents' independent effects. One can observe a wide variety of effects due to synergism, antagonism or other forms of interactions, and the common concepts of concentration addition and independent action often come short. Despite the expected and indicated differences in modes of action of the two chemicals, we observed less evident changes in T-RFLP patterns in soils treated with the chemicals in mixture, as compared to the single chemical treatments. Our results showed that statistically significant different T-RFLP patterns, for both the soil bacterial and fungal communities, resulted from treatments with picoxystrobin, 4-n-nonylphenol, and a mixture of these. However, the changes in T-RFLP patterns evolved more slowly and were less pronounced for the mixture than for the single chemical treatments, as compared to the solvent control. Further, the negative effects of low levels of picoxystrobin on respiration were apparently remediated when mixing with high concentration

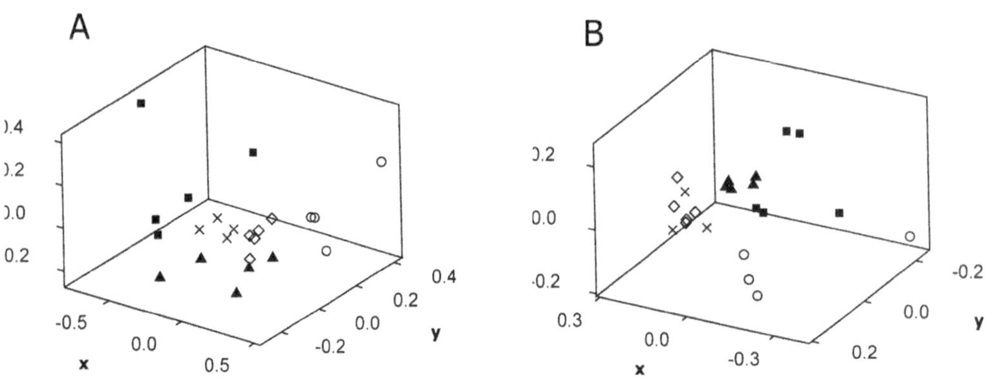

Figure 4. Effects of picoxystrobin and 4-n-nonylphenol on soil microbial community structure. Effects of picoxystrobin (PI) and 4-n-nonylphenol (NP) as single chemicals and in mixture on soil bacterial (a) and fungal (b) community structure, as shown from nmMDS of data from T-RFLP analyses after 28 days of incubation. Legends: Treatment PI high (■; 10 mg kg^{-1} dry soil), NP high (▲; 10 mg kg^{-1} dry soil), Mix high (◇; 10 mg picoxystrobin and 10 mg 4-n-nonylphenol kg^{-1} dry soil), solvent control (×; no chemical added) and original soil sample (○).

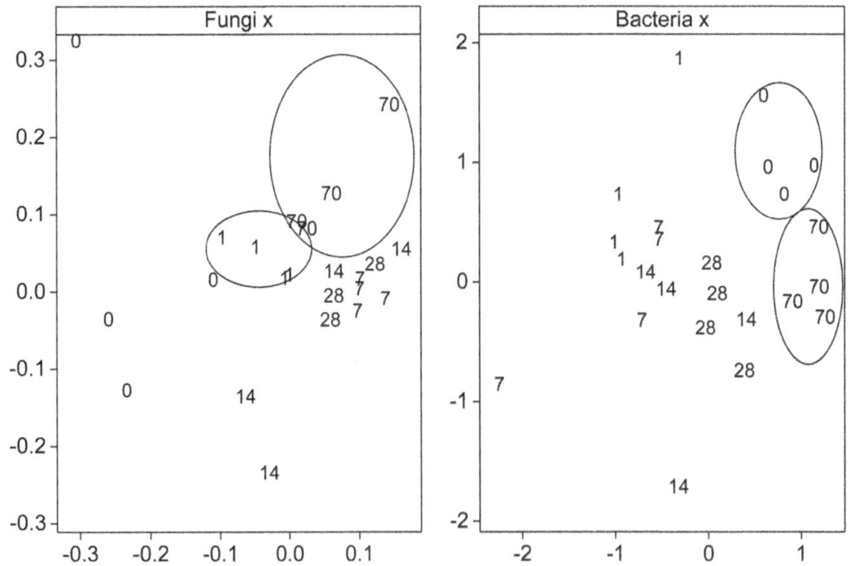

Figure 5. Duration of effects of chemicals on soil microbial community structure. Representative pattern (exemplified by the solvent control treatment) for observed clustering of T-RFLP fingerprints of soil microbial communities at day 0, 1, 7, 14, 28, and 70 after addition of chemicals, shown for fungi (left panel) and bacteria (right panel). Small distance between start (0) and end (70) of the incubation period indicates only minor differences in T-RFLP fingerprints, and imply transient effects of the chemicals.

of 4-n-nonylphenol (Mix NP high), while high levels of picoxystrobin nulled the positive effects of high levels of 4-n-nonylphenol (Mix high). In a practical mixture situation in the field, the expected environmental concentrations of nonylphenol can from this not be expected to increase the potentially adverse effects of picoxystrobin.

To assure the validity in extrapolation of the observed effects to a risk assessment situation in the field, the results need to be verified in field/semi-field/mesocosm trials. These lab studies do not take into account the additional stress the soil microbial community experience in nature from climatic conditions (drought, frost), predators, competition, or food shortage, nor the effects of alteration in the soil microbial community structure on other parts of the soil fauna. The studied mixture is relevant in an agricultural context and the results should be possible to extrapolate to a certain degree to other nonylphenols and strobilurin fungicides. However, care should be taken to consider the variety of physico-chemical and toxicological properties of the different strobilurins [6] as well as the differences in estrogenicity, sorption properties and degradation rates of different isomers of linear and branched nonylphenols [42]. Further, due to the general importance of soil properties for the sorption of chemicals, extrapolation of results from this sandy loam soil with low organic carbon content should be done with care.

Table 5. Estimated degradation half-lives (DT50) for picoxystrobin and 4-n-nonylphenol in a sandy loam soil.

Start concentrations	DT50 initial	r^2	DT50 2nd phase	r^2
2 mg picoxystrobin kg^{-1} dry soil[a]	20.0±2.5	0.87	300±170	0.74
10 mg picoxystrobin kg^{-1} dry soil[a]	26.0±4.0	0.74		
0.5 mg 4-n-nonylphenol kg^{-1} dry soil[b]	3.0±1.0	0.85	42±2	0.58
10 mg 4-n-nonylphenol kg^{-1} dry soil[a]	6.5±0.5	0.92	52±21	0.89

[a]Initial phase 0–28 days, 2nd phase 28–70 days,
[b]Initial phase 0–14 days, 2nd phase 14–70 days.
Calculations are based on results from both single chemical and mixture experiments. (The test statistic of $p<0.05$ is valid for all estimated DT50-values.).

Methodological Considerations

Our results showed that both soil microbial respiration activity and community structure was affected by the chemicals. They also illustrate the need for the use of several measures to be able to assess effects of chemicals in soil with a minimum degree of certainty. We found statistically significant negative effects of picoxystrobin on activity levels as shown by CO_2 development, but no statistically significant differences on cumulative CO_2-values (Table 4) due to too large variability between replicate samples. This was despite observable trends in the data (Fig. 1). Further, the T-RFLP results indicated a need for frequent sampling shortly after addition of chemicals, and possibly the resolution of the sampling was too low to capture the effects on the soil fungal community structure as the largest spread in these data as shown by nmMDS was found after 7 days (Fig. 2). The community fingerprints arising from T-RFLP analyses depend on the PCR-primers and restriction enzymes utilized, and cannot easily be interpreted in terms of ecological relevance of the observed effects. This was ameliorated through correspondence with effects on soil respiration activity. Our studies showed that the T-RFLP technique could be used as a valuable tool in elucidating how rapid and to what extent effects (positive or negative) of the chemicals in the soil microbial community can be expected.

Our results for the fungicide picoxystrobin show that the mere measurement of changes in soil chemical concentrations for this moderately sorbing chemical could not be used as an indication of when to expect effects, while exposure assessment in ecological risk assessment is often restricted to external exposure like concentrations in water, soil or sediment. However it has been recognized that in natural ecosystems, in particular in soils and sediments, the amount of chemicals truly available for uptake into organisms is frequently only a fraction of the total amount present, due to a

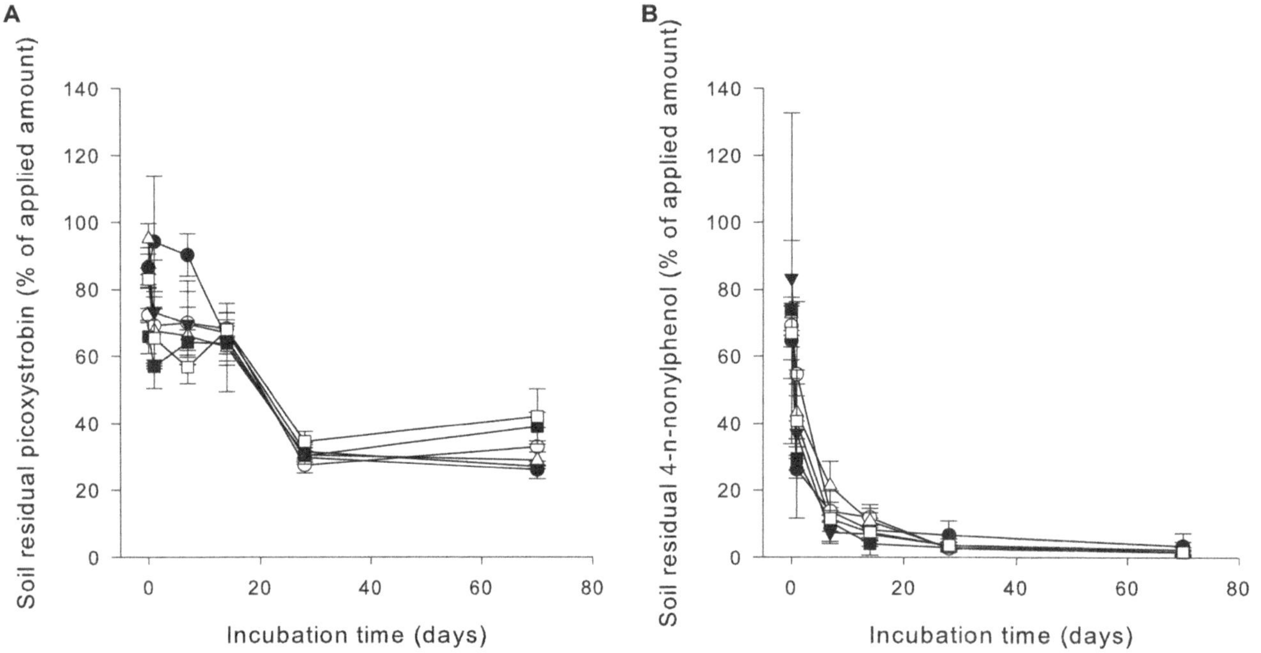

Figure 6. Degradation of picoxystrobin and 4-n-nonylphenol in soil. Degradation curve for picoxystrobin (a) and 4-n-nonylphenol (b) in soil, showing similar pattern and similar percentage degradation during the incubation period for all treatments with picoxystrobin and 4-n-nonylpyhenol, respectively, regardless of initial concentration or mixture situation. Mean values ± SD. Legends: Treatment PI low (a) and NP low (b) (●; PI = 2 mg and NP = 0.5 mg kg^{-1} dry soil), PI high (a) and NP high (b) (○; PI and NP = 10 mg kg^{-1} dry soil), Mix low (▼; PI = 2 mg and NP = 0.5 mg kg^{-1} dry soil), Mix NP high (△;PI = 2 mg and NP = 10 mg kg^{-1} dry soil), Mix PI high (■;PI = 10 mg and NP = 0.5 mg kg^{-1} dry soil), and Mix high (□; PI and NP = 10 mg kg^{-1} dry soil).

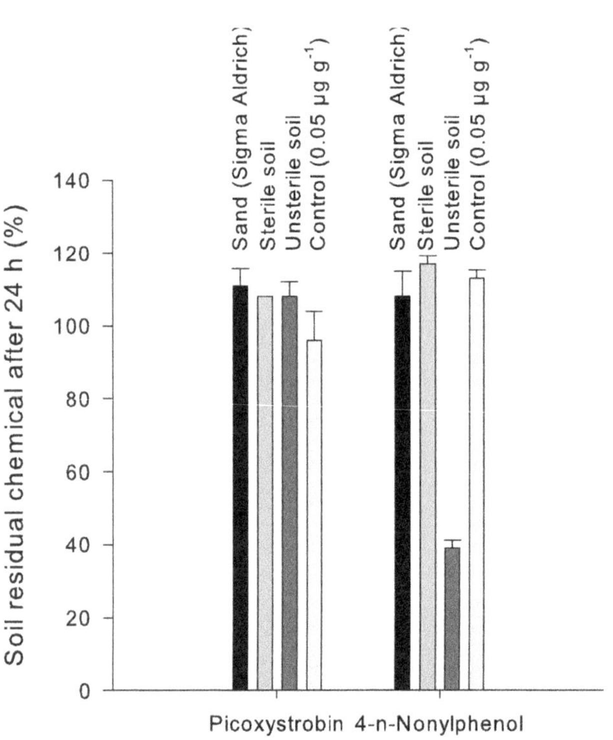

Figure 7. Microbial degradation of picoxystrobin and 4-n-nonylphenol in soil. Laboratory test results for degradation of picoxystrobin and 4-n-nonylphenol in selected media during 24 h, showing rapid degradation of 4-n-nonylphenol in unsterile soil.

complex environmental behaviour [43]. Measurements of residual chemical concentrations in our test system was, however, shown to be a valuable tool for explaining the observed changes in soil microbial community structure and respiration activity and elucidating the concept of a chemicals bioavailability, and should be included in effects assessments.

Conclusions

In summary, we have studied the degradation and effects of picoxystrobin and 4-n-nonylphenol on the soil microbial community structure and respiration activity in a sandy loam soil. The fungicide picoxystrobin was shown to decrease the soil microbial activity, while 4-n-nonylphenol caused an increase. These effects were accompanied by statistically significant changes in the T-RFLP fingerprints for the soil microbial community that were still detectable, but small, after 70 days. A mixture situation relevant for assessing the environmental risk of fungicide application in years with sewage sludge amendment of agricultural soil was tested. These results implied that a mixture of picoxystrobin and 4-n-nonylphenol will not have more adverse effects on soil respiration or the soil microbial community structure than the single chemicals. They indicated that the chemicals affected different parts of the soil microbial community, and resulted in a low net effect of the chemicals in mixture. The presented results do, however, not allow us to propose an explanation for the mechanisms causing this response.

In conclusion, our results imply that the application of picoxystrobin and nonylphenol to agricultural soil in environmentally relevant and worst-case concentrations will not cause irreversible effects on soil microbial respiration and community structure. Further, a mixture situation with the fungicide picoxystrobin and nonylphenol from sewage sludge application, will not increase any potentially adverse effects of picoxystrobin on the soil microbial community. In a wider perspective, these results illustrate that there is a need and possibility to refine today's risk assessment procedures to encompass the study of both soil bacterial and fungal communities, both broad-scale measures and genomics, and interpret the results in relation to residual soil chemical concentrations.

Supporting Information

Figure S1 Effects of picoxystrobin on soil bacterial community structure. Effects of picoxystrobin (PI) on soil bacterial community structure, as shown from nmMDS of data from T-RFLP analyses. Legends: Treatment PI low (□; 2 mg kg^{-1} dry soil) and PI high (■; 10 mg kg^{-1} dry soil), solvent control (X; no chemical added) and original sample (○).

Figure S2 Effects of picoxystrobin on soil fungal community structure. Effects of picoxystrobin (PI) on soil fungal community structure, as shown from nmMDS of data from T-RFLP analyses. Legends: Treatment PI low (□; 2 mg kg^{-1} dry soil) and PI high (■; 10 mg kg^{-1} dry soil), solvent control (X; no chemical added) and original sample (○).

Figure S3 Effects of 4-n-nonylphenol on soil bacterial community structure. Effects of 4-n-nonylphenol (NP) on soil bacterial community structure, as shown from nmMDS of data from T-RFLP analyses. Legends: Treatment NP low (Δ; 0.5 mg kg^{-1} dry soil) and NP high (▲; 10 mg kg^{-1} dry soil), solvent control (X; no chemical added) and original sample (○).

Figure S4 Effects of 4-n-nonylphenol on soil fungal community structure. Effects of 4-n-nonylphenol (NP) on soil fungal community structure, as shown from nmMDS of data from T-RFLP analyses. Legends: Treatment NP low (Δ; 0.5 mg kg^{-1} dry soil) and NP high (▲; 10 mg kg^{-1} dry soil), solvent control (X; no chemical added) and original sample (○).

Figure S5 Effects of picoxystrobin and 4-n-nonylpyhenol on soil bacterial community structure. Effects of picoxystrobin (PI) and 4-n-nonylphenol (NP) on soil bacterial community structure, as shown from nmMDS of data from T-RFLP analyses. Legends: Treatment PI high (■; 10 mg kg^{-1} dry soil), NP high (▲; 10 mg kg^{-1} dry soil), Mix high (◇; 10 mg picoxystrobin and 4-n-nonylphenol kg^{-1} dry soil), solvent control (X; no chemical added) and original sample (○).

Figure S6 Effects of picoxystrobin and 4-n-nonylpyhenol on soil fungal community structure. Effects of picoxystrobin (PI) and 4-n-nonylphenol (NP) on soil fungal community structure, as shown from nmMDS of data from T-RFLP analyses. Legends: Treatment PI high (■; 10 mg kg^{-1} dry soil), NP high (▲; 10 mg kg^{-1} dry soil), Mix high (◇; 10 mg picoxystrobin and 4-n-nonylphenol kg^{-1} dry soil), solvent control (X; no chemical added) and original sample (○).

Table S1 Summary of results from nmMDS (STRESS; k = 3 dimensions, Bray-Curtis distance measure) and ANOSIM (overall R) of T-RFLP fingerprints for the soil microbial community after different chemical treatments.

Acknowledgments

Monica Skogen, Hege Særvold Steen, Grete Lund and Elameen Abdelhameed, Norwegian Institute for Agricultural and Environmental Research, are acknowledged for technical assistance in the lab. Many thanks to Tim Daniell, James Hutton Institute, for advice on the experimental set-up, and analyses of T-RFLP-data. We thank the anonymous reviewers for a thorough review and helpful comments.

Author Contributions

Conceived and designed the experiments: MS SSK OME. Performed the experiments: MS HRN. Analyzed the data: MS HRN. Contributed reagents/materials/analysis tools: MS SSK HRN OME. Wrote the paper: MS SSK HRN OME.

References

1. JRC (2001). Organic contaminants in sewage sludge for agricultural use. Study coordinated by European Commission and Joint Research Centre and the Institute for Environment and Sustainable Soil and Waste Unit. Report prepared by UMEG Center for Environmental Measurements, Environmental Inventories and Product Safety, 18 October 2001. 73 p. Available: http://ec.europa.eu/environment/waste/sludge/pdf/organics_in_sludge.pdf. Accessed 21 May 2013.
2. Norwegian Scientific Committee for Food Safety (2009). Risk assessment of contaminants in sewage sludge applied on Norwegian soils – Opinion from the Panel on Contaminants in the Norwegian Scientific Committee for Food Safety. Oslo: VKM. 208 p. Available: http://www.vkm.no/dav/2ae7f1b4e3.pdf. Accessed 21 May 2013.
3. Hansen AB, Lassen P (2008). Screening of phenolic substances in the Nordic environments. TemaNord 2008:530. Copenhagen: Nordic Council of Ministers. 145 p.
4. Soares A, Guieysse B, Jefferson B, Cartmell E, Lester JN (2008). Nonylphenol in the environment: A critical review on occurrence, fate, toxicity and treatment in wastewaters. Environ Int 34, 1033–1049.
5. EC (2000). Directive 2000/60/EC of the European Parliament and of the Council of 23 October 2000 establishing a framework for Community action in the field of water policy.
6. Bartlett DW, Clough JM, Godwin JR, Hall AA, Hamer M, Parr-Dobrzanski B (2002). The strobilurin fungicides. Pest Manag Sci 58: 649–62.
7. BCPC (2011). ePesticide Manual V5.2 2011–2012.
8. EC (2003). Review report for the active substance picoxystrobin. SANCO/10196/2003-Final. Available: http://ec.europa.eu/food/plant/protection/evaluation/newactive/list1_picoxystrobin_en.pdf. Accessed 21 May 2013.
9. Norwegian Food Safety Autority (2005). Vurdering av plantevernmidlet Acanto Prima – pikoksystrobin+cyprodinil – vedrørende søknad om godkjenning. (In Norwegian). Available: http://www.mattilsynet.no/mattilsynet/multimedia/archive/00014/Plantevernmidler__Ra_14615a.pdf. Accessed 21 May 2013.
10. Coleman DC (2008). From peds to paradoxes: Linkages between soil biota and their influences on ecological processes. Soil Biol Biochem 40: 271–289.
11. Wu L, Wang H, Zhang Z, Lin R, Zhang Z, et al. (2011). Comparative metaproteomic analysis on consequtively *Rehmannia glutinosa*-monocultured rhizosphere soil. PloS ONE 6(5): e20611. Doi:10.1371/journal.pone.0020611.
12. Winding A, Hund-Rinke K, Rutgers M (2005). The use of microorganisms in ecological soil classification and assessment concepts. Ecotox Environ Safe 62: 230–248.
13. EFSA (2010). EFSA Panel on Plant Protection Products and their Residues (PPR); Scientific Opinion on the development of specific protection goal options for environmental risk assessment of pesticides, in particular in relation to the revision of the Guidance Documents on Aquatic and Terrestrial Ecotoxicology (SANCO/3268/2001 and SANCO/10329/2002). EFSA Journal 8: 1821. 55 p. Available: http://www.efsa.europa.eu/en/efsajournal/doc/1821.pdf. Accessed 21 May 2013.
14. Imfeld G, Vuilleumier S (2012). Measuring the effects of pesticides on bacterial communities in soil: A critical review. Eur J Soil Biol 49: 22–30.
15. Øvreås L, Torsvik V (1998). Microbial diversity and community structure in two different agricultural soil communities. Microbial Ecol 36: 303–315.
16. Bending GD, Rodrígues-Cruz MS, Lincoln SD (2007). Fungicide impacts on microbial communities in soils with contrasting management histories. Chemosphere 69: 82–88.
17. Griffiths RI, Thomson BC, James P, Bell T, Bailey M, et al. (2011). The bacterial biogeography of British soils. Environ Microbiol 13: 1642–1654.
18. OECD (2000). Test No. 217: Soil Microorganisms: Carbon Transformation Test, OECD Guidelines for the Testing of Chemicals, Section 2, OECD Publishing. doi: 10.1787/9789264070240-en.

19. OECD (2000). Test No. 216: Soil Microorganisms: Nitrogen Transformation Test, OECD Guidelines for the Testing of Chemicals, Section 2, OECD Publishing. doi: 10.1787/9789264070226-en.
20. MacDonald CA, Clark IM, Zhao F-J, Hirsch PR, Singh BK, et al. (2011). Long term impacts of zinc and copper enriched sewage sludge additions on bacterial, archeal and fungal communities in arable and grassland soils. Soil Biol Biochem 43: 932–941.
21. Schütte UME, Abdo Z, Bent SJ, Shyu C, Williams CJ, et al. (2008). Advances in the use of terminal restriction fragment length polymorphism (T-RFLP) analysis of 16S rRNA genes to characterize microbial communities. Appl Microbiol Biot 80: 365–380.
22. Aiken JT (2011). Terminal restriction fragment length polymorphism for soil microbial community profiling. Soil Sci Soc Am J 75: 102–111.
23. Chaussod R, Nicolardot B, Catroux G (1986). Mesure en routine de la biomasse microbienne des sols par la méthode de fumigation au chloroforme. Science du Sol - Bulletin AFES 2: 201–211.
24. Singh BK, Nunan N, Ridgeway KP, McNicol J, Peter W, et al. (2008). Relationship between assemblages of mycorrhizal fungi and bacteria on grass roots. Environ Microbiol 10: 534–541.
25. Anderson IC, Cairney JWG (2004). Diversity and ecology of soil fungal communities: increased understanding through the application of molecular techniques. Environ Microbiol 6: 769–779.
26. Gardes M, Bruns TD (1993). ITS primers with enhanced specificity for basidiomycetes: application to the identification of mycorrhizae and rusts. Mol Ecol 2: 113–118.
27. White TJ, Bruns T, Lee S, Taylor J (1990). Amplification and direct sequencing of fungal ribosomal RNA genes for phylogenetics. In: Innis MA, Gelfand DH, Sninsky JJ, White TJ, editors. PCR Protocols: a guide to methods and applications. New York: Academic Press. 315–322.
28. R Core Team (2012). R: A language and environment for statistical computing. R Foundation for Statistical Computing, Vienna, Austria. ISBN 3-900051-07-0, URL http://www.R-project.org/. Accessed 28 February 2013.
29. Oksanen J, Blanchet FG, Kindt R, Legendre P, Minchin PR, et al. (2013). vegan: Community Ecology Package. R package version 2.0-6. Available: http://CRAN.R-project.org/package=vegan. Accessed 28 February 2013.
30. Venables WN, Ripley BD (2002). Modern Applied Statistics with S. Fourth Edition. New York: Springer. ISBN 0-387-95457-0.
31. Kruskal JB (1964a), Multidimensional Scaling by Optimizing Goodness of Fit to a Nonmetric Hypothesis. Psychometrika 29: 1–27.
32. Kruskal JB (1964b). Nonmetric multidimensional scaling: a numerical method. Psychometrika 29: 115–129.
33. Hammer Ø, Harper DAT, Ryan PD (2001). PAST: Paleontological statistics software package for education and data analysis. Palaeontol Electron 4(1): art. 4. 9 p. Available: http://palaeo-electronica.org/2001_1/past/issue1_01.htm. Accessed 28 Februart 2013.
34. Kollmann A, Brault A, Touton I, Dubroca J, Chaplain V, et al. (2003). Effect of nonylphenol surfactants on fungi following the application of sewage sludge on agricultural soils. J Environ Qual 32: 1269–1276.
35. Corvini PFX, Chäffer A, Schlosser D (2006). Microbial degradation of nonylphenol and other alkylphenols – our evolving view. Appl Microbiol Biot 72: 223–243.
36. PPDB (2009). The Pesticide Properties Database (PPDB) developed by the Agriculture & Environment Research Unit (AERU), University of Hertfordshire, funded by UK national sources and the EU-funded FOOTPRINT project (FP6-SSP-022704). http://sitem.herts.ac.uk/aeru/footprint/en/index.htm. Accessed 21 May 2013.
37. Defra (2010). Development of guidance on the implementation of aged soil sorption studies into regulatory exposure assessments. Research report for DEFRA project PS 2235. The Food and Environment Research Agency and Alterra.
38. Hesselsøe M, Jensen D, Skals K, Olesen T, Moldrup P, et al. (2001). Degradation of 4-Nonylphenol in homogeneous and nonhomogeneous mixtures of soil and sewage sludge. Environ Sci Technol 35: 3695–3700.
39. Topp E, Starratt A (2000). Rapid mineralization of the endocrine-disrupting chemical 4-nonylphenol in soil. Environ Toxicol Chem 19: 313–318.
40. Loewe S (1927). Die Mischarznei. Versuch einer allgemeinen Pharmakologie der Arzneikombinationen. Klin Wochenschr 6: 1077–1085.
41. Bliss CI (1939). The toxicity of poisons applied jointly. Ann Appl Biol 26: 585–615.
42. Shan J, Jiang B, Yu B, Li C, Sun Y, et al. (2011). Isomer-Specific Degradation of Branched and Linear 4-Nonylphenol Isomers in an Oxic Soil. Environ Sci Technol 45: 8283–8289.
43. Alexander M (1995). How toxic are toxic chemicals in soil? Environ Sci Technol 29: 2713–2717.

9

Cross-Species Extrapolation of Prediction Model for Lead Transfer from Soil to Corn Grain under Stress of Exogenous Lead

Zhaojun Li[1]*, Hua Yang[1,2], Yupeng Li[1,3], Jian Long[2], Yongchao Liang[1]

1 Institute of Agricultural Resources and Regional Planning, Chinese Academy of Agricultural Sciences, Key Laboratory of Plant Nutrition and Fertilizer, Ministry of Agriculture, Beijing, China, **2** Guizhou Key Laboratory of Mountain Environment, Guizhou Normal University, Guiyang, China, **3** College of City and Environmental Sciences, Shanxi Normal University, Linfen, China

Abstract

There has been increasing concern in recent years regarding lead (Pb) transfer in the soil-plant system. In this study the transfer of Pb (exogenous salts) was investigated from a wide range of Chinese soils to corn grain (Zhengdan 958). Prediction models were developed with combination of the Pb bioconcentration factor (BCF) of Zhengdan 958, and soil pH, organic matter (OM) content, and cation exchange capacity (CEC) through multiple stepwise regressions. Moreover, these prediction models from Zhengdan 958 were applied to other non-model corn species through cross-species extrapolation approach. The results showed that the soil pH and OM were the major factors that controlled Pb transfer from soil to corn grain. The lower pH and OM could improve the bioaccumulation of Pb in corn grain. No significant differences were found between two prediction models derived from the different exogenous Pb contents. When the prediction models were applied to other non-model corn species, the ratio ranges between the predicted BCF values and the measured BCF values were within an interval of 2-fold and close to the solid line of 1:1 relationship. Moreover, the prediction model i.e. Log[BCF] = −0.098 pH-0.150 log[OM] −1.894 at the treatment of high Pb can effectively reduce the measured BCF intra-species variability for all non-model corn species. These suggested that this prediction model derived from the high Pb content was more adaptable to be applied to other non-model corn species to predict the Pb bioconcentration in corn grain and assess the ecological risk of Pb in different agricultural soils.

Editor: Xianglin Shi, University of Kentucky, United States of America

Funding: This work was jointly supported by the Special Fund of Public Industry in China (Agriculture, project no. 200903015, http://www.moa.gov.cn/) and the Natural Science Foundation of Institute of Agricultural Resources and Regional Planning, Chinese Academy of Agricultural Sciences (2013-9, http://www.iarrp.cn/SoisWeb/home/index.html). The funders had no role in study design, data collection and analysis, decision to publish, or preparation of the manuscript.

Competing Interests: The authors have declared that no competing interests exist.

* E-mail: lizhaojun@caas.cn

Introduction

Lead (Pb) is one of the most toxic heavy metals for people in the world [1–3]. In children, Pb has been known to cause decreases in intelligence quotient (IQ) scores, retardation of physical growth, and hearing problems. In individuals of all ages, it has been known to cause anemia brain damage, impaired function of the peripheral nervous system, high blood pressure, reproductive abnormalities, development defects, abnormal vitamin D metabolism, and in some situations death of people [4–5]. Soil-to-plant transfer of heavy metals is the major exposure pathway for humans with regards to soil contamination [6]. Therefore, the accumulation of heavy metals in plants in agricultural soils, especially in edible part of plants, is great concern regarding its posing potential threats [7–8]. The efficiency of the plants for absorbing metals is mainly evaluated by plant uptake or the metal transfer factors from soil to plant [9]. Evaluating the metal transfer factors from soil to plant especially to edible parts can further provide the information about accumulation of contaminants into the food chain [10].

The Pb transfer factor from soil to plant mainly depends on its availability in soils. Soil properties can determine the availability of Pb to crops, including soil pH, organic matter (OM), cation exchange capacity (CEC), redox potential, clay minerals, Fe and Mn oxides, and calcium carbonate content [11–13]. In addition, it has been reported that soil pH is a key factor in controlling the cation mobility and regulating the solubility of heavy metals in soils [14–15]. The heavy metals tended to be more available in acid soil [16]. The low soil pH is favorable for the transference of Pb from soil to plant [17–19]. As well as, soil OM has also been found to play an important role in influence Pb adsorption processes in soils [20–21]. Pb in soils with low OM content was difficult to be adsorbed but was easy to be uptaken by plants [22]. The importance for assessing the ecological risk of heavy metals in complex soil types has urged a lot of prediction models to be developed to predict heavy metals transfer from soil to plant [23]. The heavy metal transfer factor from soil to plant also depends on plant species. It has been observed that significant differences exist in heavy metal uptake among cultivars [24]. Due to the expenses and efforts involved, these models were limited to specific contaminated sites or single species [25–27]. In addition, most of these models were developed to describe the transfer of heavy metals from soil to plant leaves, vegetables, earthworms by using the log-transformed concentrations [28–29]. A few attempts were only made to describe heavy metals uptake by rice, wheat and barley grain directly from soil with combination of total concentrations of heavy metal and soil basic properties [30–31]. Little is reported about the prediction models developed to describe the transfer of heavy metals, especially Pb, from soils to corn grain. Intake of heavy metals from dietary sources may represent a significant exposure pathway for human populations. Prediction of heavy metals including Pb concentrations in edible part of crops based on their concentration in soil and other environmental factors are urgently required for human risk assessment and derivation of soil environmental benchmark croplands.

The species sensitivity distributions (SSD) have been proved to be successful and were increasingly used in ecological risk assessment procedures [32–33]. However, the toxicology data from the soils with different soil properties need to be normalized using prediction models to eliminate the influence of soil factors [34]. It can improve the accuracy of the sensitivity distribution of species and environmental quality standard values [35]. These models can also provide great opportunities to carrying out ecological risk assessments and establishing soil quality criteria for heavy metals in croplands. Due to the fact that significant differences existed among cultivars for heavy metal uptake, it is ideal that separate species-specific models should be developed for each single species. However, this is unrealistic due to the expenses and efforts involved. Thus the cross-species extrapolation of toxicity data models was often applied in some cases of pollutants risk assessment practice. For example, Deleebeeck et al. [36–37] have applied biotic bigand model (BLM) from *O. mykiss* to other fish species and applied BLM from D. *magna* to other cladocera species. Schlekat et al. [38] applied chronic nickel BLMs developed for the cladocera such as Daphnia magna and Ceriodaphnia dubia to predict chronic toxicity of nickel to three other invertebrates including snail *(Lymnaea stagnalis)*, insect (*Chironomus tentans*), and rotifer (*Brachionus calyciflorus*). Yang et al. [16] applied corn grain Cd BCF predication models developed for Zhengdan 958 to predict other non-model corn species and wheat grain Cd BCF. However, little effort has been put on establishing models to describe the relationship between Pb accumulation in corn grain and soil properties. The objectives of this study are: 1) to establish the prediction model for Pb accumulation in corn grain on seventeen soils with greatly different properties; 2) to assess the feasibility of applying these models to other non-model corn species, and 3) to investigate their accuracy in predicting accumulation of Pb in non-model corn species.

Materials and Methods

None of these 17 soil samples were collected from national parks or other protected areas. It is confirmed that no specific permissions were required for the soils sampling activities in the 17 locations in China. It is also can be confirmed that the field studies did not involve endangered or protected species. No tested corn species are under first- or second-class state protection, and they are not listed in the Inventory of Rare and Endangered Plants of China (http://zrbhq.forestry.gov.cn/portal/zrbh/s/3053/

Table 2. Soil Environmental Quality Standards of China (GB15618-1995) and the added contents of exogenous Pb (mg.kg^{-1}).

pH	<6.5	6.5~7.5	>7.5
Grade Two Standard	250	300	350
Low Pb	125	150	175
High Pb	250	300	350

Table 1. Selected soil properties of the 17 soils used in the lead bioconcentration factor test.

Soil NO.	Location*	pH	OM (g.kg^{-1})	CEC (cmol.kg^{-1})	TN (g.kg^{-1})	TP (mg.kg^{-1})	TK (g.kg^{-1})	Background Pb (mg.kg^{-1})
S1	Hunan	4.90	15.52	10.85	1.14	0.47	15.26	23.12
S2	Chongqing	5.74	17.48	21.34	1.00	0.55	22.61	29.72
S3	Liaoning	5.74	25.84	12.19	1.00	0.73	23.94	27.94
S4	Yunnan	5.92	34.26	11.10	2.01	0.81	4.77	24.58
S5	Jiangxi	6.01	11.69	8.70	0.51	0.52	9.96	33.48
S6	Anhui	6.25	20.04	19.08	0.99	0.35	15.41	29.34
S7	Heilongjiang	6.27	35.69	28.59	1.74	0.48	24.70	33.99
S8	Jilin	6.82	32.85	31.11	1.75	0.35	24.58	31.01
S9	Jiangsu	6.93	47.69	26.20	2.44	0.69	21.03	36.81
S10	Shaanxi	7.90	16.49	22.37	1.36	0.98	24.37	37.30
S11	Hebei	7.98	8.57	8.12	0.68	0.53	24.22	31.96
S12	Henan	8.07	17.79	16.01	1.07	0.75	19.86	35.89
S13	Xinjiang	8.12	19.43	25.25	1.32	0.78	25.49	31.73
S14	Shanxi	8.24	23.17	16.80	1.13	0.95	23.70	37.97
S15	Tianjin	8.29	22.02	24.67	1.42	0.92	24.63	34.18
S16	Gansu	8.37	19.27	11.23	1.05	0.74	23.62	35.22
S17	Shandong	8.65	11.84	13.09	0.93	0.97	21.37	34.55

*Soil sample locations are listed in the order of increasing pH.

content-457748.html), or the Key Protected Inventory of Wild Plants of China (http://zrbhq.forestry.gov.cn/uploadfile/zrbh/2010-10/file/2010-10-14-bb296addeaa047798d6b6c476aaa1da9.doc). These corn species were used for only scientific research as permitted by the Ministry of Agriculture of China.

Soil Samples

A set of 17 soils covering a wide range of soil properties was sampled from typical locations in China. In each sample location, the soils were collected from the top 20 cm of the soil profile. For analysis of their phys-chemical characteristics, the soils were air dried and sorted to pass a 2-mm sieve. Soil pH was measured in deionized water (soil: solution ratio, 1:5) [39]. CEC was determined by the unbuffered silver-thiourea method [40]. OM was measured by dry combustion [41]. The total phosphorus in soils (TP) was measured by colorimetric method [42]. The total nitrogen in soils was determined by the Kjeldahl's method [43]. The background Pb content in the soils was determined by aqua regia (1:3 fresh mixture of concentrated HNO_3 and HCl) digestion [44]. The selected properties of the 17 soil samples were shown in Table 1.

Experimental Design

Bioaccumulation factors for Pb by corn grain from seventeen soils. According to Grade Two Standard for Pb in the Soil Environmental Quality Standards of China (GB15618-1995), the tested levels of Pb added in soils were shown in Table 2. To obtain the tested levels, a certain amount of exogenous Pb i.e. $Pb(NO_3)_2$ was added to the 8 kg air-dried soils. All the soils were thoroughly mixed and placed into pots (ϕ×h, 23 cm×26.5 cm). Then the soils were moistened with deionized water to the 60% of field moisture capacity. Each tested level of Pb was replicated three times. Then the soils containing Pb at different levels were covered with plastic film with six small holes for air ventilation. All the pots were kept for 3 months for Pb aging at a greenhouse at the temperature of 25±3°C during the daytime and 20±3°C at night with a natural light photoperiod. During the period of aging, water contents of each soil sample was maintained by adding deionized water every 3 days. After 3 months' aging, corn seeds of Zhengdan 958 were sown in each soil containing Pb. During the experimental period, soil moisture content was maintained at approximately 60% of water holding capacity through weighing method and adding deionized water to the pots. The corn with grain were sampled and harvested till the corn maturity.

Bioaccumulation factors for Pb of eight corn species from two soils. Two soils with typical properties sampled from Jiangxi and Shanxi provinces were selected for this experiment. The tested levels of Pb and exogenous Pb addition were the same as those described in the experiment of bioaccumulation factors for Pb by corn from different soils (Table 2). Three replications were conducted on each tested level of Pb. After three months' aging of soils, the corn seeds, such as Jingketian 183, Chuandan 30, Liaodan 565, Tunyu 88, Zhongdan 808, and Nongda 84,were surface sterilized with 1% (v/v) NaOCl, rinsed, and soaked in distilled water for 24 h at 33°C in dark. Then these seeds were sown in each soil containing Pb. All these pots were placed in the greenhouse at the temperature of 25±3°C during the daytime and 20±3°C at night with a natural light photoperiod.

Soil Sampling and Analysis

After three months' aging, soils in the pots of the above two experiments were sampled, air-dried, passed through a <0.20-mm sieve, and then thoroughly mixed. 0.5 g of the soil sample was weighed and added to digestion tubes containing 9 ml HNO_3 and 3 ml HF.

The corn grains were harvested and oven-dried at 105°C for 30 minutes and then kept at 70°C until the weight of the grains was stable. 0.5 g of corn grain samples oven-dried were weighed and digested in 6 ml concentrated HNO_3 (70% w/v) and 3 ml H_2O_2 in a CEM Mars X microwave oven (CEM Mars) at a pressure of 3.1 MPa. The Pb concentration in the digestion solution from soil and grain was determined by inductively coupled plasma-mass spectrometry (ICP-MS, Agilent 7500a, Agilent Technologies Co. Ltd., USA).

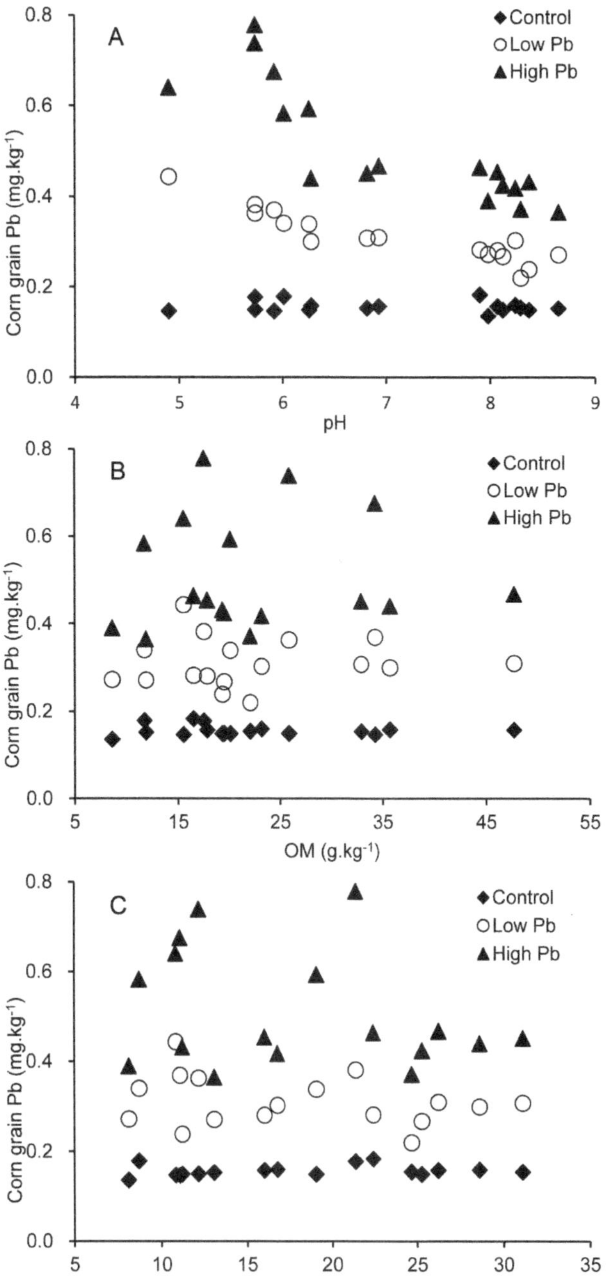

Figure 1. Relationships between Pb in corn grain and soil properties.

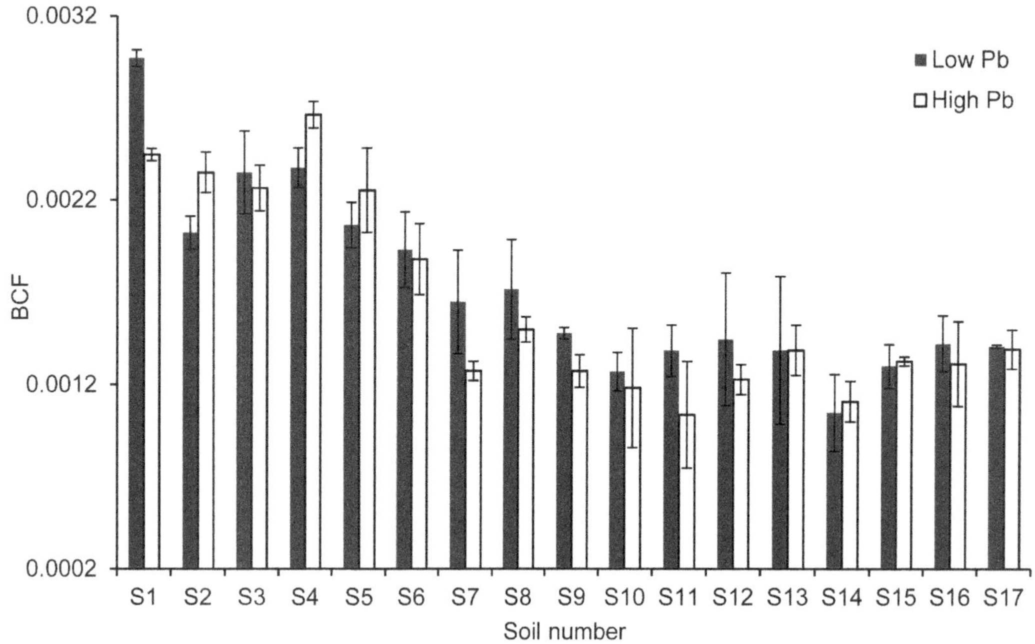

Figure 2. Effects of different soil types on the bioconcentration factor under stress of exogenous Pb.

Data Analysis

The bioconcentration factor (BCF) was calculated as the ratio of the content of Pb in the corn grain to that in the soil (equation 1) [45–46].

$$BCF = C_{grain}/C_{soil} \quad (1)$$

Where C_{grain} is the Pb concentration in the corn grain and C_{soil} is Pb concentration in the soil.

SPSS 16.0 for Windows® 107 (SPSS Inc, Chicago, IL, USA) was used for the regression analysis and statistical analysis of the significant differences. Origin 8.0 (OriginLab Co., Northampton, MA, USA) was employed for figure rendering.

Model

The prediction models for Pb transfer were established through multiple stepwise regression of the Pb BCF from Zhengdan 958 in a wide range of soil properties. The basis of these models is on the basis of equation 2 [38].

$$Log[BCF] = a \times pH + b \times \log[OM] + c \times \log[CEC] + k \quad (2)$$

Where Log [BCF], Log [OM], and Log [CEC] were the logarithm base 10 of the BCF values, the content of organic matter ($g.kg^{-1}$) in soils, and the cation exchange capacity ($cmol.kg^{-1}$) of soils, respectively? The slope of soil property parameters such as a, b, and c indicate the impacts of soil properties on Pb accumulation in corn grain. The intercept k is the intrinsic sensitivity that characterizes the ability of corn species to absorbing Pb.

Cross-species Extrapolation

In the process of cross-species extrapolation, an interim alternative was to assume that interactions between Pb accumulation in corn grain and soil pH, OM, CEC were the same among related corn species. In other words, the models stability constants including a, b, c were assumed the same among related species, and the only difference between related species was assumed to be their intrinsic sensitivity (k) [38]. The variation of intrinsic sensitivities within a species among plants reflects residual variation [34]. Under the condition of the squared error with the lowest value between predicted BCF value and the measured BCF value calculated as $\sum_{i=1}^{n}(measured\,BCF_i - predicited\,BCF_i)^2$, the intercept ($k$) for different models corresponding to various species were obtained through Excel Solver for linear optimization [47].

The accuracy of the model predictions was evaluated by comparing the measured BCF of the other non-model corn species with the predicted BCF from the model. The predictions for other non-model corn species were calculated with the prediction model derived from corn species i.e. Zhengdan 958.

Table 3. Prediction models for the different Pb levels.

Model No.	Lead levels	Prediction models	R^2	*P*
Model 1	Low Pb	Log[BCF] = −0.098 pH-0.150 log[OM] −1. 894	0.9085	<0.001
Model 2	High Pb	Log[BCF] = −0.108 pH-0.178 log[OM] −1. 806	0.8603	<0.001

Table 4. Intrinsic sensitivity (*k*) for non-model species fitted by models from Zhengdan 958.

Prediction Model	Intrinsic sensitivities (*k*) Jingketian 183	Chuandan 30	Liaodan 565	Tunyu 88	Zhongdan 808	Nongda 84
Model 1	−1.823	−1.881	−1.924	−2.295	−2.288	−2.152
Model 2	−1.681	−1.785	−1.846	−1.857	−1.794	−1.717

Analysis of the Reduction of Intra-species Variability

The Pb BCF values of non-model corn species were normalized to the specific soil conditions through the obtained model for Zhengdan 958 and the intra-species variability was computed by equation 3 [47].

$$f=\sqrt{\frac{\sum_{i=1}^{n}(BCFsi-\overline{BCFs})^2}{(n-1)\times(\overline{BCFs})^2}} \qquad (3)$$

Where the BCF of i-th conditions for specific corn species was normalized to specific soil conditions. *BCFsi* is the value that Pb BCF specific corn species were normalized to i-th soil conditions. $\overline{BCFs}$ is the mean of n BCFs. *n* is the number of different conditions for specific corn species, *f* is the intra-species variability.

The BCFs that Pb for different corn species was normalized by the prediction models to a set soil condition should be equal. Therefore, the decreases in intra-species variability indicate that normalization processing eliminated soil properties to a certain extent.

Results

Major Factors Affecting Pb Accumulation in Corn Grain in Different Soils

The relationships between Pb concentration in corn grain at different Pb levels and soil pH, OM and CEC were shown in Figure 1. At the treatment of low concentration Pb, the Pb concentration in corn grain decreased with the increasing pH of soil. The similar trend was also found at the treatment of high concentration Pb. There was no significant difference was found among Pb concentration in corn grain with the change of pH at the treatment of control. The significantly negatively direct correlations were observed between Pb accumulation in corn grain and soil pH at low concentration Pb, and high concentration Pb, respectively. However, no significantly direct correlations between Pb concentration in corn grain and soil OM, CEC were found.

Effects of Different Soil Types on Pb Transfer from Soil to Corn Grain

As shown in Figure 2, the BCF values from Zhengdan 958 under stresses of exogenous Pb significantly differed in 17 soils with a wide range of soil properties in China. The variation tendency of BCF values from Zhengdan 958 was generally consistent at different Pb levels of treatments. These BCF values decreased with

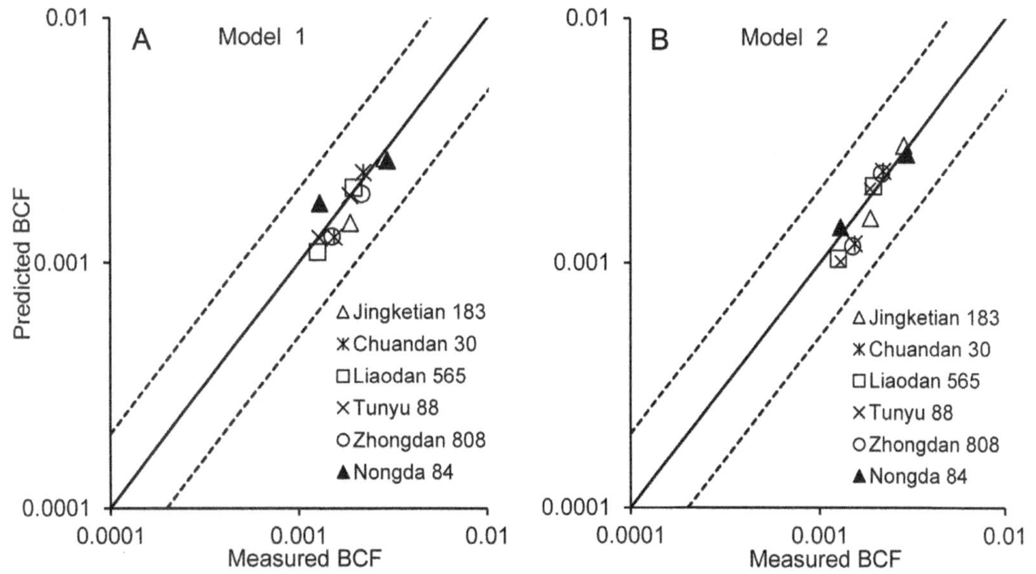

Figure 3. Relativity between measured and predicted BCF values for Pb in non-model corn grains. The predicted BCF values were estimated by model 1 and model 2 in Table 3; the solid line represents a 1:1 relationship; the interval between two dashed lines indicates a 2-fold prediction interval between the predicted and measured values.

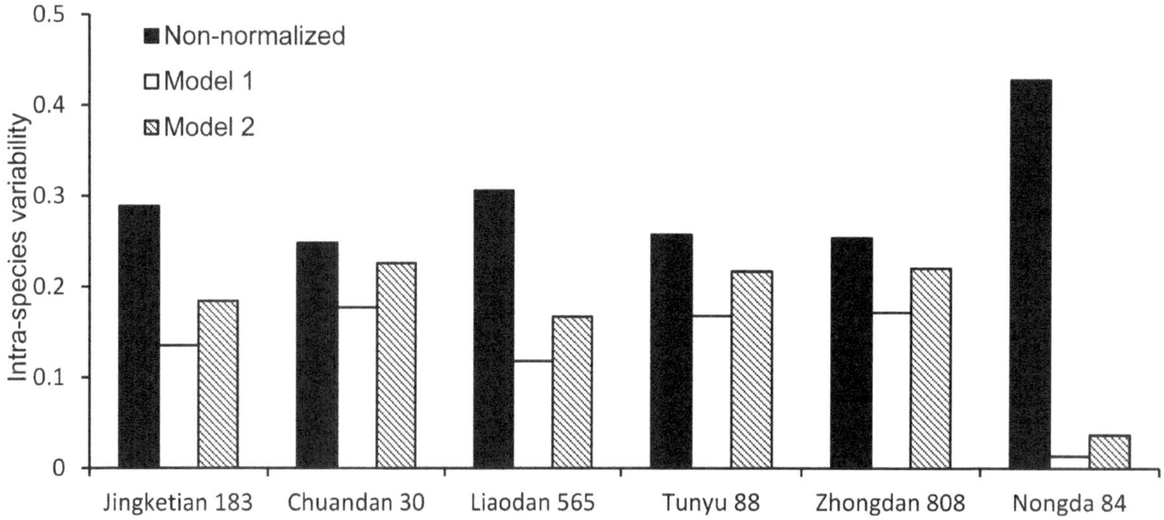

Figure 4. Intra-species variability of Pb BCF. The data from all non-model corn species were normalized with the models listed in Table 3.

the increases of soil pH. The BCF values detected from S1 to S9 (pH from 4.90 to 6.93) were significantly higher than those from S10 to S17 (pH from 7.90 to 8.80). At the treatment of low Pb, the maximum BCF (0.0030) among the 17 soils was observed in S1, and the minimum value for BCF (0.0010) was found in S14. The maximum value was 3.0 folds the minimum. At the same time, the highest BCF (0.0024) values were once again found in S1, and the lowest BCF (0.0010) was found in S11 at the treatment of high Pb, The maximum was 2.40 folds the minimum. These results indicate that Pb under acidic soil conditions is more highly bioavailable and leads to increased absorption by the plants.

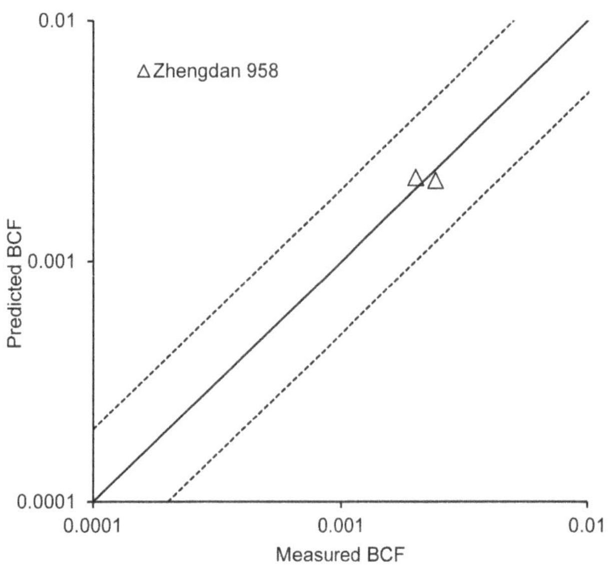

Figure 5. Relativity between measured and predicted BCF values for Pb in corn grains. The measured BCF were obtained from references. The predicted BCF values in Figure 5 were estimated by model 1. The solid line represents a 1:1 relationship; the dashed lines indicate a 2-fold prediction interval between the predicted and measured values.

Prediction Models

The prediction models established by stepwise regression were shown in Table 3. The result showed that pH and OM were the major factors affecting Pb accumulation in corn grain. A significantly negative correlation was observed between the Log[BCF] and soil pH and Log[OM] at the different treatments including low Pb, and high Pb, with the R^2 values of 0.8227, and 0.8955 ($P<0.001$), respectively. There was no significant difference between the two equations at the different treatment of Pb levels.

Cross-species Extrapolation

The intercept (k) can indicate the sensitivity of corn species to Pb accumulation. The intercept (k) for different corn species deduced by two different models were shown in Table 4. In terms of the same corn cultivar, no significant differences were observed among the k values for all tested non-model corn species deduced by the same model including model 1, and 2. However, the variation of k values among corn species deduced by the model 2 was stronger than that by the model 1. For example, the k values of corn species were different with the range from −1.888 to −1.736 deduced by model 1 and with range from −2.295 to −1.823 deduced by the model 2.

On the basis of the k values, the BCF values for non-model corn species were predicted by different models developed for Zhengdan958. The relationship between the predicted BCF values and the measured BCF values for non-model corn species was shown in Figure 3. The ratio between the predicted BCF values and the measured BCF values was within 2-fold interval and close to the solid line of 1:1 relationship. It indicates that the two models from Zhengdan 958 can be applied to predict the Pb BCF for non-model corn species.

Reduction of Intra-species Variability

Figure 4 shows the intra-species variability of Pb BCF for non-model species which were normalized with the prediction models listed in Table 3. The two prediction models can reduce the measured BCF intra-species variability for all non-model corn species. The intra-species variability of Pb BCF values fitted by model 1 was found to be significantly lower than those by model 1. It indicates that model 1 is more effective in reducing the uncertainty caused by the soil property differences than model 2.

Table 5. Soil properties and BCF values for Pb from reference.

Cultivation soil	cultivar species	pH	OM (g.kg-1)	CEC (cmol.kg^{-1})	Pb (g.kg^{-1})	BCF	Reference
Soil	Zhengdan 958	6.3	17.6	22.6	30.62	0.0020	[61]
Soil	Zhengdan 958	5.9	38.6	–	30.72	0.0024	[62]

This novel result also indicate that model 1 can be more reasonable and to be applied to other non-model corn species to establish the soil Pb SSD curves and to provide an ecological benchmark in China in the future.

Discussion

The bioavailable concentration of heavy metals is a key factor affecting its uptake and accumulation in plants [48]. In the present study, the results showed that corn grain taken up Pb from soil was mainly governed by soil pH and OM content. The Pb accumulation in the corn grain increased with the decreasing soil pH. This mainly attributed to pH played an important role in Pb uptake by plants from soil, this process through diverse reactions such as adsorption, ionic exchange, redox reactions, and precipitation-dissolution. These reactions are mainly affected by soil pH because of pH affecting Pb speciation and the charge of soil surface groups [49–50]. Christensen [51] and Naidu R et al [52] have found that the pH also influence on the solution activity of Pb and the distribution of Pb between the soil phase and the solution phase. The increased sorption of Pb at high pH values could reduce the Pb speciation in soils. In the present study, the BCF values observed in acidic soils (S1 to S9) with low pH values were significantly higher than those in alkaline soils (S10 to S17) with high pH values (Figure 2). These new results suggest that Pb is much more bioavailable under acidic conditions and is easily accumulated in corn grain. In the present paper, it is newly found that the Pb accumulation in corn grain was negatively related with OM. This is alike to the Cu accumulation in plant shoots [53], however it is different from Cd accumulation in corn grain [17]. This may be attributed to organic matter action as a primary sorbent of Pb because of its high sorption affinity [54–56].

In the present paper, no significant differences were observed between model 1 and model 2 at the different exogenous Pb concentration. In addition, there were also no significantly differences among Pb BCF values of corn grain at different exogenous Pb concentration, either. This was consistent with the findings of Rezvani and Zaefarian [57].

More importantly, we confirmed that the Pb BCF prediction models from Zhengdan 958 could be used to predict Pb BCF of other non-model corn species according to the soils pH and OM through cross-species extrapolation approach. Further, it was more accurate to apply model 2 to predict the BCF values of other corn species than model 1. Although many prediction models were developed to describe the relationship between plants uptake heavy metals and soil properties in terrestrial ecosystems, these models can't be extrapolated to non-model plant species, being only for single species [58–59]. Only a few studies were carried out to develop the prediction models for one species in aquatic ecosystems and then they are applied to other non-model species [36,38]. Few prediction models were developed for Pb accumulation in corn grain [60].

In order to further validate the feasibility of the prediction modes developed in the present study, the measured Pb BCF values of corn grains from soils were selected from the references [61,62], which were shown in Table 5. Although the corn species was also the same as that used to develop the model 1, the properties of the soils used in these references including OM, pH, and so on were different from those in the present paper. Therefore it is suitable that these data were used to test the prediction accuracy of model 1. The relativity between measured and predicted BCF values for Pb is shown in Figure 5. The result showed that the ratios between the predicted BCF values and the measured BCF values were also within 2 folds interval and close to 1:1 relationship. This indicates that the model 1 developed from Zhengdan 958 in the present study is feasible. It could be applied to other non-model corn species.

Conclusions

The soil pH and OM were the most major factors that can control Pb uptake in corn grain. The Pb accumulation in corn grain was negatively related with pH and Log [OM], respectively. The Pb BCF in corn grain can be well predicted by the model 1, which could be applied to other non-model corn species.

Acknowledgments

We are indebted to Prof. Lifang Hong for her help in soils collection, and Prof. Yibin Ma and Dr. Xiaoqing Wang for their help in the data processing and model derivation.

Author Contributions

Conceived and designed the experiments: ZL Y. Liang. Performed the experiments: Y. Li. Analyzed the data: HY. Contributed reagents/materials/analysis tools: Y. Li JL. Wrote the paper: ZL HY.

References

1. Cecchi M, Dumat C, Alric A, Felix-Faure B, Pradere P, et al (2008) Multi-metal contamination of a calcic cambisol by fallout from a lead-recycling plant. Geoderma 144: 287–298.
2. Grover P, Rekhadevi P, Danadevi K, Vuyyuri S, Mahboob M, et al (2010) Genotoxicity evaluation in workers occupationally exposed to lead. Int J Hyg Environ Health 213: 99–106.
3. Shahid M, Pinelli E, Pourrut B, Silvestre J, Dumat C (2011) Lead-induced genotoxicity to *Viciafaba L.* roots in relation with metal cell uptake and initial speciation. Ecotoxicol Environ Saf 74: 78–84.
4. Hrudey SE, Chen W, Rousseaux CG (1995) Bioavailability in environmental risk assessment. Lewis Publ, Boca Raton, FL.
5. USEPA (1992) Common chemicals found at Superfund sites. EPA540/R-94/044. U S Gov Print Office, Washington, DC.
6. Khan S, Cao Q, Zheng YM, Huang YZ, Zhu YG (2008) Health risks of heavy metals in contaminated soils and food crops irrigated with wastewater in Beijing, China. Environmental Pollution 152: 686–692.
7. Li ZJ, Xu JM, Tang CX, Wu JJ, Muhammad A, et al. (2006) Application of 16S rDNA-PCR amplification and DGGE fingerprinting for detection of shift in microbial community diversity in Cu-, Zn-, and Cd-contaminated paddy soils. Chemosphere 26: 1374–1380.
8. Chien CC, Kuo YM, Chen CC, Hung CW, Yeh CH, et al. (2008) Microbial diversity of soil bacteria in agricultural field contaminated with heavy metals. Journal of Environmental Sciences 20: 359–363.

9. Rattan RK, Datta SP, Chhonkar PK, Suribabu K, Singh AK (2005) Long-term impact of irrigation with sewage effluents on heavy metal content in soils, crops and groundwater-a case study. Agriculture Ecosystems and Environment 109: 310–322.
10. Chopin EIB, Alloway BJ (2007) Distribution and mobility of trace elements in soils and vegetation around the mining and smelting areas of Tharsis, Ríotinto and Huelva, Iberian Pyrite Belt, SW Spain. Water, Air, and Soil Pollution 182: 245–261.
11. Jung MC (2008) Heavy metal concentration in soils and factors affecting metal uptake by plants in the vicinity of a Korean Cu-W mine. Sensors 8: 2413–2423.
12. Wang J, Zhang CB, Jin ZX (2009) The distribution and phytoavailability of heavy metal fractions in rhizosphere soils of paulowniu fortune (seem) Hems near a Pb/Zn smelter in Guangdong, PR China. Geodema 148: 299–306.
13. Chapman EEV, Dave G, Murimboh JD (2013) A review of metal (Pb and Zn) sensitive and pH tolerant bioassay organisms for risk screening of metal-contaminated acidic soils. Environmental Pollution 179: 326–342.
14. Rodrıguez JA, Nanos N, Grau JM, Gil L, Lopez-Arias M (2008) Multiscale analysis of heavy metal contents in Spanish agricultural topsoils. Chemosphere 70: 1085–1096.
15. Bang J, Hesterberg D (2004) Dissolution of trace element contaminants from two coastal plain soils as affected by pH. Journal of Environmental Quality 33: 891–901.
16. Yang H, Li ZJ, Lu L, Long J, Liang YC (2013) Cross-species extrapolation of prediction models for cadmium transfer from soil to corn grain. PLoS ONE 8(12): e80855.doi: 10.1371/journal.pone.0080855.
17. Wang AS, Angle JS, Chaney RL, Delorme TA, Reeves RD (2006) Soil pH effects on uptake of Cd and Zn by Thlaspi caerulescens. Plant and Soil 281: 325–337.
18. Kirkham MB (2006) Cadmiumin plantson polluted soils: Effect of soil factors, hyperaccumulation and amendments. Geoderma 137: 19–32.
19. Zeng F, Ali S, Zhang H, Quyang Y, Qiu B, et al. (2001) The influence of pH and organic matter content in paddy soil on heavy metal availability and their uptake by rice plants. Environmental Pollution 159: 84–91.
20. Hettiarachchi GM, Ryan JA, Chaney RL, La Fleur CM (2003) Sorption and desorption of cadmium by different fractions of biosolids-amended soils. Journal of Environmental Quality 32: 1684–1693.
21. Antoniadis V, Robinson JS, Alloway BJ (2008) Effects of short-term pH fluctuations on cadmium, nickel, lead, and zinc availability to ryegrass in a sewage sludge-amended field. Chemosphere 71: 759–764.
22. Zeng FR, Shafaqat Ali, Zhang HT, Ouyang YN, Qiu BY, et al. (2011)The influence of pH and organic matter content in paddy soil on heavy metal availability and their uptake by rice plants. Environmental Pollution 159: 84–91.
23. Ge Y, Murray P, Hendershot WH (2000) Trace metal speciation and bioavailability in urban soils. Environ Pollut 107: 137–144.
24. Chen WP, Li LQ, Chang AC, Wu LS, Chaney RL, et al. (2009) Characterizing the solid-solution partitioning coefficient and plant uptake factor of As, Cd, and Pb in California croplands. Agriculture, Ecosystems and Environment 129: 212–220.
25. Zarcinas BA, Ishak CF, McLaughlin MJ, Cozens G (2004) Heavy metals in soils and crops in southeast Asia. 1. Peninsular Malaysia. Environmental Geochemistry and Health 26: 343–357.
26. Simmons RW, Noble AD, Pongsakul P, Sukreeyapongse O, Chinabut N (2008) Analysis of field-moist Cd contaminated paddy soils during rice grain fill allows reliable prediction of grain Cd levels. Plant and Soil, 302: 125–137.
27. Kuo S, Huang B, Bembenek R (2004) The availability to lettuce of zinc and cadmium in a zinc fertilizer. Soil Science, 169: 363–373.
28. Hough RL, Breward N, Young SD, Crout NMJ, Tye AM, et al. (2004) Assessing potential risk of heavy metal exposure from consumption of home-produced vegetables by urban populations. Environmental Health Perspectives, 112: 215–221.
29. Nahmani J, Hodson ME, Black S (2007) A review of studies performed to assess metal uptake by earthworms. Environmental Pollution, 145: 402–424.
30. Zhang HZ, Luo YM, Song J, Zhang HB, Xia JQ, et al. (2011) Predicting As, Cd and Pb uptake by rice and vegetables using field data from China. Journal of Environmental Sciences 23: 70–78.
31. Adamsa ML, Zhao FJ, McGratha SP, Nicholsonb FA, Chambersb BJ (2004) Predicting cadmium concentrations in wheat and barley grain using soil properties. Journal of Environmental Quality 33: 532–541.
32. Zhao XJ, Ma RC, Li ZJ (2013) Non-target organism species sensitivity distributions to oxytetracycline in water. Research Journal of Chemistry and Environment 17: 74–81.
33. Larras F, Bouchez A, Rimet FR, Montuelle B (2012) Using bioassays and species sensitivity distributions to assess herbicide toxicity towards benthic diatoms. PloS ONE 8: 1–9.
34. Van Sprang PA, Verdonck FAM, Van Assche F, Regoli L, De Schamphelaere KAC (2009) Environmental risk assessment of zinc in European freshwaters: a critical appraisal. Sci Tot Environ 407: 5373–5391.
35. Allard P, Fairbrother A, Hope BK, Hull RN, Johnson MS, et al. (2010) Recommendations for the development and application of wildlife toxicity reference values. Intergrated Environmental Assessment and Management 6: 28–37.
36. Deleebeeck NME, De Schamphelaere KAC, Janssen CR (2007a) A bioavailability model predicting the toxicity of nickel to rainbow trout (Oncorhynchus mykiss) and fathead minnow (Pimephales promelas) in synthetic and natural waters. Ecotoxicol Environ Saf 67: 1–13.
37. Deleebeeck NME, Muyssen BTA, De Laender F, Janssen CR, De Schamphelaere KAC (2007b) Comparison of nickel toxicity to cladocerans in soft versus hard surface waters. Aquat Toxicol 84: 223–35.
38. Schlekat CE, Genderen EV, Schamphelaere KACD, Antunes PMC, Rogevich EC, et al. (2010). Cross-species extrapolation of chronic nickel biotic ligand models. Science of the Total Environment 408: 6148–6157.
39. Sparks DL, Page AL, Helmke PA, Loeppert RH, Soltanpour PN, et al. (1996) Methods of soil analysis. Part 3 - chemical methods. Soil Science Society of America, Inc, Madison. WI.
40. Dohrmann R (2006) Cation exchange capacity methodology II: A modified silver-thiourea method. Applied Clay Science 34: 38–46.
41. Ball DF (1964) Loss-on-ignition as an estimate of organic matter and organic carbon in noncalcareous soils. Journal of Soil Science 15: 84–92.
42. Bray RH, Kurtz LT (1945) Determination of total, organic, and available forms of phosphorus in soils. Soil Science 59: 39–46.
43. Bremner JM (1996) Nitrogen total Methods of soil analysis. Part 3: Chemical methods, American Society of Agronomy-Soil Science Society of America, 1996, PP 1085–1121.
44. Zarcinas BA, McLaughlin MJ, Smart MK (1996) The effect of acid digestion technique on the performance of nebulisation systems used in inductively coupled plasma spectrometry. Commun Soil Sci Plant Anal 27: 1331–1354.
45. Baker AJM, Reeves RD, Hajar ASM (1994) Heavy metal accumulation and tolerance in British populations of metallophyte Thlaspi caerulescens. New Phytol 127: 61–68.
46. Raskin I, Kumar PBAN, Dushenkov S, Salt DE (1994) Bioconcentrationof heavy metals by plants. Curr Opin Biotechnol 5: 285–290.
47. Wang XQ (2012) Ecological Thresholds for Copper and Nickel in Chinese Agricultural Soils, Beijing, China Mining University 1–50.
48. Sarwar N, Saifullah, Malhi SS, Zia MH, Naeem A, et al. (2010) Role of mineral nutrition in minimizing cadmium accumulation by plants. J Sci Food Agric 90: 925–937.
49. Martınez-Villegas N, Flores-Velez LM, Domınguez O (2004) Sorption of lead in soil as a function of pH:a study case in Mexico. Chemosphere 57: 1537–1542.
50. Smical AI, Hotea V, Oros V, Juhasz J, Pop I (2008) Studies on transfer and bioaccumulation of heavy metals from soil into lettuce. Environmental Engineering and Management Journal 7: 609–615.
51. Christensen TH (1984) Cadmium soil sorption at low concentrations: I. Effect of time, cadmium load, pH, and calcium. Water. Air. Soil Pollut 21: 105–114.
52. Naidu R, Bolan NS, Kookana RS, Tiller KG (1994) Ionic-strength and pH effects on the sorption of cadmium and the surface charge of soils. European Journal of Soil Science 45: 419–429.
53. Moreno JL, Garcia C, Hernandez T, Pascual JA (1996) Transference of heavy metals from a calcareous soil amended with sewage sludge compost to barley plants. Bioresource Technology 55: 251–258.
54. Ge Y, Hendershot W (2005) Modeling sorption of Cd, Hg and Pb in soils by the NICA (non-ideal competitive adsorption)-Donnan model. Soil Sediment Contam 14: 53–69.
55. Sauve S, Manna S, Turmel MC, Roy AG, Courchesne F (2003) Solid solution partitioning of Cd, Cu, Ni, Pb, and Zn in the organic horizons of a forest soil. Environ Sci Technol 37: 5191–5196.
56. Puschenreiter M, Horak O, Friesl Wand Hartl W (2005) Low-cost agricultural measures to reduce heavy metal transfer into the food chain – a review. Plant Soil Environ 51: 1–11.
57. Rezvani M, Zaefarian F (2011) Bioaccumulation and translocation factors of cadmium and lead in Aeluropus littoralis. Australian Journal Agricultural Engineering 2: 114–119.
58. Gandoisa L, Probstb A, Dumata C (2010) Modelling trace metal extractability and solubility in French forest soils by using soil properties. European Journal of Soil Science 61: 271–286.
59. Wang XP, Shan XQ, Zhang SZ, Wen B (2004) A model for evaluation of the phytoavailability of trace elements to vegetables under the field conditions. Chemosphere 55: 811–822.
60. Rainbow PS (2007) Trace metal bioaccumulation: Models, metabolic availability and toxicity. Environment international 33: 576–582.
61. Cao Y, Liu Y, Wang GJ, Han Y, Wang SB (2009). Differences in Lead and Cadmium Concentrations among Plant Tissues of Maize Cultivars under the Combined Stress of Lead and Cadmium. Journal of Maize Sciences 17: 80–85.
62. Li YF, Liu L, Chen X, Hu LL, Tang JJ (2010) Plant Growth,Lead Uptake and Partitioning of Maize (*Zea mays L.*) Under Simulated Mild/Moderate Lead Pollution Stress. Journal of Agro-Environment Science 29: 2260–2267.

10

Analysis of the Diversity of Substrate Utilisation of Soil Bacteria Exposed to Cd and Earthworm Activity Using Generalised Additive Models

Selene Muñiz, Juan Lacarta, María P. Pata, Juan José Jiménez, Enrique Navarro*

Department of Biodiversity & Conservation and Ecosystem Restoration, Pyrenean Institute of Ecology, Spanish National Research Council, Zaragoza, Spain

Abstract

Biolog EcoPlates^{TM} can be used to measure the carbon substrate utilisation patterns of microbial communities. This method results in a community-level physiological profile (CLPP), which yields a very large amount of data that may be difficult to interpret. In this work, we explore a combination of statistical techniques (particularly the use of generalised additive models [GAMs]) to improve the exploitation of CLPP data. The strength of GAMs lies in their ability to address highly non-linear relationships between the response and the set of explanatory variables. We studied the impact of earthworms (*Aporrectodea caliginosa* Savigny 1826) and cadmium (Cd) on the CLPP of soil bacteria. The results indicated that both Cd and earthworms modified the CLPP. GAMs were used to assess time-course changes in the diversity of substrate utilisation (DSU) using the Shannon-Wiener index. GAMs revealed significant differences for all treatments (compared to control -S-). The Cd exposed microbial community presented very high metabolic capacities on a few substrata, resulting in an initial acute decrease of DSU (i.e. intense utilization of a few carbon substrata). After 54 h, and over the next 43 h the increase of the DSU suggest that other taxa, less dominant, reached high numbers in the wells containing sources that are less suitable for the Cd-tolerant taxa. Earthworms were a much more determining factor in explaining time course changes in DSU than Cd. Accordingly, Ew and EwCd soils presented similar trends, regardless the presence of Cd. Moreover, both treatments presented similar number of bacteria and higher than Cd-treated soils. This experimental approach, based on the use of DSU and GAMs allowed for a global and statistically relevant interpretation of the changes in carbon source utilisation, highlighting the key role of earthworms on the protection of microbial communities against the Cd.

Editor: Volker Thiel, Kantonal Hospital St. Gallen, Switzerland

Funding: This work was supported by Spanish Ministry of Economy and Competitiveness (National Research Plan, ref. BFU2010-22053, http://www.mineco.gob.es/portal/site/mineco/). The ARAID Foundation (http://www.araid.es/) is also acknowledged for support provided to J.J. JimÃ©nez. The funders had no role in study design, data collection and analysis, decision to publish, or preparation of the manuscript.

Competing Interests: The authors have declared that no competing interests exist.

* E-mail: enrique.navarro@ipe.csic.es

Introduction

Biolog Ecoplates

Biolog MicroPlates were developed in the late 1980s to assist in the identification of bacterial strains [1]. These 96-well plates contained carbon sources and a tetrazolium violet redox dye that turned purple if inoculated microorganisms utilised these sources. By comparing the obtained carbon substrate utilisation patterns with databases, it was possible to establish a probable identification [2]. Later, microbial ecologists used Biolog plates to investigate patterns at the community level. Therefore, a new plate specifically designed for community analysis and microbial ecological studies was created; this new plate was referred to as the EcoPlate [3]. The EcoPlate contains 31 of the most useful carbon sources for soil community analysis [3], allowing for community-level physiological profiling (CLPP) of heterotrophic bacterial assemblages. This technique has also been widely used to assess the toxicological impacts of different pollutants [4], including different heavy metals [5–9].

The use of EcoPlates results in a CLPP, yielding a very large amount of data that may be difficult to interpret. These difficulties are related to the alteration of the original microbial communities due to soil sampling and pre-treatment as well as bacterial extraction [4,10]. Moreover, the Biolog procedure can be considered a culture method in which the originally inoculated community is altered. The contribution of a certain species population to the colour profile will depend on its culturability in the Biolog wells and on its interactions with other species (differential growth and competition). The tetrazolium dye also introduces some bias in the profile because not all bacteria are able to reduce it [4]. Thus, the Biolog method is more useful for comparing soil microbial communities than for community characterisation [4]. CLPPs provide little insight regarding the function of the community *in situ* unless they are combined with other microbial methods that do not rely on the culturing of the soil microflora [11,12].

The pattern of positive and negative responses as well as substrate oxidation rate and extent are highly reproducible for simple microbial communities, particularly if the inoculum densities are similar [10]. Nonetheless, data corrections are required if inoculum density is not controlled. A kinetic approach must be used to extract data from the plates to avoid problems related to incubation time. However, there is no consensus regarding which statistics are most appropriate, and the interpretation of their meaning remains unclear [4].

Earthworms and cadmium (Cd) on soils

Earthworms are key members of the soil macrofauna in temperate soils [13] that directly or indirectly modulate resource availability (quality, quantity, and distribution) for other organisms. Earthworms are considered ecosystem engineers because they modify, maintain, or create habitats [14] as they build soil biostructures, which consist of aggregates and macropores [15–17]. The soil zone influenced by earthworm burrowing and casting was first termed the drilosphere by Bouché, 1975 [18,19], where microbial biomass is greater than in the surrounding soil [20]. In general, the mechanical and biological activities of earthworms favour organic matter mineralisation and humification and stimulate microbial activity [13].

Cd levels in the environment vary widely, and the average natural abundance of this element in the earth's crust varies from 0.1 to 0.5 mg kg^{-1}. However, human activities, such as mining, metal smelting, urban and industrial emissions, waste incineration, coal combustion, traffic dust, and particularly the use of phosphate fertilisers and sewage sludges, are quoted as the primary reason for the increase in soil Cd content over the last several years in Europe [21]. As heavy metals cannot be degraded, they tend to accumulate in soils. Their disappearance indicates that they have leached into deep soil layers and transferred to subterranean waters or transported to other locations by runoff [22]. The fate of Cd in soils will depend on a balance between adsorption, leaching, and biological uptake. In most soils, more than 99% of Cd is found in the solid phase, and less than 1% is present in the soil solution [21].

Earthworm species belonging to the three main ecological groups, i.e. epigeics, endogeics and anecics, have been shown to increase the mobility and bioavailability of some heavy metals in soils [23,24]. As a result, soil ecosystem function can be altered by heavy metal pollution through changes in microbial communities and earthworms in the drilosphere [25]. These changes in microbial community composition may lead to increased heavy metal tolerance in the bacterial community [26,27]. However, a decrease in microbial diversity also leads to less soil resilience to natural or human-induced perturbations [28]. Thus, studies on the joint effect of earthworm engineering and soil pollution on the bacterial community may help to prevent and remediate human impacts at the ecosystem level.

Experimental approach: the use of generalised additive models (GAMs)

Our goal is to demonstrate that the joint action of earthworm activity and Cd toxicity induce changes in the bacterial community that are measurable by changes in the CLPP. To this end, we have assessed and measured these changes using a combination of descriptive and multivariate statistics together with GAMs [29,30]. GAMs are semiparametric extensions of generalised linear models that allow for the flexible modelling of covariate effects, avoiding the limitations of parametric models.

This methodology is appropriate when the relationship between the variables is expected to be complex and not easily fitted by standard linear or non-linear models, which is the case in most biological studies. One of the most frequent uses of GAMs in ecology is the modelling of species distribution and species abundance [31], related to environmental and geographical variables [32], being plant ecology and fisheries the two fields where the use of GAMs have been generalized in the last decade. Studies of air pollution and health are another domain where these models have been widely used as standard methods since its flexibility in modelling spatial and temporal processes [33]. One of the great advantages of these models is its ability to determine the nature of the relationship between the response and explanatory variables rather than assuming a fixed relationship, allowing for fitting complex relationships in a straightforward manner.

Methods

Soil and earthworm sampling

The earthworm *Aporrectodea caliginosa* is commonly used in soil ecotoxicological studies [34]. This earthworm is present in the natural soils surrounding our facilities, is easy to culture and sensitive enough against different toxicants, and being a key actor of the matter cycling and soil formation, the results obtained with it are environmentally relevant.

A total of 102 individuals were hand-sorted in an experimental field near the Pyrenean Institute of Ecology (Montañana, Zaragoza) in June 2011. Experimental soils were collected in an experimental field near Jaca (northern Spain). No specific permissions were required for these activities since both experimental fields from which the soil and the earthworms were collected belong to the Spanish National Research Council. The activities neither involve vertebrates nor endangered or protected species.

The soil pH was determined from extracts obtained with a 1:2.5 ratio (g soil:g distilled H_2O [pH_w] or in 0.1 M KCl [pH_{KCl}]). Organic matter (OM) content, cation exchange capacity (CEC), and soil texture were measured by calcination, atomic absorption spectrometry, and aerometry, respectively [35]. Field capacity (0.3 bar) and wilting point (15 bar) were determined in two pressure chambers [36]. The pH values at the earthworm sampling site were 7.82 (water) and 7.43 (KCl). The prevailing soil conditions (pH, OM content, and CEC) are provided in Table 1. The soil was

Table 1. Soil pH, organic matter (OM), and cation exchange capacity (CEC) at day 32.

Treatment	pH_w	pH_{KCl}	OM	CEC	[Cd] μg g^{-1}	[Cd] mg L^{-1}
			%	meq 100 g^{-1}	in soils	in pore water
S	8.02	7.89	5.33	17.7	< 0.5	-
Cd	7.8	7.88	5.28	18.3	329	0.78
Ew	7.84	7.91	5.25	19.4	< 0.5	-
EwCd	7.63	7.84	5.16	18.7	311±76	0.49±.002

Soil pH was determined in water (pH_w) and KCl (pH_{KCl}). Prior to Cd analysis in soils, six replicates from Cd treatments were integrated, whereas EwCd replicates were analysed separately (accordingly, standard deviation is available).

loamy with 44.2% sand, 38.1% silt, and 17.7% clay. The bulk density of dried and sieved soil was 1,270.83 kg m^{-3}. The field capacity was 30.3% (w/w), and the wilting point was 14.5% (w/w).

All earthworms remained in a terrarium with the same soil from which they were collected for two weeks (maintenance period). Then, they were moved to another terrarium that contained the experimental soil sieved at < 2 mm. The duration of this preconditioning period was 16 days, and its purpose was to allow the earthworms to empty their guts of the previously ingested soil.

Design and duration of the experiment

We tested the effect of Cd and earthworms on the ability of bacteria to degrade the 31 different sources of carbon included in Biolog EcoPlates. Accordingly, a bifactorial experiment with four treatments was designed (S: soil control; Cd: soil irrigated with Cd; Ew: soil engineered by earthworms; EwCd: soil engineered by earthworms and irrigated with Cd). A detailed schema of the design is provided in Figure 1. Six replicates (pots) per treatment were filled with 1.5 L of sieved soil. The soil surface exposed to air in each pot was approximately 180 cm^2. The earthworms were exposed to a natural light/dark cycle from June to August. The air humidity as well as air and soil temperatures were controlled. The soil moisture was measured by a Tektronix 1502C TDR Cable Tester reflectometer (Bracknell, Berkshire, United Kingdom) [37]. After the preconditioning period, specimens of a similar size and weight were selected and introduced into each pot (three individuals). Each pot was subsequently watered with distilled water to reach field capacity (i.e., 30% w/w).

All pots were watered with 200 mL of distilled water from Days 1–16. From Day 17 onward, the pots were watered with 200 mL of distilled water or Cd solution depending on the treatment (Figure 1). Cd was added to the soil as $CdCl_2 + 2.5\ H_2O$ (Panreac), and the pots were watered with a 343 mg L^{-1} solution. The irrigation was designed to achieve soil concentrations below the median lethal concentration (LC_{50}) to *A. caliginosa, which* has been estimated in 540 μg Cd g-1 for a 56-day exposure [38]. The percolated water was collected daily 24 h after the watering that generated it and before watering the pots again. The soils were exposed to both Cd and earthworms for 16 days (see Figure 1).

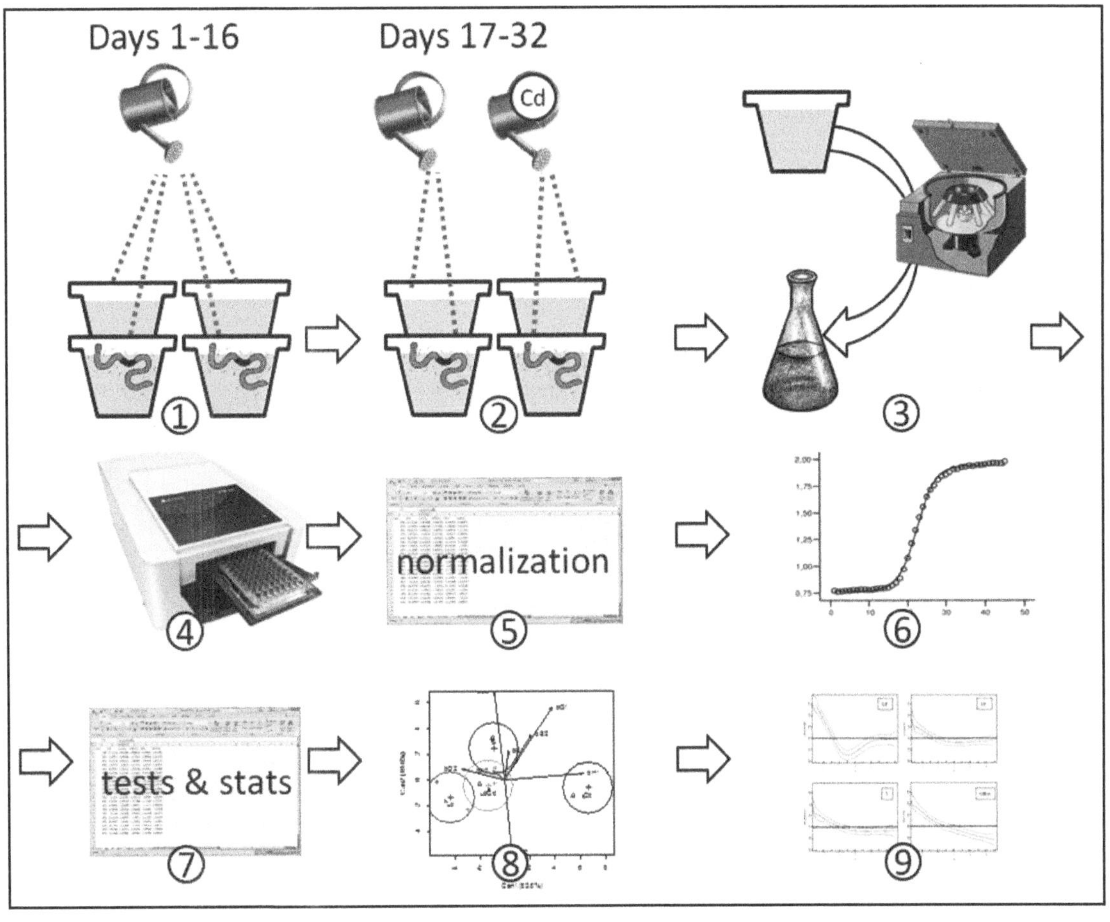

Figure 1. Diagram showing experimental design and data processing. During the first days (1), soil bacteria were exposed to earthworms and irrigated with clean water. Later, soils were irrigated with Cd or clean water depending on the treatment (2). At the end of the experimental period, bacteria were extracted by centrifugation and suspended in liquid media (3). This media was inoculated in Biolog EcoPlates (4). The results of colour development over 167 h were recorded and normalised (5). The results were fitted to a sigmoidal model (6), and their parameters were calculated (slope, time to reach 50% of the maximum colour intensity, and maximum intensity reached). Parameters were compared to assess the effects of the different treatments (7), and the carbon sources that were more informative (accumulating more discriminating power) were selected for multivariate analysis (8). Finally, GAMs were fitted to assess the effect of the different treatments on the use of diverse carbon sources over time, determined by the Shannon-Weaver index (9).

Measurement of heavy metal

Following Hobbelen et al. [39], three soil sample types were obtained on Day 32. First, pore water was collected by placing 40 g samples of fresh soil into 50 mL Falcon tubes and centrifuging for 40 min (2,000 g). The supernatants were filtered with a Büchner funnel and cellulose filters (0.45 μm pore size; Sartorius Stedim Biotech) and fixed with 69% HNO_3 (Hiperpur; Panreac) at a final concentration of 4% (v/v). Next, the soil was air dried. To obtain $CaCl_2$ extracts, a 5:1 mixture (5 g of 0.01 M $CaCl_2$:1 g of soil) was prepared and shaken for 2 h at 200 rpm. The extracts were filtered and fixed. Finally, the soil samples were acid-digested in a microwave (Milestone digester) by adding 4 mL of HNO_3 (PA-ISO 69%) and 1 mL of H_2O_2 (33%) to each 0.1 g of dried soil (20 min at 220°C, final temperature). All samples were stored at 4°C until further analysis. The Cd concentration in the soil and percolated water was determined by inductively coupled plasma-optical emission spectrometry (ICP-OES).

Bacterial community-level physiological profiling (CLPP)

The microbial community population structure and changes in carbon source usage were assessed by live/dead staining and Biolog tests, respectively. Bacterial cells were extracted from soils following Katayama [40] and modified as described in Ikeda [41]. Briefly, 10 g of fresh soil was diluted in 95 mL of Milli-Q water. The samples were homogenised for 15 min, and 10 mL of the soil suspension was pipetted into a 50 mL Falcon tube. The sample was sonicated for 1 min and subsequently centrifuged at 1,000 g for 10 min to obtain 9.5 mL of sample. The remaining 0.5 mL was resuspended in 10 mL and submitted to a new dispersion-centrifuge cycle to obtain additional microbial inocula. This cycle was repeated five times until a 47.5 mL sample was obtained, which was refrigerated at 4°C until analysis.

Bacterial extracts were later filtered using 60 μm cellulose filters to reduce interference during OD measurements (for Biolog). Viable and dead cells (live/dead staining) were stained following the LIVE/DEAD protocol [42]. The cells in 10 randomly selected fields per sample were counted using NIS-ELEMENTS software (Nikon ©). The number of viable cells was calculated as total - dead.

For CLPP, integrated soil samples were used (sampling was performed at different depths). One Biolog™ EcoPlate was inoculated per treatment and incubated at 25°C in a dark chamber. Each plate contained three sets (replicates) of carbon sources, and each plate was filled with samples from one pot. An Anthos 2010 microplate reader (Biochrom) was used to read the absorbance in each well at 590 nm (OD_{590}) at even intervals from 0 to 167 h.

Diversity of substrate utilisation (DSU)

DSU was determined by the Shannon-Weaver index (H), which is one of the most commonly used diversity measures [43,44]. This index considers both the number of species and their abundance. In this case, the H value describes the ability of the bacterial community to degrade more or fewer types of carbon sources, thus being an index of physiological diversity of the bacterial community. Microbial communities that are able to degrade more substrates or/and to degrade them with similar efficiency would have higher values of H. The index was calculated as follows:

$$H = -\sum p_i(\ln p_i)$$

where p_i is the normalised value of OD_{590} (see point 2.6). This equation is commonly used to process Biolog data [45,46].

Statistical analysis

The carbon degradation data from every well were normalised by being divided by the initial OD_{590} at time 0 after averaging the three wells containing the same carbon source. This procedure was performed to reduce the variability due to differences in inocula densities [47,48].

The results of colour development showed sigmoidal relationship between the OD_{590} and time for all carbon sources. Therefore, the curves were fitted to a non-linear model [47,49] with the four-parameter log-logistic function given by the formula:

$$f(x_i,(b,c,d,e)) = c + \frac{d-c}{1+\exp(b(\log(x_i)-e))} + \varepsilon_i$$

Where parameter b is the slope of the curve (S), *c* is the lower asymptote of the sigmoidal curve, *d* is the upper asymptote of the curve representing the maximal degradation increase (M), and parameter *e* is the time to reach 50% of maximum degradation (TM_{50} or ED_{50}). T-tests were performed to evaluate which ratios between parameters from two different treatments were significantly ($p<0.05$) different from 1; p-values were adjusted using Bonferroni correction for multiple test.

For the CLPP analysis, the carbon sources shown to be more informative regarding the effects of heavy metals on bacterial communities were selected. Other sources (such as D-xylose and D-cellobiose) identified in previous studies were also considered [50–52]. Canonical discriminant analysis (CDA) was performed on the selected carbon sources (β-Methyl-D-Glucoside, i-Erythritol, D-Mannitol, α-Cyclodextrin, N-Acetyl-D-Glucosamine, α-D-Lactose, and the two previously mentioned).

Thin-plate regression splines [53] were used as smoothers, with optimal effective degrees of freedom chosen automatically using the generalised cross-validation criteria [30].

All analyses were performed with the free statistical software R [54]. The *drm and compParm* functions of the *drc* package [49] were used to estimate and compare kinetic parameters, the *candisc* package [55] was used for CDA, and the *gam* function of the *mgcv* library [30] was used for GAM modelling.

Results

Bacterial cell density and Cd in soils

Cd-treated soils contained significantly fewer bacterial cells (1.1e7 cells g^{-1} DW) than control soils (paired t-tests, p = 0.03) and a smaller proportion of dead cells (12%). Control soils (S) contained more bacterial cells (4.5e7 cells g^{-1} DW), and earthworm-treated soils had intermediate values (3.0–3.6e7 cells g^{-1} DW). The control and earthworm treatments contained very similar proportions of dead cells (17–20%).

The final Cd concentration in the soils was lower than 0.5 μg $g^{-1}{}_{dry\ soil}$ in the S and Ew treatments, 328.54 μg $g^{-1}{}_{dry\ soil}$ in the Cd treatment, and 310.93±75.58 μg $g^{-1}{}_{dry\ soil}$ (average value, n = 6) in the EwCd treatment. The Cd concentration in the pore water was 0.78 mg L^{-1} in the Cd treatment and 0.49±0.002 mg L^{-1} (n = 2) in the EwCd treatment. The amount of $CaCl_2$-extractable metal was 2.18 μg $g^{-1}{}_{dry\ soil}$ in the Cd treatment and 2.2±0.55 μg $g^{-1}{}_{dry\ soil}$ (n = 3) in the EwCd treatment.

Table 2. Comparison of kinetic parameters.

Code	Carbon sources	S/Ew	S/Cd	S/EwCd	Cd/Ew	Cd/EwCd	Ew/EwCd
A2	β-Methyl-D-Glucoside				M:1.7	M:2.08	
A3	D-Galactonic acid α-Lactone	S:0.26			M:1.61	M:1.86	
		(−0.38,0.90)			(1.46,1.75)	(1.68,2.04)	
B1	Pyruvic Acid Methyl Ester						M:0.78
							(0.68,0.87)
B3	D-Galacturonic Acid		M:0.75		M:1.3		
			(0.67,0.83)		(1.22,1.37)		
C2	i-Erythritol					M:1.59	
						(1.37,1.81)	
C3	2-Hydroxy Benzoic Acid						S:0.14
							(−0.27,0.55)
C4	L-Phenylalanine		S:0.23			M:2	
			(−0.12,0.58)			(1.88,2.12)	
D2	D-Mannitol		M:0.68		M:1.71		
			(0.51,0.85)		(1.56,1.85)		
D3	4-Hydroxy Benzoic Acid	T:0.51	M:0.7			M:1.79	
		(0.12,0.90)	(0.55,0.85)			(1.61,1.97)	
			T:0.55				
			(0.15,0.94)				
D4	L-Serine		M:0.7		M:1.53		
			(0.41,0.99)		(1.42,1.63)		
E1	α-Cyclodextrin					M:1.56	
						(1.38,1.74)	
E2	N-Acetyl-D-Glucosamine		M:0.67		M:1.67		
			(0.41,0.93)		(1.44,1.89)		
F1	Glycogen			M:1.59		M:1.7	
				(1.29,1.89)		(1.50,1.90)	
F2	D-Glucosaminic Acid		T:0.68			M:1.61	
			(0.36,0.98)			(1.36,1.86)	
F3	Itaconic Acid					M:0.75	
						(0.65,0.85)	
F4	Glycyl-L-Glutamic Acid				M:1.68	M:1.72	
					(1.49,1.86)	(1.53,1.91)	
G2	Glucose-1-Phosphate	M:1.53			M:1.36		
		(1.33,1.73)			(1.28,1.43)		
G3	α-Ketobutyric Acid		S:0.27			M:1.6	
			(−0.06,0.60)			(1.50,1.70)	
G4	Phenylethyl-amine	S:0.15					
		(−0.18,0.48)					
H1	α-D-Lactose	S:0.35		S:0.31	S:0.29	S:0.29	
		(−0.06,0.76)		(−0.14,0.76)	(−0.04,0.62)	(−0.32,0.66)	
		M:1.51		M:1.59	M:1.25	M:1.32	
		(1.29,1.73)		(1.35,1.83)	(1.10,1.40)	(1.16,1.48)	
H2	D,L-α-Glycerol Phosphate	M:1.84		M:1.64			
		(1.55,2.13)		(1.33,1.95)			
H3	D-Malic Acid	M:1.65	T:1.93	M:1.5	M:1.72		
		(1.46,1.84)	(1.28,2.57)	(1.31,1.69)	(1.54,2.71)		
H4	Putrescine	S:0.19			M:2.18	M:2.2	
		(−0.20,0.58)			(1.64,2.71)	(1.72,2.68)	

Table 2. Continued.

To assess the effect of the different treatments in the degradation of the different carbon sources, the parameters from the 4 parameters log-logistic models, were divided and the resulting ratio compared to 1 using t-tests. The letter indicates the parameter compared (T = TM_{50}, S = slope, M = maximum). Values between brackets are 95% CI. Table shows only those comparisons with significant p-values (adjusted using Bonferroni correction for multiple test), meaning that these ratios significantly ($p<0.05$) different from 1 and thus the parameters compared were significantly different. As example, the first value in the column S/Ew indicates that the slope of the log-logistic model of degradation for A3 in treatment S was four times lower than in treatment Ew.

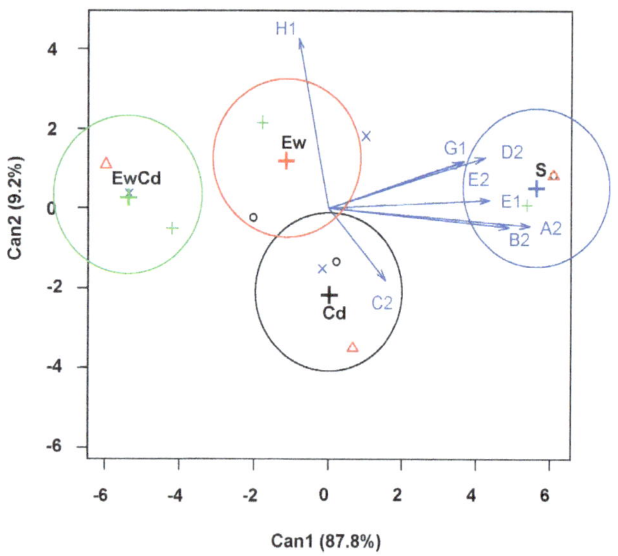

Figure 2. Plot of the two canonical discriminant functions for TM_{50}. TM_{50} refers to the time to reach 50% of the maximum colour intensity. Centroid locations of each treatment are indicated by "+".

Kinetic parameters

The results of colour development revealed sigmoidal relationships between the OD_{590} and time for all carbon sources (see point 6 in Figure 1); thus, the results were adjusted to the four parameters log-logistic model [47]. Three parameters of this model were selected for their usefulness in comparing treatments: the maximum value attained (M), the time at which the colour development is half of its maximum value (TM_{50}), and the slope (S). Those parameters presenting significant differences between treatments are shown in Table 2.

This process is the first step of the experimental approach proposed in this study; the carbon sources exhibiting significant differences between treatments (i.e., discriminant sources) would be potential candidates for the subsequent data analysis (see Figure 1) with the aim of increasing the discriminant power of the procedure.

Treatments provoked differences in the kinetic parameters of 23 of the 31 carbon sources. Specifically, the largest differences in CLPP were promoted by the activity of earthworms in Cd-treated soils, which accounted for the differences in the kinetics of 13 carbon sources. On the other hand, the addition of Cd to earthworm-treated soils only resulted in differences in two carbon sources (Table 2).

Canonical discriminant analysis (CDA)

Figure 2 shows the first two CDA axes based on the TM_{50} values of the sigmoidal models (see section 3.1). The cumulative percentage of the variance explained by the first two canonical

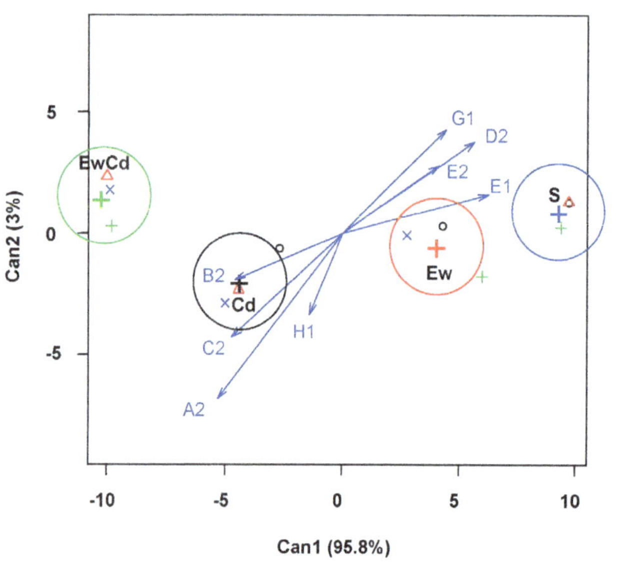

Figure 3. Plot of the two canonical discriminant functions for the maximum colour intensity (M). Centroid locations of each treatment are indicated by "+".

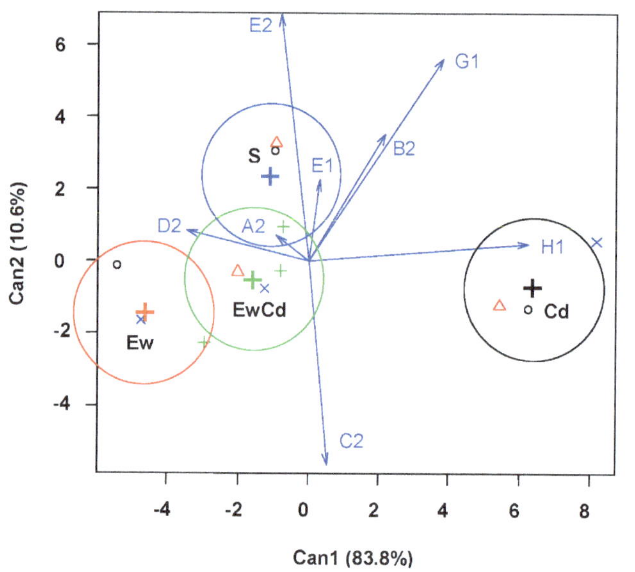

Figure 4. Plot of the two canonical discriminant functions for slope (S). Centroid locations of each treatment are indicated by "+".

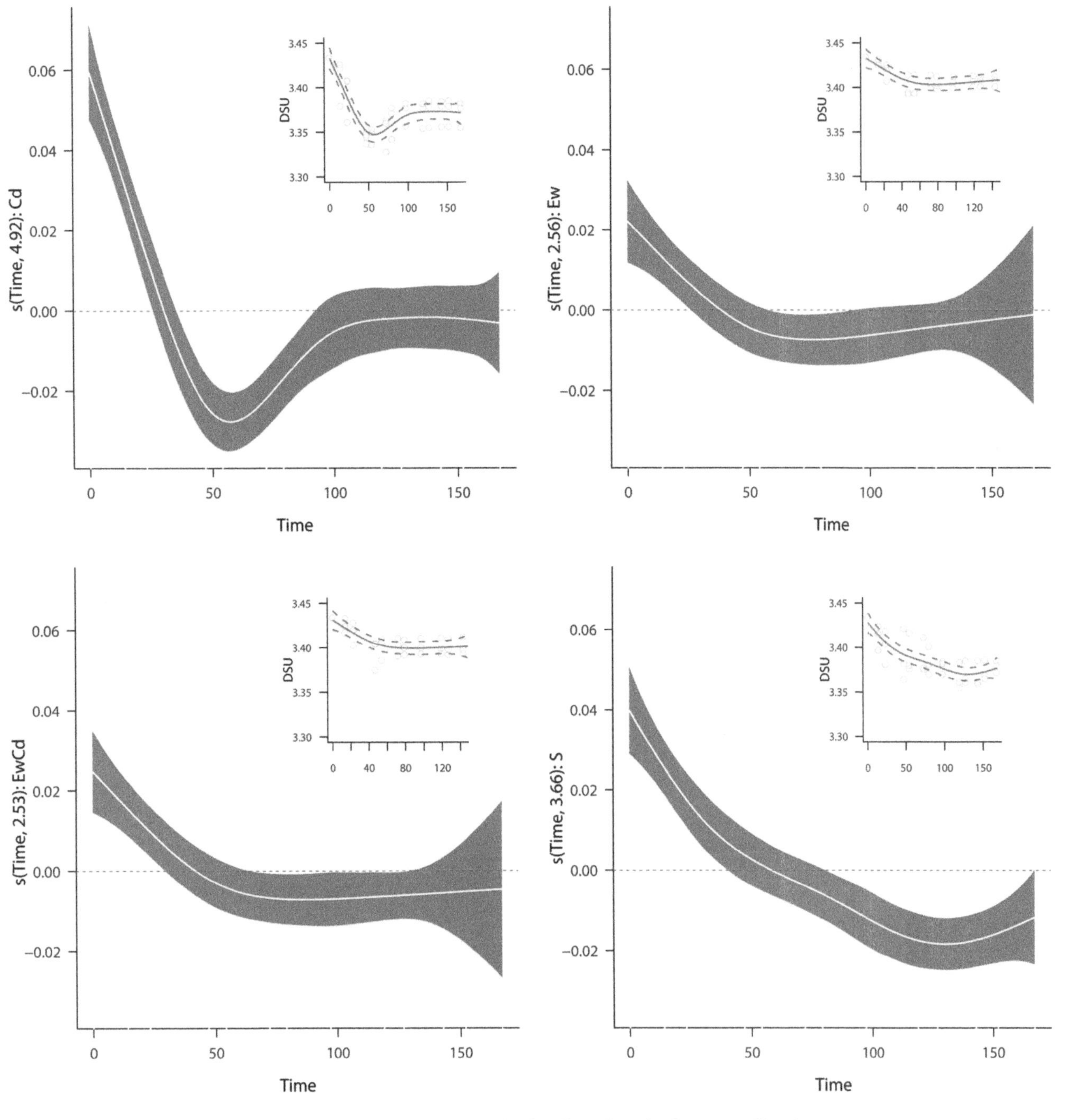

Figure 5. Estimated flexible effects of time-related changes in the diversity of substrate utilisation (DSU). The x axis represents the changes on DSU values over time (hours), obtained from the GAM with the 95% point-wise confidence band (grey coloured) and centred at zero (red dotted line) to allow comparisons. The effective degrees of freedom obtained for each predictor are indicated on the y-axis. The estimated curves in the response scale along with raw data are located in the upper right corner of each plot.

functions is 97%. The first axis accounts for 87% of the variance and is strongly influenced by D-cellobiose. The second function is strongly influenced by D-cellobiose and is also influenced by N-acetyl-D-glucosamine. Figure 3 presents the CDA based on M values. The cumulative percentage of the first two functions is very high (approximately 99%). In this case, the sources with the most influence are also N-acetyl-D-glucosamine and D-cellobiose for both axes. Figure 4 presents the CDA based on slopes. Although the cumulative percentage is also very high (94%), in this case, the standardised coefficients indicate that the weight of the different sources is more similar than that in the previous CDA.

Assessing the time course changes in diversity of substrate utilization (DSU) based on the Shannon-Wiener index using generalised additive models (GAMs)

Results from GAM model shows that treatments exhibit significantly different time course changes in DSU, with a explained deviance of 87%. The effective degrees of freedom, *chi*

Table 3. Estimated non-parametric components of GAM model, with the corresponding effective degrees of freedom, F-statistic and P value.

Smooth effect of variable	edf	F-statistic	P value
f(Time)EwCd	2.525	8.519	< 0.001
f(Time)Cd	4.92	23.868	< 0.001
f(Time)Ew	2.555	6.772	< 0.001
f(Time)S	3.664	18.483	< 0.001

square statistic and p-values of estimated flexible effects from the model are shown in table 3. In Figure 5, the red dotted line represents a hypothetical community with a DSU that was constant as a function of time; in other words, a community exhibiting such a time-course trend would degrade all carbon sources with similar affinity (S) and capacity (M). Thus, this red line can aid in our interpretation of the results. All treatments caused a decrease in DSU during the first 54 h. This decrease was more acute in the bacterial community from the Cd-treated soils. Thereafter, differences were evident between treatments. The communities from soils containing earthworms (Ew and EwCd) presented similar values until the end of the experiment (143 h) regardless the presence of Cd. On the other hand, the community from the Cd-treated soil exhibited an increase in DSU during the next 43 h and did not change thereafter. The community exhibiting the most diverse behaviour was that obtained from control soils (S). In this community, the decline in DSU was constant over time. The model validation did not show any violation of variance homogeneity (i.e. no patterns in residuals) or normality.

Discussion

Kinetic parameter analysis

Cd exposure resulted in the enhanced degradation of eight carbon sources, as shown by increases in S, M, and/or TM_{50} (see column 2 in Table 2). These results indicate the Cd tolerance of soil microbial communities. It has been shown that long-term exposure to Cd (months to years) may cultivate a more tolerant community [26,27], exhibiting enhanced enzyme activities [56]. However, our data indicate that the acclimatisation of the communities to Cd may be achieved after only a few days, which has also been shown in communities exposed to mercury [6]. The low percentage of dead cells in Cd-treated soils also supports this observation.

On the other hand, earthworm activity had a modulating effect on Cd impact. Although earthworm activity allowed for the survival of a higher number of bacterial cells, it also caused reductions in bacterial catabolic activity. As shown in column Cd/EwCd in Table 2, earthworm activity resulted in lower M values for 12 carbon sources. Because EwCd contained three times more active bacteria than Cd, this result indicates a marked reduction in their enzymatic activity. The reduction of bacterial growth and activity by earthworms has also been recently described in the absence of toxicants [57].

Canonical discriminant analysis (CDA)

In all three CDAs (based on TM_{50}, M and S), treatments resulted in different CLPPs, as shown by the centroids. The centroids are the weighted position of each treatment in the multidimensional space created by the eight carbon sources used in the analysis (Figs. 2, 3, and 4); arrows indicate the effect of each source in the discriminatory space. The clear separation between centroids means that these sources gathered a sufficient amount of information to differentiate the effect of each treatment on the CLPP. The CDA illustrates how the CLPP is affected by both Cd and earthworms. Compared with Cd, earthworms had a greater effect on the physiological profile of bacterial communities. The three parameters used (TM_{50}, M, and S) allow for a clear discrimination between the effects of the four treatments; however, these parameters have different sensitivities for each experimental factor.

CDA on TM_{50} (Figure 2). This parameter allowed for discrimination of the control soils and can be used to detect the interactive effects of Cd and earthworms. The EwCd treatment is characterised by faster degradation of D-cellobiose (G1), α-cyclodextrin (E1), and β-methyl-D-glucoside (A2), whereas Cd is characterised by faster degradation of N-acetyl-D-glucosamine (E2) and D-xylose (B2).

CDA on M (Figure 3). Cd provoked clear differences in M values; its presence reduced the capacity for degrading α-D-lactose (H1), β-methyl-D-glucoside (A2), D-xylose (B2), and i-erytritol (C2), whereas the degradation of D-mannitol (D2), α-cyclodextrin (E1), N-acetyl-D-glucosamine (E2), and D-cellobiose (G1) was enhanced.

CDA on S (Figure 4). Slopes also allow for a clear discrimination of Cd impacts. Thus, the Cd treatment appeared clearly separated from the other treatments. The S values of α-D-lactose (H1) and D-cellobiose (G1) were enhanced by Cd.

Sources indicative of specific impacts. The increase in TM_{50} for the degradation of D-cellobiose (code G1 in Biolog plates and in Figure 2) appears to be a promising indicator for earthworm activity. Changes in the slope of α-D-lactose are also indicative of earthworm activity regardless of the presence of Cd. In contrast, an increase in S for 2-hydroxybenzoic acid may be indicative of Cd impacts on bacterial-earthworm consortia.

Time course changes in diversity of substrate utilization (DSU) by GAM analysis

GAMs revealed significant differences for all treatments (compared to control -S-) in the time trend of DSU. However, all treatments shown a common initial decline (the first 54 h) in the DSU; that is generally attributed to changes in the community inoculated [11], which becomes less diverse. The CLPP would be dominated by fast growing species that are positively selected by high nutrient levels [4]. Other mechanisms related to metal tolerance and earthworm's effects would explain the differences shown beginning at 54 h and continuing onward.

The acute initial DSU decrease shown by Cd treated soils, suggests the dominance of a few Cd-tolerant microbial taxa [26,27]. That community presented very high metabolic capacities on a few substrata (as indicated by the results shown in Table 2), and that may explain the acute decrease of DSU (i.e. intense utilization of a few carbon substrata). After 54 h, and over the next 43 h the increase of the DSU suggest that other taxa, less dominant, may reach high numbers in the wells containing sources that are less suitable for the Cd-tolerant taxa.

Earthworms were a much more determining factor in explaining time course changes in DSU than Cd. Accordingly, Ew and EwCd soils presented similar trends and values, regardless the presence of Cd. Moreover, both treatments presented similar number of bacteria (between 3 and 3.6e7 cells g^{-1} DW) and higher than Cd-treated soils (1.1e7). Altogether indicates that earthworm activity provided some protection to the bacterial community against Cd. The bacterial community promoted by earthworms activity during the first 16 days of the experiment, maintained its degrading diversity even after the exposure to a new stressor (Cd). Differently to other studies [58], Cd did not exerted an additional selection pressure. This experimental approach, based on the use of DSU and GAMs allowed for a global and statistically relevant interpretation of the changes in carbon source utilisation, highlighting the key role of earthworms on the protection of microbial communities against the cadmium.

Author Contributions

Conceived and designed the experiments: SM JL JJ EN. Performed the experiments: SM JL. Analyzed the data: SM JL MP EN. Contributed reagents/materials/analysis tools: JJ EN. Wrote the paper: SM EN.

References

1. Garland JL (1997) Analysis and interpretation of community-level physiological profiles in microbial ecology. FEMS Microbiol Ecol 24: 289–300.
2. Klingler JM, Stowe RP, Obenhuber DC, Groves TO, Mishra SK, et al. (1992) Evaluation of the Biolog Automated Microbial Identification System. Appl Environ Microbiol 58: 2089–2092.
3. Insam H (1997) A new set of substrates proposed for community characterization in environmental samples. In: Insam H, Rangger A, editors. Microbial Communities Functional versus structural approaches. Heidelberg: Springer Verlag. Springer. pp. 260–261.
4. Preston-Mafham J, Boddy L, Randerson PF (2002) Analysis of microbial community functional diversity using sole-carbon-source utilisation profiles: a critique. FEMS Microbiol Ecol 42 1–14.
5. Rutgers M, Van't Verlaat IM, Wind B, Posthuma L, Breure AM (1998) Rapid method for assessing pollution-induced community tolerance in contaminated soil. Environ Toxicol Chem 17: 2210–2213.
6. Müller AK, Rasmussen LD, Sorensen JS (2001) Adaptation of the bacterial community to mercury contamination. FEMS Microbiol Lett 204: 49–53.
7. Van Beelen P, Wouterse M, Posthuma L, Rutgers M (2004) Location-specific ecotoxicological risk assessment of metal-polluted soils. Environ Toxicol Chem 23: 2769–2779.
8. Niklińska M, Chodak M, Laskowski R (2005) Characterization of the forest humus microbial community in a heavy metal polluted area. Soil Biol Biochem 37: 2185–2194.
9. Kamitani T, Oba H, Kaneko N (2006) Microbial biomass and tolerance of microbial community on an aged heavy metal polluted floodplain in Japan. Water Air Soil Pollut 172: 185–200.
10. Haack SK, Garchow H, Klug MJ, Forney LJ (1995) Analysis of Factors Affecting the Accuracy, Reproducibility, and Interpretation of Microbial Community Carbon Source Utilization Patterns. Appl Environ Microbiol 61: 1458–1468.
11. Smalla K, Wachtendorf U, Heuer H, Liu WT, Forney L (1998) Analysis of BIOLOG GN substrate utilization patterns by microbial communities. Appl Environ Microbiol 64: 1220–1225.
12. Classen AT, Boyle SI, Haskins KE, Overby ST, Hart SC (2003) Community-level physiological profiles of bacteria and fungi: plate type and incubation temperature influences on contrasting soils. FEMS Microbiol Ecol 44: 319–328.
13. Lavelle P, Decaens T, Aubert M, Barot S, Blouin M, et al. (2006) Soil invertebrates and ecosystem services. Eur J Soil Biol 42: S3–S15.
14. Jones CG, Lawton JH, Shachak M (1994) Organisms as ecosystem engineers. Oikos 69: 373–386.
15. Joschko M, Diestel H, Larink O (1989) Assessment of earthworm burrowing efficiency in compacted soil with a combination of morphological and soil physical measurements. Biol Fertil Soils 8: 191–196.
16. Shipitalo MJ, Protz R (1989) Chemistry and micromorphology of aggregation in earthworm casts. Geoderma 45: 357–374.
17. Marinissen JCY, Dexter AR (1990) Mechanisms of stabilization of earthworm casts and artificial casts. Biol Fertil Soils 9: 163–167.
18. Jégou D, Schrader S, Diestel H, Cluzeau D (2001) Morphological, physical and biochemical characteristics of burrow walls formed by earthworms. Appl Soil Ecol 17: 165–174.
19. Bouché MB (1975) Action de la faune sur les états de la matière organique dans les écosystèmes. In: Gilbertus K, Reisinger O, Mourey A, Cancela da Fonseca JA, editors. Biodegradation et Humification. Sarreguemines: Pierson. pp. 157–168.
20. Zhang H, Schrader S (1993) Earthworm effects on selected physical and chemical properties of soil aggregates. Biol Fertil Soils 15: 229–234.
21. Mc Laughlin MJ, Singh BRE (1999) Cadmium in Soils and Plants. DordrechtThe Netherlands: Kluwer Academic Publishers. 271 p.
22. Violante A, Huang PM, Gadd GME (2008) Biophysico-chemical processes of heavy metals and metalloids in soil environments. New York: John Wiley & Sons. 658 p.
23. Sizmur T, Hodson ME (2009) Do earthworms impact metal mobility and availability in soil?: a review. Environ Pollut 157: 1981–1989.
24. Udovic M, Lestan D (2007) The effect of earthworms on the fractionation and bioavailability of heavy metals before and after soil remediation. Environ Pollut 148: 663–668.
25. Lavelle P, Spain A (2001) Soil Ecology. DordrechtThe Netherlands: Kluwer Academic Publishers. 654 p.
26. Almås ÅR, Bakken LR, Mulder J (2004) Changes in tolerance of soil microbial communities in Zn and Cd contaminated soils. Soil Biol Biochem 36: 805–813.
27. Diaz-Ravina M, Baath E (1996) Development of metal tolerance in soil bacterial communities exposed to experimentally increased metal levels. Appl Environ Microbiol 62: 2970–2977.
28. Degens BP, Schipper LA, Sparling GP, Duncan LC (2001) Is the microbial community in a soil with reduced catabolic diversity less resistant to stress or disturbance? Soil Biol Biochem 33: 1143–1153.
29. Hastie T, Tibshirani R (1990) Exploring the Nature of Covariate Effects in the Proportional Hazards Model. Biometrics 46: 1005–1016.
30. Wood SN (2006) On confidence intervals for generalized additive models based on penalized regression splines. Aust N Z J Stat 48: 445–464.
31. Guisan A, Edwards TC, Hastie T (2002) Generalized linear and generalized additive models in studies of species distributions: setting the scene. Ecol Model 157: 89–100.
32. Yee TW, Mitchell ND (1991) Generalized Additive-Models in Plant Ecology. J Veg Sci 2: 587–602.
33. Dominici F, McDermott A, Hastie TJ (2004) Improved semiparametric time series models of air pollution and mortality. J Amer Statist Assoc 99: 938–948.
34. Römbke J (2006) Tools and techniques for the assessment of ecotoxicological impacts of contaminants in the terrestrial environment. Hum Ecol Risk Assess 12: 84–101.
35. Ministry-of-Agriculture (1975) Orden de 5 de diciembre de 1975, del Ministerio de Agricultura, por la que se aprueban como oficiales los métodos de análisis de suelos y aguas. Boletín Oficial del Estado (B.O.E.) 78: 6458–6491.
36. ASTM Standard D2216 (1998) Standard test method for laboratory determination of water (moisture) content of soil and rock by mass. West Conshohocken, PA: ASTM International. DOI: 10.1520/D2216-98. Available: http://www.astm.org.
37. Moret-Fernández D, Vicente J, Lera F, Latorre B, López MV, et al. (2011) TDR-Lab version 1.2. Users' Guide. Available: http://digital.csic.es/bitstream/10261/35790/3/TDR-Lab%20V.1.2.3_User-Guide.pdf.
38. Khalil MA, Abdel-Lateif HM, Bayoumi BM, van Straalen NM, van Gestel CAM (1996) Effects of metals and metal mixtures on survival and cocoon production of the earthworm Aporrectodea caliginosa. Pedobiologia 40: 548–556.
39. Hobbelen PH, Koolhaas JE, van Gestel CA (2006) Bioaccumulation of heavy metals in the earthworms Lumbricus rubellus and Aporrectodea caliginosa in relation to total and available metal concentrations in field soils. Environ Pollut 144: 639–646.
40. Katayama A, Kai K, Fujie K (1998) Extraction efficiency, size distribution, colony formation and [H-3]-thymidine incorporation of bacteria directly extracted from soil. Soil Sci Plant Nutr 44: 245–252.
41. Ikeda H, Wang X, Inoue K, Someya T (2004) Extraction of bacterial cells from soil by centrifugation. Proceeding of the International Symposium on Identification and Use of Microbial Resources for Sustainable Agriculture. pp. 169.
42. Gregori G, Citterio S, Ghiani A, Labra M, Sgorbati S, et al. (2001) Resolution of viable and membrane-compromised bacteria in freshwater and marine waters based on analytical flow cytometry and nucleic acid double staining. Appl Environ Microbiol 67: 4662–4670.
43. Shannon CE (1948) A Mathematical Theory of Communication. The Bell System Technical Journal 27: 623–656.
44. Tuomisto H (2010) A consistent terminology for quantifying species diversity? Yes, it does exist. Oecologia 164: 853–860.
45. Tam L, Derry AM, Kevan PG, Trevors JT (2001) Functional diversity and community structure of microorganisms in rhizosphere and non-rhizosphere Canadian arctic soils. Biodivers Conserv 10: 1933–1947.
46. Wang Y, Ouyang ZY, Zheng H, Wang XK, Chen FL, et al. (2011) Carbon metabolism of soil microbial communities of restored forests in Southern China. J Soils Sediments 11: 789–799.
47. Hackett CA, Griffiths BS (1997) Statistical analysis of the time-course of Biolog substrate utilization. J Microbiol Methods 30: 63–69.
48. Insam H, Goberna M (2004) Use of Biolog for the Community Level Physiological Profiling (CLPP) of environmental samples. In: Kowalchuk GA, de Bruijn FJ, Head IM, Akkermans AD, van Elsas JD, editors. Molecular Microbial Ecology Manual. Boston: Kluwer Academic Publishers. p. 1780.
49. Ritz C, Streibig JC (2005) Bioassay analysis using R. J Stat Softw 12: 1–22.
50. Jiang W, Wang JJ, Tang JS, Hou F, Lu YT (2010) Soil bacterial functional diversity as influenced by cadmium, phenanthrene and degrade bacteria application. Environ Earth Sci 59: 1717–1722.

51. Akmal M, Xu JM, Li ZJ, Wang HZ, Yao HY (2005) Effects of lead and cadmium nitrate on biomass and substrate utilization pattern of soil microbial communities. Chemosphere 60: 508–514.
52. Yao HY, Xu JM, Huang CY (2003) Substrate utilization pattern, biomass and activity of microbial communities in a sequence of heavy metal-polluted paddy soils. Geoderma 115: 139–148.
53. Wood SN (2003) Thin plate regression splines. J R Stat Soc Series B Stat Methodol 65: 95–114.
54. R_Development_Core_Team (2012) R: A language and environment for statistical computing. Vienna, Austria: R Foundation for Statistical Computing. ISBN 3-900051-07-0. Available: http:www.r-project.org.
55. Friendly M, Fox J (2013) Candisc: Visualizing Generalized Canonical Discriminant and Canonical Correlation Analysis. R package version 0.6–3.
56. Karaca A, Cetin S, Turgay O, Kizilkaya R (2010) Effects of Heavy Metals on Soil Enzyme Activities. In: Sherameti I, Varma A, editors. Soil Heavy Metals. Berlin Heidelberg: Springer Verlag. pp. 237–262.
57. Gomez-Brandon M, Lores M, Dominguez J (2012) Species-Specific Effects of Epigeic Earthworms on Microbial Community Structure during First Stages of Decomposition of Organic Matter. Plos One 7 (2): e31895. doi:10.1371/journal.pone.0031895.
58. Nakatsu CH, Carmosini N, Baldwin B, Beasley F, Kourtev P, et al. (2005) Soil microbial community responses to additions of organic carbon substrates and heavy metals (Pb and Cr). Appl Environ Microbiol 71: 7679–7689.

11

LEGO® Bricks as Building Blocks for Centimeter-Scale Biological Environments: The Case of Plants

Kara R. Lind[1], Tom Sizmur[1,2¤], Saida Benomar[1,2], Anthony Miller[1,3], Ludovico Cademartiri[1,2,4*]

1 Department of Materials Science & Engineering, Iowa State University, Ames, Iowa, United States of America, **2** Ames Laboratory, US Department of Energy, Iowa State University, Ames, Iowa, United States of America, **3** Department of Agronomy, Iowa State University, Ames, Iowa, United States of America, **4** Department of Chemical & Biological Engineering, Iowa State University, Ames, Iowa, United States of America

Abstract

LEGO bricks are commercially available interlocking pieces of plastic that are conventionally used as toys. We describe their use to build engineered environments for cm-scale biological systems, in particular plant roots. Specifically, we take advantage of the unique modularity of these building blocks to create inexpensive, transparent, reconfigurable, and highly scalable environments for plant growth in which structural obstacles and chemical gradients can be precisely engineered to mimic soil.

Editor: Matthias Rillig, Freie Universität Berlin, Germany

Funding: The work was funded by Iowa State University through a startup grant to LC. The funders had no role in study design, data collection and analysis, decision to publish, or preparation of the manuscript.

Competing Interests: The authors have declared that no competing interests exist.

* Email: lcademar@iastate.edu

¤ Current address: Department of Sustainable Soils and Grassland Systems, Rothamsted Research, Harpenden, Heartfordshire, United Kingdom

Introduction

Microfluidics[1,2,3,4] and other engineered environments[5,6] can produce highly controlled micrometer-scale environments for the study of organismal model systems (e.g., mammalian cells). However, scientists or engineers interested in manipulating the environment of cm-scale organisms (e.g., plants) have remarkably few convenient tools at their disposal[7,8]. This paucity is partly due to the demanding design requirements associated with larger scales (e.g., cost). This liability is particularly evident in the study of plants and their root systems.

The development of plants in soil is an important subject of investigation. The provision of food to the global human population is under severe pressure (our supply of food is predicted to be far below demand by 2050[9]) and depends on plant roots[10] (97.6% of global calorie consumption is derived from plants[11]). Roots influence a plant's yield and whether a plant will survive stresses. We know that root growth is strongly affected by its environment, soil, but our mechanistic understanding of these effects is imperfect[10,12] and strongly limited by technical challenges.

Root development is a difficult process to study experimentally. (i) Plants display highly variable root systems, even when genetically identical[13]. (ii) Roots are remarkably sensitive to a variety of stimuli (e.g., gravity, light, touch, moisture, nutrients, oxygen, temperature, trauma, electric fields[14]). (iii) Any volume of soil is unique and impossible to replicate exactly[15,16]. (iv) Its heterogeneity makes it opaque to most forms of radiation[17]. (v) Its structural and chemical characteristics (i.e., porosity, surface chemistry, nutrient gradients, oxygen gradients, bulk composition, soil biota) cannot be independently manipulated.

One approach to avoid this complexity is to characterize the growth of plants in soil-less media, e.g., hydrogels, paper, glass beads, sand. These systems are less inhomogeneous and irreproducible than soil and can be modified – usually to a limited extent – to mimic soil properties such as chemical composition [18], physical structure [19,20], water availability [21], refractive index [22], or mechanical strength [23]. However, the lack of modularity, versatility, structural precision, and the very limited control over structural and chemical heterogeneities in these systems severely limits the type, complexity, and reproducibility of the experiments they can perform. Microfluidic approaches offer fascinating capabilities for the study of plant roots, but are subjected to limitations in their throughput and in the size of the plants they can host [4,24,25].

We here demonstrate that LEGO bricks are highly convenient and versatile building blocks for building cm-scale engineered environments for plant roots. Their modularity enables the fabrication of environments with highly controlled structural and chemical heterogeneities that are suitable for convenient quantitative studies of environmental effects on plant phenotypes[26].

System Design

A convenient experimental platform for the study of root development in controlled environments must satisfy a demanding set of design constraints. LEGO bricks, while conceived and sold as toys, satisfy these constraints.

Modularity

Modular systems can produce many structurally distinct environments from a few different components. Features can be added or removed without remanufacturing the entire experimental setup. LEGO structures are modular. The smallest bricks are 8*x*8*x*6 mm. The largest are 48*x*8*x*50 mm. The number of different structures that can be made with these units is staggering: six identical bricks can form almost a billion different structures[27].

Scalability

Confinement can affect the physiology of an organism[28]. The ability to create experimental platforms of a range of sizes enables researchers to study any plant and their ensembles. LEGO structures can be easily scaled to accommodate different plant species: the smallest enclosed environment that can be produced with LEGO bricks measures 0.35 cm^3 in volume, and it is theoretically possible to create LEGO structures capable of containing the largest plant species.

Structurally precise

Roots are sensitive to the physical structure of their environment. For example, the study of root thigmotropism (the response of a root to touch) requires structures that are of an exact size and shape. The molds used to produce LEGO bricks are accurate to within 5 μm[29], which is comparable to the diameter of a root hair and to the resolution of 3D printing (minimum layer thickness is ~50 μm in some of the best current models).

Capable of increasing levels of complexity

A good model system allows for the controlled introduction of experimental variables. LEGO bricks can be used – as shown below – for the generation of physical barriers, air pockets, chemical gradients, and interconnecting chambers to control the growth environment of a plant.

Simplicity

Simple setups reduce the risk of operator-induced systematic errors. Differently from microfluidic approaches, the assembly of structures from LEGO bricks does not require technical training so undergraduate students can perform LEGO brick-based plant experiments from their first day in the laboratory. Simple experiments that demonstrate fundamental principles of plant growth (e.g., tropisms) or encourage experimental creativity can be conducted by school children of all ages during science education classes[30].

Reproducibility

Plant root experimental platforms (e.g. sand columns, rhizotrons, split-root pots) are typically made from scratch. Their reproducibility between labs or across continents cannot be guaranteed. The unique selling point of LEGO bricks is that bricks bought in separate batches are essentially identical and backward- and forward-compatible with each other. Experiments created from LEGO bricks can be accurately replicated anywhere in the world.

Affordability

The more expensive each experiment is, the fewer experiments can be conducted with finite resources. This fact is especially meaningful in developing nations[31] and in research fields, like plant science, where throughput is an essential parameter. Individual LEGO bricks cost between $0.10 and $1.00 and are sold worldwide. A LEGO structure capable of growing a plant costs $3.1 and is reusable: some LEGO bricks in our lab have been in near-constant use for two years.

High throughput

The ability to run a large number of experiments at the same time is essential for the establishment, for example, of genotype-environment-phenotype relationships[32]. A LEGO structure like the one shown in Figure 1 can be assembled in less than a minute.

Transparency

Twenty eight different LEGO bricks are made from transparent polycarbonate which can be assembled into transparent structures for the real-time monitoring of plant roots over time.

Autoclavable

Tissue cultures require sterile conditions. Transparent LEGO bricks (with the exception of large base plates) are autoclavable due to their polycarbonate composition: they still fit together in the same way as they do prior to autoclaving and are still transparent after more than 50 autoclave cycles. Opaque LEGO bricks are made from acrylonitrile-butadiene-styrene block copolymer (ABS), and can be sterilized with ethanol or bleach.

Three-dimensionality

While 2D platforms offer significant advantages in terms of visualization and practicality[33], 3D mediums are more representative of the natural environment of roots[34]. LEGO bricks allow for the creation of nearly arbitrary 3D structures.

Chemical inertness

Legislative standards ensure the safety to children of LEGO bricks sold in the USA and EU. These standards include maximum soluble levels of toxic or hazardous substances.

Compatibility with existing growth environments

Tools that integrate with existing experimental platforms are often the most useful. The modularity of LEGO structures enables them to integrate with laboratory protocols e.g., LEGO structures can hold gel, beads, sand, soil, 3D-printed elements, or be structurally precise elements in other setups[35].

Results and Discussion

Figure 1 shows a flow diagram of the design, assembly, disassembly, and re-assembly of an experiment based on LEGO bricks. The website of the LEGO Group (www.lego.com) provides a free software (LEGO Digital Designer, LDD) for the CAD-like design of structures using any available LEGO brick. The software outputs a step-by-step assembly guide and a list of the required parts. Individual bricks can be purchased through the "Pick a Brick" section of www.lego.com or other outlets (e.g. local LEGO stores, EBay). Sterilization of LEGO bricks can be performed before or after assembly. The preservation of sterility requires the structure to be maintained in a sterile container during the course of an experiment.

The simplest example of a plant germination and growth environment based on LEGO bricks is shown in Figure 1. The LEGO bricks are assembled into a container that contains a root growth medium on which a seed is germinated and grown: Figure 1, for example, shows a *Brassica rapa*, Wisconsin Fast Plant, Astroplant, *dwf1*, growing on a transparent hydrogel, Gellan gum. While gel media for root growth are very commonly used in

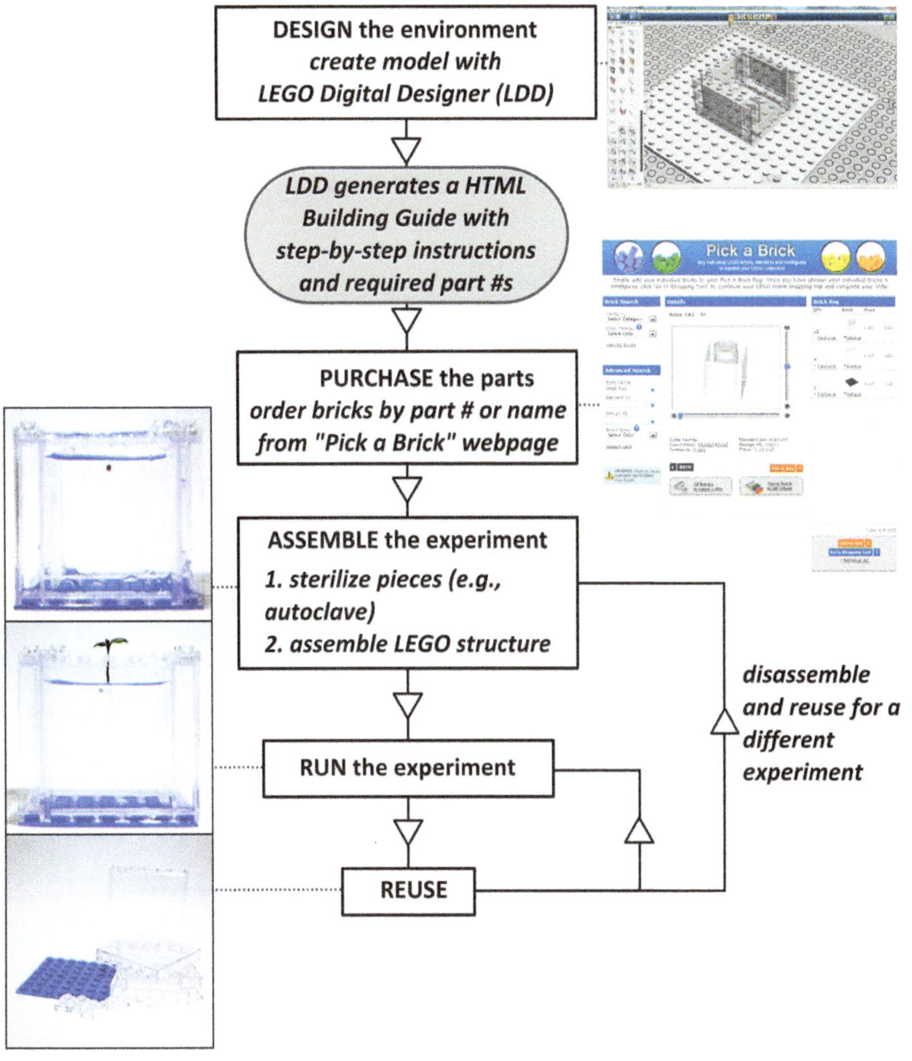

Figure 1. Scheme of the process of carrying out a plant growth experiment using LEGO bricks as building blocks. The same process can be used to prototype and fabricate other biological experiments.

experiments[26], they are not the best mimic of soil: root architectures grown in an homogeneous media will not match those of plants grown in real soil[36]. However, gel media allows us to demonstrate three essential capabilities of LEGO-based biological environments: their ability to hold liquids, their compatibility with real-time observation and root structure analysis, and their use in generating reconfigurable environments that include controlled heterogeneities. Furthermore, LEGO environments are not limited to gel media: the environment shown in Figure 1 can hold other media of choice, e.g., sand, perlite, soil.

Since structures built from LEGO bricks are not waterproof, their use to hold gels requires some stratagems (see Supporting Information for details and Movie S1 for a demonstration). The LEGO structure must be chilled in a freezer before the cool gel solution is poured in it just prior to setting. Using this approach, leakage of the gel solution was minimal. These basic environments can be easily scaled to match the dimensions of the organism under consideration and the time the organism is allowed to grow. Figures 2a, 2b, and 2c show the use of LEGO bricks to create containers with very different dimensions ($5\times5\times5$ cm, $10\times10\times5$ cm, and $20\times20\times10$ cm) for the growth of Fast Plants, *Triticum polonicum* (Wheat), and *Zea mays* (Corn).

The transparency and flat walls of LEGO bricks allows for good quality real time imaging of the development of the root system. Figure 2d shows time-lapse imaging of *Lepidium sativum* (Garden cress) roots over the course of ~48 hrs from germination in a LEGO-based environment. The plant was chosen for its relatively fine roots (~350 μm thickness) that would have been hard to image in a poorly transparent system.

The reversible nature of the mechanical bond between the bricks provides two important capabilities: the creation of reconfigurable biological environments, and of highly controlled heterogeneities (i.e., solid obstacles, air pockets, and chemical and soil biota gradients) in an otherwise homogeneous growth medium. Figure 2e demonstrates a reconfigurable plant growth environment. Two Fast Plants were grown in gel in separate containers assembled on the same base plate. The LEGO brick walls separating the two containers were removed and reconfigured to make one larger container. The volume separating the two plants was then filled with more gel, fluidically connecting the two plants. Figure 3 demonstrates the generation of controlled

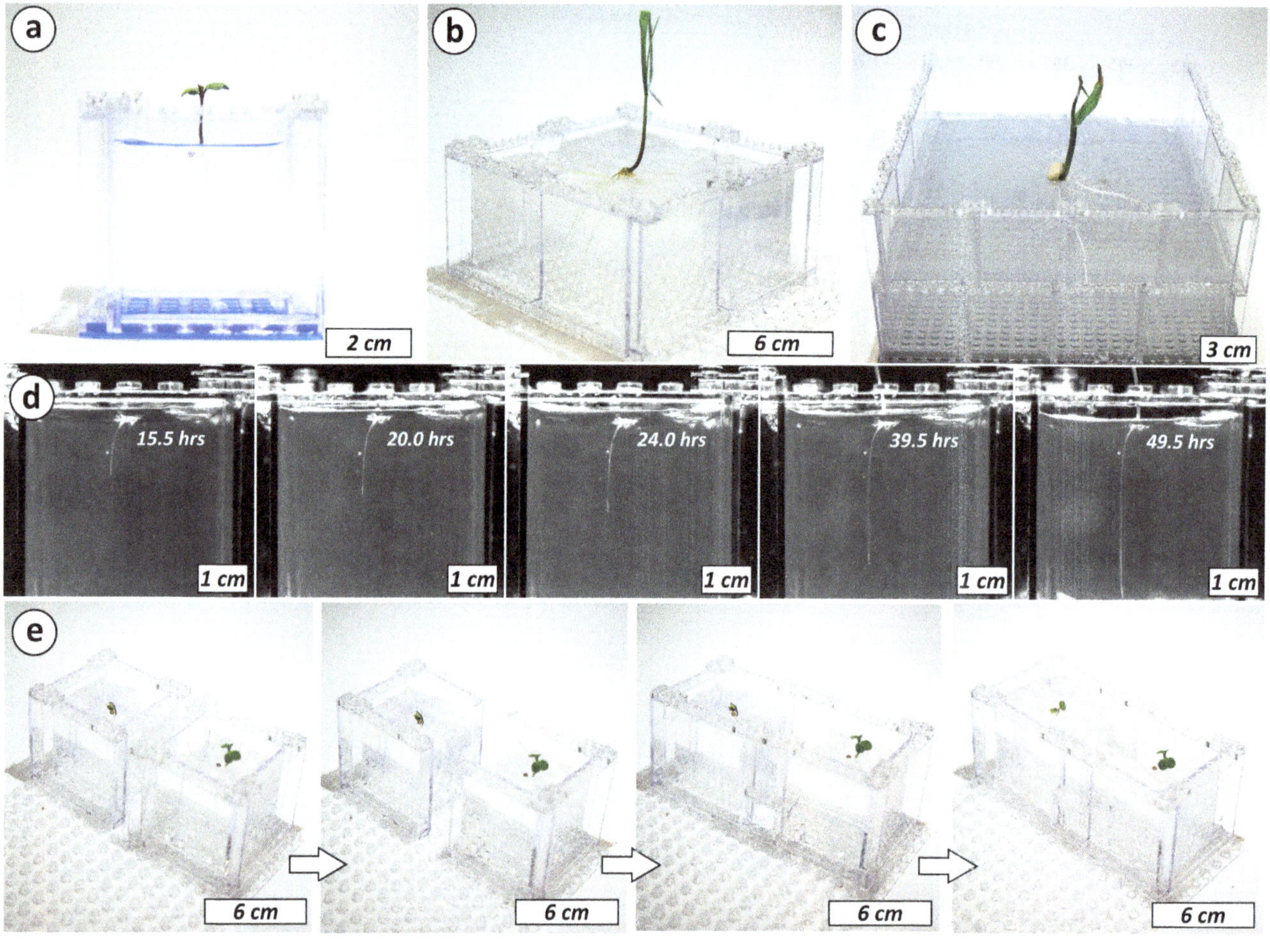

Figure 2. Versatility, transparency, and modularity of the LEGO-based environments for plant growth. a-c) pictures of basic LEGO-based environments growing Fast Plants, Wheat and Corn. The size of the environments can be controlled to match the size of the organism under consideration. d) Timelapse imaging of *Lepidium sativum* root development through the walls of a LEGO-based environment. The images indicate the time since germination. e) Examples of a LEGO-based system that allows for the dynamic change of the environment of a plant. Two plants (Fast Plants) are grown in isolated environments. The environment is then modified, during growth, to allow the two plants to share the same environment and interact.

heterogeneities in a homogeneous gel medium for plant growth by a simple templating strategy borrowed from the materials science "toolbox". A gelling mixture was poured into a LEGO-based mold. LEGO-based features in the mold can be used as solid heterogeneities to study the physical interaction of plant roots with solid objects (thigmotropism). After gelation, LEGO-based molds could be removed, leaving behind precisely positioned air pockets that would serve as sources of oxygen gradients into the gel. These pockets could be then refilled with a hydrogel containing a desired chemical to generate precisely positioned one-dimensional (Figure 3, bottom left panel) or two-dimensional (Figure 3, bottom right panel) nutrient gradients. The above process can be combined to create environments with solid heterogeneities, air pockets (i.e., oxygen gradients), and chemical (e.g., nutrients, toxins, signaling molecules) gradients simultaneously (see Appendices S1).

Conclusions

In summary we demonstrated that LEGO-based environments can (i) scale to the size of the organism under consideration, (ii) allow for real time monitoring of root systems in 3D, (iii) be structurally reconfigured to change the environment of an organism during its development, and (iv) generate precisely controlled heterogeneities (i.e., solid barriers, air pockets, chemical and soil biota gradients) in an otherwise homogeneous growing medium.

This manuscript also proposes a broader concept: the use of reusable and mechanically interlocking building blocks for the construction of biological environments for cm-scale organisms and systems of organisms. Modular and reusable building blocks can alleviate the challenges associated with the large scales of plant science experiments, while providing new capabilities (e.g., controlled heterogeneities, reconfigurable environments) for the study of environmental effects on biosystem development. Furthermore, this concept provides materials chemists and engineers with two stimulating opportunities: (i) to creatively engage with the synthesis or development of increasingly capable cm-scale biological environments for important organisms such as plants, and (ii) to use these environments to test hypothesis concerning plants that are compatible with their skillset. Compelling opportunities lie in extending our approach to chemically synthesized bricks, LEGO-compatible 3D-printed bricks and objects, and commercial bricks from other manufacturers. Our

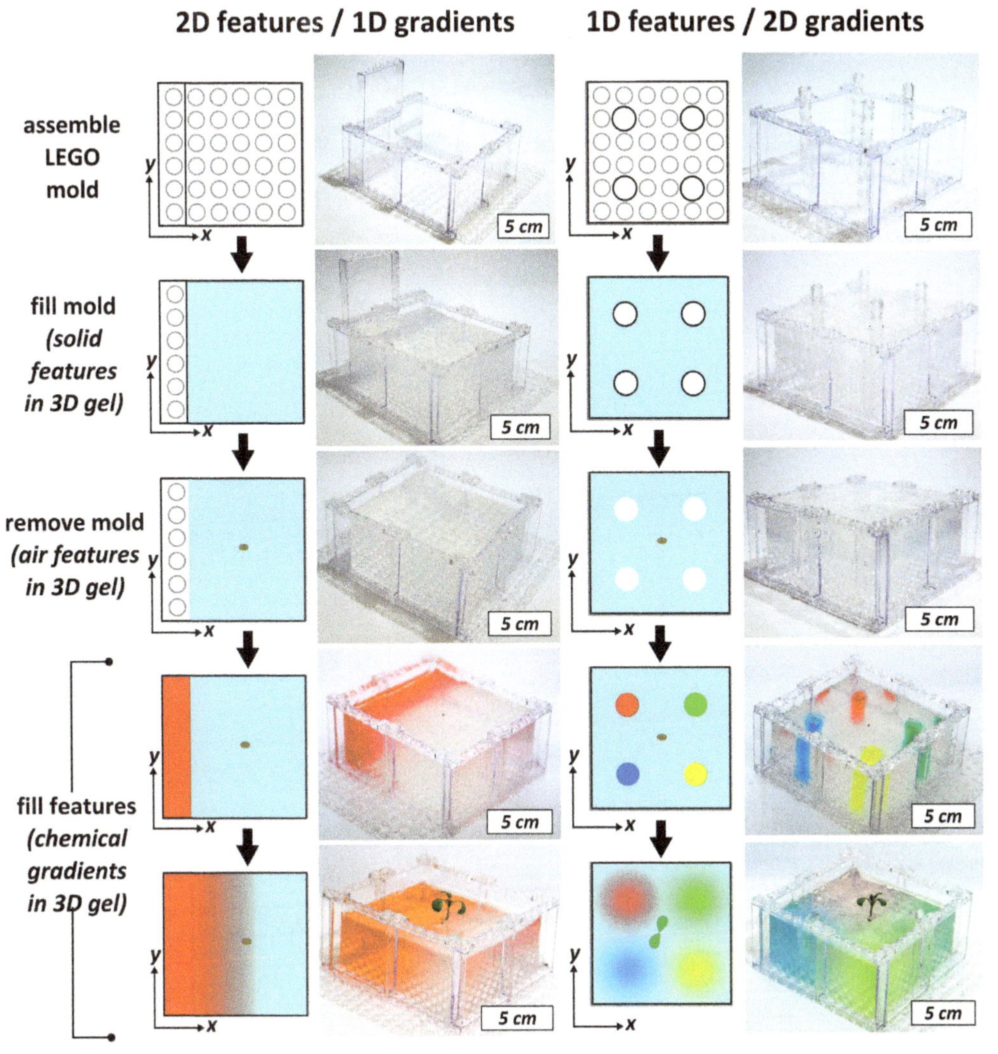

Figure 3. Fabrication of controlled heterogeneities in plant growth environments. Sequence of diagrams and corresponding images illustrating the generation of a 1D and 2D heterogeneities (solid features, air pockets, and chemical gradients) across a developing root system of a Fast Plant. In the bottom panels, the red linear gradient is of MS nutrients (dye is added for visibility), while the radial gradients are from potassium phosphate (green), potassium nitrate (yellow), calcium chloride (red), and magnesium sulfate (blue).

laboratory will be introducing a set of integrated tools for the fabrication of frugal but sophisticated[37] cm-scale environments for the study of plants and other organisms[35].

Supporting Information

Appendices S1 Materials, methods, and procedures for the generation of (i) basic LEGO-based environments, (ii) LEGO-based environments with linear chemical gradients, (iii) LEGO-based environments with cylindrical chemical gradients, (iv) larger scale LEGO-based environments. Demonstration of a LEGO-based environment combining controlled obstacles, air pockets, and multiple chemical gradients. Calculation of the smallest possible LEGO-based environment. Limitations, open questions, and failed experiments.

Figure S1 Summary snapshots of the assembly of a basic LEGO-based plant growth environment.

Figure S2 Summary snapshots of steps for root analysis using WinRhizo of two *brassica rapa* roots grown in LEGO-based plant growth environment.

Figure S3 Snapshots of the procedure to produce linear features (solid obstacles, air pockets and chemical gradients) in a homogeneous gel by using LEGO bricks.

Figure S4 Snapshots of the procedure to produce 2-dimensional features (solid obstacles, air pockets and cylindrical chemical gradients) in a homogeneous gels by using LEGO bricks.

Figure S5 Photograph of a 3D plant growth environment based on LEGO bricks featuring three different types of heterogeneities: a solid barrier (top left), an air pocket (bottom right) and two different cylindrical chemical gradients (top right and bottom left).

Figure S6 Depiction of the smallest LEGO-based environment.

Movie S1 Assembly of a basic LEGO-based environment.

Movie S2 Plant harvesting procedure from a basic LEGO-based environment.

Table S1 Preparation of salt solutions for cylindrical dye experiment.

Acknowledgments

We thank Dr. Kuloth V. Shajesh for valuable discussions and William Rekemeyer for help in the laboratory.

Author Contributions

Conceived and designed the experiments: LC. Performed the experiments: KRL TS SB AM. Analyzed the data: TS LC. Contributed reagents/materials/analysis tools: LC. Contributed to the writing of the manuscript: LC TS SB KRL.

References

1. Hulme SE, Shevkoplyas SS, Apfeld J, Fontana W, Whitesides GM (2007) A microfabricated array of clamps for immobilizing and imaging c-elegans. Lab Chip 7: 1515–1523.
2. Lucchetta EM, Lee JH, Fu LA, Patel NH, Ismagilov RF (2005) Dynamics of drosophila embryonic patterning network perturbed in space and time using microfluidics. Nature 434: 1134–1138.
3. Lutolf MP, Hubbell JA (2005) Synthetic biomaterials as instructive extracellular microenvironments for morphogenesis in tissue engineering. Nat. Biotechnol. 23: 47–55.
4. Parashar A, Pandey S (2011) Plant-in-chip: Microfluidic system for studying root growth and pathogenic interactions in arabidopsis. Appl. Phys. Lett. 98.
5. Derda R, Laromaine A, Mammoto A, Tang SKY, Mammoto T, et al. (2009) Paper-supported 3d cell culture for tissue-based bioassays. Proc. Natl. Acad. Sci. U. S. A. 106: 18457–18462.
6. Tibbitt MW, Anseth KS (2009) Hydrogels as extracellular matrix mimics for 3d cell culture. Biotechnol. Bioeng. 103: 655–663.
7. Clark KA, Krysan PJ (2007) Protocol: An improved high-throughput method for generating tissue samples in 96-well format for plant genotyping (ice-cap 2.0). Plant Methods 3.
8. Krysan P (2004) Ice-cap. A high-throughput method for capturing plant tissue samples for genotype analysis. Plant Physiol. 135: 1162–1169.
9. Ray DK, Mueller ND, West PC, Foley JA (2013) Yield trends are insufficient to double global crop production by 2050. PLoS ONE 8.
10. Lynch J (1995) Root architecture and plant productivity. Plant Physiol. 109: 7.
11. FAO (2009) Food balance sheets. Food and Agriculture Organization of the United Nations.
12. de Dorlodot S, Forster B, Pagès L, Price A, Tuberosa R, et al. (2007) Root system architecture: Opportunities and constraints for genetic improvement of crops. Trends Plant Sci. 12: 474–481.
13. Sultan SE (2000) Phenotypic plasticity for plant development, function and life history. Trends Plant Sci. 5: 537–542.
14. Gilroy S, Masson PH (2008) Plant tropisms: Wiley Online Library.
15. Nunan N, Wu K, Young IM, Crawford JW, Ritz K (2002) In situ spatial patterns of soil bacterial populations, mapped at multiple scales, in an arable soil. Microb. Ecol. 44: 296–305.
16. Young IM, Crawford JW (2004) Interactions and self-organization in the soil-microbe complex. Science 304: 1634–1637.
17. Tracy SR, Roberts JA, Black CR, McNeill A, Davidson R, et al. (2010) The x-factor: Visualizing undisturbed root architecture in soils using x-ray computed tomography. J. Exp. Bot. 61: 311–313.
18. Zhang H, Forde BG (1998) An arabidopsis mads box gene that controls nutrient-induced changes in root architecture. Science 279: 407–409.
19. Bengough AG, Hans J, Bransby MF, Valentine TA (2010) Piv as a method for quantifying root cell growth and particle displacement in confocal images. Microsc. Res. Techniq. 73: 27–36.
20. Massa GD, Gilroy S (2003) Touch modulates gravity sensing to regulate the growth of primary roots of arabidopsis thaliana. Plant J. 33: 435–445.
21. Emmerich W, Hardegree S (1991) Seed germination in polyethylene glycol solution: Effects of filter paper exclusion and water vapor loss. Crop Sci. 31: 454–458.
22. Downie H, Holden N, Otten W, Spiers AJ, Valentine TA, et al. (2012) Transparent soil for imaging the rhizosphere. PLoS ONE 7: e44276.
23. Whalley WR, Dodd IC, Watts CW, Webster CP, Phillips AL, et al. (2013) Genotypic variation in the ability of wheat roots to penetrate wax layers. Plant Soil 364: 171–179.
24. Grossmann G, Guo WJ, Ehrhardt DW, Frommer WB, Sit RV, et al. (2011) The rootchip: An integrated microfluidic chip for plant science. Plant Cell 23: 4234–4240.
25. Meier M, Lucchetta EM, Ismagilov RF (2010) Chemical stimulation of the arabidopsis thaliana root using multi-laminar flow on a microfluidic chip. Lab Chip 10: 2147–2153.
26. Fiorani F, Schurr U (2013) Future scenarios for plant phenotyping. In: Merchant SS, editor. Annual review of plant biology, vol 64. Palo Alto: Annual Reviews. pp. 267–291.
27. Abrahamsen M, Eilers S (2011) On the asymptotic enumeration of lego structures. Exp. Math. 20: 145–152.
28. Poorter H, Bühler J, van Dusschoten D, Climent J, Postma JA (2012) Pot size matters: A meta-analysis of the effects of rooting volume on plant growth. Functional Plant Biology.
29. The LEGO Group (2013) The lego group company profile.
30. McNamara S, Cyr M, Rogers C, Bratzel B (1999) Lego brick sculptures and robotics in education. ASEE Annual Conference.
31. Whitesides G (2012) The frugal way: The promise of cost-conscious science. The Economist: The world in 2012.
32. Ingram PA, Zhu J, Shariff A, Davis IW, Benfey PN, et al. (2012) High-throughput imaging and analysis of root system architecture in brachypodium distachyon under differential nutrient availability. Philos. T. R. Soc.B 367: 1559–1569.
33. Balvin M, Sohn E, Iracki T, Drazer G, Frechette J (2009) Directional locking and the role of irreversible interactions in deterministic hydrodynamics separations in microfluidic devices. Phys. Rev. Lett. 103.
34. Pampaloni F, Reynaud EG, Stelzer EH (2007) The third dimension bridges the gap between cell culture and live tissue. Nat. Rev. Mol. Cell Biol. 8: 839-845.
35. Sizmur T, Lind KR, Benomar S, VanEvery H, Cademartiri L (2014) A simple and versatile 2-dimensional platform to study plant germination and growth under controlled humidity. PLoS ONE 9: e96730.
36. Hargreaves C, Gregory P, Bengough AG (2009) Measuring root traits in barley (hordeum vulgare ssp. Vulgare and ssp. Spontaneum) seedlings using gel chambers, soil sacs and x-ray microtomography. Plant Soil 316: 285–297.
37. Whitesides GM (2013) Cool, or simple and cheap? Why not both? Lab Chip 13: 11–13.

12

Soil Nutrient Content Influences the Abundance of Soil Microbes but Not Plant Biomass at the Small-Scale

Kadri Koorem[1*¤], Antonio Gazol[1], Maarja Öpik[1], Mari Moora[1], Ülle Saks[1], Annika Uibopuu[1], Virve Sõber[2], Martin Zobel[1]

1 Department of Botany, Institute of Ecology and Earth Sciences, University of Tartu, Tartu, Estonia, **2** Department of Zoology, Institute of Ecology and Earth Sciences, University of Tartu, Tartu, Estonia

Abstract

Small-scale heterogeneity of abiotic and biotic factors is expected to play a crucial role in species coexistence. It is known that plants are able to concentrate their root biomass into areas with high nutrient content and also acquire nutrients via symbiotic microorganisms such as arbuscular mycorrhizal (AM) fungi. At the same time, little is known about the small-scale distribution of soil nutrients, microbes and plant biomass occurring in the same area. We examined small-scale temporal and spatial variation as well as covariation of soil nutrients, microbial biomass (using soil fatty acid biomarker content) and above- and belowground biomass of herbaceous plants in a natural herb-rich boreonemoral spruce forest. The abundance of AM fungi and bacteria decreased during the plant growing season while soil nutrient content rather increased. The abundance of all microbes studied also varied in space and was affected by soil nutrient content. In particular, the abundance of AM fungi was negatively related to soil phosphorus and positively influenced by soil nitrogen content. Neither shoot nor root biomass of herbaceous plants showed any significant relationship with variation in soil nutrient content or the abundance of soil microbes. Our study suggests that plants can compensate for low soil phosphorus concentration via interactions with soil microbes, most probably due to a more efficient symbiosis with AM fungi. This compensation results in relatively constant plant biomass despite variation in soil phosphorous content and in the abundance of AM fungi. Hence, it is crucial to consider both soil nutrient content and the abundance of soil microbes when exploring the mechanisms driving vegetation patterns.

Editor: Kurt O. Reinhart, USDA-ARS, United States of America

Funding: This research was funded by grants from the Estonian Science Foundation (9050, 9157), targeted financing (IUT 20-28), European Union's Seventh Framework Programme for research, technological development and demonstration under grant agreement no GA-2010-267243 – PLANT FELLOWS, ERMOS programme grant 14 (co-funded by Marie Curie Actions), and the European Regional Development Fund (Centre of Excellence FIBIR). The funders had no role in study design, data collection and analysis, decision to publish, or preparation of the manuscript.

Competing Interests: MM and MÖ (co-authors of this study) are PLOS ONE Editorial Board members. This does not alter our adherence to PLOS ONE Editorial policies and criteria.

* E-mail: kadri.koorem@ut.ee

¤ Current address: Department of Terrestrial Ecology, Netherlands Institute of Ecology, NIOO-KNAW, Wageningen, The Netherlands

Introduction

Resource heterogeneity in space and time is a ubiquitous phenomenon in natural ecosystems and plays a fundamental role in determining plant community dynamics [1,2]. Small-scale resource heterogeneity can be especially important in determining plant productivity since plants can concentrate their root biomass in areas of high nutrient content [3–5]. However, plant growth can also be strongly influenced by soil microorganisms [6]. Symbiotic microorganisms, such as mycorrhizal fungi, contribute to plant nutrient uptake, thereby affecting plant growth and biomass allocation [7]. Similarly, bacteria play an important role in increasing nutrient availability for plants [6]. Despite increasing recognition of the roles played by soil nutrients and especially soil microbes in determining vegetation patterns, information about how soil nutrients, microbes and plant biomass covary in natural ecosystems remains limited.

The spatial distribution of soil microbes is heterogeneous at various scales [8,9] and can be closely related to soil fertility [7,10–12]. Less is known about how temporal variation in soil nutrient content [1] influences soil microbial communities, though the abundance of fungi and bacteria has been shown to display significant temporal variation, with maximum abundance in spring and autumn [13]. Indeed, the abundance of soil microbes is expected to be high in spring and autumn when nutrient availability is high and decline as the plant growth season proceeds and nutrient availability subsequently decreases [14]. However more evidence is needed to better understand these dynamics and the mechanisms driving them.

Studies have found significant small-scale patchiness in the abundance in AM fungi, ectomycorrhizal fungi, other fungi and bacteria [15–17] but knowledge about the simultaneous dynamics of different microbial groups remains limited. Both positive and negative associations in the abundance of co-occuring microbes have been reported: in saprotrophic fungi and bacteria [18] and AM fungi and bacteria [19,20]. Positive interactions may be the result of one group providing substrate for the other [19], while negative interactions may indicate competition for nutrients [20]. A recent study demonstrated that high N:P ratio in soil can provide an advantage, in the form of plant assimilated carbon, to

AM fungi, in competition with non-mycorrhizal fungi that also inhabit plant roots [21].

Mycorrhizal fungi contribute to plant nutrient acquisition from the soil [7] and their abundance has been found to decrease with increasing nutrient availability [10,11,22]. Hence, mycorrhizal fungi and other soil microbes are expected to play an important role in determining plant productivity in ecosystems with low soil nutrients [6]. However, it is unknown whether the effect of soil microbes on plant productivity remains dependent on soil nutrient content at small scales. A recent experimental study did not find any change in the effect of AM fungi on plant productivity in conditions where local soil fertility varied [23], but that study did not consider the effect of microbes besides AM fungi.

We simultaneously measured the temporal and spatial variation of soil nutrient content, soil microbial biomass and plant biomass in a natural herb-rich boreonemoral spruce forest. We focused on a scale that corresponds to the approximate size of adult plant individuals in the understory (i.e. small scale, 15 cm^2). The study site is characterized by heterogeneous light and soil conditions, patchy understory vegetation [24], and remarkably taxon rich AM fungal communities in the roots of herbaceous species [25,26]. We ask:

i) Does soil nutrient content and microbial abundance vary during the first half of the growing season when plant growth is rapid?

ii) Are there negative associations in the small-scale spatial distribution of soil microbial groups?

iii) Is the spatial distribution of soil microbes and plant biomass related to soil nutrient content?

Materials and Methods

Study site and sampling design

The study was carried out in a boreonemoral forest near Koeru, Estonia (58°58′N; 26°03′E). The climate of the study site is transitional between maritime and continental. The mean annual precipitation in the area is 665 mm and the mean annual air temperature is 5.0 °C (−5.4 °C in January and 16.6 °C in July on average; Estonian Meteorological and Hydrological Institute, 2013). The study site is a 130 ha patch of *Hepatica* type forest on a calcareous cambisol (cf. [24]). The soil of the area is relatively uniform in terms of pH, ranging from 5.2–5.7, but exhibits some variation in nutrient content [27,28]. The tree layer is dominated by *Picea abies* (L.) Karst. with individuals of *Fraxinus excelsior* L. and *Acer platanoides* L. also present. The shrub layer is dominated by *Corylus avellana* L. Common species in the herb layer are *Oxalis acetosella* L., *Hepatica nobilis* Schreb., *Paris quadrifolia* L. and *Viola mirabilis* L. The forest can be classified as relatively undisturbed old-growth forest, but clear cutting of approximately 1–2 ha patches has taken place repeatedly in parts of the study site. Detailed descriptions of the area are provided by Moora *et al.* [24] and Zobel *et al.* [27]. This study complies with the laws of Estonia in which no permits are required to carry out research on public land, unless specified otherwise in legislation. No regulation applies to the Koeru site and no protected species were sampled during the study.

Three 105×105 cm plots (A, B, C) were positioned at 5 m intervals in an area of homogenous forest in terms of vegetation cover and light conditions as estimated by eye. All plots were divided into 49 (15×15 cm) quadrats. During preliminary measurements conducted in summer 2007, five plant individuals of the five most common understory species (*Oxalis acetosella, Hepatica nobilis, Fragaria vesca* L., *Galeobdolon luteum* Huds., *Viola mirabilis*) were excavated to a depth of 15 cm, and the root systems of those plants measured with a ruler to determine the size of an average adult plant root system. Soil cores studied during the preliminary study harboured most of the roots in the upper 10 cm of the soil and the average size of plant roots systems was 15×15 cm (data not presented). The size of quadrats used in the main experiment therefore corresponds to the average size of the root system of an adult plant (the average of the species listed above), and is expected to harbour the majority of the biomass of herbaceous plant roots and microbes present in the rhizosphere. Plot A was used to address temporal variation, and plots A, B and C were used to address spatial variation in soil nutrient content and soil microbial activity at small scales. Plot A was sampled twice, at the beginning (26.05.2008) and in the middle (21.07.2008) of the growing season. The small spatial scale of the study meant that it was not possible to address temporal variation in the abundance of soil nutrients and microbes over a longer time period, as additional sampling would have necessitated considerable disturbance. Soil samples (50 g and 10 g) were collected from the soil surface (depth 10 cm) at both sampling times for soil chemical, and fatty acid analysis respectively. Soil samples collected at beginning of the plant growing season (in plot A only) were excavated with a soil corer (2 cm diameter and 10 cm deep) at the corner of each quadrat (n = 49) in order to minimize disturbance to the plots. The hole was filled with soil taken from nearby, to minimize sampling effects on soil conditions. Collected soil was mixed (to avoid any effect of vertical structure on measured parameters) and subjected to soil chemical and fatty acid analysis (see below). Soil samples collected in the middle of the growing season (July) were collected during destructive harvesting (see below). One core was taken from each quadrat; in plot A, the locations where the first samples were collected were avoided. The soil corer used during destructive harvesting had a diameter of 15 cm and a depth of 10 cm. All soil cores were transported to the laboratory where, after the removal of soil samples (50 and 10 grams as before) from all of the soil cores (n = 147), the shoots and roots of herbaceous plant species were separated carefully by hand from the samples from plots B and C (n = 98). Plant biomass was dried to a constant weight at 70 °C, and weighed. As the vegetation in plot A was disturbed by the sampling at the beginning of the growing season, biomass was not collected from that plot.

Soil chemical and fatty acid analysis

Soil samples for chemical analyses were kept at −80°C prior to analysis, in order to prevent the loss of soil nutrients. Analysis of soil nutrient content followed International Organization for Standardization (ISO) protocols. Specifically, determination of soil ammonium (NH_4) and nitrate (NO_3) content (mg/kg) was based on ISO 14256–2: 2005. Organic carbon (C) and total nitrogen (N) content (%) were determined based on ISO 10694: 1995 and ISO 13878: 1998, respectively. As soil NH_4, NO_3 and C content were highly correlated with total N content, only the latter was used in further analysis. The Mehlich III procedure was used to determine soil phosphorus (P) and potassium (K) content (mg/kg): P was extracted by reaction with acetic acid and fluoride compounds; exchangeable K was extracted by the action of ammonium nitrate and nitric acid [29]. Chemical analyses were performed at the laboratory of the Agricultural Research Centre in Saku, Harjumaa, Estonia.

Ester-linked fatty acid (ELFA) content in soil samples was used as a proxy for the biomass of major microbial groups: AM fungi, other fungi and bacteria. Fatty acid profiling is widely used to measure microbial biomass and coarse microbial community composition [10,30–35]. The abundance of soil microbes can also

Table 1. Summary statistics of the Linear Mixed-Effect Models analysis fitted to study the influence of sampling time (May or July) on soil nutrient content and ester-linked fatty acid (ELFA) biomarkers of arbuscular mycorrhizal (AM) fungi, other fungi and bacteria measured from plot A.

Variable	Num DF	Den DF	F	P
Nitrogen (%)	1	45	3.43	0.07
Phosphorus (mg/kg)	1	45	0.15	0.70
Potassium (mg/kg)	1	45	5.20	0.03
AM fungi (ELFA 16:1ω5c μg g−1 soil)	1	45	4.87	0.03
Other fungi (ELFA 18:2ω6,9 μg g−1 soil)	1	45	1.16	0.29
AM fungi: other fungi	1	45	0.79	0.38
Bacteria (ELFA μg g−1 soil)	1	45	6.53	0.01

The degrees of freedom of the numerator (Num DF) and denominator (Den DF), F statistic (F) and associated probability (P) for the sampling time (i.e. fixed factor) are presented.

be estimated by quantifying their DNA by qPCR [35] but the advantage of fatty acid profiling is that it enables the biomass of different microbial groups to be determined simultaneously. Three grams of fresh soil (randomly taken from collected 10 g) was lyophilized immediately after separation and stored at room temperature. Phospholipids were converted to ester-linked fatty acids and extracted from the soil following the protocol of Schutter & Dick [36], which methylates only ester-linked and not free fatty acids. Briefly, 15 ml of 0.2 M KOH in methanol was added to 2 g of dry soil, mixed and incubated with shaking at 200 rpm at 37°C for 1 h. The sample was neutralized with 3 ml of 1 M acetic acid. Fatty acid methyl esters were extracted by adding 10 ml of hexane, the sample was mixed by vortexing and left to settle into two layers. The upper hexane layer was transferred to a clean tube and evaporated under a vacuum in a Jouan RC1022 Vacuum Concentrator. Fatty acids were re-suspended in 100 μl hexane and transferred to a GC vial for analysis using gas chromatography.

The ester-linked fatty acid peak areas were normalized against the internal standard and expressed as μg per g of dry soil. Fatty acid nomenclature follows Ranneklev & Bååth [37]. Ester-linked fatty acid 16:1ω5c was used as a biomarker for AM fungi. This marker is widely used as an indicator for AM fungi [30,32], though the different fatty acid fractions of cells may be of different indicative value for AM fungi. Namely, NLFA (neutral lipid fatty acids, storage lipids) 16:1ω5c occurs only in AM fungi, while PLFA (phospholipid fatty acids, structural lipids in cell membranes) 16:1ω5c can also occur in some Gram-negative bacteria [38,39]. NLFA is therefore proposed as a more suitable indicator of AM fungi in samples with potentially high-bacterial content, such as soils [39,40]. However, ELFA 16:1ω5c has been shown to correlate very well with AM fungal inoculation in an experiment containing both AM fungi and microbes from the AM fungal inocula, where ELFA 16:1ω5c was absent in microbial wash-treated, but AM fungal-free soils [41]. Therefore, ELFA measurements used in this paper represent whole-cell fatty acids, i.e. they contain both the PLFA and NLFA fraction. Fatty acid 18:2ω6,9 was used as a biomarker for fungi other than AM fungi, as it is a dominant fatty acid in most fungi [31] but is only found at very low levels in AM fungi [30]. We calculated the ratio between the levels of fatty acids 16:1ω5c and 18:2ω6,9 (hereafter AM fungi: other fungi ratio) to estimate the relative abundance of AM fungi in the soil fungal community. Fatty acids i15:0, a15:0, 15:0, i16:0,

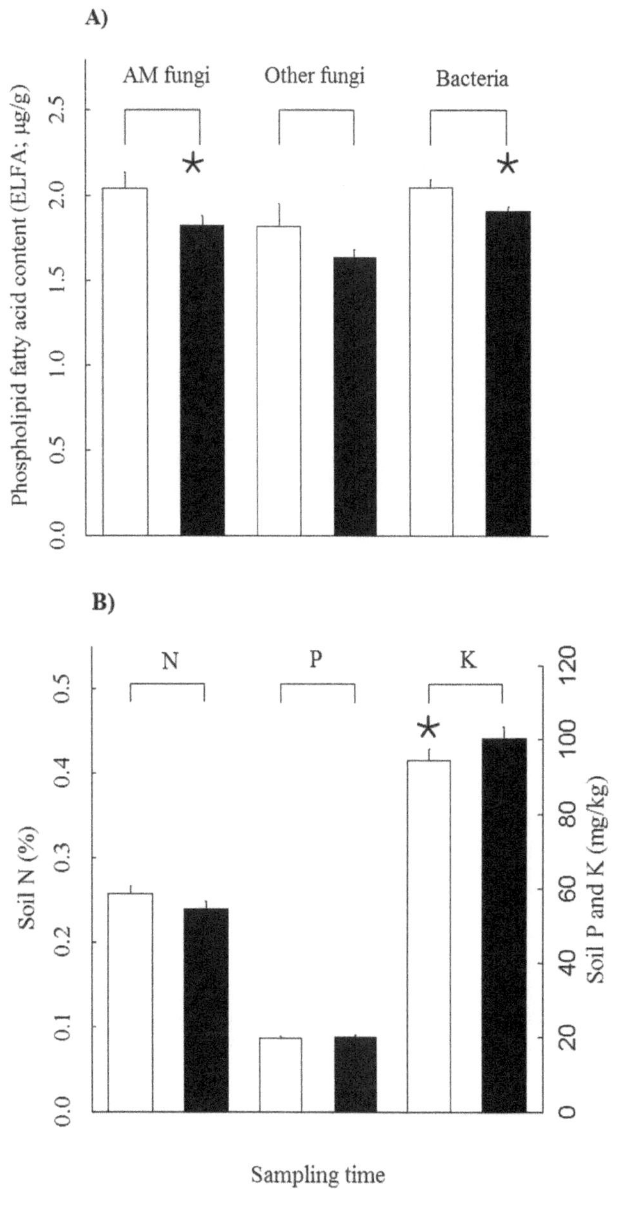

Figure 1. The abundance of soil microbes (A) and nutrients (B) in May (open bars) and July (closed bars). The abundance of arbuscular mycorrhizal (AM) fungi, other fungi and bacteria was estimated as the content of respective ester-linked fatty acid (y-axis on graph A, see Methods for details). Soil N concentration was measured in % (left y-axis of graph B) whereas P and K content were measured as mg/kg (right y-axis of graph B). Significant differences (Linear Mixed-Effect Models; $p<0.05$) in the measured parameters over time are marked with *.

16:1ω9c, br17:0, 10Me16:0, i17:0, a17:0, 17:0, cy17:0, br18:0, 10Me17:0, 18:1ω7c, 10Me18:0, cy19:0 ω 9, cy19:0 ω 7 were used as bacterial biomarkers [31,42].

Statistical analyses

Linear Mixed-Effect Models (LMM) [43] were used to address temporal differences in soil nutrient content (N, P, K) and soil microbial biomass (fatty acid biomarkers for AM fungi, other fungi, AM fungi: other fungi ratio, bacteria). The abundance of other fungi and bacteria as well as the AM fungi: other fungi ratio

Table 2. Results of the modified t-Test for correlation performed to study the relationship between the abundance of arbuscular mycorrhizal (AM) fungi, other fungi and bacteria in the three plots (A, B, C).

		Other fungi	Bacteria
AM fungi	Plot A	0.44**	0.53**
	Plot B	0.54**	0.36*
	Plot C	0.81**	0.56**
Other fungi	Plot A	-	0.41**
	Plot B	-	0.54**
	Plot C	-	0.55**

For each plot, the Pearson's moment correlation coefficient is presented, accompanied by its significance corrected by the presence of spatial autocorrelation. * P<0.05; ** P<0.01.

was log transformed prior to analysis to meet the assumptions of parametric tests. For each variable, a model using time (sampling time: May or July) as a fixed factor and quadrat identity as a random factor were fitted. The homogeneity assumption for model residuals was evaluated graphically [44]. When any sign of heterogeneity was detected in the residuals, it was tested if keeping the variance fixed between the two time periods resulted in homogeneity for the residuals (i.e. including a fixed variance structure in the model; [44]). Models were fitted using Restricted Maximum Likelihood (REML) in order to compare models with different variance structures [44].

Moran's I was used to determine the presence of spatial autocorrelation in the distribution of soil microbes. Moran's I calculates the degree of correlation between observations as a function of the spatial distance separating them [45]. A correlogram is a graphical representation of the Moran's I statistic against distance classes, and provides information about the spatial autocorrelation of a variable and the size of patches. We constructed a spatial correlogram for each group of soil microbes (AM fungi, other fungi and bacteria) in each plot (A, B and C). Moran's I was evaluated at different distance classes (0 to 50 cm) and its significance was assessed by performing 9999 unrestricted permutations. Since the quadrats were distributed regularly, all correlograms were constructed using distance classes with increments of 0.075 m up to a maximum distance of 0.75 m. The first distance class (0.000–0.075) was discarded since it did not contain any pairs of quadrats.

The Modified t-Test for correlation [46] was used to study pairwise relationships between the abundance of soil microbes: AM fungi, other fungi and bacteria in each plot (A, B and C). This method evaluates the correlation between two spatial patterns using the Pearson's product-moment correlation, but corrects the significance of the correlation for the presence of spatial autocorrelation.

Generalized least square models (GLS) [43] were used to determine which factors best explain variations in soil microbial abundance (in plots A, B, C) and both above- and belowground biomass of herbaceous plants (in plots B, C). The set of predictor variables were different depending on the response variable considered. Specifically, in the models of soil microbial abundance (AM fungi, other fungi, bacteria and AM fungi: other fungi ratio), the influence of plot identity (A, B and C) and soil nutrients (N, P, K) were considered. In the models of above- and belowground biomass of herbaceous plants, the influence of plot identity, soil nutrients and microbial abundance were considered. In order to select the most parsimonious model for each variable, the protocol proposed by Zuur *et al.* [44] was followed. First, a model which included the full set of covariates as potential predictors was fitted for each variable. After that, the homogeneity of the residuals of the model was evaluated graphically. In addition, due to the nested structure of the sampling design, within plot spatial autocorrelation of residuals was studied. When spatial autocorrelation was present, the inclusion of correlation structures of different form were evaluated [47]. Finally, models were fitted with Maximum Likelihood (ML) and the best set of predictor variables was selected using Akaike's Information Criterion (AIC). If several models showed similar AIC values (difference lower than two), the most parsimonious one, i.e. the one with fewer fixed components, was chosen [44]. The final model was then refitted using REML to obtain estimates of factor effects and their significance.

LMM, GLS and modified t-Tests and correlograms were performed in the R environment [48], using the packages nlme [49], SpatialPack [50] and ncf [51]. Maps of the spatial patterns of AM fungal abundance and soil P content were created using ArcGIS 9.3 software (ESRI Redlands, NY, USA). For each variable, the "geostatistical analyst" extension of ArcGIS 9.3 was used to generate semivariograms of density and create kriging maps.

Results

Temporal variation

After the removal of outliers, data from 46 quadrats from plot A was subjected to analysis of temporal variation in soil nutrient content and soil microbial abundance. The linear mixed-effect models (LMMs) revealed significant temporal variation in the abundance of arbuscular mycorrhizal (AM) fungi, bacteria and soil K content, but not in the other variables (Table 1, Figure 1).

Table 3. Results of the generalized least square models performed to study the influence of environmental conditions on soil microbes.

	Plot B	Plot C	Nitrogen	Phosphorus	Potassium	AIC weight	R^2
AM fungi	0.206*	1.051**	2.874*	−0.074**	-	0.36	0.57
Other fungi	0.014	0.627**	-	-	0.003*	0.71	0.37
AM: other fungi	0.151**	0.166**	-	-0.024**	-	0.23	0.23
Bacteria	1.612**	3.874**	-	-	0.017*	0.36	0.36

For each soil microbial group, the coefficient associated with the explanatory variable is presented. In addition, the AIC weight of the regression model (i.e. its importance as compared to other models containing a different subset of explanatory variables) and the R^2 are shown. * P<0.05; ** P<0.01.

Table 4. The mean, standard deviation (SD) and the range of the soil nutrient content, ester-linked fatty acid (ELFA) biomarkers of AM fungi, other fungi, bacteria and vegetation characteristics measured at 1 m^2 scale.

Variable	Plot A		Plot B		Plot C	
	Mean (±SD)	Range	Mean (±SD)	Range	Mean (±SD)	range
Nitrogen (%)	0.24 ± 0.07^a	0.15–0.43	0.27 ± 0.05^b	0.15–0.41	0.24 ± 0.03^a	0.18–0.30
Phosphorus (mg/kg)	20.11 ± 3.02^a	13.00–28.00	20.63 ± 2.63^a	15.00–30.00	14.16 ± 1.64^b	12.00–18.00
Potassium (mg/kg)	100.43 ± 21.07^a	57.00–159.00	92.80 ± 22.58^a	54.00–159.00	68.59 ± 10.56^b	47.00–93.00
AM fungi (ELFA 16:1ω5c μg g-1 soil)	1.83 ± 0.42^a	1.16–2.95	2.09 ± 0.52^a	1.22–3.07	3.32 ± 0.81^b	1.56–4.98
Other fungi (ELFA 18:2ω6,9 μg g-1 soil)	1.65 ± 0.30^a	1.02–2.36	1.64 ± 0.35^a	1.02–2.67	2.18 ± 0.39^b	1.49–2.88
AM fungi: other fungi	1.13 ± 0.24^a	0.59–1.67	1.29 ± 0.24^b	0.68–1.72	1.52 ± 0.22^c	1.04–1.93
Bacteria (ELFA μg g-1 soil)	19.14 ± 1.61^a	16.08–22.38	20.58 ± 1.75^b	16.20–25.50	22.54 ± 2.30^c	15.64–26.85
Root biomass of woody species (g)	–	–	10.30±6.76	2.30–37.00	8.92±5.53	1.80–28.10
Shoot biomass of herbaceous species (g)	–	–	0.81±0.42	0.16–2.06	0.79±0.36	0.04–1.53
Root biomass of herbaceous species (g)	–	–	0.57±0.63	0.03–3.07	0.64±0.53	0.03–1.88

Due to destructive sampling, vegetation characteristics were only measured in plots B and C (see explanation in Methods). Different letters (when present) mark a significant difference among means according to Tukey HSD test ($p<0.05$).

Means of the measured variables are given in Table S1. The abundance of AM fungal and bacterial fatty acid biomarkers in the soil was higher in May than in July. Soil K content was lower at the beginning of the growing season. The model fitted to explain temporal variation in the abundance of AM fungi included a fixed variance structure as it substantially improved the model fit.

Spatial variation

After the outliers were removed 47, 46 and 44 quadrats from plots A, B and C respectively were subjected to analysis to test the spatial variation of soil nutrient content and the abundance of soil microbes. Limited evidence was found for the presence of spatial autocorrelation in the abundance of soil microbes studied in each plot, since only the abundance of bacteria showed significant autocorrelation (Figure S1). The abundance of AM fungi and soil bacteria tended to show positive spatial autocorrelation up to a distance of 20 cm, but this trend was significant only in the case of bacteria for plots A and C (Figure S1). The Modified t-Test for correlation revealed significant correlations between the abundance of AM fungi and other fungi, AM fungi and bacteria and between bacteria and other fungi (Table 2).

Generalized least square models revealed that plot identity was an important predictor of the abundance of AM fungi (Table 3); the mean abundance was significantly lower in plots A and B compared to plot C (Table 4). In addition, the abundance of AM fungi was significantly positively related to the content of soil N and negatively affected by soil P content (Table 3, Figure 2). Plot identity also played a significant role in predicting the abundance of other fungi (Table 3); the mean abundance was considerably higher in plot C compared to plots A and B (Table 4). In addition, the abundance of other fungi was significantly positively influenced by soil K (Table 3). The AM fungi: other fungi ratio was also predicted by plot identity (Table 3); the mean AM fungi: other fungi ratio differed significantly between all of the plots, being lowest in plot A and highest in plot C (Table 4). In addition, the AM fungi: other fungi ratio was negatively associated with soil P content (Table 3). The abundance of bacteria was the only variable that was best explained by a model where a spatial correlation structure was included, indicating small-scale autocorrelation in bacterial biomass. The abundance of bacteria was explained by plot identity (Table 3), as the mean differed significantly between all plots, being lowest in plot A and highest in plot C (Table 4). In addition, the abundance of bacteria was significantly positively influenced by soil K content (Table 3).

Of all the variables considered, the AM fungi: other fungi ratio was the only one with a significant and positive effect on the shoot biomass of herbaceous plants ($F = 5.29$; $P = 0.02$). However, the mean shoot biomass of herbaceous plants did not differ between plots (Table 4). None of the studied variables explained the root biomass of herbaceous plants, which also did not vary significantly between plots (Table 4).

Discussion

The small-scale abundance of soil microbes varied through time in the studied boreonemoral herb-rich spruce forest. We recorded a lower abundance of soil microbes in the middle compared to the beginning of the growing season, whereas no decrease in soil nutrient content occurred. Contrary to our expectations, the small-scale spatial abundance of different soil microbes, including bacteria, arbuscular mycorrhizal (AM) fungi and other fungi, was positively related, suggesting little or no importance of resource competition among these groups in our study system. However, the abundance of soil microbes was significantly influenced by soil nutrient content. In particular, the abundance of AM fungi was strongly negatively related to soil P content. Plant biomass on the other hand showed little variation. Based on these results, we suggest that plants potentially compensate for low soil P content with alternative resource acquisition mechanisms, for example via symbiosis with AM fungi, and therefore achieve equal biomass production despite variations in soil nutrient content and the abundance of soil microbes. Soil microbes (especially plant

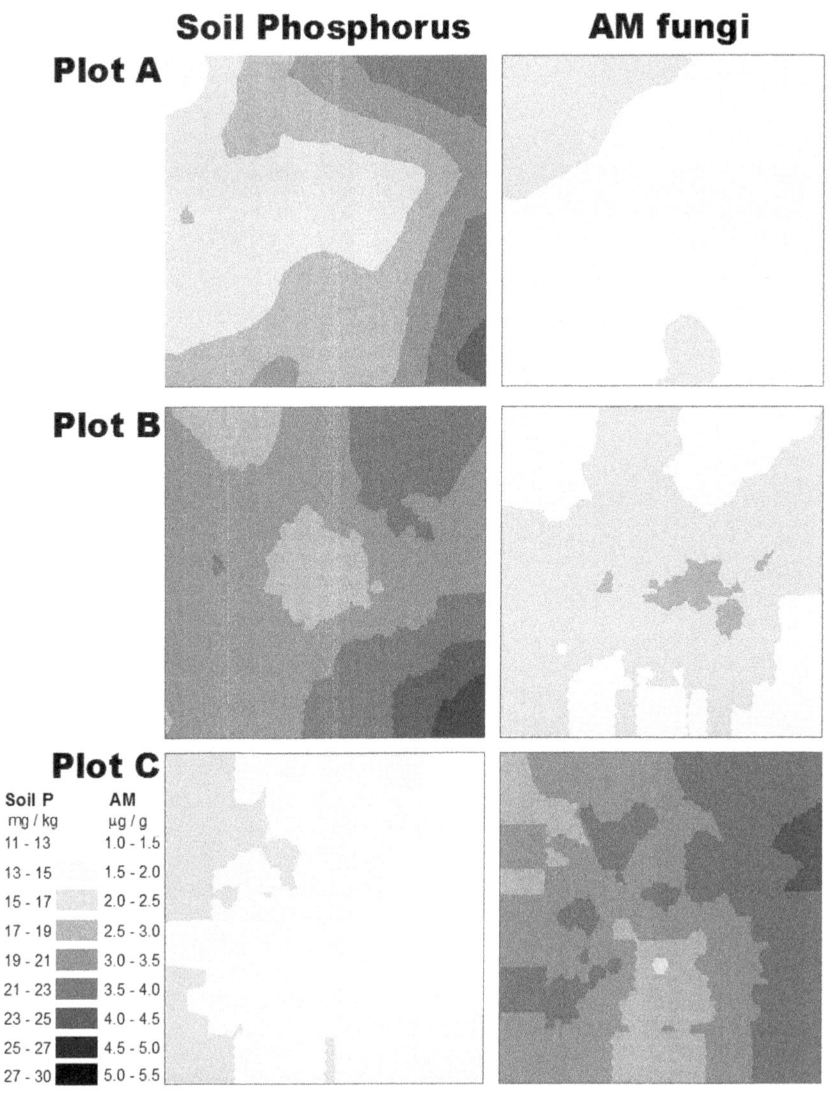

Figure 2. Spatial pattern of soil phosphorus content and abundance of arbuscular mycorrhizal fungi (ester-linked fatty acid 16:1ω5c in soil) in plots A, B and C. Maps were created by kriging, using data from soil samples collected from 15×15 cm quadrats in each 105×105 cm plot.

symbionts such as AM fungi) thus play a fundamental role in the relationship between soil nutrients and plant productivity.

The abundance of soil microbes (especially AM fungi and bacteria) was lower in the middle of the plant growing season, which matches the findings of previous studies that have addressed the temporal variation of soil microbes at larger and smaller spatial scales [13,52]. Active plant growth during the first half of the growing season did not bring about a decrease in soil nutrient content. Thus, the importance of soil nutrients in determining temporal changes in the abundance of soil microbes seem to be minor [52]. Some previous studies have reported a contrasting pattern of AM fungal abundance during the plant growing season: an initial increase with a peak in the summer, followed by a decrease in the autumn [53,54]. A potential explanation for this inconsistent variation of AM fungal abundance in soil can be the particular weather conditions of the sampling year, as soil AM fungal hyphal density is considerably affected by temperature and precipitation [54,55]. Thus, data from long-term measurements with multiple sampling times during the growing season is needed in order to draw solid conclusions about the dynamics of soil microbes.

We found a strong positive association between the abundance of bacteria, AM fungi and other fungi, suggesting a limited role of resource competition among these microbial groups in a natural ecosystem. Several previous experimental studies have reported negative associations between the abundance of ectomycorrhizal fungi, AM fungi and bacteria [20,56], as well as between the abundance of ectomycorrhizal and saprotrophic fungi [57], and AM fungi and other fungi [21,58]. On the other hand, studies conducted in natural conditions have demonstrated similar biomass dynamics of fungi other than AM fungi and bacteria during both the growing season [13] and over decades [59]. These contrasting results highlight the importance of studies addressing the pattern of soil microbes in natural environments. The strong positive relationship between the abundance of bacteria, AM fungi and other fungi found in this study suggests that the use of slightly different resources by soil microbes promotes their coexistence. Indeed, AM fungi, which are expected not to be highly saprotrophic themselves, have been shown to promote the activity of saprotrophic microbes [60]. In addition, fungal hyphae and spores can function as substrates for bacteria [19,61], and carbon released from dead bacteria can promote the growth of saprotrophic fungi [32]. Also, living roots excrete carbon compounds [62], which can be consumed by saprotrophic soil microorganisms [13]. In addition, AM fungi as well as ectomycorrhizal fungi have been shown to increase their growth in the presence of root exudates [63,64].

The abundance of all groups of soil microorganisms tested depended on the sampling location (plot identity), which confirms the small-scale spatial heterogeneity of soil microbes previously demonstrated [12]. In addition, soil nutrients played an important role in determining the abundance of soil microorganisms. The negative effect of soil P content on the abundance of AM fungi has also been reported in earlier field fertilization experiments in grasslands and forests [65,66]. As AM fungi can improve the acquisition of soil nutrients, especially P, by plants [7,67], they can play an important role when soil nutrient content is low. When soil nutrient levels increase, the cost to a plant of supporting AM fungi may outweigh the benefits, and can result in a decrease of AM fungal colonization of plant roots [7]. However, soil AM fungal fatty acid content has also been reported to be positively or insignificantly related to soil P content [68,69]. We recorded a positive effect of soil N on the abundance of AM fungi, a pattern that has also been registered at a large scale [35] and can be explained by the high N demand of AM fungi [70]. In addition, AM fungal communities have been shown to be highly influenced by soil pH [71]. However, Dumbrell *et al.* [71] focused on sites with high variation in soil pH (ranging from 4.5–8), whereas our study area is characterised by fairly uniform pH values [27,28]. Therefore, the role of soil nutrients in determining the abundance of AM fungi in current study site could be expected to be more pronounced, while the effect of small-scale variation in soil pH remains to be fully considered in future studies.

Saprotrophic and ectomycorrhizal fungi as well as bacteria were positively influenced by soil K content. A recent study demonstrated an increase in bacterial biomass, and growth of extra-matrical mycelia of ectomycorrhizal fungi in response to K addition in peatland soils with severe K deficiency [72]. Our results suggest that K also has an important role to play in determining the abundance of soil microbes in soils where K content is not extremely low. On the other hand, the observed positive relationship can also reflect increased weathering and K influx resulting from active growth of bacteria and saprotrophic fungi [73,74].

Plants have been shown to concentrate their biomass into patches with high nutrient content [3–5], but in the current study, conducted in an old and stable ecosystem, we found no direct effect of nutrients on plant biomass. One explanation for the absence of the expected pattern is that plants are continuously foraging, while the pattern of biomass measured in studies such as this one reflects dynamics that have operated over numerous years. Greenhouse experiments have shown that nutrient poor patches contain higher plant biomass when next to nutrient rich patches [4,75] but those experiments were short-term. In the longer term, there will be natural turnover of nutrient rich patches (both temporal and spatial), and active foraging can therefore result in relatively uniform plant biomass. Secondly, the natural variation in nutrient content could be too low to generate distinct patterns. Indeed, understory plant biomass in this study area has not shown changes in response to induced variation in soil nutrient content either [23]. Thirdly, the effect of soil nutrients can be mitigated by the influence of soil microbes such as AM fungi. AM fungi seemed to be the only group of soil microbes to affect understory plant growth, as the relative increase of AM fungi compared to other fungi resulted in increased plant shoot biomass. As AM fungi can greatly improve plant P acquisition [7,67], the presence of AM fungi can be more important for plant growth when soil P content is low and less beneficial for plants when soil nutrient content is high. In accordance with this, a decrease of AM fungal hyphal density in soil and AM colonization in plant roots has been reported in response to increased P availability [22,76]; whereas the nutrient content of plant tissues and plant biomass did not exhibit any change [22,68]. Our study, based on data from a natural environment, demonstrates stable plant biomass in conditions of low abundance of AM fungi and high P content and *vice versa.* We support the suggestion of Beauregard *et al.* [68] that plants can compensate for low available nutrient concentration with more efficient symbiosis with soil microbes. Hence, it is crucial to consider both the soil nutrient content and the abundance of soil microbes when exploring the mechanisms driving vegetation patterns.

Supporting Information

Figure S1 Spatial correlograms of microbes. Spatial correlograms of arbuscular mycorrhizal fungi (a–c), other fungi (d–f), and bacteria (g–i) in plot A (a, d, g), B (b, e, h) and C (c, f, i). Distances with significant spatial autocorrelation are marked with solid circles.

Table S1 Temporal variation in the abundance of soil nutrients and microbes. Summary statistics of the soil nutrient content and ester-linked fatty acid (ELFA) biomarkers of arbuscular mycorrhizal (AM) fungi, other fungi and bacteria measured from plot A at the beginning (May) and in the middle (July) of the growing season. The mean, standard deviation (SD) and the range are presented for each variable.

Acknowledgments

We thank Sandra Caul at The James Hutton Institute, Dundee, UK, for help with fatty acid analyses. We are grateful to Jodi Price and John Davison for their comments on earlier drafts of this manuscript.

Author Contributions

Conceived and designed the experiments: MZ MM MÖ ÜS AU KK. Performed the experiments: KK MÖ MM ÜS AU VS MZ. Analyzed the data: AG MZ MÖ MM KK. Contributed reagents/materials/analysis tools: AG MM MZ MÖ. Wrote the paper: KK AG MÖ MM ÜS AU VS MZ.

References

1. Farley RA, Fitter AH (1999) Temporal and spatial variation in soil resources in a deciduous woodland. J. Ecol. 87: 688–696.
2. Pausas JG, Carreras J, Ferre A, Font X (2003) Coarse-scale plant species richness in relation to environmental heterogeneity. J. Veg. Sci. 14: 661–668.
3. Hutchings MJ, Wijesinghe DK (1997) Patchy habitats, division of labour and growth dividends in clonal plants. Trends Ecol. Evol. 12: 390–394.
4. Day KJ, Hutchings MJ, John EA (2003) The effects of spatial pattern of nutrient supply on yield, structure and mortality in plant populations. J. Ecol. 91: 541–553.
5. Hodge A (2004) The plastic plant: root responses to heterogeneous supplies of nutrients. New Phytol. 162: 9–24.

6. van der Heijden MGA, Bardgett RD, van Straalen NM (2008) The unseen majority: soil microbes as drivers of plant diversity and productivity in terrestrial ecosystems. Ecol. Lett. 11: 296–310.
7. Smith SE, Read DJ (2008) Mycorrhizal Symbiosis. Third edition. Academic Press, Amsterdam
8. Ettema CH, Wardle DA (2002) Spatial soil ecology. Trends Ecol. Evol. 17: 177–183.
9. Baldrian P, Vetrovsky T (2012) Scaling down the analysis of environmental processes: monitoring enzyme activity in natural substrates on a millimeter resolution scale. Appl. Environ. Microbiol. 78: 3473–3475.
10. Olsson PA, Bååth E, Jakobsen I (1997) Phosphorus effects on the mycelium and storage structures of an arbuscular mycorrhizal fungus as studied in the soil and roots by analysis of fatty acid signatures. Appl. Environ. Microbiol. 63: 3531–3538.
11. Nilsson LO, Bååth E, Falkengren-Grerup U, Wallander H (2007) Growth of ectomycorrhizal mycelia and composition of soil microbial communities in oak forest soils along a nitrogen deposition gradient. Oecologia 153: 375–384.
12. Franklin RB, Mills AL (2009) Importance of spatially structured environmental heterogeneity in controlling microbial community composition at small spatial scales in an agricultural field. Soil Biol. Biochem. 41: 1833–1840.
13. Kaiser C, Koranda M, Kitzler B, Fuchslueger L, Schnecker J, et al. (2010) Belowground carbon allocation by trees drives seasonal patterns of extracellular enzyme activities by altering microbial community composition in a beech forest soil. New Phytol. 187: 843–858.
14. Paul EA, Clark FE (1996) Soil Microbiology and Biochemistry. Second edition. Academic Press, San Diego.
15. Klironomos JN, Rillig MC, Allen MF (1999) Designing belowground field experiments with the help of semi-variance and power analyses. Appl. Soil Ecol. 12: 227–238.
16. Mummey DL, Rillig MC (2008) Spatial characterization of arbuscular mycorrhizal fungal molecular diversity at the submetre scale in a temperate grassland. FEMS Microbiol. Ecol. 64: 260–270.
17. Baldrian P, Merhautova V, Cajthaml T, Petrankova M, Snajdr J (2010) Small-scale distribution of extracellular enzymes, fungal, and bacterial biomass in *Quercus petraea* forest topsoil. Biol. Fertil. Soils 46: 717–726.
18. Rousk J, Bååth E (2011) Growth of saprotrophic fungi and bacteria in soil. FEMS Microbiol. Ecol. 78: 17–30.
19. Lecomte J, St-Arnaud M, Hijri M (2011) Isolation and identification of soil bacteria growing at the expense of arbuscular mycorrhizal fungi. FEMS Microbiol. Lett. 317: 43–51.
20. Leigh J, Fitter AH, Hodge A (2011) Growth and symbiotic effectiveness of an arbuscular mycorrhizal fungus in organic matter in competition with soil bacteria. FEMS Microbiol. Ecol. 76: 428–438.
21. Chagnon P-L, Bradley RL (2013) Evidence that soil nutrient stoichiometry controls the competitive abilities of arbuscular mycorrhizal *vs.* root-borne non-mycorrhizal fungi. Fungal Ecol. 6: 557–660.
22. McNaughton SJ, Oesterheld M (1990) Extramatrical mycorrhizal abundance and grass nutrition in a tropical grazing ecosystem, the Serengeti National Park, Tanzania. Oikos 59: 92–96.
23. Koorem K, Saks Ü, Sõber V, Uibopuu A, Öpik M, et al. (2012) Effects of arbuscular mycorrhiza on community composition and seedling recruitment in temperate forest understory. Basic Appl. Ecol. 13: 663–672.
24. Moora M, Daniell TJ, Kalle H, Liira J, Püssa K, et al. (2007) Spatial pattern and species richness of boreonemoral forest understorey and its determinants – a comparison of differently managed forests. For. Ecol. Manage. 250: 64–70.
25. Öpik M, Metsis M, Daniell TJ, Zobel M, Moora M (2009) Large-scale parallel 454 sequencing reveals host ecological group specificity of arbuscular mycorrhizal fungi in a boreonemoral forest. New Phytol. 184: 424–437.
26. Davison J, Öpik M, Daniell TJ, Moora M, Zobel M (2011) Arbuscular mycorrhizal fungal communities in plant roots are not random assemblages. FEMS Microbiol. Ecol. 78: 103–115.
27. Zobel M, Kalamees R, Püssa K, Roosaluste E, Moora M (2007) Soil seed bank and vegetation in mixed coniferous forest stands with different disturbance regimes. For. Ecol. Manage. 250: 71–76.
28. Koorem K, Moora M (2010) Positive association between understory species richness and a dominant shrub species (*Corylus avellana*) in a boreonemoral spruce forest. For. Ecol. Manage. 260: 1407–1413.
29. Sen Tran T, Simard RR (1993) Mehlich III-extractable elements. In: Carter MR, editor. Soil sampling and methods of analysis. pp. 43–48.
30. Graham JH, Hodge NC, Morton JB (1995) Fatty acid methyl ester profiles for characterization of Glomalean fungi and their endomycorrhizae. Appl. Environ. Microbiol. 61: 58–64.
31. Frostegård A, Bååth E (1996) The use of phospholipid fatty acid analysis to estimate bacterial and fungal biomass in soil. Biol. Fertil. Soils 22: 59–65.
32. Hinojosa MB, Carreira JA, Garcia-Ruiz R, Dick RP (2005) Microbial response to heavy metal-polluted soils: community analysis from phospholipid-linked fatty acids and ester-linked fatty acids extracts. J. Environ. Qual. 34: 1789–1800.
33. De Deyn GB, Quirk H, Bardgett RD (2011) Plant species richness, identity and productivity differentially influence key groups of microbes in grassland soils of contrasting fertility. Biol. Lett. 23: 75–78.
34. Schnoor TK, Martensson LM, Olsson PA (2011) Soil disturbance alters plant community composition and decreases mycorrhizal carbon allocation in a sandy grassland. Oecologia 167: 809–819.
35. Birkhofer K, Schöning I, Alt F, Herold N, Klarner B, et al. (2012) General relationships between abiotic soil properties and soil biota across spatial scales and different land-use types. PLoS ONE 7: e43292.
36. Schutter ME, Dick RP (2000) Comparison of fatty acid methyl ester (FAME) methods for characterizing microbial communities. Soil Sci. Soc. Am. J. 64: 1659–1668.
37. Ranneklev SB, Bååth E (2003) Use of phospholipid fatty acids to detect previous self-heating events in stored peat. Appl. Environ. Microbiol. 69: 3532–3539.
38. Nichols P, Stulp BK, Jones JG, White DC (1986) Comparison of fatty acid content and DNA homology of the filamentous gliding bacteria *Vitreoscilla, Flexibacter, Filibacter*. Arch. Microbiol. 146: 1–6.
39. Olsson PA (1999) Signature fatty acids provide tools for determination of the distribution and interactions of mycorrhizal fungi in soil. FEMS Microbiol. Ecol. 29: 303–310.
40. Frostegård A, Tunlid A, Bååth E (2011) Use and misuse of PLFA measurements in soils. Soil Biol. Biochem. 43: 1621–1625.
41. Walder F, Niemann H, Lehmann MF, Boller T, Wiemken A, et al. (2013) Tracking the carbon source of arbuscular mycorrhizal fungi colonizing C3 and C4 plants using carbon isotope ratios (delta 13C). Soil Biol. Biochem. 58: 341–344.
42. Zelles L (1999) Fatty acid patterns of phospholipids and lipopolysaccharides in the characterisation of microbial communities in soil: a review. Biol. Fertil. Soils 29: 111–129.
43. Pinheiro JC, Bates DM (2000) Mixed-Effect Models in S and S-PLUS. Springer.
44. Zuur AF, Ieno EN, Walker NJ, Saveliev AA, Smith GM (2009) Mixed effects models and extensions in ecology with R. Springer, New York.
45. Legendre P, Legendre L (1998) Numerical Ecology. Second English edition. Elsevir Science, Amsterdam.
46. Dutilleul P, Clifford P, Richardson S, Hemon D. (1993) Modifying t test for assessing the correlation between two spatial processes. Biometrics 49: 305–314.
47. Dormann CF, McPherson JM, Araujo MB, Bivand R, Bolliger J, et al. (2007) Methods to account for spatial autocorrelation in the analysis of species distributional data: a review. Ecography 30: 609–628.
48. R Development Core Team (2011) R: A language and environment for statistical computing. R Foundation for Statistical Computing, Vienna, Austria. ISBN 3-900051-07-0, URL http://www.R-project.org/ Accessed 2013 February 27
49. Pinheiro J, Bates D, DebRoy S, Sarkar D, the R Development Core Team (2011) Nlme: linear and nonlinear mixed effects models. R package version 3: 1–98.
50. Osorio F, Vallejos R, Cuevas F (2012) SpatialPack: package for analysis of spatial data.
51. Bjornstad ON (2012) Ncf: spatial nonparametric covariance functions. R package version 1.1–4.
52. Hamel C, Hanson K, Selles F, Cruz AF, Lemke R, et al. (2006) Seasonal and long-term resource-related variations in soil microbial communities in wheat-based rotations of the Canadian prairie. Soil Biol. Biochem. 38: 2104–2116.
53. Kabir Z, O'Halloran IP, Fyles JW, Hamel C (1998) Dynamics of the mycorrhizal symbiosis of corn (*Zea mays* L.): effects of host physiology, tillage practice and fertilization on spatial distribution of extra-radical mycorrhizal hyphae in the field. Agric. Ecosyst. Environ. 68: 151–163.
54. Staddon PL, Thompson K, Jakobsen I, Grime JP, Askew AP, et al. (2003) Mycorrhizal fungal abundance is affected by long-term climatic manipulations in the field. Glob. Chang. Biol. 9: 186–194.
55. Hu Y, Rillig MC, Xiang D, Hao Z, Chen B (2013) Changes of AM fungal abundance along environmental gradients in the arid and semi-arid grasslands of Northen China. PLoS ONE 8: e57593.
56. Olsson PA, Chalot M, Bååth E, Finlay RD, Söderström B (1996) Ectomycorrhizal mycelia reduce bacterial activity in a sandy soil. FEMS Microbiol. Ecol. 21: 77–86.
57. Pereira E, Coelho V, Tavares RM, Lino-Neto T, Baptista P (2012) Effect of competitive interactions between ectomycorrhizal and saprotrophic fungi on *Castanea sativa* performance. Mycorrhiza 22: 41–49.
58. Olsson PA, Francis R, Read DJ, Söderström B (1998) Growth of arbuscular mycorrhizal mycelium in calcareous dune sand and its interaction with other soil microorganisms as estimated by measurement of specific fatty acids. Plant Soil 201: 9–16.
59. Campbell CD, Cameron CM, Bastias BA, Chen CR, Cairney JWG (2008) Long term repeated burning in a wet sclerophyll forest reduces fungal and bacterial biomass and responses to carbon substrates. Soil Biol. Biochem. 40: 2246–2252.
60. Cheng L, Booker FL, Tu C, Burkey KO, Zhou L, et al. (2012) Arbuscular mycorrhizal fungi increase organic carbon decomposition under elevated CO_2. Science 337: 1084–1087.
61. Roesti D, Ineichen K, Braissant O, Redecker D, Wiemken A, et al. (2005) Bacteria associated with spores of the arbuscular mycorrhizal fungi *Glomus geosporum* and *Glomus constrictum*. Appl. Environ. Microbiol. 71: 6673–6679.
62. Jones DL, Hodge A, Kuzyakov Y (2004) Plant and mycorrhizal regulation of rhizodeposition. New Phytol. 163: 459–480.
63. Buee M, Rossignol M, Jauneau A, Ranjeva R, Becard G (2000) The pre-symbiotic growth of arbuscular mycorrhizal fungi is induced by a branching factor partially purified from plant root exudates. Mol. Plant Microbe Interact. 13: 693–698.

64. Kikuchi K, Matsushita N, Suzuki K, Hogetsu T (2007) Flavonoids induce germination of basidiospores of the ectomycorrhizal fungus *Suillus bovinus*. Mycorrhiza 17: 563–570.
65. Groffman PM, Fisk MC (2011) Phosphate additions have no effect on microbial biomass and activity in a northern hardwood forest. Soil Biol. Biochem. 43: 2441–2449.
66. Liu YJ, Shi GX, Mao L, Cheng G, Jiang SJ, et al. (2012) Direct and indirect influences of 8 yr of nitrogen and phosphorus fertilization on Glomeromycota in an alpine meadow ecosystem. New Phytol. 194: 523–535.
67. Smith SE, Smith FA, Jakobsen I (2003) Mycorrhizal fungi can dominate phosphate supply to plants irrespective of growth responses. Plant Physiol. 133:16–20.
68. Beauregard MS, Hamel C, Nayyar A, St-Arnaud M (2010) Long-term phosphorus fertilization impacts soil fungal and bacterial diversity but not AM fungal community in alfalfa. Microb. Ecol. 59: 379–389.
69. Thoms C, Gattinger A, Jacob M, Thomas FM, Gleixner G (2010) Direct and indirect effects of tree diversity drive soil microbial diversity in temperate deciduous forest. Soil Biol. Biochem. 42: 1558–1565.
70. Hodge A, Fitter AH (2010) Substantial nitrogen acquisition by arbuscular mycorrhizal fungi from organic material has implications for N cycling. Proc. Natl. Acad. Sci. U.S.A. 107: 13754–13759.
71. Dumbrell AJ, Nelson M, Helgason T, Dytham C, Fitter AH (2010) Relative roles of niche and neutral processes in structuring a soil microbial community. ISME J. 4: 337–345.
72. Potila H, Wallander H, Sarjala T (2009) Growth of ectomycorrhizal fungi in drained peatland forests with variable P and K availability. Plant Soil 316: 139–150.
73. Hoffland E, Giesler R, Jongmans AG, van Breemen N (2003) Feldspar tunneling by fungi along natural productivity gradients. Ecosystems 6: 739–746.
74. Calvaruso C, Turpault MP, Frey-Klett P (2006) Root-associated bacteria contribute to mineral weathering and to mineral nutrition in trees: a budgeting analysis. Appl. Environ. Microbiol. 72: 1258–1266.
75. Gazol A, Tamme R, Price JN, Hiiesalu I, Laanisto L, et al. (2013) A negative heterogeneity–diversity relationship found in experimental grassland communities. Oecologia 173: 545–555.
76. Nayyar A, Hamel C, Forge T, Selles F, Jefferson PG, et al. (2008) Arbuscular mycorrhizal fungi and nematodes are involved in negative feedback on a dual culture of alfalfa and Russian wildrye. Appl. Soil Ecol. 40: 30–60.

13

Carbon Sequestration and Fertility after Centennial Time Scale Incorporation of Charcoal into Soil

Irene Criscuoli[1,7]*, Giorgio Alberti[2,3], Silvia Baronti[4], Filippo Favilli[5], Cristina Martinez[1,4], Costanza Calzolari[6], Emanuela Pusceddu[4], Cornelia Rumpel[7], Roberto Viola[8], Franco Miglietta[1,4]

1 FoxLab, Forest & Wood Science, Fondazione E.Mach, San Michele all'Adige (Trento), Italy, **2** MountFor Project Centre, European Forest Institute, San Michele all'Adige (Trento), Italy, **3** Department of Agricultural and Environmental Sciences, University of Udine, Udine, Italy, **4** Institute of Biometeorology, National Research Council, Firenze, Italy, **5** European Academy of Bolzano, Bolzano, Italy, **6** Institute of Research for Hydrogeological Protection, National Research Council, Sesto Fiorentino (FI), Italy, **7** Biogéochimie et écologie des milieux continentaux, CNRS-INRA-ENS-Paris 6, Thiverval-Grignon, France, **8** Centre for Research and Innovation, Fondazione E.Mach, San Michele all'Adige (Trento), Italy

Abstract

The addition of pyrogenic carbon (C) in the soil is considered a potential strategy to achieve direct C sequestration and potential reduction of non-CO_2 greenhouse gas emissions. In this paper, we investigated the long term effects of charcoal addition on C sequestration and soil physico-chemical properties by studying a series of abandoned charcoal hearths in the Eastern Alps of Italy established in the XIX century. This natural setting can be seen as an analogue of a deliberate experiment with replications. Carbon sequestration was assessed indirectly by comparing the amount of pyrogenic C present in the hearths (23.3 ± 4.7 kg C m^{-2}) with the estimated amount of charcoal that was left on the soil after the carbonization (29.3 ± 5.1 kg C m^{-2}). After taking into account uncertainty associated with parameters' estimation, we were able to conclude that $80 \pm 21\%$ of the C originally added to the soil via charcoal can still be found there and that charcoal has an overall Mean Residence Time of 650 ± 139 years, thus supporting the view that charcoal incorporation is an effective way to sequester atmospheric CO_2. We also observed an overall change in the physical properties (hydrophobicity and bulk density) of charcoal hearth soils and an accumulation of nutrients compared to the adjacent soil without charcoal. We caution, however, that our site-specific results should not be generalized without further study.

Editor: Ben Bond-Lamberty, DOE Pacific Northwest National Laboratory, United States of America

Funding: This study was financially supported by the European Commission through the EuroChar project (FP7-ENV-478 2010ID-265179) and AgroPyroGas funded by Regione Toscana, Italy. I.C. was supported by a Ph.D. grant from Fondazione Edmund Mach, Trento. The funders had no role in study design, data collection and analysis, decision to publish, or preparation of the manuscript.

Competing Interests: The authors have declared that no competing interests exist.

* E-mail: irene.criscuoli@fmach.it

Introduction

Thermo-chemical conversion of organic material under limited oxygen supply, within a certain range of temperatures (200–1200°C), transforms biomass into bio-oil and syngas, which may be used as an energy source, and produces a carbonaceous co-product (i.e. biomass-derived Pyrogenic-C or charcoal or biochar [1]) which has been proposed as a tool to mitigate climate change and improve soil fertility [2]. A recent study [3] quantified the theoretical carbon (C) sequestration potential of biochar following its incorporation in agricultural soils at a maximum rate of 50 Mg C ha^{-1} to a depth of 0.15 m as 1.8 Gt CO_2-$C_{equivalent}$ per year. This estimate corresponds to 12% of current global anthropogenic C emissions and includes: a) direct C sequestration, associated with the burial of recalcitrant organic C forms [4]; (b) potential reduction of N_2O and CH_4 emissions from soils associated with biochar application [5]; and (c) CO_2 emissions avoided due to fossil fuel substitution by the energy released by biomass during pyrolysis and gasification. Moreover, several studies have shown that the addition of biochar to both poor and fertile agricultural soils may have beneficial effects on plant yields, thus amplifying its environmental benefit. These effects are associated with improvements in soil physical [6] and chemical properties [7], microbiological activity [8], temperature increase due to changes in surface albedo [9], hormesis (i.e. favorable biological responses to low exposures; [10]), as well as combinations of several of these different drivers [8].

However, while short-term studies have confirmed the potential of biochar to increase C storage and to improve soil physico-chemical properties in the short-term, the long-term effects of incorporating large amounts of pyrogenic C into the soil remain rather elusive. The actual ability of biochar to act as a C sink into the soil remains controversial due to uncertainties related to its long term stability [11]. Thousand-year old charcoal residues identified in archeological sites and areas interested by wildfires have been considered a demonstration of its long term stability in soils [12] [13] even though some studies have outlined the fact that the amount of Pyrogenic Carbon measured in soils is much lower than what would have been expected according to other paleontological and archeological artifacts [14] or the frequency and intensity of fires [15]. Rapid transformations of charcoal in soils by abiotic and biotic oxidation can occur [16] and its stability varies according to the initial feedstock, the charring conditions, and the environmental characteristics of the burial site [17]. Very ancient charcoal deposits, such as Terra Preta de Indio in

Figure 1. Soil profile at the control site (Panel a) and the charcoal hearth (Panel b). The letters indicate different pedologic horizons. In the charcoal hearth the dark anthropogenic layer (Acoal; 0–10 cm) can be easily identified.

Amazonia [18] and Bronze Age human settlements of Terramare in the Po Valley in northern Italy [19], are still rich in C and are still fertile substrates [20] [21].

The present study aimed to explore the centennial time scale stability of pyrogenic C incorporated as charcoal in soil. To do this, we used Alpine areas where charcoal, produced in traditional charcoal piles, was added to the soil more than one century ago and was not mixed with other organic sources. Moreover, we were able to assess the effect on physio-chemical soil properties after char addition to soil.

Materials and Methods

Ethics Statement

Collection of soil samples was authorized by the Stelvio National Park and Trentino Forest Service.

Site Description and Soil Sampling

The study site is located in Val di Pejo (Trentino, Northern Italy; 46°20′16.18″ N, 10°36′07.02″ E) at an elevation ranging from 2120 to 2170 m a.s.l. The mean annual temperature at the site is 3.5°C and the mean annual precipitation is 903 mm [22]. The lowest precipitations are registered in January, while the highest are distributed between April and November.

Starting from the 16th century the area was subject to intensive wood resource exploitation for larch charcoal production which was subsequently used in the local iron industry. Production ceased in 1858 when a severe fire event destroyed the major iron foundry in the valley [13] [23]. Charcoal production was based on large forest clear-cuts, wood chopping and downhill transportation to artificial flat terraces (charcoal hearths) with an elliptical shape. Some of these hearths are still identifiable today as terraces where

Table 1. Parameters, coefficients and variables used to distinguish and quantify the different carbon pools in charcoal soil layer (means ± standard error, n = 3).

Parameter/coefficient/ variable	Definition	Unit	Value	Source(s)
$\delta^{13}C_{TOT}$	Isotopic signature of the bulk anthropogenic soil layer in hearths	‰	−24.72±0.14	Measured with IRMS
$\delta\ ^{13}C_{CHAR}$	Isotopic signature of charcoal fragments extracted from the anthropogenic soil layer in hearths	‰	−24.53±0.01	Measured with IRMS
$\delta\ ^{13}C_{SOM}$	Isotopic signature of soil organic matter of control soils	‰	−26.28±0.30	Measured with IRMS
$\delta\ ^{13}C\text{-}CO_{2HEARTHS}$	Isotopic signature of respired CO_2 from incubated hearth soils	‰	−25.22±1.13	Measured with CRDS and Keeling plot
$\delta\ ^{13}C\text{-}CO_{2CONTROL}$	Isotopic signature of respired CO_2 from incubated control soils	‰	−24.81±0.31	Measured with CRDS and Keeling plot
C_{TOT}	Carbon content of anthropogenic soil layer in hearths	kg C m^{-2}	26.2±5.3	Measured and calculated
C_{CHAR}	Pyrogenic carbon content of anthropogenic soil layer in hearths	kg C m^{-2}	23.3±4.7	Calculated with mass balance
C_{SOM}	Soil organic matter content of anthropogenic soil layer in hearths	kg C m^{-2}	2.9±0.6	Calculated with mass balance

Table 2. Total carbon and nutrient stocks in the control soils and charcoal hearths and the estimated amount added by carbonization calculated according to Eq. [2].

Element	control soils	charcoal hearths	p-value	Input by carbonization
C (kg m^{-2})	8.1±0.3	26.2±5.3	0.03	29±5
P (g m^{-2})	45.8±3.1	95±7	0.003	117±21
K (g m^{-2})	231±34	285±70	0.53	112±20
Ca (g m^{-2})	136±23	368±80	0.10	229±41
Mg (g m^{-2})	127±88	62±5	0.51	59±10
N (g m^{-2})	582±84	500±29	0.42	80±15

Mean ± standard error (n = 3). Results of the comparison between control and charcoal hearth (p-value) are also reported.

wood piles were prepared and subsequently covered by soil and tree branches [24]. The relatively large size and elliptical shape of the hearths suggest that more than one pile of wood was carbonized at a particular time as it was already well known that only circular piles could ensure a uniform and high quality charcoal production [25]. Wood carbonization required between four and ten days according to the dimension of the wood pile. After a two-day cooling down period, the charcoal was finally transported to the foundry.

Three hearths and three adjacent control areas with southerly aspects and unaffected by significant geo-morphological dynamics (erosion or sedimentation) or recent anthropogenic disturbances were selected. These are flat (2% slope) and have an average surface area (*s*) of 94 m^2. Soils within the control areas are shallow to moderately deep (35–70 cm), sandy-loam brown acid soils and with restricted areas of podzols (Lithic Dystrudept and Entic Haplorthod according to USDA, 2010 [26]) with an approximate 25% slope. Soils in the charcoal hearths show a truncated profile, with a shallow surface organic horizon approximately 2 cm deep covering a thicker (19.3±2.8 cm) black anthropogenic layer. This horizon contains a large amount of charcoal fragments and fine particles left after carbonization which have been subsequently incorporated and well mixed with the pre-existing soil in response to bioturbation [27] and freeze-thaw processes [28] (Figure 1).

Both control soils and charcoal hearths are, nowadays, covered by the same herbaceous vegetation dominated by *Nardus stricta L.* while the surrounding forest is dominated by *Larix decidua* L. and *Picea abies* L.

The exact date of charcoal production was assessed using a dendro-anthracological approach. This method relies on cross-dating tree ring widths in charcoal fragments with known tree chronologies and has been used previously by [29], who showed that the oldest and youngest tree rings identified in charcoal fragments at our study area were dated 1530 and 1858 respectively. This last date corresponds to the year in which a wildfire event down in the valley caused the destruction of industrial plants thus determining the interruption of the local iron industry and charcoal production in the area [13].

Soil Sampling and Chemical Analysis

The anthropogenic layer within the charcoal hearths was sampled using a manual soil corer at five different sampling points in each hearth. Similarly, the soil at approximately the same depth was sampled at five points in each control area. Soil samples were dried for 72 hours at 35°C and sieved to 2 mm. In the case of the charcoal hearths, the fraction of soil >2 mm was further separated into two subsamples, one including charcoal fragments and one including plant debris, roots and stones. All further analysis was completed on the five sampling points separately.

Soil pH was measured in a soil/water solution (1:2.5 ratio). Soil C and N contents were determined by dry combustion using a CHN elemental analyzer (© Perkin Elmer 2400 series II CHNS/O elemental analyzer). Total Ca, K, Mg, Na, P, and available Ca, K, Mg, P concentrations were determined for subsamples oven-dried at 105°C for 24 h according to the EPA method 3052 [30] and the filtered solutions were analyzed using an ICP-OES spectrophotometer (Varian Inc., Vista MPX). A further set of subsamples was used to assess NO_3^-N according to the method proposed by [31] and NH_4^+-N according to the method proposed by [32]. Hydrophobicity was measured following the method of [33]. Such term defines the affinity for soils to water controlling infiltration or wetting. It can be caused by coating of long-chained hydrophobic organic molecules on individual soil particles in response to the decay of organic matter but also to the diversity of soil micro-organisms. Increased hydrophobicity is normally observed after wildfires that leaves charcoal fragments at the soil surface.

C and N content, and total and available Ca, K, Mg, Na, P concentrations of >2 mm charcoal individual fragments were determined using the same methodology described above. Micrographs of those charcoal fragments were made using a Scanning Electron Microscope, XL 20 FEI SEM, with CRYO-GATAN ALTO 2100 technology on samples dried under vacuum, following standard procedures [34].

To enable a comparison between old and fresh charcoal, fragments of larch wood were carbonized in a muffle furnace at 400°, 500°, 600°, 860°C. The time needed to complete the carbonization corresponds to the time needed for the sample to stabilize its weight loss. C and N contents were determined on samples using the methodology described above.

Pyrogenic C Determination

The relative contribution of charcoal-C ($C_{CHAR} : C_{TOT}$) to total soil carbon (C_{TOT}) was estimated using a mass balance method [35] based on the δ^{13}C values of charcoal fragments excavated from the anthropogenic soil layer ($\delta\,^{13}C_{CHAR}$), the mean $\delta\,^{13}$C of the entire layer ($\delta^{13}C_{TOT}$) and the $\delta\,^{13}$C of the SOM contained in the adjacent control soils ($\delta^{13}C_{SOM}$) (Table 1):

$$\frac{C_{CHAR}}{C_{TOT}} = \frac{\delta^{13}C_{TOT} - \delta^{13}C_{SOM}}{\delta^{13}C_{CHAR} - \delta^{13}C_{SOM}} \tag{1}$$

Table 3. Total and available nutrient concentrations (mg kg^{-1} dry matter ± standard error; n = 3) measured in control soils, charcoal hearths, old and new charcoal fragments and larch wood.

	Control soils		Charcoal hearths		Old charcoal fragments	Fresh charcoal fragments	Larch wood
Element	**Total concentration**	**Available**	**Total concentration**	**Available**	**Total concentration**	**Total concentration**	**Total concentration**
Ca^{2+}	993±135	278±35	3300±185	1006±158	3438±275	5920±70	5334±53
K^+	1603±116	147±30	2463±69	279±120	885±66	2899±39	1415±77
Mg^{2+}	2739±73	80±12	2378±384	245±4	1215±76	1533±32	1158±12
Na^+	297±50	34±1	86±8	33±26	216±1	207±4	190±4
P	321±7	7±2	921±206	12±4	346±81	3005±53	2716±46
N-NO_3^-	1.96±0.74	–	2.34±0.89	–	–	–	–
N-NH_4^+	3.93±1.00	–	3.73±1.80	–	–	–	–

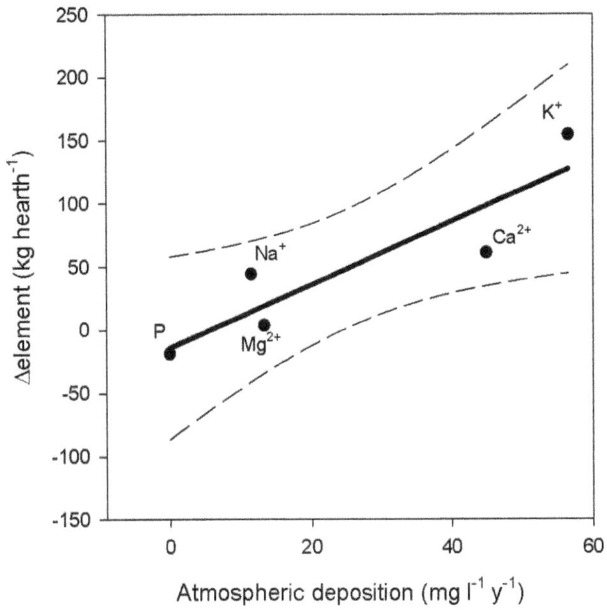

Figure 2. Correlation between average annual atmospheric deposition of P, K^+, Ca^{2+}, Mg^{2+}, Na^{2+} (mg l^{-1} y^{-1}) and the difference between the input of the same elements due to charcoal application in 1858 and the amount found today in hearth's soils (Δelement, kg $hearths^{-1}$) ($y = 2.50 \times -14.31$, $R^2 = 0.82$, $p = 0.035$). Dashed lines represent 95% confidence interval.

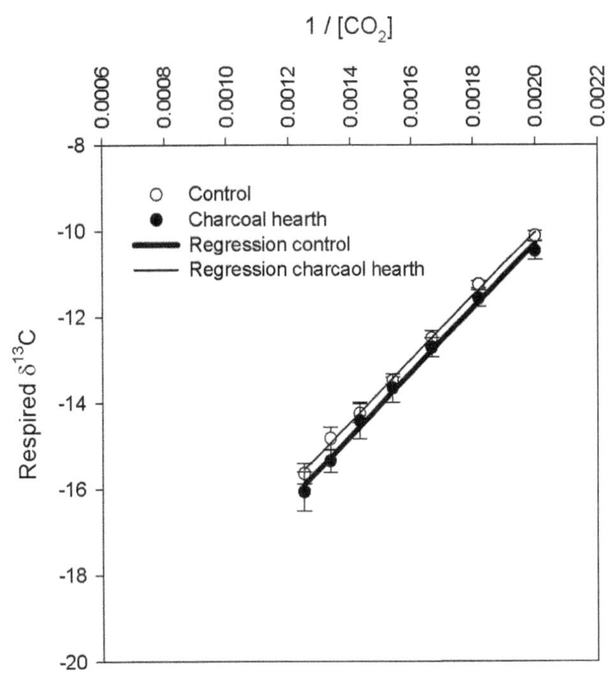

Figure 3. Keeling plots measured by CRDS showing the $\delta^{13}C$ of respired CO_2 fluxes versus the reciprocal of CO_2 concentration for control and charcoal hearth incubated soils ($\delta^{13}C_{CONTROL} = 7353*[CO_2]^{-1} - 24.8$, $R^2 = 0.99$; $\delta^{13}C_{CHARCOAL\ HEARTH} = 7467*[CO_2]^{-1} - 25.2$, $R^2 = 0.99$). Horizontal and vertical bars indicate standard deviations (n = 3).

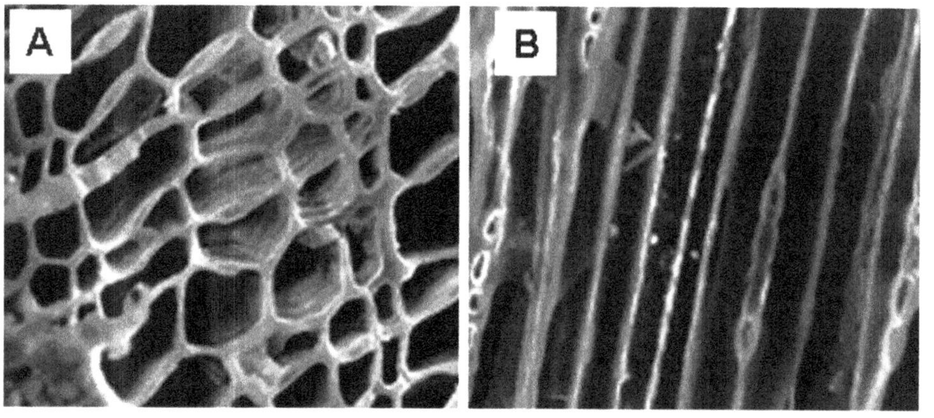

Figure 4. SEM micrographs showing the inner morphology of charcoal fragments and the absence of any microbes or plant debris. a) is a radial section b) a longitudinal section.

Stable C isotope ratio ($\delta^{13}C$) measurements were made on the fine fraction (<2 mm) of representative subsamples of control and hearth soils using an Isotope Ratio Mass Spectrometer (© Thermo Fischer Scientific, Delta V Plus) following total combustion in an elemental analyser (© EA Flash 1112 ThermoFinnigan).

The $\delta^{13}C$ signature of respired CO_2 from incubated charcoal hearth and control soils was measured using the Picarro G2131-*i* $\delta^{13}C$ High-precision Isotopic CO_2 Cavity Ring Down Spectrometer (CRDS) and Keeling plot method [36]. Representative subsamples (~250 cm^3; n = 3) were incubated in Erlenmeyer flasks at 40°C for 15 minutes. Air was continuously circulated from the flask to a pump and then back into the flask at a rate of 0.8–1.0 l min^{-1}. The CRDS was connected to the pump inlet tube and the air sub-sampled at 0.015 l min^{-1} for measurements of CO_2 concentration and $\delta^{13}C$. Sampling frequency was 0.5 Hz. To determine the $\delta^{13}C$ signature of the respired CO_2, the Keeling method was applied [36]. The intercept of the linear regression with the y-axis represents the isotopic signature of the source of the flux. Regression coefficients were calculated on averaged data at each 50 ppm interval of CO_2 concentration, starting from 450 to 800 ppm to establish a steady mixing within the flask. Finally, mean and standard deviation values of $\delta^{13}C$ were computed for both charcoal hearth and control soils (n = 3). We assumed that any difference in the $\delta^{13}C$ of SOM in control and charcoal hearths would also be reflected by a difference in the $\delta^{13}C$ of the respired CO_2.

Net C Sequestration

Ancillary information that is required to estimate the net C sequestration in hearths' soils was gathered from different sources:

- the amount of wood that was harvested and used for the carbonization process (W_s) was estimated on the basis of forest stand volume at three sub-alpine larch forests of varying ages. One forest was located on the slopes above the hearths of Val di Pejo, one in the nearby Val di Rabbi and one in Val Comasine. All forests had the same altitude (2000 m) and the latter two sites are known to be old-growth forests where 600 year old larch trees can still be found. Stand volume (m^3 ha^{-1}) was estimated using Light Detection And Ranging (LiDAR) measurements processed according to [37] and converted into biomass using an average wood basal density of 860 kg m^{-3}. For each forest, maximum wood stock was calculated using a fixed number of pixels;
- forest surface area that supplied each hearth during charcoal production (h), was determined by geo-referencing the charcoal hearths and drawing a 26 ha polygon on a digital orthophoto (Terraitaly™; © Compagnia Generale Ripreseaeree S.p.A. – Parma). h was calculated according to the following two criteria: i) the wood collection area should be located above each charcoal hearth up to the tree line; ii) the lateral boundaries of the area corresponded to the mean distance between two charcoal hearths (~100 m). Finally, the total area was divided by the total number of hearths identified in the area (n = 7).
- carbonization efficiency, i.e. the ratio between produced charcoal and used wood (q), was assumed to be equal to 20% according to [38];
- the fraction of charcoal that was left on the soil surface at the end of each carbonization cycle after char was removed by charcoal makers (w), was experimentally estimated using a modern analogue. A charcoal hearth that is currently in use

Table 4. Description of larch forests considered as analogues of the larch forest harvested for charcoal production in Val di Pejo (mean ± standard error).

Forest site	Coordinates	Elevation (m a.s.l.)	Forest age	Number of plots	Wood stock (W, t ha^{-1})
Val di Pejo	46°20′16.18″ N, 10°36′07.02″ E	2152	150	51	288±41
Val di Rabbi	46°26′39.45″ N, 10°45′59.56″ E	1864	500	20	213±27
Val Comasine	46°20′02.00″ N, 10°39′58.78″ E	2119	650	40	258±25

Data were determined using LiDAR measurements [37].

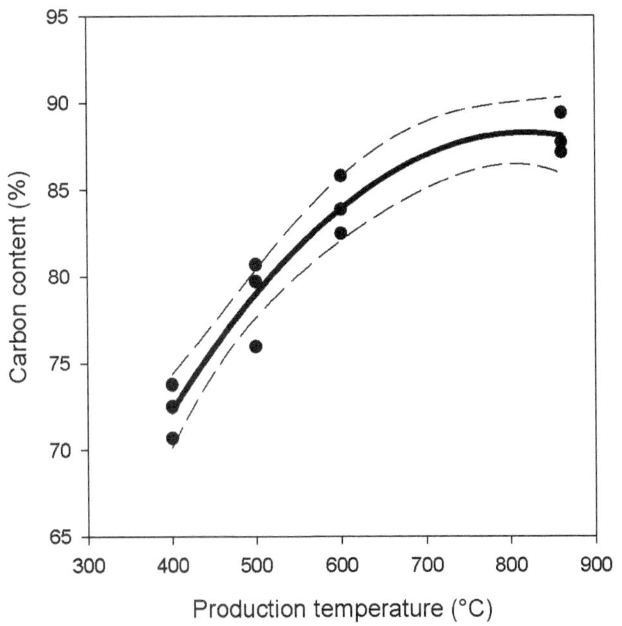

Figure 5. Carbon content of charcoal produced from larch wood at different temperatures. Wood samples were collected in close proximity to the hearths. Charcoal was produced in a muffle furnace at 400°, 500° 600° and 860°C. Dashed lines represent 95% confidence interval. ($Y = 26.9 + 0.15\ X - 9.2\ 10^{-5}\ X^2$; $r^2 = 0.94$; $p < 0.0001$).

was identified at short distance from the Val di Pejo, where expert charcoal makers repeat traditional charcoal production mainly for didactical purposes. They also recorded, year by year, the exact amount of wood used and of charcoal produced. The amount of pyrogenic C left on hearth soil was quantified using the loss-on-ignition (LOI) method [39] using four replicates randomly selected within the hearth area. w was determined as the ratio between the sum of C contained in the charcoal that was made over the last ten years and the corresponding amount of C that was found in the soil. The uncertainty was estimated as the standard error of the mean;

- C content of freshly produced larch wood charcoal (r_0) was also assessed experimentally. Known amounts of larch wood taken from wood disc collected in the proximity of Val di Peio charcoal hearths were pyrolized at different temperatures in a muffle furnace. The C concentration was measured on each charcoal sample using a CHN elemental analyzer. Measured C content data were fitted to production temperature using a second order polynomial relationship and C content at a reference temperature of 450°C was assumed to be an analog of the charcoal originally produced [40].

Uncertainty and Sensitivity Analysis

All data in the text and in the tables are reported as mean ± standard error (n = 3) if not differently indicated.

Gaussian error propagation technique (GEP) was used in error analysis to analytically determine the error or uncertainty produced by multiple and interacting measurements or variables. For this, the uncertainty associated to each measurement was calculated as standard error (se) of the mean (se = standard deviation/root square of the number of samples) and the classical error propagation theory and equations were used [41]. Error sensitivity analysis was also made by constructing an error budget [42], thus enabling further understanding of the error structure, i.e. the relative contribution of the errors associated with each parameter to the overall error estimate. Such sensitivity indices provide in this way a measure of the percentage rate of change in an output variable produced per unit percentage change in its input variable, an information that may be used critically to identify where error reduction of estimates may lead to lower uncertainty.

Table 5. Parameters, coefficients and variables used to estimate charcoal stability in soil (mean ± standard error; n = 3).

Parameter/ coefficient/variable	Definition	Unit	Value	Source(s)
q	Carbonization efficiency	–	0.20	[38]
w	Fraction of charcoal left on the ground at the end of carbonization	–	0.02±0.017	Measured with LOI
r_0	Carbon content of larch wood charcoal produced at 450°C	$g\ g^{-1}$	0.76±0.040	Measured with CHN and extrapolation
r_{155}	Carbon content of charcoal fragments found in the hearth's soil	$g\ g^{-1}$	0.60±0.032	Measured with CHN
W_s	Larch wood stock of the forest in Val di Pejo	$t\ ha^{-1}$	288±41	Measured (LiDAR) and calculated to convert from $m^3\ ha^{-1}$ to $t\ ha^{-1}$
h	Forest area for wood collection for charcoal production per hearth	ha $hearth^{-1}$	3.7	Measured on ortophotos
s	Surface area of charcoal hearths	m^2	94	Measured
C_{IN}	Pyrogenic carbon left on the ground at the time of the carbonization	$Kg\ C\ m^{-2}$	29.3±5.1	Calculated using Eq. [2]
–	Fraction of pyrogenic carbon lost since the time of the carbonization calculated with Eq. [2]	–	0.20±0.28	Calculated
–	Fraction of pyrogenic carbon lost since the time of the carbonization calculated with r_0 and r_{155}	–	0.21±0.04	Calculated
k_{CHAR}	Annual pyrogenic carbon decay rate in hearths	$years^{-1}$	0.0015±0.0003	Calculated
MRT	Mean residence time of charcoal in the soil	years	650±139	Calculated

Table 6. Uncertainty and sensitivity analysis results for the setimation of equation 3 parameters.

Variable	Variable uncertainty (standard error)	Relative contribution to the overall uncertainty in ΔC_{CHAR} in Eq.3
W_s	41	16%
w	0.002	11%
r_0	0.004	1%
Soil organic C content	0.03	18%
Soil bulk density	71	14%
Anthropogenic soil layer depth	0.04	26%
$\delta^{13}C_{TOT}$	0.14	11%
$\delta^{13}C_{CHAR}$	0.31	3%
$\delta^{13}C_{SOM}$	0.01	1%

Results and Discussion

Soil Bulk Density and Hydrophobicity

The anthropogenic layer of the charcoal hearth soils has a lower bulk density than control soils (0.60 ± 0.08 vs. 0.87 ± 0.12 Mg m^{-3}; $p<0.01$). Soil bulk density is an important indicator of physical soil quality, being linked to air capacity, resistance to root growth and capacity of storing and transmitting water [43]. Such a decrease in bulk density is associated with a 97.3% ± 1.6 decrease in hydrophobicity. This is in line with the water infiltration data from charcoal production sites in Ghana [44], but in contrast with short term observations following biochar applications to soil [45] [46] that showed small but consistent increases in soil hydrophobicity which is largely controlled by the surface chemistry of fresh biochar particles [47]. We speculate that a prolonged residence time of charcoal caused substantial leaching or degradation of hydrophobic compounds [48], a shift in soil texture ([49], [44]), and a microbially-driven creation of functional groups [50]. Decreased hydrophobicity is known to increase water availability for plants and is also important for nutrient cycling, as it favors water infiltration into the soil and reduces runoff, thus preventing lateral nutients losses.

Nutrients and Carbon

Nutrient content (total and plant available fractions) is higher in the hearths than in the control soils (Table 2; Table 3). In particular, the total P-stock is 107% larger in hearths than in the control (95 ± 6 vs. 46 ± 3 g m^{-2}, $p = 0.003$), while the plant available P is 24% higher (1.2 ± 0.04 g m^{-2} vs. 0.9 ± 0.3 g m^{-2}). The higher P content is not surprising as charcoal contains at least 20% of P originally contained in the wood (Table 2). If we assume that charcoal made in the middle of the XIX century had a P content comparable to that of modern charcoal (3.0 ± 0.05 g P kg^{-1}; Table 3), we may estimate that carbonization events led to the addition of 117 ± 21 g P m^{-2} (Table 2). In the absence of grass mowing, a virtually closed P-cycle can be hypothesized for these soils as P-leaching does not usually occur if the overall concentrations are low so that P can be considered "virtually immobile" [51] [52]. Nevertheless, large herbivores are known to be net P-exporters in alpine grasslands as they may preferentially graze in P-enriched areas and then release P as dung elsewhere [53]. This export largely depends on the grazing pressure, but it is unlikely to exceed the maximum value of 0.07 g P m^{-2} y^{-1} that was assessed experimentally in the Swiss Alps [53]. When scaled to the time that charcoal was added to the soil the amount of P exported would not have exceeded 11.5 g m^{-2}. Atmospheric deposition may have also contributed to the P balance as a result of long-range desert dust transport and as a consequence of atmospheric pollution, including biomass combustion [54]. Although the latter is known to be variable in time and space, this input is not large in the Eastern Alps, with less than 0.01 g P m^{-2} y^{-1} [55]. When measured today, the total amount of P contained in the anthropogenic layer of the hearth's soil is only 19% less than what was initially added during the carbonization events (95 ± 6 vs. 117 ± 21 g m^{-2}; Table 2). This highlights the fact that charcoal production was indeed a long-lasting source of P in an otherwise P-limited environment. P-fertilization persisted on a centennial time-scale and it is also interesting to observe that both organic and inorganic (extractable) fractions of P that were added to the soil are now mostly contained in the non-pyrogenic fraction of the SOM, as the P contained in old charcoal fragments is only 12% of that of modern laboratory-produced larch charcoal (Table 3).

Other nutrients, such as potassium (K) and calcium (Ca) are also more abundant in the charcoal hearth soils even though there are no significant differences with control soils due to the large spatial variability (Table 2). Furthermore K and Ca are 155% and 61% higher than what was added to the soil with charcoal (Table 2). This excess may be attributed to an higher retention of atmospheric K and Ca depositions which have been previously reported to be relevant in the Alpine region [56] and can be related to the higher cation exchange capacity (CEC) of charcoal [57]. A strong correlation between current atmospheric deposition rates and element excesses found in the hearth soil seems to confirm such hypotheses (Figure 2).

Total N content is not significantly different between hearth and control soils ($p = 0.42$; Table 2) and no significant difference was found in the concentrations of mineral N (NO_3^-: $p = 0.31$; NH_4^+: $p = 0.92$; Table 3).

Total C content of the anthropogenic layer at the charcoal hearths is three times higher than that of the adjacent control soils ($p = 0.03$; Table 1). The amount of C that is now contained in the anthropogenic soil layer (C_{TOT}, kg C m^{-2}) is the sum of pyrogenic (C_{CHAR}, kg C m^{-2}) and non-pyrogenic components (C_{SOM}, kg C m^{-2}) as carbonates are absent due to the low pH (4.2 ± 0.3 and 4.6 ± 0.3 in charcoal hearths and control areas, respectively). Charcoal fragments larger than 2 mm represent approximately 4.1 ± 1.7% (by weight) of the entire mass of the anthropogenic layer within the deeper soil horizons where no charcoal debris can be identified (Figure 1).

Any meaningful evaluation of net C sequestration achieved in charcoal hearths firstly requires a precise separation of C fractions contained in the charcoal (C_{CHAR}) and in the rest of the soil, followed by a accurate estimation of the C input at the time of the carbonization event.

Assessing the exact ratio between C_{CHAR} and C_{TOT} using sable C isotopes is problematic because:

- Equation 1 assumes a net difference in the isotopic signature of old charcoal and modern SOM. This is supported by the observation that wood formed before 1900 is on average ~1‰ less negative than that formed after 1950 [58] due to the recent rapid rise in $\delta^{13}C$-depleted atmospheric CO_2 concentrations as a result of fossil fuel burning [59]. In addition, it has been reported that branch or stem wood of C_3 plants is generally enriched by 1–3‰ compared to leaves [60]. Despite the fact that limited ^{13}C-depletion may occur during wood carbonization at temperatures above 300°C [61], it could be expected that the charcoal fragments found in the hearth's soils are significantly enriched in the heavier C isotope compared to the more recent SOM pools that are mainly derived from the decomposition of litter formed more recently;
- Equation 1 assumes that $\delta^{13}C_{SOM}$ in the hearth's soils is equal to $\delta^{13}C_{SOM}$ in the control soils. Such equivalence cannot be assumed *a priori*, but CRDS-based $\delta^{13}C$ flux measurements demonstrated that the signature of respired CO_2 was not significantly different between control and charcoal hearth incubated soil ($p = 0.75$; Figure 3 and Table 1). Similarly, it may be assumed that the $\delta^{13}C$ of the less recalcitrant (non-pyrogenic) SOM fractions is the same in both hearth and control soils. Such an equivalence is further confirmed by the linearity observed by plotting the reciprocal of the C content of charcoal fragments, hearth and control soils versus their respective stable C isotopic ratios (slope = −11.3; intercept = −24.3; $r^2 = 0.98$; $n = 9$; $p < 0.0001$). The fact that the data points fall within close proximity to a straight line indicates that the two C pools (pyrogenic and non-pyrogenic C) are distinguished in the hearth's soils;
- The $\delta^{13}C$ of the charcoal fragments ($\delta^{13}C_{CHAR}$ in Equation 1) could be affected by the presence of organic debris or micro-organisms in charcoal pores. The analysis of charcoal fragments with a Scanning Electron Microscope showed that, despite being exposed in soil for over 150 years, no organic or inorganic debris were present in the inner portion of the fragments ([34]; Figure 4). A recent study [62] showed that, in spite of the large empty space in charcoal pores, these were only sparsely populated by microorganisms three years after application to soil. This was attributed to the adsorption of inorganic and organic compounds which may cause blockage of the charcoal pores and thus prevent microorganisms penetrating the inner portion of fragments. Our data support this observation, showing that charcoal, also in the long term, does not provide a habitat for microbes and, accordingly, the $\delta^{13}C$ of the charcoal fragments is unaffected.

Based on the above assumptions and considerations, the fraction of pyrogenic C (C_{CHAR}) contained in the overall soil C (C_{TOT}) was finally estimated using Equation 1 to be equal to 89±9%, corresponding to an absolute amount of 23.3±4.7 kg of pyrogenic C m^{-2} (Table 1).

Carbon Sequestration

Once the amount of pyrogenic C is known, a reliable estimation of the C sequestration achieved in the hearth soil over centennial timescale requires a proper estimation of the initial C input (C_{IN}, kg C m^{-2}). This can be obtained according to equation:

$$C_{IN} = \frac{W_s \times h \times q \times w \times r_0}{s} * 1000 \qquad (2)$$

Solving Equation 2 is problematic mainly because of the uncertainty associated with the determination of parameters W_s, w and r_0:

- the total amount of wood biomass (W_s) that was used for carbonization cannot be directly estimated but requires the use of a proxy. The assumption can be made that the forest standing biomass one and half centuries ago was comparable to what is currently found in our study area (288±41 t ha^{-1}; Table 4). This assumption is partly confirmed by the fact that LIDAR-based standing volumes of two additional old-growth forest sites (>200 years) are comparable to the wood stock of the study area. The fact that tree volume is independent of stand age is not surprising and is confirmed by the usually reported plateau of forest wood stocks over centennial times scales [63] in alpine larch forests [64];
- the fraction of charcoal left on the soil after carbonization can only be estimated indirectly as there a no reliable sources reporting such a value. We assumed that an experimental assessment using modern charcoal hearth is the most reliable proxy of ancient charcoal hearths. The value obtained in this way ($w = 0.02 \pm 0.002$) was not far from judgment of two expert charcoal makers who unanimously said that no more than 2% charcoal fragments are normally left over the soil at the end of each carbonization cycle;
- the C content of new charcoal (r_0) is known to vary substantially with production temperature ($r^2 = 0.94$, $p < 0.0001$; Figure 5). Here again there are not historical data on temperature during charcoal production but FAO reports that the average temperature in traditional carbonization wood piles is around 450°C [40]. Using this value, r_0 was estimated to be equal to 0.76±0.004.

Based on the above assumptions and considerations, C_{IN} could be calculated to be equal to 29.3±5.1 kg C m^{-2} (Table 5) and the overall pyrogenic C lost was then quantified as:

$$\Delta C_{CHAR} = \frac{(C_{IN} - C_t)}{C_{IN}} \qquad (3)$$

where C_t is the actual pyrogenic C in the soil and t is time since the last char production at the charcoal hearth (153 years). Thus, the fraction of pyrogenic C lost since the time of carbonization was equal to 0.20±0.28 of C_{IN} (Table 5). This value is given by the sum of the direct charcoal degradation by biotic and abiotic factors and the lateral transport due to surface runoff and erosion occurring during and after char production, before a new soil layer covered the anthropogenic layer [65]. The presence of small amounts of minute charcoal fragments predominantly down-slope of the hearths seems to confirm the occurrence of such lateral flows.

Recent studies have used changes of the C-content of individual charcoal fragments buried in the soil as a proxy of the fraction of carbon lost from charcoal over long time scales [66]. Such method is certainly questionable, as it cannot distinguish between actual C-losses due to oxidation and C-dilution effects associated to the deposition of inorganic salts and minerals on those fragments.

Nevertheless, it is noteworthy, here, to highlight that the C content of individual charcoal fragments found in the charcoal hearth's soils (r_{155} = 0.60±0.03 gC g^{-1}) is consistently lower than the C content of modern larch charcoal made in the laboratory (r_0 = 0.76±0.04 gC g^{-1}). The fact that the relative difference between r_0 and r_{155} (0.21) almost exactly matches our estimate of the fraction of C lost on centennial time scale (Equations 2 and 3) is likely coincidental but suggests that more detailed investigations on the oxidation of ancient charcoal fragments ([67]) and on mineral deposition on old charcoal fragments are needed.

The decay rate (k, years^{-1}) can be finally calculated as:

$$k = \frac{\ln\left(\frac{1}{1-\Delta C_{CHAR}}\right)}{t} \quad (4)$$

and is equal to 0.0015±0.0003. Such a value is less than half of the value estimated in a recent meta-analysis of charcoal decomposition [68] and corresponds to a Mean Residence Time (MRT) of 650±139 years (Table 5).

Error sensitivity analysis performed on the results of Equation 2 and 3 illustrates the relative contribution of errors to the overall uncertainty of the result (Table 6). The data show that the largest contribution to the overall uncertainty is in the error associated to the anthropogenic soil layer depth. Such large error suggests that variance among the three replicated hearths may not be simply random but reflects, instead, some systematic effects possibly associated to differences in the amount of wood that was carbonized in each hearth. This is indeed a critical aspect, possibly requiring a re-analysis of the simplifying assumption of an even distribution of carbonization intensity (amount of wood carbonized) among the replicated hearths. Such re-analysis would in fact enable a substantial reduction of the uncertainty, while not affecting the means. Other important sources of error are related to the estimation of the C-content of the anthropogenic layers (C_{tot}) and in the estimated amount of wood that was used for the carbonization process (W_s). Error reduction would have been possible, in the first case, by increasing the number of replicates and more unlikely by increasing samplings in each replicate. Nevertheless, a relative large uncertainty would remain associated to the second case, as W_s was inferred on the basis of large simplification and that by no means new information could be retrieved to finally reduce its error.

Conclusions

This study presents two key messages. First, it provides novel insights into the long-term decomposition of pyrogenic C in the soil, by demonstrating that charcoal addition to soil is indeed a way to obtain substantial C-sequestration. Despite some inevitable uncertainty, we have shown that 23.3±4.7 kg C m^{-2} of pyrogenic C are still present in the soil after an addition of 29.3±5.1 kg C m^{-2} that was made in the middle of the XIX century. Carbon sequestration was estimated as 80±21% of the original added C. Secondly, our investigation provides substantial evidence that the avaliability of macro- and some micro-nutrients is higher in charcoal hearth soils over centennial timescales. This supports the common observation that the addition of various forms of pyrogenic C (biochar) increases soil fertility and plant yield, even in the long-term. The persistence of enhanced nutrient avaliability over centennial timescales is likely associated with mechanisms favoring their accumulation and improving soil water relations. Overall, this strongly supports the idea that the addition of biochar to soil is indeed a feasible, effective, and sustainable strategy to both sequester atmospheric C and to enhance crop yields.

However, we call for some caution on excessive generalization of our results: the hearths within the Val di Pejo are located in a mountainous area, at high elevations, and are exposed to peculiar climatic conditions, that are certainly different to the vast majority of areas where croplands are concentrated. Even if soil freeze-thaw cycles associated with large seasonal temperature fluctuations are likely to favour decomposition and C oxidation, it is not certain if and how the C decomposition rates that we observed are greater or smaller than in other climates. Additional caution is required since that the charcoal added to the studied hearths came from a single feedstock (larch wood) and was produced in traditional charcoal production systems. The fate and the effects of other feedstocks and of other production processes such as slow and fast pyrolysis, pyrogasification, and hydrothermal conversion may indeed create totally different biochar types, possibly behaving in different ways in the soil.

Charcoal hearths in Val di Pejo are certainly a unique resource for investigating the long term effect of pyrogenic C addition to the soil. The number of replications which are available, the long residence time of the charcoal in the soil, the accuracy of charcoal dating, the reliability of ancillary information on the sequence of events that were associated to charcoal production and its sudden cessation, contribute to the scientific value of these sites. A number of questions, not considered in the present study, could be addressed in the near future, such as, for instance, the effect of long-term charcoal burial on (i) its physio-chemical properties; (ii) non-CO_2 greenhouse gas fluxes (CH_4 and N_2O); (iii) plant productivity; and (iv) shifts and changes in soil biodiversity.

Acknowledgments

The authors would like to thank Federica Camin and Luca Ziller, Matteo Girardi, Michele Dalponte and Damiano Gianelle (E. Mach Foundation, San Michele a/Adige) Luisa Andrenelli, Adriano Baglio, Fabio Zuliani (Stelvio National Park) for their support in field and laboratory work, as well as Christa Backmeroff, Fabio Angeli, Dino Buffoni, Jean Christophe Lata, Marie Alexis and Katell Quenea for useful discussions.

Author Contributions

Conceived and designed the experiments: IC GA CR RV FM. Performed the experiments: IC GA SB FF CM CC EP FM. Analyzed the data: IC GA CC FM. Wrote the paper: IC GA CM FM.

References

1. Lehmann J, Gaunt J, Rondon M (2006) Bio-char Sequestration in Terrestrial Ecosystems – A Review. Mitigation and Adaptation Strategies for Global Change 11: 395–419. Available: http://link.springer.com/10.1007/s11027-005-9006-5. Accessed 23 May 2013.
2. Laird DA (2008) The Charcoal Vision: A Win–Win–Win Scenario for Simultaneously Producing Bioenergy, Permanently Sequestering Carbon, while Improving Soil and Water Quality. doi:10.2134/agronj2007.0161.
3. Woolf D, Amonette JE, Street-Perrott FA, Lehmann J, Joseph S (2010) Sustainable biochar to mitigate global climate change. Nature communications 1: 56. Available: http://www.pubmedcentral.nih.gov/articlerender.fcgi?artid=2964457&tool=pmcentrez&rendertype=abstract. Accessed 29 May 2013.
4. Joseph SD, Downie A, Munroe P, Crosky A, Lehmann J (2007) Biochar for Carbon Sequestration, Reduction of Greenhouse Gas Emissions and Enhancement of Soil Fertility; A Review of the Materials Science: 130–133.
5. Yanai Y, Toyota K, Okazaki M (2007) Effects of charcoal addition on N 2 O emissions from soil resulting from rewetting air-dried soil in short-term laboratory experiments. Soil Science and Plant Nutrition 53: 181–188. Available: http://www.tandfonline.com/doi/abs/10.1111/j.1747-0765.2007.00123.x. Accessed 28 May 2013.

6. Peng X, Ye LL, Wang CH, Zhou H, Sun B (2011) Temperature- and duration-dependent rice straw-derived biochar: Characteristics and its effects on soil properties of an Ultisol in southern China. Soil and Tillage Research 112: 159–166. doi:10.1016/j.still.2011.01.002.
7. Oguntunde PG, Fosu M, Ajayi AE, Van de Giesen N (2004) Effects of charcoal production on maize yield, chemical properties and texture of soil. Biology and Fertility of Soils 39: 295–299. Available: http://link.springer.com/10.1007/s00374-003-0707-1. Accessed 29 May 2013.
8. Lehmann J, Rillig MC, Thies J, Masiello C, Hockaday WC, et al. (2011) Biochar effects on soil biota – A review. Soil Biology and Biochemistry 43: 1812–1836. Available: http://linkinghub.elsevier.com/retrieve/pii/S0038071711001805. Accessed 25 May 2013.
9. Genesio L, Miglietta F, Lugato E, Baronti S, Pieri M, et al. (2012) Surface albedo following biochar application in durum wheat. Environmental Research Letters 7: 014025. Available: http://stacks.iop.org/1748-9326/7/i=1/a=014025?key=crossref.4c8f29d72864c065a47107c0fadd5c71. Accessed 29 May 2013.
10. Graber ER, Meller Harel Y, Kolton M, Cytryn E, Silber A, et al. (2010) Biochar impact on development and productivity of pepper and tomato grown in fertigated soilless media. Plant and Soil 337: 481–496. Available: http://link.springer.com/10.1007/s11104-010-0544-6. Accessed 29 May 2013.
11. Gurwick NP, Moore L a, Kelly C, Elias P (2013) A systematic review of biochar research, with a focus on its stability in situ and its promise as a climate mitigation strategy. PloS one 8: e75932. Available: http://www.pubmedcentral.nih.gov/articlerender.fcgi?artid=3786913&tool=pmcentrez&rendertype=abstract. Accessed 17 December 2013.
12. Schmidt MWI, Skjemstad JO, Jäger C (2002) Carbon isotope geochemistry and nanomorphology of soil black carbon: Black chernozemic soils in central Europe originate from ancient biomass burning. Global Biogeochemical Cycles 16. doi:10.1029/2002GB001939.
13. Favilli F, Cherubini P, Collenberg M, Egli M, Sartori G, et al. (2010) Charcoal fragments of Alpine soils as an indicator of landscape evolution during the Holocene in Val di Sole (Trentino, Italy). The Holocene 20: 67–79. doi:10.1177/0959683609348850.
14. Kaal J, Martínez Cortizas A, Eckmeier E, Costa Casais M, Santos Estévez M, et al. (2008) Holocene fire history of black colluvial soils revealed by pyrolysis-GC/MS: a case study from Campo Lameiro (NW Spain). Journal of Archaeological Science 35: 2133–2143. doi:10.1016/j.jas.2008.01.013.
15. Masiello Ca (2004) New directions in black carbon organic geochemistry. Marine Chemistry 92: 201–213. Available: http://linkinghub.elsevier.com/retrieve/pii/S0304420304002051. Accessed 27 May 2013.
16. Zimmermann M, Bird MI, Wurster C, Saiz G, Goodrick I, et al. (2012) Rapid degradation of pyrogenic carbon. Global Change Biology 18: 3306–3316. Available: http://doi.wiley.com/10.1111/j.1365-2486.2012.02796.x. Accessed 29 May 2013.
17. Knicker H (2011) Pyrogenic organic matter in soil: Its origin and occurrence, its chemistry and survival in soil environments. Quaternary International 243: 251–263. doi:10.1016/j.quaint.2011.02.037.
18. Smith N (1980) Anthrosols and human carrying capacity in Amazonia. Annals of the Association of American Geographers 70: 553–566.
19. Cremaschi M, Pizzi C, Valsecchi V (2006) Water management and land use in the terramare and a possible climatic co-factor in their abandonment: The case study of the terramara of Poviglio Santa Rosa (northern Italy). Quaternary International 151: 87–98. doi:10.1016/j.quaint.2006.01.020.
20. Glaser B, Haumaier L, Guggenberger G, Zech W (2001) The "Terra Preta" phenomenon: a model for sustainable agriculture in the humid tropics. Naturwissenschaften 88: 37–41. Available: http://link.springer.com/10.1007/s001140000193. Accessed 24 May 2013.
21. Mercuri AM, Accorsi CA, Mazzanti MB, Bosi G, Cardarelli A, et al. (2006) Economy and environment of Bronze Age settlements – Terramaras – on the Po Plain (Northern Italy): first results from the archaeobotanical research at the Terramara di Montale. Vegetation History and Archaeobotany 16: 43–60. doi:10.1007/s00334-006-0034-1.
22. Di Piazza A, Eccel E (2012) Analisi di serie giornaliere di temperatura e precipitazione in Trentino nel periodo 1958–2010. Provincia Autonoma di Trento e Fondazione E. Mach.
23. Sonna R (n.d.) Le miniere di Comasine. Studi e Ricerche: 1–10.
24. Rizzi G (2010) Sentiero Val di Sole, le sfumature del verde. Trento: Tipografia Editrice Temi s.a.s.
25. Schenkel Y, Bertaux P, Vanwijnbserghe S, Carre J (1998) An Evaluation of the Mound Kiln Carbonization Technique. Biomass and Bioenergy 14: 505–516.
26. Smith RT, Atkinson K (1975) Techniques in pedology. London: Paul Elek (Scientific Books) Ltd.
27. Eckmeier E, Gerlach R, Skjemstad JO, Ehrmann O, Schmidt MWI, et al. (2007) Minor changes in soil organic carbon and charcoal concentrations detected in a temperate deciduous forest a year after an experimental slash-and-burn: 377–383.
28. Carcaillet C (2001) Are Holocene wood-charcoal fragments stratified in alpine and subalpine soils? Evidence from the Alps based on AMS 14C dates. The Holocene 11: 231–242. doi:10.1191/095968301674071040.
29. Backmeroff C (2013) Precisely dating iron-ore mining and its effects on an alpine valley: Summary of a dendrochronological investigation of charcoal hearths and relict woodland stands. In: Silvertant J ed., editor. Mining and Cultural Landscape –8th International Symposium on Archaeological Mining History. pp. 218–251.
30. USEPA (1996) Method 3052, Microwave assisted aced digestion of siliceous and organically based matrices: 1–20.
31. Vendrell P, Zupancic J (1990) Determination of soil nitrate by transnitration of salicylic acid. Communications in Soil Science and Plant Analysis 21.
32. Willis RB, Schwab GJ, Gentry CE (1993) Elimination of interferences in the colorimetric analysis of ammonium in water and soil extracts. Communications in Soil Science and Plant Analysis 24.
33. Letey J (1969) Measurement of contact angle, water drop penetration time, and critical surface tension. In: University of California, editor. Water repellent soils. Proceedings of the Symposium on Water Repellent Soils. Riverside. pp. 43–47.
34. Pusceddu E, Criscuoli I, Miglietta F (2013) Morphological investigation and physical characterization of ancient fragments of pyrogenic carbon. Journal of Physics: Conference Series 470: 012003. Available: http://stacks.iop.org/1742-6596/470/i=1/a=012003?key=crossref.442c46e432fd91c4a0ddb5660ebc00b8. Accessed 28 December 2013.
35. Del Galdo I, Six J, Peressotti A, Cotrufo F (2003) Assessing the impact of land-use change on soil C sequestration in agricultural soils by means of organic matter fractionation and stable C isotopes. Global change biology: 1204–1213.
36. Keeling CD (2011) The Concentration and Isotopic Abundances of Carbon Dioxide in the Atmosphere. Tellus A. Available: http://journals.sfu.ca/coaction/index.php/tellusa/article/view/9366.
37. Tonolli S, Dalponte M, Vescovo L, Rodeghiero M, Bruzzone L, et al. (2010) Mapping and modeling forest tree volume using forest inventory and airborne laser scanning. European Journal of Forest Research 130: 569–577. Available: http://link.springer.com/10.1007/s10342-010-0445-5. Accessed 29 May 2013.
38. Mantovani P (2006) Ricordi di un carbonèr. Da Bondone a Bissina. Memorie autobiografiche. Antolini, editor Antolini.
39. Ball DF (1964) Loss-on-ignition as an estimate. 15.
40. FAO (1987) Simple technologies for charcoal making. Rome: Mechanical Wood Products Branch Forest Industries Division FAO Forestry Department.
41. Lehrter JC, Cebrian J (2010) Uncertainty propagation in an ecosystem nutrient budget. Ecological applications: a publication of the Ecological Society of America 20: 508–524. Available: http://www.ncbi.nlm.nih.gov/pubmed/20405803.
42. Lo E (2005) Gaussian error propagation applied to ecological data: post-ice-storm-downed woody biomass. Ecological Monographs 75: 451–466. Available: http://www.esajournals.org/doi/abs/10.1890/05-0030. Accessed 28 December 2013.
43. Reynolds WD, Drury CF, Tan CS, Fox C a., Yang XM (2009) Use of indicators and pore volume-function characteristics to quantify soil physical quality. Geoderma 152: 252–263. doi:10.1016/j.geoderma.2009.06.009.
44. Oguntunde PG, Abiodun BJ, Ajayi AE, Van de Giesen N (2008) Effects of charcoal production on soil physical properties in Ghana. Journal of Plant Nutrition and Soil Science 171: 591–596. Available: http://doi.wiley.com/10.1002/jpln.200625185. Accessed 29 May 2013.
45. Scott DF, Van Wyk DB (1990) The effects of wildfire on soil wettability and hydrological behaviour of an afforested catchment. Journal of Htdrology 121: 239–256.
46. Lane PNJ, Croke JC, Dignan P (2004) Runoff generation from logged and burnt convergent hillslopes: rainfall simulation and modelling. Hydrological Processes 18: 879–892. doi:10.1002/hyp.1316.
47. Kinney TJ, Masiello C a., Dugan B, Hockaday WC, Dean MR, et al. (2012) Hydrologic properties of biochars produced at different temperatures. Biomass and Bioenergy 41: 34–43. Available: http://linkinghub.elsevier.com/retrieve/pii/S0961953412000438. Accessed 27 May 2013.
48. Briggs CM, Breiner J, Graham R (2005) Contributions of Pinus Ponderosa charcoal to soil chemical and physical properties. The ASA-CSSA-SSSA International Annual Meetings (November 6–10, 2005). Salt Lake City. pp. 1–13.
49. Glaser B, Lehmann J, Zech W (2002) Ameliorating physical and chemical properties of highly weathered soils in the tropics with charcoal - a review. Biology and Fertility of Soils 35: 219–230. Available: http://link.springer.com/10.1007/s00374-002-0466-4. Accessed 23 May 2013.
50. Lehmann J, Liang B, Solomon D, Lerotic M, Luizão F, et al. (2005) Near-edge X-ray absorption fine structure (NEXAFS) spectroscopy for mapping nano-scale distribution of organic carbon forms in soil: Application to black carbon particles. Global Biogeochemical Cycles 19: n/a–n/a. Available: http://doi.wiley.com/10.1029/2004GB002435. Accessed 22 May 2013.
51. Hesketh N, Brookes PC (2000) Development of an Indicator for Risk of Phosphorus Leaching. Journal of Environmental Quality 29: 105–110.
52. Flueck WT (2009) Biotic Translocation of Phosphorus: The Role of Deer in Protected Areas. Sustainability 1: 104–119. doi:10.3390/su1020104.
53. Schütz M, Risch AC, Achermann G, Thiel-Egenter C, Page-Dumroese DS, et al. (2006) Phosphorus Translocation by Red Deer on a Subalpine Grassland in the Central European Alps. Ecosystems 9: 624–633. doi:10.1007/s10021-006-0091-4.
54. Bergametti G, Remoudaki E, Losno R, Steiner E, Chatenet B (1992) Source, transport and deposition of atmospheric phosphorus over the northwestern Mediterranean. Journal of Atmospheric Chemistry 14: 501–513.
55. Mahowald N, Jickells TD, Baker AR, Artaxo P, Benitez-Nelson CR, et al. (2008) Global distribution of atmospheric phosphorus sources, concentrations and

deposition rates, and anthropogenic impacts. Global Biogeochemical Cycles 22. doi:10.1029/2008GB003240.

56. Tait D, Thaler B (2000) Atmospheric deposition and lake chemistry trends at a high mountain site in the eastern Alps. 59: 61–71.
57. Basso AS, Miguez FE, Laird D a., Horton R, Westgate M (2013) Assessing potential of biochar for increasing water-holding capacity of sandy soils. GCB Bioenergy 5: 132–143. doi:10.1111/gcbb.12026.
58. Saurer M, Cherubini P, Reynolds-Henne CE, Treydte KS, Anderson WT, et al. (2008) An investigation of the common signal in tree ring stable isotope chronologies at temperate sites. Journal of Geophysical Research 113: G04035. Available: http://doi.wiley.com/10.1029/2008JG000689. Accessed 29 May 2013.
59. Friedli H, Lötscher H, Oeschger H, Stauffer U, Siegenthaler B (1986) Ice core record of the 13C/12C ratio of atmospheric CO2 in the past two centuries. Nature 324: 237–238.
60. Badeck F-W, Tcherkez G, Nogués S, Piel C, Ghashghaie J (2005) Post-photosynthetic fractionation of stable carbon isotopes between plant organs–a widespread phenomenon. Rapid communications in mass spectrometry: RCM 19: 1381–1391. doi:10.1002/rcm.1912.
61. Bird MI, Ascough PL (2012) Isotopes in pyrogenic carbon: A review. Organic Geochemistry 42: 1529–1539. Available: http://linkinghub.elsevier.com/retrieve/pii/S0146638010002305. Accessed 22 May 2013.
62. Quilliam RS, Glanville HC, Wade SC, Jones DL (2013) Life in the "charosphere" – Does biochar in agricultural soil provide a significant habitat for microorganisms? Soil Biology and Biochemistry 65: 287–293. Available: http://linkinghub.elsevier.com/retrieve/pii/S0038071713002149. Accessed 17 October 2013.
63. Mencuccini M, Martínez-Vilalta J, Vanderklein D, Hamid H a, Korakaki E, et al. (2005) Size-mediated ageing reduces vigour in trees. Ecology letters 8: 1183–1190. doi:10.1111/j.1461-0248.2005.00819.x.
64. Poda A (1963) Tavola alsometrica locale del larice cresciuto in fustaia coetanea del bosco diel comune di Cesana Torinese (Torino). Piano di assestamento, decennio 1963–1972. In: Castellani CE, editor. Tavole Stereometriche Ed Alsometriche Costruite Per Boschi Italiani. Trento: ISAFA. p. 780.
65. Jaffé R, Ding Y, Niggemann J, Vähätalo A V, Stubbins A, et al. (2013) Global charcoal mobilization from soils via dissolution and riverine transport to the oceans. Science (New York, NY) 340: 345–347. Available: http://www.ncbi.nlm.nih.gov/pubmed/23599492. Accessed 27 May 2013.
66. Cheng C-H, Lehmann J, Engelhard MH (2008) Natural oxidation of black carbon in soils: Changes in molecular form and surface charge along a climosequence. Geochimica et Cosmochimica Acta 72: 1598–1610. Available: http://linkinghub.elsevier.com/retrieve/pii/S0016703708000306. Accessed 27 May 2013.
67. Naisse C, Alexis M, Plante A, Wiedner K, Glaser B, et al. (2013) Can biochar and hydrochar stability be assessed with chemical methods? Organic Geochemistry 60: 40–44. Available: http://linkinghub.elsevier.com/retrieve/pii/S0146638013000946. Accessed 28 December 2013.
68. Singh N, Abiven S, Torn MS, Schmidt MWI (2012) Fire-derived organic carbon in soil turns over on a centennial scale. Biogeosciences 9: 2847–2857. Available: http://www.biogeosciences.net/9/2847/2012/. Accessed 27 May 2013.

14

Biochar Decelerates Soil Organic Nitrogen Cycling but Stimulates Soil Nitrification in a Temperate Arable Field Trial

Judith Prommer[1], Wolfgang Wanek[1]*, Florian Hofhansl[1], Daniela Trojan[2], Pierre Offre[2], Tim Urich[2], Christa Schleper[2], Stefan Sassmann[3], Barbara Kitzler[4], Gerhard Soja[5], Rebecca Clare Hood-Nowotny[1]*

1 Department of Microbiology and Ecosystem Science, University of Vienna, Vienna, Austria, **2** Department of Ecogenomics and Systems Biology, University of Vienna, Vienna, Austria, **3** Core Facility of Cell Imaging and Ultrastructure Research, University of Vienna, Vienna, Austria, **4** Institute of Forest Ecology and Soil, Federal Research and Training Centre for Forests, Natural Hazards and Landscape, Vienna, Austria, **5** Department of Health and Environment, Austrian Institute of Technology, Tulln, Austria

Abstract

Biochar production and subsequent soil incorporation could provide carbon farming solutions to global climate change and escalating food demand. There is evidence that biochar amendment causes fundamental changes in soil nutrient cycles, often resulting in marked increases in crop production, particularly in acidic and in infertile soils with low soil organic matter contents, although comparable outcomes in temperate soils are variable. We offer insight into the mechanisms underlying these findings by focusing attention on the soil nitrogen (N) cycle, specifically on hitherto unmeasured processes of organic N cycling in arable soils. We here investigated the impacts of biochar addition on soil organic and inorganic N pools and on gross transformation rates of both pools in a biochar field trial on arable land (Chernozem) in Traismauer, Lower Austria. We found that biochar increased total soil organic carbon but decreased the extractable organic C pool and soil nitrate. While gross rates of organic N transformation processes were reduced by 50–80%, gross N mineralization of organic N was not affected. In contrast, biochar promoted soil ammonia-oxidizer populations (bacterial and archaeal nitrifiers) and accelerated gross nitrification rates more than two-fold. Our findings indicate a de-coupling of the soil organic and inorganic N cycles, with a build-up of organic N, and deceleration of inorganic N release from this pool. The results therefore suggest that addition of inorganic fertilizer-N in combination with biochar could compensate for the reduction in organic N mineralization, with plants and microbes drawing on fertilizer-N for growth, in turn fuelling the belowground build-up of organic N. We conclude that combined addition of biochar with fertilizer-N may increase soil organic N in turn enhancing soil carbon sequestration and thereby could play a fundamental role in future soil management strategies.

Editor: Shibu Jose, University of Missouri, United States of America

Funding: The authors highly appreciate the funding through FWF Project V138-B17 (to RCHN) as well as the FFG Project 825438 (to GS). The funders had no role in study design, data collection and analysis, decision to publish, or preparation of the manuscript.

Competing Interests: The authors have declared that no competing interests exist.

* E-mail: rebecca.clare.hood-nowotny@univie.ac.at (RCHN); wolfgang.wanek@univie.ac.at (WW)

Introduction

Carbon farming opportunities in temperate agricultural systems are limited to no-till farming and conservation agriculture strategies. In these systems careful carbon management is required to avoid problems of nutrient, particularly N, immobilization by the soil microbial community, which can lead to substantial crop yield depressions. Biochar application to temperate agricultural soils offers the prospect for substantial carbon sequestration without the concomitant yield losses or greenhouse gas emissions associated with other strategies [1–4]. The pyrolysis derived aromatic-macromolecular structure of biochar [5] makes it more recalcitrant to microbial decomposition, which means it does not immobilize significant quantities of valuable nutrients when added to soil, rendering biochar a viable long-term carbon sink option for temperate agriculture [6–7]. Promoting novel carbon farming options in a wider agronomic context requires facing risk-benefit assessments by the soil custodians, namely farmers and other environmentally concerned stakeholders. As a result, an in-depth understanding of the impact of biochar on the soil N cycle, the most economically relevant biogeochemical cycle to farmers in mid to high latitudes, will be required to facilitate decision-making and allow for improved biochar management and subsequent adoption as a carbon sequestering strategy.

There is increasing evidence suggesting that biochar addition to soil impacts a number of processes of the soil N cycle [8] and that the impacts of biochar addition are strongly dependent on the biochar feedstock and the pyrolysis conditions, in addition to the respective soil properties and local environmental and climatic conditions [9]. For instance, biochar was shown to decrease nitrate leaching [10–11], to increase soil N immobilization [12–13], to increase ammonification, the conversion of organic N into ammonium [14–15], to increase nitrification, the oxidation of ammonium to nitrate [16–17], to promote volatilization, the loss of ammonia to the atmosphere [18], and to increase plant N uptake [19–21]. The mechanisms posited to explain the differences in N cycling processes when biochar is added to soil are mostly direct ones, resulting from the unique properties of the biochar such as its highly porous structure, the large surface area and its ion-exchange capacity [22], with associated beneficial effects on nutrient retention and water holding capacity. Some N cycle processes like NH_3 volatilization can be stimulated by inherent biochar properties such as the potential of alkaline biochar to increase soil pH [23]. Moreover biochars can have high nutrient contents and increase soil nutrient availability directly or through the effects of priming which may increase the bio-availability of soil nutrients [24–25]. In addition, indirect mechanisms associated with changes in the soil microbial community can be triggered by the properties specific to biochar and thereby have strong implications for soil microbial N processing [16,26]. To date the majority of biochar studies were carried out as short-term pot experiments, many of these focusing on poor, acidic and nutrient depleted soils where the addition of biochar has the potential to cause largest effects including fundamental changes in soil nutrient cycles, often resulting in pronounced increases in crop production [27–28]. Long-term field trials examining the effects of biochar on intensively managed and fertile agricultural soils in temperate regions are therefore urgently needed to advance our predictive capabilities of biochar - soil N cycle interactions [8,20].

The impacts of biochar on soil N transformation processes are extremely important, given the dominant role N nutrition plays in regulating crop production in temperate regions, as it has to be replenished either as inorganic N fertilizer or it is cycled back to the soil in a complex organic form such as crop residues. In the past two decades there has been a major paradigm shift in our conceptualization of the soil N cycle [29]. It places greater emphasis on the organic turnover processes in the soil N cycle such as the depolymerization of organic matter, which is now regarded as the rate limiting step of dissolved organic N production and subsequently of soil N mineralization and nitrification. As a result of the biochar characteristics described above biochar addition to soil could have major impacts on the organic N turnover component of the soil N cycle. The multitude of simultaneously occurring soil N processes (Figure 1) make measuring and understanding soil N transformation rates difficult and thus requires isotopic tools [30–31]. Using an array of parallel established and novel ^{15}N isotope pool dilution techniques [32–33], it should be possible to study each key transformation process independently. These techniques in consortium allow determining the impact of biochar on multiple soil processes in a synchronized manner and with a high degree of resolution and reliability. In the context of biochar addition this issue of complexity is of particular relevance as biochar addition could have conflicting impacts on different control points of the soil N cycle.

Our major objective therefore was to investigate the effects of biochar on key inorganic and organic N transformation processes, on soil C and N pools and on the soil microbial community. A hardwood-based biochar (pyrolysis at 500°C for two hours) was added in March 2011 to an arable field on calcareous Chernozem in Traismauer, Austria. The field trial consisted of four treatments replicated four times in circular plots of 6.5 m, comprising NPK fertilized plots as controls combined with biochar addition rates of 0, 24 and 72 t dry mass ha^{-1}. As agricultural systems rely on fertilization to grant crop production, fertilized plots were taken as controls (NPK, no biochar) and compared to the two biochar addition rates i.e. 24 t BC ha^{-1} with NPK amendment (BC1N) and 72 t BC ha^{-1} with NPK amendment (BC3N). A treatment with 72 t BC ha^{-1} lacking NPK amendment was also setup (BC3) to check for nutrient immobilization effects of BC-alone treatments. In this study we focused on the control (NPK) and the highest biochar treatment (BC3N). Soil samples were taken

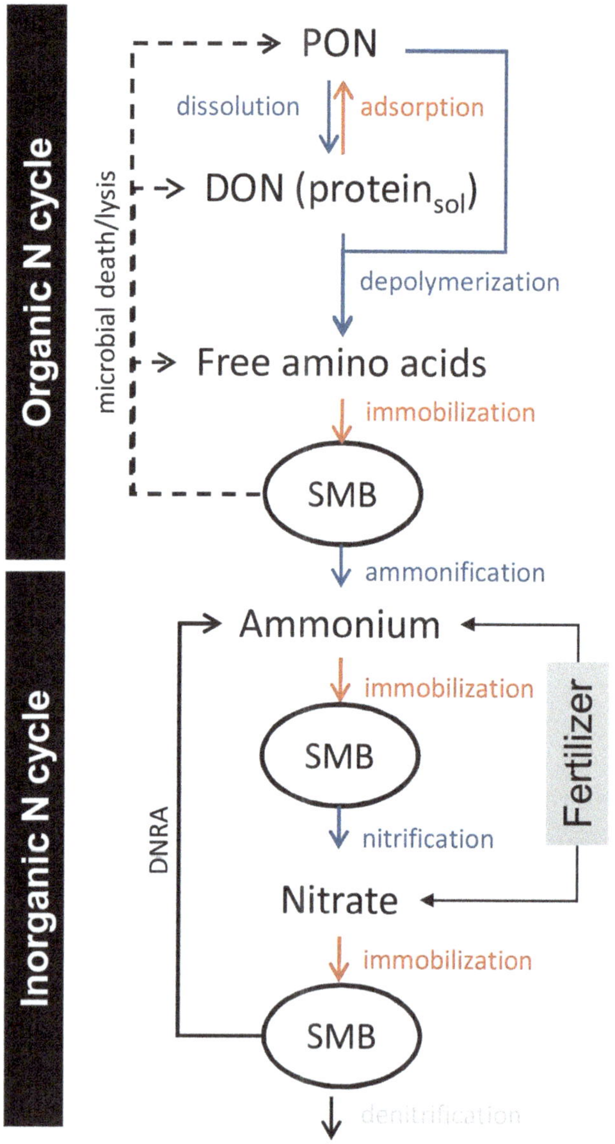

Figure 1. Overview of the agricultural soil N cycle. The organic N and inorganic N cycles are highlighted. DNRA...dissimilatory nitrate reduction to ammonium, DON...dissolved organic N, PON...particulate organic N such as detritus and soil organic matter, SMB...soil microbial biomass. Not shown - heterotrophic nitrification which results from microbial uptake of e.g. free amino acids, oxidation of reduced N compounds and release of nitrate into the soil.

four months after biochar application and sampling was repeated several times during a period of fourteen months. We applied a novel array of isotope pool dilution assays to measure gross rates of protein, amino acid, ammonium and nitrate dynamics (Figure 1) and molecular techniques to investigate nitrifier (ammonium oxidizer) population sizes. In addition we used scanning electron microscopy to examine the physical structure of the biochar in order to understand its possible role in shaping the soil microbial community.

Results

Biochar Porosity

The cellular structure of wood was well conserved in the biochar as seen in Figure S1. The mean pore coverage of xylem cross sections was 47.6±5.9% (mean±SD) for the biochar applied in this field trial. Overall pore size number distribution showed a maximum at approximately 0.7 μm diameter (related to simple xylem pits), and a second and third peak which are closely related to bordered xylem pits and parenchyma cells (around 3 μm) and xylem vessels (15–25 μm; Fig. 2, Fig. S2). Beech has a diffuse-porous wood structure [34] and early-wood and late-wood had distinctly different mean vessel diameters (24±12 and 15±9 μm, respectively), resulting in a decrease in mean pore coverage from 61.8±6.1% for early-wood to 40.2±4.1% for late-wood. Secondary cortex cells showed a wide range of diameters, with a peak ranging from 3 to 5 μm (data not shown). On an area basis the contribution of small pores decreased and the major contribution to total porosity shifted towards larger pores with diameters >10 μm i.e. xylem vessels (Fig. 2b).

Effect of Biochar on General Soil Properties

The effect of biochar on soil pH and cation exchange capacity was minimal. Soil pH decreased by 0.1 units from 7.5 to 7.4 in biochar-treated soils (Table 1). Soil cation-exchange capacity decreased slightly from an initial 22.5 to 20.8 cmolc kg^{-1} in biochar-treated soils but in both cases was dominated by Ca^{2+} (~90%) and Mg^{2+} (~7%), with a trend towards an increase of K^+ saturation from 1.6 to 4.8% (measured on one composite soil sample per treatment, Table 1).

Bulk Soil C and N Contents and Soil Water Relations

Biochar amendment resulted in increased soil organic C and there was a trend of enhanced total soil N concentration with biochar addition. Soil bulk densities and gravimetric soil water contents declined while water-filled pore space remained unaltered by biochar addition (two-way mixed ANOVA for effects of treatment and time, Table 2, Figure 3). Time affected most of these and the following soil parameters (Table 2), but emphasis here is given on treatment effects and treatment × time interactions; the latter might indicate a decreasing or increasing biochar-effect on soil properties or fluxes. However, all time × treatment interactions effects investigated in this study were either non-significant or not related to increasing or decreasing biochar-effects over time. Direct time effects were most likely caused by seasonal changes, agricultural management and subsequent plant growth effects. Soil organic C was significantly higher in biochar-treated soils (47.4 mg C g^{-1}) than in non-biochar controls (18.7 mg C g^{-1}; $p=0.001$) (Table 2). Total soil N responded analogously, with a slight but non-significant increase due to biochar-treatment (1.47 to 1.64 mg N g^{-1}; $p=0.062$). Concurrent with increases in soil organic C the soil bulk density decreased from 1.17 to 1.04 g cm^{-3} following biochar amendment ($p<0.001$). Along the same lines biochar amendment caused an increase in soil porosity from 56.0 to 60.6% ($p<0.001$) and in gravimetric soil water content from 0.140 to 0.163 g H_2O g^{-1} dry soil ($p<0.001$). As a result of increased soil water content and soil porosity with biochar addition, water-filled pore space (and air-filled pore space) did not change between controls (29.8%) and biochar-treated soils (28.9%, $p=0.229$). Interestingly, by comparing all four treatments we found the biochar-effect to be amount-dependent i.e. the magnitude of the effects increased or decreased with the amount of biochar added in the field (Table S1). Soil organic C increased with biochar rate, bulk density decreased and soil porosity and soil water content increased with biochar rate (Table S1).

Extractable C and N Concentrations

Dissolved organic C in soil extracts was slightly but significantly lower in the biochar treatment compared to the controls (111 vs. 123 μg C g^{-1} respectively; $p=0.010$) (Figure 3, Table 2). Concentrations of extractable ammonium were low compared to extractable nitrate and not significantly different between the control and biochar treatments, with 1.66 and 1.55 μg NH_4^+-N g^{-1} respectively ($p=0.383$). In contrast soil nitrate concentrations in the biochar (BC3N) treated soils were significantly lower than those from the control soils ($p=0.001$), with 8.40 and 14.4 μg NO_3^--N g^{-1} respectively. Dissolved organic N concentrations in the soil extracts decreased, though not significantly, from 9.62 in biochar-treated soils to 8.35 μg N g^{-1} in control treatments ($p=0.187$). The ratio of dissolved organic C: dissolved organic N varied around 16:1 and was not affected by biochar addition ($p=0.909$). Nitrate dominated the extractable N pool, accounting for between 67% (BC3N) and 76% (controls). The decrease in soil nitrate therefore translated into a significant decline of total dissolved N from 25.0 to 17.0 μg N g^{-1} in the biochar treatment ($p=0.001$). Free amino acids were low in concentration (around 0.7 μg N g^{-1}) and were not affected by treatment ($p=0.110$). Measured only once (July 2011, four months after the start of the field experiment) we found no significant differences in the concentrations of different size classes of dissolved organic N between the two treatments, with extractable high-molecular weight organic N concentrations around 3.3 μg N g^{-1} and low-molecular weight organic N concentrations of around 6 μg N g^{-1} (Table 2). Free amino acids accounted for approximately 12% of the low molecular weight organic N pool and comprised 8% of dissolved organic N in both treatments. Around 40% of the dissolved organic N and all the high-molecular weight organic N fraction (>3 kDa; 105–107%) was protein, although this measurement is based only on the July 2011 sampling. The negative effect of biochar on concentrations of dissolved organic C, total dissolved N and nitrate was consistent even when all four treatments were analyzed (Table S1), with decreases in dissolved organic C, nitrate and free amino acids with increasing biochar addition rate (Table S1).

Soil N Transformation Rates

The full set of gross N transformation processes was measured once, and one-way ANOVA results are presented in Table 3 and Figure 4; nitrate fluxes were measured on three occasions throughout the seasons (Table 3, Figure 4, see circles). Measured rates of protein influx into the extractable organic N pool were significantly ($p=0.016$) higher in the NPK treatment (80.0 μg N g^{-1} d^{-1}) and almost double the rates in the biochar treatment (44.2 μg N g^{-1} d^{-1}). Biochar showed a trend ($p=0.097$) of depressing protein efflux rates out of the extractable organic N pool, with rates in the biochar treatment (53 μg N g^{-1} d^{-1}) being half that of the control treatment (115 μg N g^{-1} d^{-1}). The decline in amino acid transformation rates was even more pronounced. Amino acid influx rates into the organic N pool were significantly ($p=0.0004$) lower in biochar treated soils (5.4 μg N g^{-1} d^{-1}), more than fivefold lower than those in the control treatment (28.5 μg N g^{-1} d^{-1}) and so too were amino acid consumption rates. Amino acid consumption rates declined more than six-fold from 50.3 μg N g^{-1} d^{-1} in the control treatment to 7.9 μg N g^{-1} d^{-1} in the biochar treatment ($p=0.039$). Gross mineralization rates were not significantly different between the two treatments ($p=0.616$) nor were ammonium consumption rates ($p=0.526$;

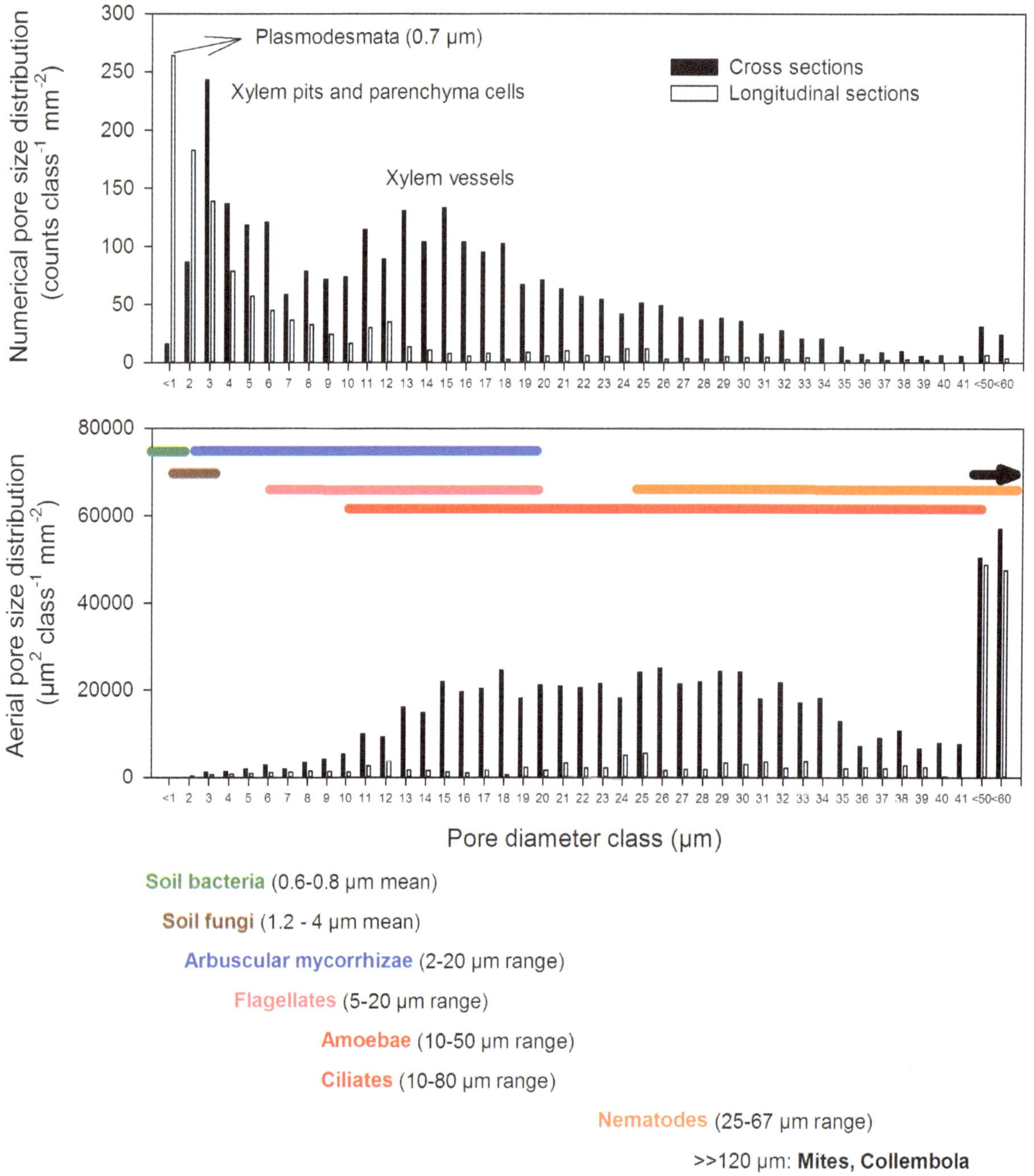

Figure 2. Pore size distribution of beech wood biochar as determined by scanning electron microscopy. In total, mean diameters of 7715 pores were measured in 6 longitudinal and 7 cross sections of charcoal pieces, each image covering 0.35 mm^2. Pore size distributions are presented (1, top panel) as counts mm^{-2} per 1 µm size class (top) and (2, middle panel) on an area basis, i.e. µm^2 mm^{-2} per 1 µm size class. Typical mean diameters (soil bacteria, soil fungi) and diameter ranges (arbuscular mycorrhizal fungi; protozoan grazers such as flagellates, amoebae, ciliates; and mesofaunal grazers and predators such as nematodes, mites and collembola) are presented color-coded (bottom panel) and with range lines in the middle panel.

Figure 3), and both processes ranged between 1.5 and 2.1 µg N g^{-1} d^{-1}. In contrast, gross nitrification and gross nitrate immobilization were strongly enhanced by biochar application. Gross nitrification rates in the biochar treatment (10.2 µg N g^{-1} d^{-1}) were fourfold higher than the rates in the control treatment (2.4 µg N g^{-1} d^{-1}), and significantly different ($p<0.001$). A similar trend ($p=0.0573$) was evident for nitrate consumption rates which almost doubled from 6.8 to 11.3 µg N g^{-1} d^{-1} in the biochar treatment.

Rates of nitrification and nitrate consumption were re-measured twice. Over the course of time nitrification rates were consistently higher (threefold) in the biochar treated compared to the control plots ($p<0.001$, Table 3). The same pattern was apparent for nitrate consumption rates which were consistently negatively affected by biochar amendment ($p=0.001$). On the sampling date

Table 1. Soil properties in control (NPK) and biochar-treated (BC3N) soils of the Traismauer biochar field trial.

Parameter	Unit	Control (NPK)	Biochar (BC3N)
pH ($CaCl_2$)		7.5	7.4
$CaCO_3$	%	15.8	15.2
Humus	%	2.4	18.1
Total N	%	0.148	0.203
P (CAL)	mg kg^{-1}	49	84
P_{tot} (acid digest)	g kg^{-1}	5.46	5.54
Sand	%	18.3	n.d.[1]
Silt	%	57.2	n.d.
Clay	%	24.5	n.d.
CEC	cmol kg^{-1}	22.5	20.8
Ca (CEC)	cmol kg^{-1}	20.7	18.2
Mg (CEC)	cmol kg^{-1}	1.46	1.53
K (CEC)	cmol kg^{-1}	0.36	0.99
Na (CEC)	cmol kg^{-1}	<0.04	<0.04
Al (CEC)	cmol kg^{-1}	<0.06	<0.06
Fe (CEC)	cmol kg^{-1}	<0.01	<0.01
Mn (CEC)	cmol kg^{-1}	<0.01	<0.01
H (CEC)	cmol kg^{-1}	0.002	0.002
Fe (EDTA)	mg kg^{-1}	40	67
Mn (EDTA)	mg kg^{-1}	107	128
Cu (EDTA)	mg kg^{-1}	7.2	7.1
Zn (EDTA)	mg kg^{-1}	2.3	7.5

All parameters were measured on composite samples collected in July 2011 from the two treatments (n = 1). n.d. - not determined.
[1]Soil texture is not shown for biochar treated soils. At high contents humus biases the particle size distribution and is therefore conventionally decomposed by prolonged H_2O_2 (or similar oxidant) treatment prior to soil texture analysis; biochar/charcoal is largely resistant against this oxidant treatment.

where we had gross nitrification rates measured for all four treatments nitrification rates were significantly higher in all the biochar treatments ($p<0.001$, Table S1) with rates increasing significantly with increasing rates of biochar application.

Microbial Community Size and Abundance of Archaeal and Bacterial Ammonia-oxidizers

Soil DNA concentration was significantly higher in the biochar treatment (18.0 µg DNA g^{-1} soil) compared to the controls (12.4 µg DNA g^{-1}; $p = 0.042$) (Figure 5, Table 2). Ammonia-oxidizing archaea (AOA) outnumbered bacterial ammonia-oxidizers (AOB) by a factor of 3.2 to 3.6, in both treatments ($p = 0.402$). At the same time archaeal amoA gene copy numbers (AOA) and bacterial amoA gene copy numbers (AOB) expressed on a dry soil mass basis increased through biochar application (AOA: $P = 0.045$, AOB: $p = 0.020$). AOA increased 1.5-fold from 1.68×10^7 to 2.47×10^7 amoA copy numbers g^{-1} soil and AOB 1.7-fold from 5.01×10^6 to 8.45×10^6 amoA copy numbers g^{-1} soil through biochar amendment. As a result of increasing soil DNA concentration and amoA copy numbers per gram soil we did not find a significant effect of biochar on the abundance of AOA ($p = 0.637$) and AOB ($p = 0.436$) expressed on a DNA basis, i.e. the relative share of ammonia oxidizers in the microbial community did not change according to treatment. Across all four treatments we did not find an effect of biochar rate on either soil DNA concentration or any measure of AOA or AOB abundance (Table S1). However, archaeal and bacterial amoA gene copy numbers were strongly positively related ($R^2 = 0.83$, $p = 0.0114$; data not shown). Moreover, we found a trend towards a positive relationship between both, AOA or AOB numbers on a dry soil mass basis and gross nitrification rates (AOA: $R^2 = 0.65$, $p = 0.051$, AOB: $R^2 = 0.60$, $p = 0.072$; Figure 6). The "weakness" of this relationship derived mainly from the control treatment where nitrification rates were relatively constant between the two measurement dates while amoA copy numbers varied more strongly. Nitrification rates were significantly positively correlated with the total number of amoA copy numbers on a soil dry mass basis (AOA plus AOB; $R^2 = 0.66$, $p = 0.049$, data not shown).

Discussion

Soil Organic N Cycling

Biochar addition had a strong negative effect on soil organic N cycling rates as it reduced protein production and protein consumption and concurrently decreased free amino acid production and consumption (Figure 4). This is important given that protein depolymerization plays the predominant role in organic N recycling with approximately 50% of soil organic matter and >90% of plant detritus being composed of protein [35–36], and with proteins representing around 40% of dissolved organic N in this study. Proteins are therefore considered the most important substrate for microbial N mobilization and N use [33], and thus any change in their dynamics strongly impacts whole-system N cycling. The influx into the K_2SO_4 extractable protein pool (representing a labile part of the total protein pool in soils) is mainly fuelled by microbial turnover and protein desorption processes while efflux from this pool relates to protein depolymerization or protein adsorption (Figure 1).The concomitant reductions in free amino acid turnover were most likely a direct consequence of the reduced protein-N cycling in biochar amended soils. Amino acid production is driven by depolymerization of accessible proteins through extracellular proteases and peptidases which release oligopeptides and free amino acids that are then taken up by microbes (free amino acid consumption) [33]. Though oligopeptides also represent a significant source of N for microbes and plants [37] we did not measure their dynamics, which would have been expected to change accordingly. We offer two possible explanations for the observed negative effects of biochar on organic N (protein) dynamics related to the surface properties of biochar and its unique pore structure.

(1) The observed decline in protein production rates might result from protein adsorption to biochar surfaces. The reactive surface of biochar has been shown to adsorb numerous low and high molecular weight organic molecules [38], including proteins and enzymes [39]. Hence, the reduced motility and accessibility of proteins in the biochar treated soils could explain the decrease in protein influx rates observed with resultant implications for amino acid dynamics. Very strong H-bonding between proteins and polyphenols are well known [40], and the abundant phenolic groups on low temperature biochars [41] provide ample ligands for strong sorption of proteins and peptides [9].

(2) Biochar could also provide structural protection, by the entrapment of organic matter within its inherent pore network. The "mesopore protection hypothesis" by Zimmerman et al. [42] suggests that small organic molecules can be occluded within mineral mesopores (2–50 nm diameter),

Table 2. Results of ANOVA of soil physicochemical data for the factors treatment and time in control (NPK) and biochar (BC3N) treatments.

Pools	Unit	NPK	BC3N	Treatment		Time		Interaction	
		mean	mean	F	P	F	P	F	P
C_{org}	mg C g^{-1} DW	18.66	47.43	(1, 6) 43.7	0.001	(1, 6) 0.0	0.951	2.5	0.165
N_{tot}	mg N g^{-1} DW	1.47	1.64	(1, 6) 5.3	0.062	(2, 12) 37.5	<0.001	0.3	0.778
BD	g cm^{-3}	1.17	1.04	(1, 6) 48.4	<0.001	(1, 6) 55.3	<0.001	0.3	0.595
Grav. WC	g H_2O g^{-1} DW	0.14	0.17	(1, 6) 46.7	<0.001	(6, 36) 218.7	<0.001	6.2	<0.001
Porosity	%	56.03	60.61	(1, 6) 48.6	<0.001	(1, 6) 55.3	<0.001	0.3	0.594
WFPS	%	29.79	28.92	(1, 6) 1.8	0.229	(6, 36) 215.4	<0.001	2.3	0.059
AFPS	%	70.21	71.08	(1, 6) 1.8	0.229	(6, 36) 215.4	<0.001	2.3	0.059
DOC	µg C g^{-1} DW	122.56	111.21	(1, 6) 13.6	0.010	(5, 30) 205	<0.001	2.9	0.028
DOC:DON		16.10	15.92	(1, 6) 0.0	0.909	(1.6, 9.9) 14.4	0.002	0.8	0.454
DON	µg N g^{-1} DW	9.62	8.35	(1, 6) 2.2	0.187	(2.3, 13.9) 28.5	<0.001	0.3	0.782
NO_3^-	µg N g^{-1} DW	14.43	8.40	(1, 6) 39.6	0.001	(6, 36) 95.9	<0.001	8.6	<0.001
NH_4^+	µg N g^{-1} DW	1.66	1.55	(1, 6) 0.9	0.383	(2.6, 15.5) 34.8	<0.001	1.0	0.417
FAA	µg N g^{-1} DW	0.77	0.70	(1, 6) 3.5	0.110	(1.6, 9.8) 3.2	0.091	0.9	0.429
TDN	µg N g^{-1} DW	24.95	17.03	(1, 6) 33.3	0.001	(5, 30) 97.2	<0.001	1.3	0.285
HMW N_{org}	µg N g^{-1} DW	3.24	3.47	(1, 23) 1.1	0.306				
LMW N_{org}	µg N g^{-1} DW	6.78	5.68	(1, 23) 0.7	0.401				
Protein	µg N g^{-1} DW	3.46	3.65	(1, 23) 1.4	0.253				
DNA	µg g^{-1} DW	12.40	18.04	(1, 6) 6.6	0.042	(1.1, 6.9) 0.1	0.774	2.6	0.154
AOA/DNA	amoA copy no $µg^{-1}$ DNA	1.41E6	1.36E6	(1, 6) 0.2	0.637	(2, 12) 8.2	0.006	3.3	0.070
AOA/soil	amoA copy no g^{-1} soil DW	1.68E7	2.47E7	(1, 6) 6.3	0.045	(1.1, 6.6) 3.5	0.103	0.9	0.378
AOB/DNA	amoA copy no $µg^{-1}$ DNA	419075	461583	(1, 6) 0.7	0.436	(1.2, 7) 7.2	0.028	1.6	0.246
AOB/soil	amoA copy no g^{-1} soil DW	5.01E6	8.45E6	(1, 6) 9.9	0.020	(2, 12) 4.5	0.035	1.3	0.319
AOA:AOB		3.64	3.19	(1, 6) 0.8	0.402	(2, 12) 0.4	0.681	0.0	0.970

Data were analyzed by two-way mixed ANOVA. Data for protein, HMW-N_{org} and LMW-N_{org} were available only for one sampling date and therefore were analyzed by one-way ANOVA. Abbreviations: C_{org}, soil organic C; N_{tot}, total soil N; BD, soil bulk density; Grav. WC, gravimetric soil water content; WFPS, water filled pore space; AFPS, aire filled pore space; DOC, dissolved organic C; DON, dissolved organic N; NO_3^-, nitrate; NH_4^+, ammonium; FAA, total free amino acids; TDN, total dissolved N; HMW-N_{org}, high molecular weight organic N; LMW-N_{org}, low molecular weight organic N; DNA, soil DNA content; AOA/DNA, archaeal amoA copy numbers on DNA basis; AOA/soil, archaeal amoA copy numbers on dry soil basis; AOB/DNA, bacterial amoA copy numbers on DNA basis; AOB/soil, bacterial amoA copy numbers on dry soil basis.

physically protecting organic molecules such as proteins (<10 nm) and amino acids (<2 nm) from breakdown by excluding decomposers and their extracellular enzymes. Mesopore surface area and volume are mainly determined by gas sorption analysis but are not targeted by scanning electron microscopy, and therefore have not been measured in this study. Nevertheless, low-temperature wood biochars produced at <650°C, just as the Romchar used in this study, have been shown to have a low mesopore volume and surface area as measured by gas sorption [39]. Physical protection of organic matter in mesopores by occlusion from decomposing microbes may thus have played a minor role in this study.

Just like the decrease in protein production, the decline in free amino acid production in biochar treated soils could also be triggered by protease and peptidase adsorption to or occlusion in biochar. Enzyme adsorption to solid support materials (e.g. soil matrix or biochar) has been reported to decrease soil enzymatic activity [43], and adsorbed substrates such as proteins and oligopeptides are also less accessible and bioavailable for further breakdown to free amino acids. Decreased aminopeptidase activity has been observed in experiments with fast pyrolysis biochar of switchgrass and was ascribed to sorption reactions between substrate and biochar [44]. Another study found accelerated N-mineralization of the recalcitrant soil organic N pool in arable soils treated with maize biochar [14]. The inconsistent effects of biochar on soil enzymes may be the result of several factors such as biochar feedstock, pyrolysis conditions, soil type as well as the respective enzyme and/or substrate [44]. For example, contrasting results obtained in our experiment and that of Nelissen et al. (2012) [14] could be explained by structural differences between the biochars used in these two studies, leading to variable enzyme diffusion impedance. Indeed, the Romchar employed here maintained its native wood-like structure according to the scanning electron microscopy images (Figure S1) while non-woody materials such as maize biochar have an amorphous structure with little evidence of retained pores [45].

Soil N Mineralization

Given that biochar application significantly decelerated the production and consumption of low molecular weight organic N in our field experiment, we would have expected a decrease in soil N mineralization rates. Somewhat surprisingly biochar had no significant effect on gross N mineralization and gross ammonium immobilization rates in our study, in line with a laboratory incubation of cultivated black Chernozem soil with wheat biochar

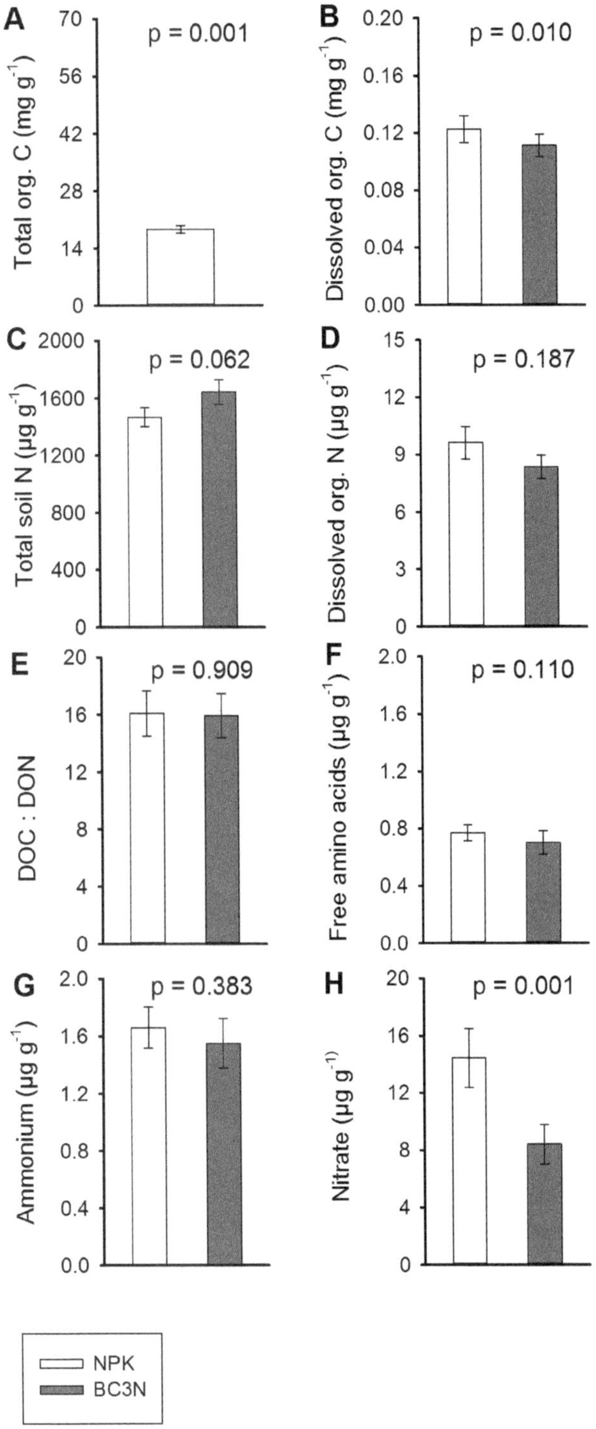

Figure 3. Soil C and N pool sizes for control (NPK) and biochar (BC3N) treatments. Bars represent means ±1 SE (n = 4) from seven measurements between July 2011 and September 2012 for extractable compounds, and from two measurements (September 2011 and 2012) for soil organic C and total soil N. Open bars, control treatment (NPK); grey bars, biochar treatment (BC3N). Units in mg C g^{-1} dry soil and μg N g^{-1} dry soil. P values are from two-way mixed ANOVA (see Table 2).

[46], but in contrast to observed accelerated rates of gross N mineralization of the recalcitrant soil organic N pool in an arable soil treated with maize biochar [14]. Interestingly mineralization (respiratory use) of amino acids and peptides, measured as the uptake of ^{14}C-labelled organic N and subsequent $^{14}CO_2$ release, was not affected by application of a wood-based biochar [15]. Divergent, positive, neutral or negative effects of biochar on soil N mineralization may be the upshot of different biochar types along with impacts on the C and N status of the soil microbial community. Carbon-rich biochar, poor in nitrogen, such as beech wood biochar (C:N ratio ~200:1) might promote N limitation, leading to the retention of catabolically produced ammonium in the N-limited microflora, which therefore results in a decrease of the amount of ammonium released to the soil [29]. Conversely, nitrogen-rich biochars with low C:N ratios such as maize-biochars [14] or manure-biochars [47] promote microbial C limitation, causing the excess of N to be mineralized and therefore N mineralization rates to increase. In our case the ammonium released (N mineralization) relative to organic N taken up by microbes (free amino acid consumption) was higher in the biochar treated soils than in controls (Figure 4), indicating that biochar addition did not lead to microbial N limitation as a consequence of the increased organic C content in soil. This suggests that the majority of organic C in the biochar was not metabolized or assimilated by the soil microbial community, and confirms that biochar is an efficient tool for long-term C sequestration in soil, with half-lifes >100 years [48]. The absence of microbial N limitation in the biochar treatment might be directly related to the field applications of fertilizer and might also be related to the slightly lower concentrations of dissolved organic C in the biochar treated plots.

Soil Nitrification

Apart from the negative effect that biochar had on soil organic N cycling, the most striking result was a strong and persistent (1.5 years) stimulation of gross nitrification. Several studies have examined the impact of biochar on soil net nitrification [8] but only few have reported gross nitrification rates, which allow the simultaneous analysis of both, productive and consumptive processes for nitrate. Biochar addition increased gross nitrification rates in two forest soils and one arable soil [14,26,49] but had no effect in other experiments with grassland and arable soils [46,49]. Several hypotheses have been raised to explain biochar-related (positive) effects on nitrification rates.

One of the dominant theories to explain the increase in nitrification rates following biochar amendment is the liming effect [14,26]. Biochar commonly has a neutral to alkaline pH [50] and therefore may increase the pH of acidic soils with a concurrent positive response of soil nitrifiers whose pH optimum is slightly acidic to neutral pH [51–52]. However the pH of the soil used in this study was slightly alkaline and did not increase but rather decreased in response to the addition of biochar.

Equally we can disregard the tenet that biochar addition leads to an increase in nitrification due to adsorption of nitrification inhibitors. In contrast to forest and grassland soils [49,53–54], intensively managed agricultural soils usually contain low amounts of naturally occurring nitrification inhibitors [55] and their effects might have been of minor importance in this experiment.

Nitrification rates have been shown to be strongly controlled by the ammonium supplied by soil N mineralization in a wide variety of ecosystem types including woodland, grassland and agricultural soils [56]. In this study, biochar amendment neither had an effect on gross ammonium production rates nor on the pool size of extractable ammonium, which obviates a simple substrate-driven explanation of increased nitrification rates in this field experiment. Whereas nitrification rates in the control soils were of the same order of magnitude as the mineralization rates, biochar treated

Table 3. Results of ANOVA of soil N transformation rates for the factor treatment in control (NPK) and biochar (BC3N) treatments.

Processes	Unit	NPK	BC3N	Treatment		Time		Interaction	
		mean	mean	F	P	F	P	F	P
One date									
Prot_GP	mg N kg^{-1} d^{-1}	79.99	44.15	(1, 11) 8.4	0.016				
Prot_GC	mg N kg^{-1} d^{-1}	115.06	52.37	(1, 11) 3.4	0.097				
AA_GP	mg N kg^{-1} d^{-1}	28.54	5.36	(1, 11) 26.7	0.000				
AA_GC	mg N kg^{-1} d^{-1}	50.25	7.91	(1, 11) 5.6	0.039				
N min	mg N kg^{-1} d^{-1}	1.88	2.10	(1, 11) 0.3	0.616				
NH_4^+_GC	mg N kg^{-1} d^{-1}	1.86	1.52	(1, 11) 0.4	0.526				
Nitrification	mg N kg^{-1} d^{-1}	2.42	10.16	(1, 11) 50.6	<0.001				
NO_3^-_GC	mg N kg^{-1} d^{-1}	6.83	11.27	(1, 11) 4.6	0.057				
Three dates									
Nitrification	mg N kg^{-1} d^{-1}	3.38	9.09	(1, 6) 82.2	<0.001	(2, 12) 0.4	0.698	3.6	0.061
NO_3^-_GC	mg N kg^{-1} d^{-1}	4.92	11.01	(1, 6) 35.2	0.001	(2, 12) 1.7	0.225	0.6	0.571

All transformation rates were measured and analyzed by one-way ANOVA for July 2011 data ("one date"). Nitrification and nitrate consumption was measured three times for NPK and BC3N treatments ("three dates") and analyzed by two-way mixed ANOVA. Abbreviations: GP, gross production (influx); GC, gross consumption (efflux); Prot, protein; AA, amino acid; N min, N mineralization; NH_4^+, ammonium; NO_3^-, nitrate.

soils exhibited nitrification rates four to five times higher than mineralization rates. This dichotomy between ammonium supply and nitrification rates could point to increased importance of heterotrophic nitrification fuelled by organic N compounds in biochar-treated soils. However heterotrophic nitrification mainly occurs in grassland and forest soils and especially in acidic soils [57–58], although this process has been recently detected in acidic arable and neutral soils [59–60]. This nitrification pathway will have to be investigated in follow-up incubation experiments, using acetylene as an inhibitor of autotrophic (archaeal and bacterial) nitrification [61] which does not affect heterotrophic nitrification [62–63].

Increased nitrification rates in soils treated with biochar could be attributed to the alteration of the soil water status, including its distribution and associated changes in soil oxygenation, since nitrification is dependent on soil O_2 availability. The soil air/water balance is essential for nitrification, which has been found to peak at 60% water-filled pore space [64]. Altered soil moisture conditions have been reported in various biochar studies [17,65–66]. However, in this study air-filled pore space was highly variable and on average greater than 70%, independent of treatments; therefore higher nitrification rates (Figure 4) could not be explained by improved oxygen availability.

While none of the proposed drivers of soil nitrification could explain the observed pattern of enhanced nitrification in biochar treated soils, the size of the ammonia-oxidizing community increased with biochar amendment (Fig. 5). Recently, the oxidation of ammonia to nitrite, the first and rate limiting step of nitrification, was shown to be performed by prokaryotes in both, the bacterial and archaeal domains [67]. Both ammonia oxidizing archaea (AOA) and ammonia oxidizing bacteria (AOB) abundances were significantly higher in the biochar treated soils, AOA outnumbering AOB in both treatments. There are numerous studies showing that AOA are often significantly more abundant than AOB in soils of different origins [68–69]. However the relative importance and contribution of these different groups to the nitrification process in natural environments is still unclear and to our knowledge there are no published studies available on the interaction of biochar and AOA, at least in temperate ecosystems. AOB abundances have been found to increase after natural forest fire events most likely as a consequence of the liming effect of charcoal and its sorption of nitrification inhibitors [26]. In a pot experiment the AOB community structure was unaffected by the sole addition of *Eucalyptus* wood biochar but changed when biochar was added together with an inorganic or organic N source [70], a possible result of substrate-limited nitrification as the ammonium concentrations declined in sole biochar treatments. The stimulation of both, AOA and AOB abundances through biochar addition in this field experiment was only significant when amoA gene copy numbers were expressed on a soil mass basis but not when expressed on a soil DNA basis. The microbial community size, indicated by total soil DNA content, increased with biochar addition revealing a general positive effect of biochar on the whole microbial community rather than a specific stimulation of the autotrophic nitrifiers. This unspecific stimulation of microbes and concomitant increase of nitrifiers is corroborated by the lack of biochar effects on more specific soil nitrification controls such as soil pH, ammonia and oxygen availability, and presence of nitrification inhibitors. The biochar effect was also unspecific in terms of stimulation of AOA and AOB; both AOA and AOB abundances increased with biochar and were strongly positively related across all soil samples independent of treatment. In addition ratios of AOA: AOB were unaffected by the biochar amendment.

AOA and AOB abundances were broadly ($P<0.1$) correlated with gross nitrification rates, suggesting that bacterial and archaeal abundance increased with increasing biochar addition and providing a functional link to the process data (Fig. 6). This and the total DNA data add weight to an underlying hypothesis, i.e. that the structural nature of the biochar provides a protective habitat for bacterial and archaeal growth. It is conceivable, that biochar promotes bacterial and archaeal community size by enhancing the colonizable surfaces in the soil and providing habitable protected pores. The unique pore-size distribution typically found in biochar with pores ranging from 1 nm to 100 μm [71] is believed to provide various beneficial functions to

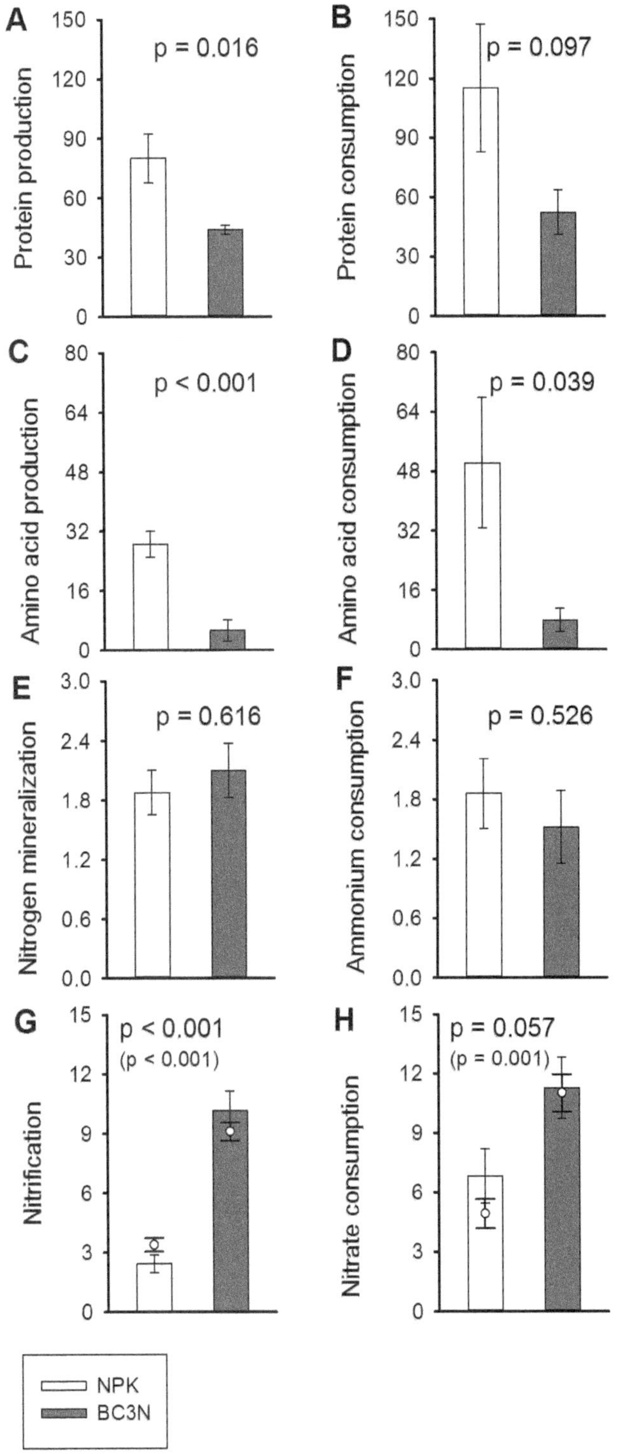

Figure 4. Soil N transformation rates for control (NPK) and biochar (BC3N) treatments. Bars represent means ±1 SE (n = 4) of measurements from July 2011. Open bars, control treatment (NPK); grey bars, biochar treatment (BC3N). Open circles represent means ±1 SE (n = 4) of nitrification rates from three measurement dates. Left panels represent gross influx rates into the target pools (production rates), right panels gross efflux rates (consumption rates). Units in μg N g^{-1} dry soil d^{-1}, equivalent to mg N kg^{-1} d^{-1}. P values are from one-way ANOVA (all rates) or two-way mixed ANOVA (nitrification for three time points, in brackets).

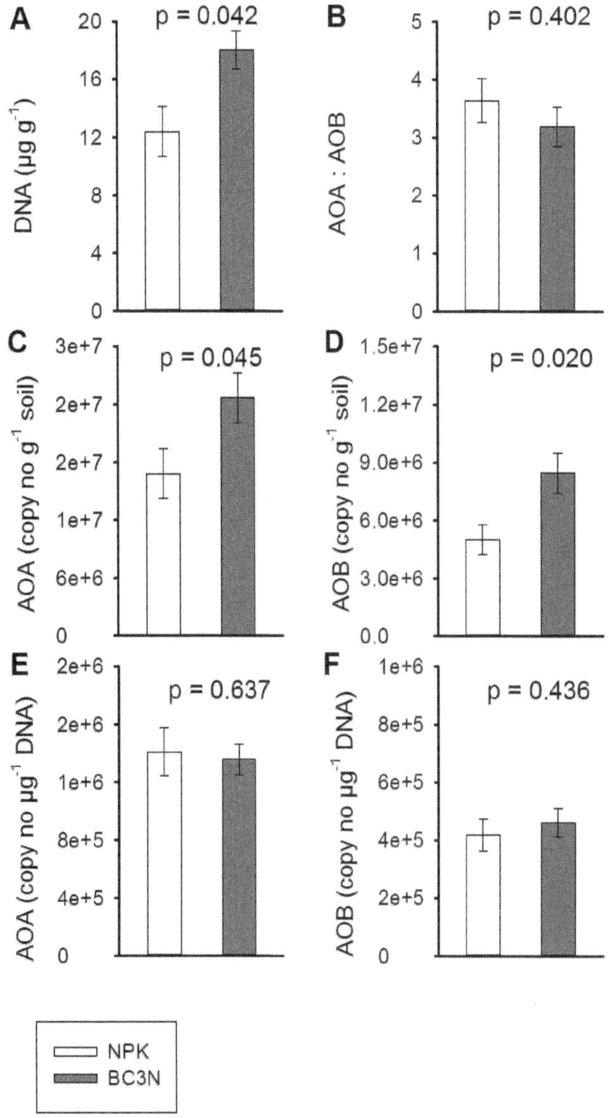

Figure 5. Soil microbial community size and ammonia-oxidizing community structure. Bars represents means ±1 SE (n = 4). Open bars, control treatment (NPK); grey bars, biochar treatment (BC3N). DNA, soil DNA content. AOA, archaeal amoA copy numbers; AOB, bacterial amoA copy numbers. AOA and AOB copy numbers are presented on soil dry mass and DNA basis. P values from two-way ANOVA (see Table 2).

soil bacteria and archaea [72]. Soil prokaryotes typically have mean cell sizes of 0.6 to 0.8 μm in diameter [73–74] with a length: diameter ratio of 1.5:1 [75] and were shown to inhabit pores greater than 0.6 μm [76]. Soil fungi have slightly larger mean diameters, of between 1.2 to 4 μm [74,77] but mycorrhizal hyphae vary enormously in diameter ranging between 2 to 20 μm [78]. As a consequence of the high proportion of colonizable pores, addition of *Eucalyptus* charcoal to a coarse textured soil increased habitable pore space surface area by 11% at a biochar rate of 25 t ha^{-1} [70] and habitable pore volume by 6.5% at 50 t ha^{-1} [79]. Indeed we found an increase of pore volume of 8% at biochar rates of 72 t ha^{-1}. More importantly it has been shown that interior surfaces and pores are rapidly colonized, microbiota showing no differences in densities between outer and internal biochar surfaces 3 years after application [79]. It has been

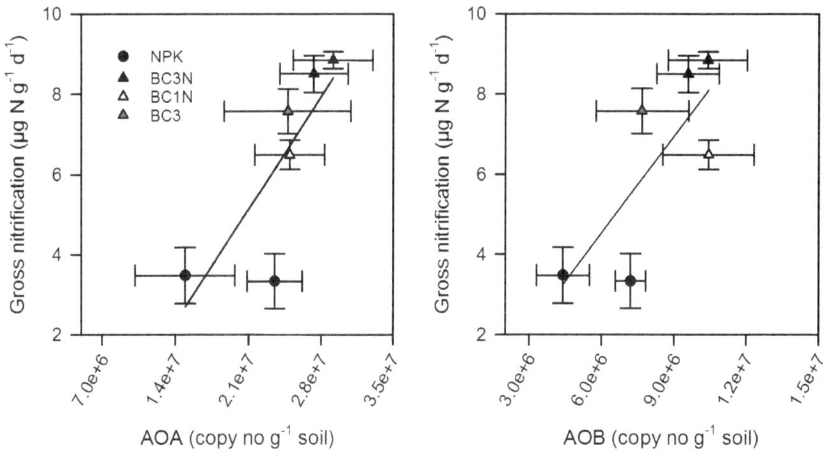

Figure 6. Relationship of archaeal (AOA) and bacterial amoA abundance (AOB) with gross soil nitrification rates. The relationships between archaeal amoA copy numbers (AOA) and bacterial amoA copy numbers (AOB) with gross nitrification rates were marginally significant for AOA ($p = 0.051$) and AOB ($p = 0.072$). Symbols represent the four treatments measured in September 2012. NPK and BC3N were also measured in June 2012. Values are means ± 1 SE ($n = 4$).

suggested that soil microbes are protected from being grazed upon their natural predators by colonizing pore habitats in biochar that are inaccessible for soil microfauna (Fig. 2). By this mechanism biochar pores can act as a safe microhabitat for various soil microbes through the exclusion of larger predators [80–81] such as nematodes, mites and collembolans [73] but also microfaunal protozoan grazers and predators, i.e. flagellates, amoebae and ciliates [82–83]. Our scanning electron microscopy data suggest that there was an abundance of colonizable pores, particularly in the 0.3 to 40 µm range with maxima around 0.7, 3 and 15 µm. The abundant sack like structures of parenchyma cells and the long cave like structures of xylem vessels (Figure S1) could therefore offer physical protection for soil prokaryotes and fungi from grazers such as nematodes and amoeba, which are just physically too big to enter the cave or pore. This in turn could reduce microfaunal grazing, which leads to a decrease in mineralization of organic N via the soil "microbial loop" [84]. This scenario provides a possible explanation for the observed increases in nitrification rate, with bacteria, fungi, AOA and AOB taking refuge in the biochar. In the biochar pores AOA and AOB have ample access to carbon dioxide and available ammonia, given the high pH of this soil; this allows them to proliferate, but protected from grazing, thus not contributing significantly to N mineralization. Interestingly this scenario would suggest possible top down control of nitrification, with predators having been shown to consume up to 60% of bacterial production in some soils [85]. In addition to the provision of protected habitat provided by biochar co-location of nutrients and water on biochar surfaces and in pores may increase substrate availability, and thus may further stimulate microbial colonization [72] and nitrification.

Conclusions and Outlook

The main findings of this field study are that application of wood biochar led to marked decreases in soil organic N transformation rates while soil nitrification was significantly stimulated in this type of agricultural soil. This field study therefore indicates a de-coupling of the intrinsic organic and inorganic soil N cycles in the short- to medium-term (four to 18 months) as a consequence of biochar addition. Assuming these effects persist, the deceleration of soil organic N cycling ultimately leads to the dichotomy of two intertwined agronomic processes; the build-up of soil organic N and the slowing-down of inorganic N release from soil organic N essential for crop growth. Addition of fertilizer-N in combination with biochar effectively decouples these processes, with plants and microbes utilizing fertilizer-N for growth, in turn fuelling the belowground build-up of soil organic N. Since soil organic N and particularly refractory soil organic matter are peptide dominated materials [86], the accumulation of soil organic N is expected to enhance soil carbon sequestration due to increases in non-biochar derived soil organic matter. However, it remains an open question whether adsorptive capacity of biochar will saturate and pore-filling of biochar slows or curtails this process, though recent investigations with aged and fresh biochar suggested that the effects of biochar on organic compound sorption may be long lasting [87].

Stimulation of soil nitrification can have positive effects on plant N availability and crop growth but concurrently increases the risk of soil N losses through nitrate leaching and gaseous losses by denitrification and nitrifier-denitrification. The decrease in soil nitrate concentration and increase in microbial nitrate utilization however point to a relatively conservative N cycling in biochar-treated arable soils. It will therefore be of great importance to study the responses of soil organic N, nitrification, soil N losses and plant growth over longer time periods and in various agricultural settings, to better understand the long-term effects of biochar amendments on soil N cycling.

Given the central role of soil organic N cycling and nitrification in controlling global terrestrial N cycles, the implications of this work are of major importance. If it is more broadly proven that addition of biochar combined with fertilizer-N leads to a greater than sum increase in non-biochar derived soil organic matter, as a consequence of the mechanisms described here, it may prove to be a fundamental factor in future soil management: as building soil organic matter through carbon farming is the key step required to kick start the next black/green revolution.

Materials and Methods

Description of Field Trial

The field experiment was established in Traismauer, Lower Austria, Austria (48° 19′ 886″ N, 15° 44′ 308″ E; 268 m a. s. l) in

March 2011 by the Austrian Institute of Technology (AIT) Tulln, Austria. The temperate climate shows strong continental influence (Pannonian climate), with mean annual precipitation of 550 mm and mean annual temperature being 10°C. The arable soil was a deep sandy to loamy silt (18.3% sand, 57.2% silt and 24.5% clay), classified as a calcareous Chernozem on loess. Soil pH of non-biochar treated soil was 7.5 in 10 mM $CaCl_2$, the carbonate content was 15.8% and humus content 2.4% (Table 1). Cation exchange capacity of the soil was 22.5 cmol kg^{-1}, dominated by Ca^{2+} (~90%) and Mg^{2+} (~7%), with a small contribution by K^+ (2%). The site was used for agriculture for >100 years, with spring barley grown in 2011 and sunflower in 2012. Hardwood-derived biochar (80% beech, rest derived from other hardwoods) was produced by pyrolysis at 500°C for two hours and was purchased from S.C. Romchar S.R.L. (Harghita, Romania). This biochar had a pH of 8.6 in water and 7.5 in 10 mM $CaCl_2$ and comprised the following elemental concentrations: 80.3% C, 9.9% O, 0.4% N, and 1.6%H contents (Table 4). Volatiles made up 11.1% and ashes 15.2% of biochar dry mass (both determined at 550°C, Table 4). Biochar was applied once to the field at a dry mass rate of 0, 24 and 72 t ha^{-1} in March 2011 and incorporated to a depth of 15 cm. NPK fertilizer was applied annually in spring providing an annual N input of 120 (2011) and 75 kg ha^{-1} (2012), with corresponding annual rates of P input of 26 (2011) and 31 kg ha^{-1} yr^{-1} (2012) and K input of 50 (2011) and 100 kg ha^{-1} yr^{-1} (2012). The four treatments were: control i.e. NPK (no biochar application, NPK added), BC1N (24 t biochar ha^{-1}, NPK added), BC3N (72 t biochar ha^{-1}, NPK added) and BC3 (72 t biochar ha^{-1}, no NPK added). The four treatments were arranged according to a Latin square and replicated four times in circular plots of 6.5 m diameter.

The field is privately owned and not protected in any way. The local winegrower association Traisen valley (Weinbauverband Traisental) gave us the written permission to conduct the field experiment at the premises owned by Rudolf Hofmann who also agreed in his role as landowner to set up the experimental plots on his land. No governmental permits were required for the described field study, and the study did not involve endangered or protected species.

Soil sampling was performed several times in 2011 and 2012 (see Table S2 for sampling dates and for the treatments sampled and the pools and processes measured at each date). Soil samples were taken from the topsoil layer (0–10 cm) using a soil corer with 8 cm diameter. From each treatment plot three soil samples were collected randomly and pooled to form one composite sample. The fresh samples were then weighed for bulk density calculation, sieved to pass 2 mm and stored at 4°C for a maximum of 2–3 days prior to analyses.

Table 4. Element contents of beech wood biochar pyrolyzed for two hours at 500°C.

Element	Content (% d.w.)
C	80.30
O	9.88
Ca	3.75
H	1.60
Si	1.58
Fe	0.79
K	0.73
N	0.40
Al	0.243
Mg	0.198
Mn	0.091
P	0.076
Si	0.053
Cl	0.010
Zn	0.009
Cr	0.005
Co	0.002
Ashes (550°C)	15.2
Volatiles (550°C)	11.1

Soil Physicochemical Analyses

General soil properties. Soil water content (SWC, in g H_2O g^{-1} dry soil) was determined gravimetrically after drying fresh soil at 105°C for 24 hours in a forced-air drying oven. Bulk density (BD, in g cm^{-3}) was calculated from the fresh weight divided by soil core volume times dry weight: fresh weight ratio. The percentage of water-filled pore space (%WFPS) was calculated as follows:

$$\%WFPS = \frac{SWC \times BD \times 100}{1 - \frac{BD}{2.65}}$$

where 2.65 g cm^{-3} represents the mean density of rock.

Soil total C and N contents were measured on oven-dried (105°C) and homogenized soil samples. Samples were ground to a fine powder with a ball mill (MM2000, Retsch, Germany) and aliquots of 10 mg weighed into tin capsules. Total soil C and N contents were determined by an elemental analyzer (EA 1110; CE Instruments, Italy). For quantification of the soil organic carbon content samples were pretreated with excess acid (2 M HCl) for 60 min and then dried at 105°C to remove carbonate from the calcareous soil. Other basic soil properties, such as soil pH, soil organic matter content, carbonate content, soil texture, total P content, cation-exchange capacity, base saturation, and exchangeable cations (see Table 1) were only measured for control (NPK) and biochar-treated soils (BC3N) at the Institute of Sustainable Plant Production, Federal Agency for Food Security (AGES) in Vienna, Austria, according to standard ÖNORM procedures (www.austrian-standards.at/en).

Biochar characterization. The biochar (beech wood charcoal) was characterized for element, volatiles and ash contents by the German Biomass Research Center (DBFZ, Leipzig, Germany) and results are presented in Table 4. Biochar pH was measured in an aqueous slurry (5% (w/v)) using a pH electrode. For measurements of the porosity (pores >300 nm diameter) and the pore size distribution of the biochar, charcoal samples were mounted on 0.5″ aluminum specimen stubs equipped with SEM-Carbon foils (Agar scientific, UK). Since the specimens were conductive no additional coating was performed after air drying at 40°C for 24 hours. Micrographs were taken on a Philips XL 20 scanning electron microscope (SEM; Philips, Eindhoven, The Netherlands) at 350× and 2000× magnification with an acceleration voltage of 15 kV. For pore size and pore coverage measurements the software Scandium Version 5.0 (Soft Imaging Systems GmbH, Olympus, Vienna, Austria) and the image analyzing software GSA Image Analyzer (GSA, Gesellschaft für Softwareentwicklung und Analytik mbH, Germany) were used. To estimate pore size distribution in total, 7715 pore measurements in

6 longitudinal and in 7 cross-sectional images covering 0.35 mm^2 and in 25 longitudinal images covering 0.0112 mm^2 were performed. Pores down to a diameter of 300 nm could be safely measured. Pore size distributions were converted to counts mm^{-2} or circular area in μm^2 mm^{-2} and summed up for each 1 μm size class.

Pool sizes of extractable organic C, inorganic N and organic N forms. For measurements of extractable C and N pools fresh soils (aliquots of 4 g) were extracted for 30 min with 30 ml 0.5 M K_2SO_4, subsequently filtered through ash-free cellulose filters and extracts kept frozen at −20°C for later analyses. Total dissolved N and dissolved organic carbon (measured as non-purgeable organic carbon) were measured by high temperature catalytic oxidation on a Shimadzu TOC-VCPH with TNM-1 unit and ASI-V autosampler (Shimadzu Austria, Korneuburg, Austria). Ammonium and nitrate were determined with colorimetric methods, NH_4^+ with a modified indophenol reaction and NO_3^- with the VCl_3/Griess assay, as described in detail by Hood et al. [88]. Amino acid concentrations were quantified by a modified fluorometric OPAME procedure based on Jones et al. [89], optimized for free amino acid measurement in protein hydrolysates. The modified OPAME reagent was made up of 10 mg o-phthaldialdehyde dissolved in 1 ml of HPLC grade methanol to which 20 μl of 3-mercaptopropionic acid was added. This reagent was mixed with 40 ml potassium borate buffer (0.2 M; pH 9.5) and left to stand overnight. Sample aliquots (50 μl) were mixed with 200 μl reagent in the microtiter plate and measured after 10 min at an excitation wavelength of 340 nm and emission wavelength of 450 nm with a fluorescence microplate reader (Tecan Infinite M200, Tecan Austria GmbH, Grödig, Austria). As NH_4^+ in the soil extracts also gives a fluorescence signal, NH_4^+ concentration had to be determined [88] and its fluorescence contribution was determined via a NH_4^+ concentration series and subtracted from the total fluorescence of the samples. Total free amino acid-N concentration was then calculated using glycine as a standard and assuming that each mol of amino acid contains 1.37 mol N. Proteins were quantified in K_2SO_4 extracts by acid hydrolysis and subsequent amino acid analysis (modification of [89–90]; see amino acid determination above). Protein hydrolysis to free amino acids was performed in 6 M HCl and heating was performed for 15 hours at 100°C in a drying oven with Bovine serum albumin (BSA containing 15.8% N) as internal standard. Samples were then dried in an N_2 stream and redissolved in deionized water. Determination of the NH_4^+ concentration to correct the amino acid fluorescence was done as in the case of amino acid determination, with the exception that the sodium nitroprusside solution (the color reagent) contained 1.5 M NaOH instead of the original 0.3 M NaOH to neutralize residual acid in the hydrolysates.

Isotope Pool Dilution Assays

Gross rates of N transformation processes were determined using the ^{15}N isotope pool dilution (IPD) technique [31,91]. This method entails the addition of an isotopic tracer to a target pool and measuring the isotopic composition and concentration over time which allows estimating gross rates of influx to (production) and efflux from (consumption, immobilization) the target pool. In order to avoid the stimulation of consumption rates due to excessive addition of ^{15}N labeled substrate to the respective target pool, each pool size was determined prior to labeling the soil samples. Thereafter to each target pool 10% of the pool size was added in the form of ^{15}N tracer (98 at% ^{15}N). Before starting the IPD assays, all soils were adjusted to 40% water holding capacity and then the ^{15}N tracer solutions were added to bring the soils to 70% water holding capacity. After addition of the tracer to the soils the vials were shaken vigorously to homogenously distribute the label. The incubations were performed at room temperature (~20°C) and terminated by adding 0.5 M K_2SO_4 and subsequent extraction for 30 min on a rotary shaker followed by filtration through ash-free filter papers into new vials. Extracts were immediately frozen and kept for further analyses. An overview of the soil preparation pipeline for these isotope pool dilution assays is presented in Figure S2.

Gross N mineralization and NH_4^+ immobilization. Each 2 g soil sub sample was weighed into disposable HDPE vials (20 ml scintillation vials) and labeled with $^{15}NH_4Cl$ (98 at% ^{15}N purchased from Isotec-Sigma Aldrich, Vienna, Austria). The incubations were terminated after 4, 24 and 48 hours by extraction with 15 ml 0.5 M K_2SO_4. Prior to isotopic analyses ammonium concentrations in the labeled extracts were measured photometrically. Ammonium was isolated from the extracts by microdiffusion [92]. In brief, this method implies the conversion of ammonium into ammonia gas which is then trapped on an acidified cellulose filter disc enclosed in a semi-permeable Teflon membrane (acid trap), where the ammonia gas dissolves and redissociates to ammonium. The interconversion of ammonium and ammonia is a consequence of the pH-dependent equilibrium between these two N forms. To each 10 ml of soil extract one acid trap and 100 mg MgO were added to raise the pH to >9.5, the vials were closed immediately and left to stand at room temperature for three days. In cases where ammonium concentrations exceeded the optimal range of N (10–25 μg NH_4^+-N 10 ml^{-1}) for elemental analyzer-isotope ratio mass spectrometry (EA-IRMS) measurement, extracts were diluted with 0.5 M K_2SO_4 prior to microdiffusion. After microdiffusion acid traps were transferred to 2 ml HDPE screw-cap vials and dried in an evacuated desiccator over concentrated sulphuric acid for two days. The cellulose discs were retrieved and transferred into tin capsules and subjected to EA-IRMS. Conversely extracts with ammonium-N concentrations less than 2 μg NH_4^+-N 10 ml^{-1} were prepared to be measured with purge-and-trap isotope ratio mass spectrometry (PT-IRMS) [93]. For this purpose acid traps were made from glass fiber filters. Acid traps to be measured with PT-IRMS were opened and the filter discs transferred to 1.5 ml HDPE snap-cap vials containing 1 ml deionized water and shaken on a Vortex to redissolve the trapped ammonium. The alkaline persulfate oxidation with subsequent VCl_3/azide reactions were then applied to convert NH_4^+ via NO_3^- to N_2O for analysis by PT-IRMS (see below) [93].

Gross nitrification and NO_3^- immobilization. Soil labeling was done with $K^{15}NO_3$ (98 at% ^{15}N purchased from Isotec-Sigma Aldrich) and carried out as a mentioned above for gross N mineralization. As in the case of gross N mineralization a reduction-microdiffusion method was used to isolate nitrate-N after conversion by Devarda's alloy to ammonium-N from the extract into acid traps [92,94] and nitrate concentrations were determined before starting the microdiffusions. In a first step ammonium was eliminated from the extracts by adding one acid trap and MgO (100 mg) to 10 ml sample, then the vials were closed and microdiffusion proceeded for three days at room temperature. After retrieving this acid trap, a new one was added together with 50 mg Devarda's alloy (45% aluminum, 50% copper and 5% zinc), a reducing catalyst converting nitrate to ammonium, to the soil extracts. After five days of microdiffusion at room temperature to collect the ammonia produced from nitrate the second set of acid traps was treated as mentioned above for gross N mineralization and isotope ratios measured via EA-IRMS (see below) to estimate gross nitrification and gross nitrate immobilization.

The organic N-IPD assays described below are novel and were applied here for the first time to measure gross protein and free amino acid fluxes in soils (but see Wanek et al. [32] for the first amino acid IPD by gas chromatography-mass spectrometry).

Gross amino acid production and immobilization. The free amino acid pool was labeled by addition of a ^{15}N-labeled algal amino acid mixture (U-^{15}N-98%, Cambridge Isotope Laboratories Europe, Radeberg, Germany). For this, each 4 g fresh soil were weighed into 50 ml HDPE centrifuge tubes and the ^{15}N tracer was added. Samples were extracted for half an hour with 30 ml 0.5 M K_2SO_4 after 5 and 30 minutes incubation time. An ultrafiltration method was used to separate low-molecular weight organic N from high molecular weight organic N. Amicon Ultra-2 prelaunch centrifugal filter devices (Millipore Corporation) with a nominal molecular weight limit (NMWL) of 3 kDa were taken to collect the low molecular weight organic N fraction (mainly amino acids and oligopeptides) together with the inorganic N forms (NO_3^- and NH_4^+) in the permeate and the high molecular weight organic N forms (mainly proteins) in the retentate. Ultrafiltration was performed on a Beckman centrifuge (Beckman J2–21, Beckmann Coulter, Krefeld, Germany). From the labeled amino acid extract aliquots of 1.7 ml were centrifuged at 4000*g* for one hour at room temperature with a swinging bucket rotor (Beckman JS-13.1). In order to get rid of inorganic N from low molecular weight organic N, each 1 ml of permeate was filled into a 2 ml cryo-vial and 20 mg Devarda's and 20 mg MgO were added together with an acid trap. The closed cryo-vials were kept at 37°C for 3 days; thereafter the acid traps were discarded. Via alkaline persulfate oxidation and VCl_3/azide reaction the low molecular weight organic N was then converted to NO_3^- and finally to N_2O for analysis by PT-IRMS (see below). Moreover free amino acid concentrations were determined in each extract using the modified OPAME procedure as detailed above, to correct the isotope enrichment of free amino acids for the contributions of other low molecular weight organic N forms using a simple two-source isotope pool mixing model [95]. Free amino acid concentrations and the corrected isotope enrichments were used for later pool dilution calculations.

Gross protein production and consumption. Soil labeling was performed as in the case of the amino acid-IPD assays but with the ^{15}N tracer being a purified algal crude protein (98 at%^{15}N purchased from Isotec-Sigma Aldrich) and with incubation times of 5 min, 30 min, 4 and 24 hours. After ultrafiltration (similar to amino acid-IPD but centrifugation parameters altered) of each 1.7 ml of the labeled extract at 7500 *g* for one hour at room temperature with a fixed angle rotor (Beckman JA-20), the permeate was put aside. The high molecular weight organic N containing retentate was washed twice with each 1 ml 50 mM K_2SO_4 via 30 min centrifugation at room temperature and 7500 *g*, to remove inorganic N and low molecular weight organic N. Thereafter the retentate was recovered by 2 min centrifugation at 1000 *g* at room temperature and diluted to 0.5 ml with 50 mM K_2SO_4. The alkaline persulfate oxidation and subsequent VCl_3/azide reactions were then used to convert the high molecular weight organic N via NO_3^- to N_2O for analysis by PT-IRMS (see below).

Alkaline persulfate oxidation (APO) of N_{org} (NH_4^+) to NO_3^-. The APO reaction was used to convert ammonium and organic N forms into nitrate [93]. At elevated temperature persulfate decomposes forming persulfate radicals that efficiently oxidize organic and inorganic compounds, in the case of organic N and ammonium producing nitrate. The persulfate reagent was prepared by dissolving 44 g $Na_2S_2O_8$, 16.8 g NaOH and 30 g H_3BO_3 in deionized water with an end volume of 1 L [96]. For digestion aliquots of 750 µl of sample were pipetted into 1.5 ml HPLC glass vials and each 750 µl of persulfate solution were added. Vials were closed immediately (to avoid losses of ammonium due to the alkaline pH) and autoclaved for one hour at 120°C. Digested samples were stored at 4°C until further analyses.

VCl_3/azide conversion of NO_3^- to N_2O. The isotope ratios of nitrate after persulfate digestion were measured by conversion of nitrate to N_2O by a coupled VCl_3/azide reaction in which in a first step, VCl_3 is used to reduce nitrate to nitrite, which in a second step is further reduced to N_2O by sodium azide [93]. For this, aliquots of 1 ml sample were pipetted into 12-ml screw cap exetainers with butyl rubber septum (IVA-VC309; IVA Analysetechnik, Merbusch, Germany) which were closed tightly and purged with helium for 10 min to get rid of air-N_2O in the headspace. Thereafter 160 µl sodium azide and finally 1 ml VCl_3 were added using gas-tight syringes by injection through the septa. Exetainers were put on a rotary shaker and kept at 37°C for 18 hours. The reaction was stopped by neutralizing and deactivating the sodium azide through injecting each 160 µl 6 M NaOH into the exetainers and N_2O was analyzed by PT-IRMS.

PT-IRMS. The exetainer vials were then loaded into a 96-slot autosampler with a double-hole needle (GC-PAL, CTC Analytics, Zwingen, Switzerland) connected via a Gasbench II headspace analyzer (Thermo Fisher, Bremen, Germany) and a PreCon cryotrapping device (Thermo Fisher) to a gas isotope ratio mass spectrometer (IRMS, Delta V Advantage, Thermo Fisher) [93]. Through the double needle of the autosampler an auxiliary helium flow purged the contents of the exetainer vials through a water and CO_2 scrubber into a sequence of two cryo-traps held at liquid N_2 temperature, and the N_2O was then released and residual CO_2 and air (O_2, N_2) separated by gas chromatography. The purified sample N_2O was then transferred into the mass spectrometer via an open split with repeated pulses of standard N_2O gas for internal calibration (Air Liquide, Vienna, Austria). For calibration of at%^{15}N respective standards varying in ^{15}N:^{14}N ratios were run throughout the sample preparation methods. For further details on PT-IRMS analyses see Lachouani et al. [93].

EA-IRMS. Acid traps and homogenized dry soils were transferred into tin capsules and directly analyzed for N content and at%^{15}N by EA-IRMS. The system consisted of an elemental analyzer (EA 1110, CE Instruments, Milan, Italy) connected via a ConFlo III interface (Thermo Fisher) to the IRMS (Delta^{PLUS}, Thermo Fisher). External calibration was performed using laboratory standards for isotope composition and N concentration [93].

Isotope pool dilution calculations. Gross production (GP) and gross consumption (GC) rates, i.e. influx into and efflux from the target pools that was labeled with ^{15}N in the course of IPD experiments was then calculated based on the isotope pool dilution equations developed by others [97]:

$$GP = \frac{C_{t_2} - C_{t_1}}{t_2 - t_1} \times \frac{\ln\left(\frac{APE_{t_1}}{APE_{t_2}}\right)}{\ln\left(\frac{C_{t_2}}{C_{t_1}}\right)}$$

$$GC = \frac{C_{t_2} - C_{t_1}}{t_2 - t_1} \times \left(1 + \frac{\ln\left(\frac{APE_{t_2}}{APE_{t_1}}\right)}{\ln\left(\frac{C_{t_2}}{C_{t_1}}\right)}\right)$$

where C_{t1}, C_{t2}, APE_{t1} and APE_{t2} are the concentrations of N (µg N g^{-1} soil dry weight) and atom percent excess of ^{15}N (APE, in %) in the respective target N pool at t_1 and t_2 (days), i.e. the times of IPD sample termination. APE was calculated as the difference between the at%^{15}N of the given N pool at time t and the background ^{15}N abundance (at%^{15}N) measured in separate non-labeled samples. Gross rates are therefore expressed in µg N g^{-1} soil dry weight d^{-1}.

Extraction of nucleic acids and quantitative PCR analyses. Nucleic acids were extracted from soil samples (0.5 g fresh weight) using a modification of the method described by Griffiths et al. 2000 [98]. Samples were loaded in 2 mL Lysing Matrix E tubes (MP Biomedicals, Illkirch, France) and cells were disrupted by bead-beating with 0.5 mL of an SDS-based lysis buffer (1% SDS, 0.7 M NaCl, 0.1 M $Na_2(SO_3)$, 0.05 M EDTA pH8 and 0.1 M Tris/HCl pH 7.5) and 0.5 mL of a phenol:-chloroform:isoamyl-alcohol (25:24:1) mixture (pH 7.5). Precipitation of nucleic acids was performed for 1 h at room temperature with 20 µg glycogen (MBI Fermentas, Vienna, Austria), 0.5 mL isopropanol and 40 µL 5 M NaCl. Precipitated nucleic acids were washed twice in 500 µL ice-cold 70% ethanol, re-suspended in 50 µL DEPC-treated water and purified by filtration through PVPP mini-columns centrifuged at 1000×*g* for 2 min at 10°C. Purity and concentration of purified nucleic acids was checked spectrophotometrically using a NanoDrop spectrophotometer (Thermo scientific) and their integrity was determined by agarose gel electrophoresis. A SybrGreen I quantitative PCR approach was used to quantify thaumarchaeal and bacterial *amoA* gene abundance. Thaumarchaeal *amoA* genes were amplified with PCR primers Arch-amoA-104F/Arch-amoA-616R [99] and bacterial *amoA* genes with primers amoA-1F/amoA-2R [100]. Duplicate dilution series (10^2–10^7 copies) of PCR products obtained with M13F and M13R vector primers from Topo® TA cloning® vectors (Invitrogen) containing either a 632 bp fragment of *Nitrososphaera viennensis* EN76 [101] *amoA* gene, obtained with PCR primers Arch-amoA-7F/Arch-amoA-638R [99], or a 491 bp fragment of *Nitrosospira multiformis* ATCC 25196 *amoA1* gene (Nmul_A0799/Nmul_A2325), obtained with primers amoA-1F/amoA-2R, were used as standards for the thaumarchaeal and bacterial *amoA* assays, respectively. Thaumarchaeal *amoA* standards were amplified with an efficiency of 93% and r^2 value of 0.997 and bacterial *amoA* standards with an efficiency of 93% and r^2 value of 0.991. All reactions (20 µL) contained 10 µL of 2× Quantifast™ SYBR® Green PCR Master Mix (Qiagen, Vienna, Austria), 0.2 mg.mL^{-1} BSA, 1 (Archaea) or 0.8 (Bacteria) µM of each primer and for the analysis of soil samples 10 ng of purified nucleic acid extracts. Amplification conditions were 95°C for 15 min, followed by 40 cycles of 95°C for 15 s, 60°C for 30 s (combined annealing and extension step) and plate read at 78°C for thaumarchaeal *amoA* genes and 35 cycles of 95°C for 15 s, 62°C for 30 s and plate read at 82°C for bacterial *amoA* genes, before a final extension step at 60°C for 10 min. Reactions were carried out in a Mastercycler® RealPlex^2 thermocycler (Eppendorf Austria GmbH, Vienna, Austria) and the specificity of the amplification was assessed by melting curve analysis and agarose gel electrophoresis.

Statistical analyses. Mixed between-within subjects ANOVAs (split-plot ANOVA) followed by Tukey HSD post hoc tests were performed in SPSS version 16.0 for Windows (SPSS Inc., USA) to test for the effects of treatment and sampling time on biological and physicochemical soil parameters. If normality (Shapiro-Wilk test) and variance homogeneity (Levene's test) were violated, data were log, square root or reciprocal transformed to meet these assumptions. Mauchly's test was used to test the assumption of sphericity and if significant, the degrees of freedom of the F-distribution were corrected using the Greenhouse-Geisser procedure as the estimated epsilon was always less than 0.75. Biochar effects on pools and processes that were only measured once were tested by oneway ANOVA. Regressions and correlations were calculated with Statgraphics Centurion XVI (StatPoint Technologies Inc., USA). Simple least squares linear regressions were calculated to estimate the amount dependency of biochar effects, with added biochar amount as independent variable. Correlations between microbial community structure parameters, or with nitrification rates were performed based on the Pearson product-moment correlation procedure.

Supporting Information

Figure S1 Structure of beech wood biochar shown by SEM analysis. Cross section across early to late wood transect in beech wood charcoal (top left), longitudinal section through beech charcoal showing open xylem vessels and xylem parenchyma cells (top right), close-up of xylem pits (bottom left) and close-up of a plasmodesmata channel (bottom right).

Figure S2 Scheme of the workflow for soil sample preparations to measure organic and inorganic N transformation rates using ^{15}N isotope pool dilution approaches.

Table S1 Results of ANOVA of soil physicochemical factors and gross nitrification rates for the factors treatment and sampling date for all four treatments.

Table S2 Soil sampling dates, treatments sampled and measurements done on these samples.

Acknowledgments

The setting up and maintenance of the field experiment by the AIT Tulln team is greatly acknowledged. All isotope analyses were run at the SILVER laboratory, Department of Microbiology and Ecosystem Science, University of Vienna, with technical assistance of Margarete Watzka. We further thank the local farmer and landowner, Rudolf Hofmann, Traismauer, Austria for allowing us to setup this biochar project.

Author Contributions

Conceived and designed the experiments: WW RCHN GS. Performed the experiments: JP DT PO FH SS. Analyzed the data: JP WW RCHN DT PO FH SS. Contributed reagents/materials/analysis tools: TU CS BK GS. Wrote the paper: JP WW RCHN PO.

References

1. Bruun EW, Muller-Stover D, Ambus P, Hauggaard-Nielsen H (2011) Application of biochar to soil and N2O emissions: potential effects of blending fast-pyrolysis biochar with anaerobically digested slurry. European Journal of Soil Science 62: 581–589.
2. van Zwieten L, Kimber S, Morris S, Downie A, Berger E, et al. (2010) Influence of biochars on flux of N2O and CO2 from Ferrosol. Australian Journal of Soil Research 48: 555–568.
3. Cayuela ML, Sanchez-Monedero MA, Roig A, Hanley K, Enders A, et al. (2013) Biochar and denitrification in soils: when, how much and why does biochar reduce N2O emissions? Scientific Reports 3.
4. Wang JY, Pan XJ, Liu YL, Zhang XL, Xiong ZQ (2012) Effects of biochar amendment in two soils on greenhouse gas emissions and crop production. Plant and Soil 360: 287–298.
5. Schmidt MWI, Noack AG (2000) Black carbon in soils and sediments: Analysis, distribution, implications, and current challenges. Global Biogeochemical Cycles 14: 777–793.
6. Baldock JA, Smernik RJ (2002) Chemical composition and bioavailability of thermally, altered *Pinus resinosa* (Red Pine) wood. Organic Geochemistry 33: 1093–1109.
7. Kuzyakov Y, Subbotina I, Chen HQ, Bogomolova I, Xu XL (2009) Black carbon decomposition and incorporation into soil microbial biomass estimated by C-14 labeling. Soil Biology & Biochemistry 41: 210–219.
8. Clough TJ, Condron LM, Kammann C, Müller C (2013) A review of biochar and soil nitrogen dynamics. Agronomy 3: 275–293.
9. Joseph SD, Camps-Arbestain M, Lin Y, Munroe P, Chia CH, et al. (2010) An investigation into the reactions of biochar in soil. Australian Journal of Soil Research 48: 501–515.
10. Chen Y, Shinogi Y, Taira M (2010) Influence of biochar use on sugarcane growth, soil parameters, and groundwater quality. Australian Journal of Soil Research 48: 526–530.
11. Ventura M, Sorrenti G, Panzacchi P, George E, Tonon G (2013) Biochar Reduces Short-Term Nitrate Leaching from A Horizon in an Apple Orchard. Journal of Environmental Quality 42: 76–82.
12. Bruun EW, Ambus P, Egsgaard H, Hauggaard-Nielsen H (2012) Effects of slow and fast pyrolysis biochar on soil C and N turnover dynamics. Soil Biology & Biochemistry 46: 73–79.
13. Zheng H, Wang Z, Deng X, Herbert S, Xing BS (2013) Impacts of adding biochar on nitrogen retention and bioavailability in agricultural soil. Geoderma 206: 32–39.
14. Nelissen V, Rütting T, Huygens D, Staelens J, Ruysschaert G, et al. (2012) Maize biochars accelerate short-term soil nitrogen dynamics in a loamy sand soil. Soil Biology & Biochemistry 55: 20–27.
15. Dempster DN, Jones DL, Murphy DV (2012) Organic nitrogen mineralisation in two contrasting agro-ecosystems is unchanged by biochar addition. Soil Biology & Biochemistry 48: 47–50.
16. Anderson CR, Condron LM, Clough TJ, Fiers M, Stewart A, et al. (2011) Biochar induced soil microbial community change: Implications for biogeochemical cycling of carbon, nitrogen and phosphorus. Pedobiologia 54: 309–320.
17. Dempster DN, Jones DL, Murphy DV (2012) Clay and biochar amendments decreased inorganic but not dissolved organic nitrogen leaching in soil. Soil Research 50: 216–221.
18. Taghizadeh-Toosi A, Clough TJ, Sherlock RR, Condron LM (2012) A wood based low-temperature biochar captures NH3-N generated from ruminant urine-N, retaining its bioavailability. Plant and Soil 353: 73–84.
19. Vaccari FP, Baronti S, Lugato E, Genesio L, Castaldi S, et al. (2011) Biochar as a strategy to sequester carbon and increase yield in durum wheat. European Journal of Agronomy 34: 231–238.
20. Jones DL, Rousk J, Edwards-Jones G, DeLuca TH, Murphy DV (2012) Biochar-mediated changes in soil quality and plant growth in a three year field trial. Soil Biology & Biochemistry 45: 113–124.
21. Ma L, Hou Z, Lv N, Ye J, Su S, et al. (2012) Effects of biochar application on wheat growth and nitrogen balance. Xinjiang Agricultural Sciences 49: 589–594.
22. Glaser B, Haumaier L, Guggenberger G, Zech W (2001) The 'Terra Preta' phenomenon: a model for sustainable agriculture in the humid tropics. Naturwissenschaften 88: 37–41.
23. DeLuca TH, MacKenzie MD, Gundale MJ (2009) Biochar Effects on Soil Nutrient Transformations. In: Lehmann J, Joseph S, editors. Biochar for Environmental Management: Science and Technology. London: Earthscan. 251–270.
24. Luo Y, Durenkamp M, De Nobili M, Lin Q, Brookes PC (2011) Short term soil priming effects and the mineralisation of biochar following its incorporation to soils of different pH. Soil Biology & Biochemistry 43: 2304–2314.
25. Fontaine S, Mariotti A, Abbadie L (2003) The priming effect of organic matter: a question of microbial competition? Soil Biology & Biochemistry 35: 837–843.
26. Ball PN, MacKenzie MD, DeLuca TH, Holben WE (2010) Wildfire and Charcoal Enhance Nitrification and Ammonium-Oxidizing Bacterial Abundance in Dry Montane Forest Soils. Journal of Environmental Quality 39: 1243–1253.
27. Liu X, Zhang A, Ji C, Joseph S, Bian R, et al. (2013) Biochar's effect on crop productivity and the dependence on experimental conditions–a meta-analysis of literature data. Plant and Soil DOI 10.1007/s11104-013-1806-x.
28. Jeffery S, Verheijen FGA, van der Velde M, Bastos AC (2011) A quantitative review of the effects of biochar application to soils on crop productivity using meta-analysis. Agriculture Ecosystems & Environment 144: 175–187.
29. Schimel JP, Bennett J (2004) Nitrogen mineralization: Challenges of a changing paradigm. Ecology 85: 591–602.
30. Barraclough D (1991) The Use of Mean Pool Abundances to Interpret N-15 Tracer Experiments.1. Theory. Plant and Soil 131: 89–96.
31. Murphy DV, Recous S, Stockdale EA, Fillery IRP, Jensen LS, et al. (2003) Gross nitrogen fluxes in soil: Theory, measurement and application of N-15 pool dilution techniques. Advances in Agronomy, Vol 79 79: 69–118.
32. Wanek W, Mooshammer M, Blochl A, Hanreich A, Richter A (2010) Determination of gross rates of amino acid production and immobilization in decomposing leaf litter by a novel N-15 isotope pool dilution technique. Soil Biology & Biochemistry 42: 1293–1302.
33. Mooshammer M, Wanek W, Schnecker J, Wild B, Leitner S, et al. (2012) Stoichiometric controls of nitrogen and phosphorus cycling in decomposing beech litter. Ecology 93: 770–782.
34. Gryc V, Vavrcik H, Gomola S (2008) Selected properties of European beech (*Fagus sylvatica* L.). Journal of Forest Science 54: 418–425.
35. Jones DL, Healey JR, Willett VB, Farrar JF, Hodge A (2005) Dissolved organic nitrogen uptake by plants - an important N uptake pathway? Soil Biology & Biochemistry 37: 413–423.
36. Kelleher BP, Simpson MJ, Simpson AJ (2006) Assessing the fate and transformation of plant residues in the terrestrial environment using HR-MAS NMR spectroscopy. Geochimica et Cosmochimica Acta 70: 4080–4094.
37. Farrell M, Hill PW, Farrar J, DeLuca TH, Roberts P, et al. (2013) Oligopeptides represent a preferred source of organic N uptake: A global phenomenon? Ecosystems 16: 133–145.
38. Kasozi GN, Zimmerman AR, Nkedi-Kizza P, Gao B (2010) Catechol and Humic Acid Sorption onto a Range of Laboratory-Produced Black Carbons (Biochars). Environmental Science & Technology 44: 6189–6195.
39. Bergeron SP, Bradley RL, Munson A, Parsons W (2013) Physico-chemical and functional characteristics of soil charcoal produced at five different temperatures. Soil Biology & Biochemistry 58: 140–146.
40. Haslam E (1974) Polyphenol-Protein Interactions. Biochemical Journal 139: 285–288.
41. Mukherjee A, Zimmerman AR, Harris W (2011) Surface chemistry variations among a series of laboratory-produced biochars. Geoderma 163: 247–255.
42. Zimmerman AR, Chorover J, Goyne KW, Brantley SL (2004) Protection of Mesopore-Adsorbed Organic Matter from Enzymatic Degradation. Environmental Science and Technology 38: 4542–4548.
43. Burns RG, DeForest JL, Marxsen J, Sinsabaugh RL, Stromberger ME, et al. (2013) Soil enzymes in a changing environment: Current knowledge and future directions. Soil Biology & Biochemistry 58: 216–234.
44. Bailey VL, Fansler SJ, Smith JL, Bolton H Jr (2011) Reconciling apparent variability in effects of biochar amendment on soil enzyme activities by assay optimization. Soil Biology & Biochemistry 43: 296–301.
45. Mullen CA, Boateng AA, Goldberg NM, Lima IM, Laird DA, et al. (2010) Bio-oil and bio-char production from corn cobs and stover by fast pyrolysis. Biomass & Bioenergy 34: 67–74.
46. Cheng Y, Cai Z-c, Chang SX, Wang J, Zhang J-b (2012) Wheat straw and its biochar have contrasting effects on inorganic N retention and N2O production in a cultivated Black Chernozem. Biology And Fertility Of Soils 48: 941–946.
47. Wang T, Arbestain MC, Hedley M, Bishop P (2012) Chemical and bioassay characterisation of nitrogen availability in biochar produced from dairy manure and biosolids. Organic Geochemistry 51: 45–54.
48. Singh BP, Cowie A, Smernik RJ (2012) Biochar carbon stability in a clayey soil as a function of feedstock and pyrolysis temperature. Environmental Science & Technology 46: 11770–11778.
49. DeLuca TH, MacKenzie MD, Gundale MJ, Holben WE (2006) Wildfire-produced charcoal directly influences nitrogen cycling in ponderosa pine forests. Soil Science Society of America Journal 70: 448–453.
50. Singh B, Singh BP, Cowie AL (2010) Characterisation and evaluation of biochars for their application as a soil amendment. Australian Journal of Soil Research 48: 516–525.
51. Prosser JI, Nicol GW (2012) Archaeal and bacterial ammonia-oxidisers in soil: the quest for niche specialisation and differentiation. Trends in Microbiology 20: 523–531.
52. Stienstra AW, Gunnewiek PK, Laanbroek HJ (1994) Repression of Nitrification in Soils under A Climax Grassland Vegetation. Fems Microbiology Ecology 14: 45–52.
53. Berglund LM, DeLuca TH, Zackrisson O (2004) Activated carbon amendments to soil alters nitrification rates in Scots pine forests. Soil Biology and Biochemistry 36: 2067–2073.
54. Gundale MJ, DeLuca TH (2006) Temperature and source material influence ecological attributes of ponderosa pine and Douglas-fir charcoal. Forest Ecology and Management 231: 86–93.

55. Clough TJ, Condron LM (2010) Biochar and the Nitrogen Cycle: Introduction. Journal of Environmental Quality 39: 1218–1223.
56. Booth MS, Stark JM, Rastetter E (2005) Controls on nitrogen cycling in terrestrial ecosystems: A synthetic analysis of literature data. Ecological Monographs 75: 139–157.
57. Nakajima Y, Ishizuka S, Tsuruta H, Iswandi A, Murdiyarso D (2005) Microbial processes responsible for nitrous oxide production from acid soils in different land-use patterns in Pasirmayang, central Sumatra, Indonesia. Nutrient Cycling in Agroecosystems 71: 33–42.
58. Wrage N, Velthof GL, van Beusichem ML, Oenema O (2001) Role of nitrifier denitrification in the production of nitrous oxide. Soil Biology & Biochemistry 33: 1723–1732.
59. Wang L, Cai Z (2008) Nitrous oxide production at different soil moisture contents in an arable soil in China. Soil Science And Plant Nutrition 54: 786–793.
60. Bateman EJ, Baggs EM (2005) Contributions of nitrification and denitrification to N2O emissions from soils at different water-filled pore space. Biology And Fertility Of Soils 41: 379–388.
61. Offre P, Prosser JI, Nicol GW (2009) Growth of ammonia-oxidizing archaea in soil microcosms is inhibited by acetylene. Fems Microbiology Ecology 70: 99–108.
62. Daum M, Zimmer W, Papen H, Kloos K, Nawrath K, et al. (1998) Physiological and molecular biological characterization of ammonia oxidation of the heterotrophic nitrifier Pseudomonas putida. Current Microbiology 37: 281–288.
63. Hynes RK, Knowles R (1982) Effect of Acetylene on Autotrophic and Heterotrophic Nitrification. Canadian Journal of Microbiology 28: 334–340.
64. Linn DM, Doran JW (1984) Effect of Water-filled Pore-space on Carbon-dioxide and Nitrous-oxide Production in Tilled and Nontilled Soils. Soil Science Society of America Journal 48: 1267–1272.
65. Busscher WJ, Novak JM, Evans DE, Watts DW, Niandou MAS, et al. (2010) Influence of Pecan Biochar on Physical Properties of a Norfolk Loamy Sand. Soil Science 175: 10–14.
66. Case SDC, McNamara NP, Reay DS, Whitaker J (2012) The effect of biochar addition on N2O and CO2 emissions from a sandy loam soil - The role of soil aeration. Soil Biology & Biochemistry 51: 125–134.
67. Schleper C (2010) Ammonia oxidation: different niches for bacteria and archaea? ISME Journal 4: 1092–1094.
68. He J, Shen J, Zhang L, Zhu Y, Zheng Y, et al. (2007) Quantitative analyses of the abundance and composition of ammonia-oxidizing bacteria and ammonia-oxidizing archaea of a Chinese upland red soil under long-term fertilization practices. Environmental Microbiology 9: 2364–2374.
69. Leininger S, Urich T, Schloter M, Schwark L, Qi J, et al. (2006) Archaea predominate among ammonia-oxidizing prokaryotes in soils. Nature 442: 806–809.
70. Dempster DN, Gleeson DB, Solaiman ZM, Jones DL, Murphy DV (2012) Decreased soil microbial biomass and nitrogen mineralisation with Eucalyptus biochar addition to a coarse textured soil. Plant and Soil 354: 311–324.
71. Downie A, Crosky A, Munroe P (2009) Physical properties of biochar. In: Lehmann J, Joseph S, editors. Biochar for Environmental Management: Science and Technology. London, UK: Earthscan.
72. Lehmann J, Rillig MC, Thies J, Masiello CA, Hockaday WC, et al. (2011) Biochar effects on soil biota - A review. Soil Biology & Biochemistry 43: 1812–1836.
73. Mulder C, Cohen JE, Setala H, Bloem J, Breure AM (2005) Bacterial traits, organism mass, and numerical abundance in the detrital soil food web of Dutch agricultural grasslands. Ecology Letters 8: 80–90.
74. Frey SD, Elliott ET, Paustian K (1999) Bacterial and fungal abundance and biomass in conventional and no-tillage agroecosystems along two climatic gradients. Soil Biology & Biochemistry 31: 573–585.
75. Mulder C, Van Wijnen HJ, Van Wezel AP (2005) Numerical abundance and biodiversity of below-ground taxocenes along a pH gradient across the Netherlands. Journal of Biogeography 32: 1775–1790.
76. Strong DT, Sale PWG, Helyar KR (1998) The influence of the soil matrix on nitrogen mineralisation and nitrification. II. The pore system as a framework for mapping the organisation of the soil matrix. Australian Journal of Soil Research 36: 855–872.
77. Baath E, Soderstrom B (1979) Significance of Hyphal Diameter in Calculation of Fungal Bio-volume. Oikos 33: 11–14.
78. Smith SE, Read DJ (2000) Mycorrhizal Symbiosis. Amsterdam, Netherlands: Academic Press.
79. Quilliam RS, Glanville HC, Wade SC, Jones DL (2013) Life in the 'charosphere' - Does biochar in agricultural soil provide a significant habitat for microorganisms? Soil Biology & Biochemistry 65: 287–293.
80. Pietikainen J, Kiikkila O, Fritze H (2000) Charcoal as a habitat for microbes and its effect on the microbial community of the underlying humus. Oikos 89: 231–242.
81. Warnock DD, Lehmann J, Kuyper TW, Rillig MC (2007) Mycorrhizal responses to biochar in soil - concepts and mechanisms. Plant and Soil 300: 9–20.
82. Finlay BJ, Black HIJ, Brown S, Clarke KJ, Esteban GF, et al. (2000) Estimating the growth potential of the soil protozoan community. Protist 151: 69–80.
83. Foissner W (1999) Soil protozoa as bioindicators: pros and cons, methods, diversity, representative examples. Agriculture Ecosystems & Environment 74: 95–112.
84. Coleman DC (1994) The Microbial Loop Concept as used in Terrestrial Soil Ecology Studies. Microbial Ecology 28: 245–250.
85. Moore JC, McCann K, Setala H, De Ruiter PC (2003) Top-down is bottom-up: Does predation in the rhizosphere regulate aboveground dynamics? Ecology 84: 846–857.
86. Knicker H (2011) Soil organic N - An under-rated player for C sequestration in soils? Soil Biology & Biochemistry 43: 1118–1129.
87. Jones DL, Edwards-Jones G, Murphy DV (2011) Biochar mediated alterations in herbicide breakdown and leaching in soil. Soil Biology & Biochemistry 43: 804–813.
88. Hood-Nowotny R, Hinko-Najera Umana N, Inselbacher E, Oswald-Lachouani P, Wanek W (2010) Alternative methods for measuring inorganic, organic, and total dissolved nitrogen in soil. Soil Science Society of America Journal 74: 1018–1027.
89. Jones DL, Owen AG, Farrar JF (2002) Simple method to enable the high resolution determination of total free amino acids in soil solutions and soil extracts. Soil Biology & Biochemistry 34: 1893–1902.
90. Roberts P, Jones DL (2008) Critical evaluation of methods for determining total protein in soil solution. Soil Biology & Biochemistry 40: 1485–1495.
91. Di HJ, Cameron KC, McLaren RG (2000) Isotopic dilution methods to determine the gross transformation rates of nitrogen, phosphorus, and sulfur in soil: a review of the theory, methodologies, and limitations. Australian Journal of Soil Research 38: 213–230.
92. Sorensen P, Jensen ES (1991) Sequential Diffusion of Ammonium and Nitrate from Soil Extracts to a Polytetrafluoroethylene Trap for N-15 Determination. Analytica Chimica Acta 252: 201–203.
93. Lachouani P, Frank AH, Wanek W (2010) A suite of sensitive chemical methods to determine the delta N-15 of ammonium, nitrate and total dissolved N in soil extracts. Rapid Communications in Mass Spectrometry 24: 3615–3623.
94. Khan SA, Mulvaney RL, Hoeft RG (2000) Direct-diffusion methods for inorganic-nitrogen analysis of soil. Soil Science Society of America Journal 64: 1083–1089.
95. Phillips DL, Gregg JW (2001) Uncertainty in source partitioning using stable isotopes. Oecologia 127: 171–179.
96. Doyle A, Weintraub MN, Schimel JP (2004) Persulfate digestion and simultaneous colorimetric analysis of carbon and nitrogen in soil extracts. Soil Science Society of America Journal 68: 669–676.
97. Kirkham D, Bartholomew WV (1954) Equations for following nutrient transformations in soil, utilizing tracer data. Soil Science Society of America Proceedings 18: 33–34.
98. Griffiths RI, Whiteley AS, O'Donnell AG, Bailey MJ (2000) Rapid method for coextraction of DNA and RNA from natural environments for analysis of ribosomal DNA- and rRNA-based microbial community composition. Applied and Environmental Microbiology 66: 5488–5491.
99. Alves RJE, Wanek W, Zappe A, Richter A, Svenning MM, et al. (2013) Nitrification rates in Arctic aoils are associated with functionally distinct populations of ammonia-ixidizing archaea. ISME Journal 7: 1620–1631.
100. Rotthauwe JH, Witzel KP, Liesack W (1997) The ammonia monooxygenase structural gene amoA as a functional marker: Molecular fine-scale analysis of natural ammonia-oxidizing populations. Applied and Environmental Microbiology 63: 4704–4712.
101. Tourna M, Stieglmeier M, Spang A, Konneke M, Schintlmeister A, et al. (2011) Nitrososphaera viennensis, an ammonia oxidizing archaeon from soil. Proceedings of the National Academy of Sciences of the United States of America 108: 8420–8425.

15

Tungsten Distribution in Soil and Rice in the Vicinity of the World's Largest and Longest-Operating Tungsten Mine in China

Chunye Lin[1]*, Ruiping Li[1], Hongguang Cheng[1], Jing Wang[2], Xiao Shao[1]

1 State Key Joint Laboratory of Environmental Simulation and Pollution Control, School of Environment, Beijing Normal University, Beijing, China, **2** Key Laboratory of Land Use, China Land Surveying and Planning Institute, Ministry of land and Resources, Beijing, China

Abstract

The objective of this study is to investigate tungsten (W) contamination in soil and its enrichment in rice in the area of the world's largest and longest-operating W mines in China. Root zone soil and rice plants were sampled at 15 sites in the agricultural field adjacent to W mines and analyzed for Al, Fe, Mn, Sc, and W contents and W chemical forms in the soil samples and W contents in the rice root, stem, leaf, and grain samples. Results showed that W content in the soil ranged from 3.99 to 43.7 mg kg^{-1}, with more than 90% of W in the residual fraction, showing its low mobility and bioavailability. Average W contents in the rice root, stem, leaf, and grain were 7.06, 2.34, 4.76, 0.17 mg kg^{-1}, respectively. In addition, they were linearly independent of W content and chemical forms in the soil. Average enrichment factor values were 0.39, 0.13, 0.28, and 0.01 for the root, stem, leaf, and grain, respectively. In can be concluded that W mining activity in the Dayu county contaminated the nearby agricultural soil and led to W bioaccumulation in the rice. This may pose a health risk to residents via food and soil ingestion, which should be a focus of scrutiny.

Editor: Jin-Song Zhang, Institute of Genetics and Developmental Biology, Chinese Academy of Sciences, China

Funding: This study was supported by the National Natural Science Foundation of China (41371441, http://www.nsfc.gov.cn/), and the Ministry of Environmental Protection Funded Project (201309044, 201109064; http://www.zhb.gov.cn/). The funders had no role in study design, data collection and analysis, decision to publish, or preparation of the manuscript.

Competing Interests: The authors have declared that no competing interests exist.

* E-mail: c.lin@bnu.edu.cn

Introduction

In recent years, the biogeochemistry of tungsten (W) has become a matter of increasing concern due to the scrutiny of a children leukemia cluster in Nevada, its toxicity to organisms, and ubiquitous presence of this element in the environment as a result of geogenic and anthropogenic processes [1–5]. Whereas accumulation of some heavy metals in soils and organisms has been well studied [6–11], the researches on W accumulation in soils and plants are extremely limited.

In general, the content of W in non-polluted soils is low, ranging from 0.4 to 5.0 mg kg^{-1} [6]. However, a few previous studies showed that the concentration of W in the soils around the W mining and/or smelting sites was elevated due to the emission of W during the mining and/or smelting processes [12–18]. For instance, high average W concentration (56 mg kg^{-1}) was reported for the topsoils in the vicinity of mining/smelting sites in North Queensland, Australia [14]. An investigation detected similar levels (10–67 mg kg^{-1}) of W content in top soils collected from four random locations in Fallon, Nevada within a few miles from a W refining plant [3]. Up to 150 mg kg^{-1} of W in the soil at an industrial production site for W trioxide in Switzerland was detected compared to geogenic background concentrations of 1 to 2.5 mg kg^{-1} [19].

Mobility and bioavailability of inorganic contaminants in soils depend on their chemical forms. However, studies on the chemical forms and bioavailability of W in soils are extremely limited. Wilson and Pyatt [15,17] investigated chemical forms and bioavailability in the soils adjacent to W mines, showing that the bioavailability and mobility of W was much lower in the acidic soils than in alkaline soils. Soils surrounding a W ore-processing plant contained total W in the range of 100 to 200 mg kg^{-1}, of which 30% is water soluble, 15% bound to Fe oxides, and 5% fixed by organic matter [6].

W uptake by plants, especially by agricultural crops, is of concern because of the potential for W to enter the food supply [20]. Land plants growing in the uncontaminated soils by W generally contain low W, being less than 0.1 mg kg^{-1} [16,19,21–23]. High concentration of W in the contaminated soils generally elevated the concentration of W in the wild land plants such as trees, shrubs, and grasses [12–18]. For example, *Calluna Vulgaris* growing in the soil of Carrock Fell Mine, UK containing 1169 mg kg^{-1} of W had 655, 48.9, and 124 mg kg^{-1} of W in its root, woody part, and leaves, respectively [15]. In addition, the concentration of W in the plants growing in the tungstate or metallic W added soils increased with the concentration of W in the soils [2,20,24–30]. For example, bioaccumulation of W into *Brassica Oleracae* leaves was dose-dependent with soil W concentration, ranging from 278 to 1420 mg kg^{-1} of W [30].

The literature findings discussed above are confusing as to W chemical forms in the soils. No one determined W concentration, distribution, and enrichment in rice, one of major food around the world.

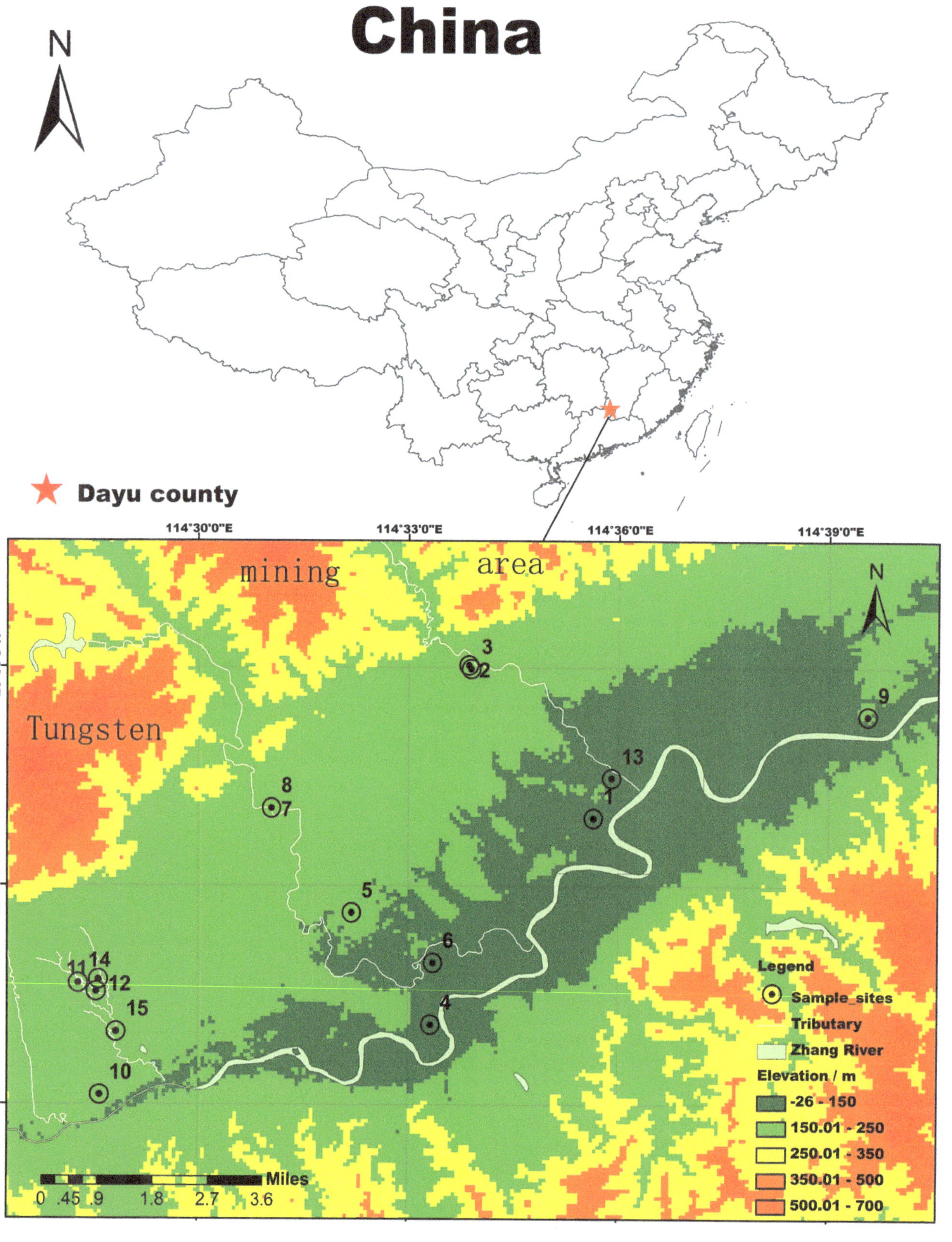

Figure 1. Schematic graph of W mine locations and soil sampling areas.

Table 1. Contents of mineral matrix elements and W in the soil.

Sample	pH	Mn	Sc	Al	Fe	W
		mg/kg	mg/kg	mg/kg	mg/kg	mg/kg
1	5.60	718.60	14.21	7.69	5.04	8.21
2	5.57	181.70	8.51	4.17	1.30	43.65
3	5.26	161.80	6.42	2.72	1.73	16.22
4	5.64	435.10	9.96	5.52	2.26	28.00
5	5.07	168.60	5.96	2.78	1.77	12.59
6	5.42	316.50	10.19	4.84	2.69	17.04
7	5.45	145.40	5.91	2.47	0.93	15.20
8	5.38	158.50	6.49	2.90	1.02	15.48
9	5.36	405.30	10.11	5.79	2.73	37.95
10	5.92	179.40	8.74	4.19	1.68	10.41
11	5.37	68.76	5.87	2.60	1.31	5.27
12	5.51	77.79	6.96	3.32	1.21	3.99
13	4.95	180.40	8.42	4.29	2.40	4.26
14	5.00	82.84	6.06	2.76	1.33	4.98
16	5.24	170.00	8.92	3.96	1.88	25.36
Mean	5.38	230.05	8.18	4.00	1.95	16.57
Median	5.38	170.00	8.42	3.96	1.73	15.20
Max	5.92	718.60	14.21	7.69	5.04	43.65
Min	4.95	68.76	5.87	2.47	0.93	3.99
Jiangxi[1]		328.00	10.31	8.60	2.88	5.28
China[1]		583.00	11.10	6.62	2.94	2.48
World[2]		1000.00	7.00	7.10	4.00	1.50

1: Data are cited from reference [34].
2: Data are cited from reference [35].

Table 2. Proportion of W in each chemical forms in the soil.

Sample	EXC	CARB	OX	OM	RES
	%	%	%	%	%
1	0.02	0.02	0.08	2.87	97.01
2	0.13	0.03	0.43	7.62	91.80
3	0.02	0.01	0.23	6.16	93.58
4	0.02	0.01	0.38	6.27	93.33
5	0.02	0.01	0.20	6.31	93.46
6	0.03	0.01	0.21	5.14	94.60
7	0.11	0.02	0.36	7.62	91.88
8	0.10	0.02	0.37	10.25	89.26
9	0.03	0.01	0.35	5.51	94.09
10	0.08	0.01	0.24	7.26	92.41
11	0.05	0.01	0.30	7.25	92.39
12	0.11	0.03	0.45	6.14	93.27
13	0.01	0.00	0.12	5.21	94.66
14	0.06	0.02	0.46	8.27	91.19
15	0.02	0.01	0.31	3.98	95.69
Mean	0.05	0.01	0.30	6.39	93.24
Median	0.03	0.01	0.31	6.27	93.33
Max	0.13	0.03	0.46	10.25	97.01
Min	0.01	0.00	0.08	2.87	89.26

China is the world's largest W producer and consumer. World tungsten supply is dominated by Chinese production and exports. In 2010, world W mine production was 61,000 t, of which 52,000 t is from China. Ganzhou in the south of Jiangxi province, being the birth place of Chinese W industry, is extremely rich in W sources. W output in Ganzhou has been at the top in China since W was discovered in Xihuashan mine in 1907 in the Dayu county, Ganzhou.

The objectives of this study were to investigate W content and distribution in the agricultural soils and rice in the vicinity of the largest W mine in China, with the longest W mining history. In addition, relationships between W contents in rice and soils were developed. This information is important to quantify W exposure to habitants via food chain.

Materials and Methods

Study site

The study site is situated in Dayu county, the southern Jiangxi province of southern China. It is characterized by subtropical monsoon climate, with average annual precipitation and temperature of 1591.5 mm and 18.5°C, respectively. Major crops are paddy rice and maize. There are three major tungsten mines in Dayu county: Xihuashan, Dangping, and Piaotang, among which Xihuashan is the first tungsten mine operated in China (Fig. 1.). Tungsten occurs in wolframite ($(Fe,Mn)WO_4$) in these mines. Local farmers usually irrigate cultivated field with river water originated from the W mining area. No specific permissions were required for the studied area (25°27′08″−25°33′02″N, 114°28′19″−114°39′33″E). This field study did not involve endangered or protected species.

The soil at Dayu county generally contains soil organic matter of 2.76% to 6.26% and is characterized with acidic pH in the range of 5.09 to 6.28 and sandy loam texture [31].

Soil and plant sampling and analysis

Root zone soil and the rice plants were sampled at 15 sites in the agricultural fields adjacent to W mines. In order to investigate the transfer of W from soils to rice plant, we think it better to collect root zone soil samples. When rice is ready for harvest, rice plants were pulled out together with their roots and root-attached soils. The plants were vigorously shaken by hand and then washed with a little deionized water to remove the root-attached soil. The removed soil and rice plants were put into the plastic bags separately and brought to lab. The soil samples were air-dried in lab, crushed, passed through 0.149 mm sieve (100 mesh), and stored in glass bottles. The rice plants were washed in deionized water and oven-dried at 70°C. Then, root, stem, leaf, and grain (without husk) of the rice plants were separated, ground, and stored in glass bottles.

The pH value of each soil sample was analyzed in a 1:10 solid/liquid ratio suspension (left for ~0.5 h) using a combination pH electrode (Orion, USA). Portions of the soil samples were digested with HNO_3–HF–$HClO_4$ [32]. The Al, Fe, and Mn in the extracts were measured with ICP-AES (IRIS Intrepid II, Thermo Electron), while W and Sc were measured with ICP-MS (X Series II, Thermo Electron). The soil mineral matrix elements (Al, Fe,

Table 3. W contents in rice root, stem, leaf, and grain and its enrichment factor in them.

Sample	W content				W enrichment factor			
	Root	Stem	Leaf	Grain	Root	Stem	Leaf	Grain
	mg/kg	mg/kg	mg/kg	mg/kg				
1	2.34	0.70	0.85	0.12	0.285	0.086	0.104	0.015
2	25.76	7.50	16.66	0.57	0.590	0.172	0.382	0.013
3	7.09	1.64	9.64	0.33	0.437	0.101	0.594	0.020
4	10.27	4.34	5.33	0.36	0.367	0.155	0.190	0.013
5	4.64	1.42	3.71	0.08	0.369	0.113	0.295	0.006
6	5.83	1.49	1.83	0.08	0.342	0.087	0.107	0.005
7	4.26	2.34	4.13	0.09	0.280	0.154	0.272	0.006
8	5.44	2.21	3.78	0.09	0.351	0.143	0.244	0.006
9	21.75	6.30	14.41	0.40	0.573	0.166	0.380	0.010
10	2.30	1.45	2.58	0.07	0.221	0.139	0.248	0.006
11	1.57	1.12	2.57	0.05	0.297	0.212	0.488	0.010
12	1.90	0.72	1.45	0.04	0.478	0.181	0.364	0.010
13	2.23	0.23	0.59	0.02	0.524	0.055	0.138	0.005
14	2.24	0.59	1.75	0.07	0.449	0.118	0.351	0.014
15	8.31	3.10	2.19	0.18	0.328	0.122	0.086	0.007
Mean	**7.06**	**2.34**	**4.76**	**0.17**	**0.393**	**0.134**	**0.283**	**0.010**
Median	**4.64**	**1.49**	**2.58**	**0.09**	**0.367**	**0.139**	**0.272**	**0.010**
Max	**25.76**	**7.50**	**16.66**	**0.57**	**0.590**	**0.212**	**0.594**	**0.020**
Min	**1.57**	**0.23**	**0.59**	**0.02**	**0.221**	**0.055**	**0.086**	**0.005**

Mn, and Sc) were determined in order to show the basic geochemical compositions of the soils. Together with digestion and measurement of our soil samples, one reference soil (GSD7a) (triplicate), provided by Institute of Geophysical and Geochemical Exploration, Chinese Academy of Geological Sciences, was digested and analyzed to check the analytical quality. Average relative errors were 3.6 (1.9′5.8)%, 4.0(3.5~4.5)%, 3.9(3.2~4.7)%, 6.5(4.1~8.2)%, and −1.1(−6.2~2.0)% for Al, Fe, Mn, Sc, and W, respectively.

Portions of the root, stem, leaf, and grain samples were digested with concentrated HNO_3 and H_2O_2, and then concentrations of W in the extracts were measured with ICP-MS (X Series II, Thermo Electron).

In order to evaluate mobility and potential bioavailability of W, chemical forms of W in the soil samples were analyzed using a selectively sequential dissolution (SSD) method developed by Tessier et al. [33]. The chemical forms were operationally fractioned into the five phases: (1) soluble and exchangeable fraction (EXC), (2) bound to carbonate minerals (CARB), (3) bound to oxides (OX) (4) bound to organic matter (OM), and (5) residual fraction (RES). The selective sequential extractions were conducted in centrifuge tubes (polypropylene, 50 mL). After each successive extraction, separation was performed by centrifuging at 10,000 rpm for 10 min using a Xiang Yi centrifuge at room temperature. The supernatants were then removed with a pipette and filtered through a 0.45-μm membrane. The W concentration in each solution was determined using ICP-MS (X Series II, Thermo Electron). In this study, only the first four forms were measured and the contents of W in the residual fraction were calculated by total content minus sum of the first four forms.

Results and Discussion

Contents of mineral matrix elements and tungsten in the soil

Table 1 summarizes levels of pH, some mineral matrix elements and W in the soils collected in the agricultural fields near the W mines. Soil pH value ranged from 4.95 to 5.92, showing its acidic property. The mineral matrix elements Mn, Sc, Al, and Fe contents in the soil ranged from 68.8 to 718.8 mg kg^{-1}, 5.87 to 14.21 mg kg^{-1}, 2.47 to 7.69%, and 0.93 to 5.04%, respectively. The mean and median contents of Mn, Sc, Al, and Fe were generally lower than their background contents in the soils of the Jiangxi province, China, and world (except Sc) [34,35]. However, mean and median contents of W in the soil were 16.57 and 15.20 mg kg^{-1}, respectively; much higher than its background contents in the soils of the Jiangxi province, China, and world [34,35]. Therefore, the agricultural soil adjacent to W mines in Dayu county is contaminated by W.

In general, soils in the areas of W mining and/or smelting contained high W, e.g., 24.7–78.4 mg kg^{-1} for Mt Carbine Mine (Queensland, Australia) [14], 10 to 67 mg kg^{-1} for Fallon (Nevada, USA) [3], 1593 mg kg^{-1} for Carrock Fell Mine (Cumbria, UK), 920 mg kg^{-1} for Trumbull Old Mine (Connecticut, USA), 116 mg kg^{-1} for Devon Great Consols (Tmar, UK) [16], and 150 mg kg^{-1} for an industrial production site for W trioxide in Switzerland [19]. In the fallowland soil down the hill from the Höllberg W mine in Germany, W content ranged 15.9 to 101.4 mg kg^{-1} [18]. W content ranged from 3.99 to 43.65 mg kg^{-1} in the agricultural top soils adjacent to W mines in Dayu county, similar to the above mentioned contents.

Table 4. Relationship between the contents of W in plants (Y, mg W kg^{-1} dry weight) and in soils (X, mg kg^{-1}) for W mining field investigations and pot experiments with W-spiked soils.

Plant	Soil W conc.	Duration	Equation	R^2	n	Ref.
	mg/kg					
Field investigation						
Rice	3.99–43.65					This study
Root			Y = 0.5692X − 2.3728	0.9141	15	
Stem			Y = 0.1703X − 0.4797	0.9496	15	
Leaf			Y = 0.3404X − 0.8781	0.7178	15	
Grain			Y = 0.0121X − 0.0306	0.8161	15	
Taraxacum officinale	15.9–101.4					[18]
Root			Y = 0.1757X − 1.229	0.8898	7	
Leaf			Y = 0.0132X+0.1979	0.8259	7	
Trifolium pratense	18.2–97.7					[18]
Root			Y = 0.1059X+2.5793	0.6139	7	
Leaf			Y = 0.0121X+0.1885	0.7048	7	
Pot experiment with W-spiked soils						
Nothofagus menziesii	0–125	8 weeks				[26]
Root			Y* = 32.205X+182.28	0.9778	12	
Stem (woody)			Y* = 0.7051X+19.313	0.7509	12	
Leaf			Y* = 0.9234X+9.8788	0.9009	12	
Sunflower	0–3900	28 days				[20]
Root			Y = 1.7098X+157.46	0.9863	5	
Leaf			Y = 0.0384X+16.234	0.9633	5	
Ryegrass	0–1000	9 months				[2]
Shoot			Y = 0.1916X+11.267	0.9935	5	
Wheat	0–120	up to fruiting				[28]
Root			Y = 0.88X − 2.66	0.9716	6	
Shoot			Y = 17.50X − 204.44	0.9288	6	
Cowpea	0–25	up to fruiting				[29]
Root			Y = 0.73X+6.3152	0.9661	6	
Shoot			Y = 0.23X+2.7685	0.9609	6	

*mg W kg^{-1} ash weight, R^2: correlation coefficient, n: number of samples, Ref.: reference.

Chemical forms of W in the soils

Generally, W in the soils was mostly bound in the RES fraction (89 to 97% of total W), while 2.9 to 10.3% of total W was associated with soil organic matter (Table 2.). Tungsten associated with EXC, CARB, and OX fraction was generally less than 0.5% of total W.

Heavy metals in soils are present in different chemical forms with a wide variety of solubility or bioavailability. Partitioning of trace metals among various soil fractions in a given soil depends on the soil fractions, affinity to the metals, soil composition and fraction concentrations in the soils, metal concentration in soil solution, and soil pH etc. [36]. Researches on W chemical forms are extremely limited. Wilson and Pyatt [15] investigated W chemical forms in the acidic spoils in the vicinity of an abandoned W mine, showing about 99% of W in the RES fraction. However, in the calcareous soils in the vicinity of an ancient metalliferour mine in the Corbières Area (Southwestern France), W content ranged from 20.5 to 77.9 mg kg^{-1}, of which about 41.2%, 23.0%, 19.1%, and 16.8% were bound in the exchangeable and carbonates, Fe/Mn oxides, soil organic matter, and residual fractions, respectively [17]. Therefore, W is more readily mobilized under alkaline conditions [17,37]. In general, mobility and bioavailability of W might be relatively low in the acidic soils of Dayu county.

Contents of W in the rice plants

The contents of W in the root, stem, leaf, and grain ranged from 1.57 to 25.76, 0.23 to 7.50, 0.59 to 16.66, and 0.02 to 0.57 mg kg^{-1}, respectively (Table 3). The W content and its enrichment factor in rice follow the order: root > leaf > stem > grain. This order is similar to that for wild *Calluna Vulgaris*, *Taraxacum officinale*, and *Trifolium pretense* reported by Wilson and Pyatt [15] and Jiang et al. [18], i.e., W content in root > W content in leaf > W content in stem. In addition, content and enrichment in the root of *Sunflowers*, *Ryegrass(Lolium perenne)*, *Nothofagus menziesii*, *wheat*, and *cowpea* growing in W-spiked soils were also higher than those in their leaves [2,12,20,28,29]. The common range of W in terrestrial plants was <0.001 to 0.15 mg kg^{-1} [6]. For example, the winter wheat grain, the spring barley grain, the wild berry grown in Sweden contained, in average, 0.006, 0.005, <0.01 mg kg^{-1} of

Table 5. Correlation matrix among mineral matrix element, W content and chemical forms in soil, and W contents in rice root, stem, leaf, and grain.

	pH	Mn	Sc	Al	Fe	TW	EXC-W	CARB-W	OX-W	OM-W	RES-W	R-W	S-W	L-W	G-W
pH	1.000	0.341	0.406	0.384	0.130	0.242	0.302	0.330	0.247	0.277	0.238	0.156	0.283	0.149	0.238
Mn		1.000	0.914*	0.933*	0.915*	0.237	−0.095	0.031	0.118	0.077	0.248	0.181	0.174	0.058	0.256
Sc			1.000	0.982*	0.903*	0.255	−0.024	0.118	0.143	0.063	0.269	0.211	0.185	0.014	0.234
Al				1.000	0.897*	0.284	−0.023	0.115	0.188	0.101	0.297	0.263	0.240	0.078	0.280
Fe					1.000	−0.002	−0.291	−0.161	−0.142	−0.205	0.014	−0.010	−0.083	−0.127	0.042
TW						1.000	0.708*	0.768*	0.969*	0.933*	1.000*	0.956*	0.975*	0.847*	0.903*
EXC-W							1.000	0.969*	0.807*	0.861*	0.692*	0.768*	0.773*	0.736*	0.680*
CARB-W								1.000	0.849*	0.869*	0.755*	0.811*	0.811*	0.742*	0.776*
OX-W									1.000	0.964*	0.964*	0.958*	0.990*	0.856*	0.897*
OM-W										1.000	0.922*	0.922*	0.951*	0.870*	0.871*
RES-W											1.000	0.953*	0.971*	0.841*	0.901*
R-W												1.000	0.968*	0.927*	0.916*
S-W													1.000	0.879*	0.899*
L-W														1.000	0.911*
G-W															1.000

Note: R-W: root W content, S-W: stem W content, L-W: leaf W content, G-W: grain W content. * Significant at $p = 0.01$ level.

W, respectively [23,38]. The content of W in the Danish onion ranged from 0.006 to 0.039 mg kg^{-1}, with an average of 0.017 mg kg^{-1} [22]. Twenty-two crop materials from Iowa contained 0 to 0.35 mg kg^{-1} of W [39]. However, plants grown in the W-contaminate soils usually contain high content of W. The maximum W levels of 50.7 mg kg^{-1} were found in cabbage, 41.2 mg kg^{-1} in onions and 12.3 mg kg^{-1} in potatoes grown in the W-contaminated soil adjacent to the W trioxide production plant [19].

Rice is the main cereals food for the local residents in Dayu county. Assuming that local residents consume 0.4 kg rice per day per person, maximal dietary exposure to W for consumers of contaminated rice was estimated to be up to 0.23 mg W d^{-1} per person corresponding to the highest W content in rice grain. On the other hand, the exposure to W for the residents via soil ingestion was estimated to be 0.002 W mg d^{-1} per person corresponding to the highest W content in the soil, assuming that soil ingestion rate is 50 mg d^{-1} per person [40]. At an industrial production site for W trioxide in Switzerland, dietary exposure to W for consumers of contaminated vegetables (cabbage, onions, and potatoes) growing in garden was estimated to be up to 1.5 mg W d^{-1} per person compared to background dietary exposure of approximately 0.01 mg W d^{-1} per person [19]. Therefore, the major way of W exposure to the residents at W contaminated sites is the dietary ingestion. Although permissible exposure limit for W in workplace atmospheric environments in USA (5 mg m^{-3}) and drinking water in the former USSR (0.05 mg L^{-1}) were established, the threshold value for W content in any plant has not been established up to now from the viewpoint of food safety [2]. Tungsten health risk caused by the rice ingestion can not be quantitatively estimated at current stage.

The enrichment factors, defined as W contents in root, stem, leaf, and grain divided by its content in the soil, were 0.221 to 0.590 for root, 0.055 to 0.212 for stem, 0.086 to 0.594 for leaf, and 0.005 to 0.020 for grain. Previous studies showed the enrichment factor of W in the wild shrub shoots growing in the W mining areas (20.5–77.9 mg W kg^{-1} soil) was generally 0.01 to 0.8 [17]. The enrichment factors of W in the root and the leaf were 0.05–0.28 and 0.01–0.03, respectively, for *Taraxacum officinale* and *Trifolium pretense* growing in the fallowland soil (15.9 mg W kg^{-1} soil) down the hill from the Höllberg W mine in Germany [18]. Therefore, the W enrichment factor values in the rice root and leaf in this studied area were higher than those for the herb *Taraxacum officinale* and *Trifolium pretense*.

The contents of W in the root, stem, leaf, and grain of rice linearly increased with the content of W in the soil (Table 4), with the determination coefficients of 0.7178 to 0.9496. Several previous studies observed the dependence of plant W content on the soil W contents [3,18,28–30]. Previous experiments in field showed W contents in the roots and leaves of maize were significantly correlated with the W content in the soil adjacent to an old tungsten mine (Pechtelsgrün) in Germany [18]. In detail, W contents in the roots and leaves of maize increased from about 1 to 16 mg kg^{-1} and 0.4 to 1.5 mg kg^{-1}, respectively, with the increase of soil W content from about 20 to 100 mg kg^{-1} [18]. In addition, W concentrations of roots and leaves both of wild *Taraxacum officinale and Trifolium pratense* were significantly correlated with soil W concentration. And there was a significant correlation between root and leaf tungsten concentrations for both species [18]. Pot experiment in greenhouse also demonstrated that W contents in root and shoot of sunflowers, ryegrass, wheat, and cowpea showed a very strong linear dependence on the W content in the W-spiked soils [3,20,28,29] (Table 4).

Correlation among elemental contents, W chemical forms, and W contents in rice

The correlation matrix in Table 5 shows the relationships between elemental contents in soil, W chemical forms in soil, and W contents in rice. pH value was positively correlated to the contents of Mn, Sc, Al, Fe, W, W chemical forms, and W in rice, whereas these correlations are not significant at $p = 0.01$ level. In general, the mineral matrix elements Mn, Sc, Al and Fe in the soil were correlated to one another, but they were not correlated to W content and chemical forms in the soil and the W contents in the rice. This might further inferred that the Mn, Sc, Al, and Fe contents in the soil mainly originated from geogenic sources, while W contents in the soil and rice were impacted by W mining activities. Tungsten contents in the soil and rice and W chemical forms in the soil were correlated to one another. These correlations showed that W mining activities in the Dayu county not only increased the total W content in soil, but also increased the W contents in each chemical form and in rice.

Plants are known to take up (possibly in anionic form, WO_4^{2-}) and accumulate W in substantial amounts [2,25]. Whereas 89% to 97% of W in the studied soil was associated to the residual fraction, non-residual content of W was 0.23 to 3.58 mg kg^{-1}, similar to W content (1.50 mg kg^{-1}) in worldwide soils [35]. In addition, W content in the non-residual fractions was positively correlated to the total content of W in the studied soil. Therefore, the contents of W in rice were positively correlated to the total content of W in the soil. Previous studies also showed that the extent of W accumulation in plants appears to be directly related to the W content of the soils whereas chemical forms of W in the soils were not measured [2,3,18,20,26,28,29].

Conclusions

About 100 years' mining W activity significantly elevated W contents in the soil adjacent to the W mines of the Dayu county. The rice grown in the soil accumulated W, with the linear dependence of W contents in the rice root, stem, leaf, and grain on the soil W content and chemical forms. The enrichment factor of W decreased in the order: root, leaf, stem, and grain. Most of W (about 93% in average) the soil was associated to the residual fraction. Tungsten contents and chemical forms in the soil and W contents in the rice root, stem, leaf, and grain were correlated to one another, but not correlated to the mineral matrix elemental (Mn, Sc, Al, and Fe) contents, due to the soil W contamination by W mining activities. The W exposure to local habitants via dietary and soil should be assessed further.

Author Contributions

Conceived and designed the experiments: CL HC JW. Performed the experiments: RL XS. Analyzed the data: CL RL HC JW. Contributed reagents/materials/analysis tools: CL RL XS. Wrote the paper: CL RL HC JW. Sample collection: CL HC XS.

References

1. inmaus C, Lu M, Todd RL, Smith AH (2004) Probability estimates for the unique childhood leukemia cluster in Fallon, Nevada, and risks near other U.S. military aviation facilities. Environ Health Perspect 112:766–771.
2. Strigul N, Koutsospyros A, Arienti P, Christodoulatos C, Dermatas D, et al. (2005) Effects of tungsten on environmental systems. Chemosphere 61:248–258.
3. Koutsospyros A, Braida W, Christodoulatos C, Dermatas D, Strigul N (2006) A review of tungsten: From environmental obscurity to scrutiny. J Hazard Mater 136:1–19.
4. Strigul N (2010) Does speciation matter for tungsten ecotoxicology?. Ecotox Environ Safe 73:1099–1113.
5. Adamakis IDS, Panteris E, Eleftheriou EP (2012) Tungsten toxicity in plants. Plants 1:82–99.
6. Kabata-Pendias A, Mukherjee AB (2007) Trace Elements from Soil to Human. Springer, Berlin.
7. Fritsch C, Coeurdassier M, Giraudoux P, Raoul F, Douay F, et al. (2011) Spatially expicit analysis to biota: Influence of soil contamination and landscape. PLoS ONE 6(5): e20682. doi:10.1371/journal.pone.0020682.
8. Zheng SA, Zheng X, Chen C (2012) Leaching behavior of heavy metals and transformation of their speciation in polluted soil receiving simulated acid rain. PLoS ONE 7(11): e49664. doi:10.1371/journal.pone.0049664.
9. Zhu H, Guo J, Chen M, Feng G, Yao Q (2012) Burkholderia dabaoshanensis sp. nov., a heavy-metal-tolerant bacteria isolated from Dabaoshan mining area soil in China. PLoS ONE 7(12): e50225. doi:10.1371/journal.pone.0050225.
10. Jiang M, Zeng G, Zhang C, Ma X, Chen M, et al. (2013) Assessment of heavy metal contamination in the surrounding soils and surface sediments in Xiawangang River, Qingshuitang District. PLoS ONE 8(8): e71176. doi:10.1371/journal.pone.0071176.
11. Huang Z, Pan XD, Wu PG, Han JL, Chen Q (2013) Health risk assessment of heavy metals in rice to the population in Zhejiang, China. PLoS ONE 8(9): e75007. doi:10.1371/journal.pone.0075007.
12. Quin BF, Brooks RR (1972) The tungsten contents of some plants from a mineralized area in New Zealand. NZ J Sci 15:308–312.
13. Quin BF, Brooks RR (1974) Tungsten concentrations in plants and soils as a means of detecting scheelite-bearing ore-badies in New Zealand. Plant Soil 41:177–188.
14. Pyatt FB, Pyatt AJ (2004) The bioaccumulation of tungsten and copper by organisms inhabiting metalliferous areas in North Queensland, Australia: an evaluation of potential health implications. J Environ Health Res 3:13–18.
15. Wilson B, Pyatt FB (2006) Bioavailability of tungsten in the vicinity of an abandoned mine in the English Lake District and some potential health implications. Sci Total Environ 370:401–408.
16. Wilson B, Pyatt FB (2009) Persistence and bioaccumulation of tungsten and associated heavy metals under different climatic conditions. Land Contam Reclam 17:93–100.
17. Wilson B, Pyatt FB (2009) Bioavailability of tungsten and associated metals in calcareous soils in the vicinity of an ancient metalliferous mine in the Corbieres area, Southwestern France. J Toxicol Env Health A 72:807–816.
18. Jiang F, Heilmeier H, Hartung W (2007) Abscisic acid relations of plants grown on tungsten enriched substrates. Plant Soil 301:37–49.
19. Brueschweiler B, Waber U, Gupta S (2009) Tungsten, a new vegetable contaminant needs further elaborated evaluation. Toxicol Letters 189:S219.
20. Johnson DR, Inouye LS, Bednar AJ, Clarke JU, Winfield LE, et al. (2009) Tungsten bioavailability and toxicity in sunflowers (Helianthus Annuus L.). Land Contam Reclam 17:141–151.
21. Bowen HJM (1960) The determination of tungsten in biological material by activation analysis. Biochem J 77:79–82.
22. Bibak A, Behrens A, Stefan Stürup S, Knudsen L, Gundersen V (1998) Concentrations of 63 major and trace elements in Danish agricultural crops measured by inductively coupled plasma mass spectrometry. 1. Onion (Allium cepa Hysam). J Agric Food Chem 46: 3139–3145.
23. Rodushkina I, Ödmana F, Holmströmb H (1999) Multi-element analysis of wild berries from northern Sweden by ICP techniques. Sci Total Environ 231:53–65.
24. Davies EB, Stockdill SMJ (1956) A pasture response to sodium tungstate on a New Zealand soil. Nature 78:866.
25. Wilson DO, Cline JF (1966) Removal of plutonium-239, tungsten-185 and lead-210 from soils. Nature 209:941–942.
26. Quin BF, Brooks RR, Reay PF (1972) The accumulation of tungsten by Nothofagus menziessi. Plant Soil 36:699–703.
27. Quin BF, Hoglund JH (1976) The effect of tungstate and nitrogen source on the dry weight and nitrogen yields, and molybdenum and tungsten content, of white clover (Trifolium Repens). Plant Soil 45:201–212.
28. Kumar A, Aery NC (2011) Effect of tungsten on growth, biochemical constituents, molybdenum and tungsten contents in wheat. Plant Soil Environ 57:519–525.
29. Kumar A, Aery NC (2012) Effect of tungsten on growth, dry-matter production, and biochemical constituents of cowpea. Commun Soil Plant Anal 43:1098–1107.
30. Kennedy AJ, Johnson DR, Seiter JM, Lindsay JH, Boyd RE, et al. (2012) Tungsten toxicity, bioaccumulation, and compartmentalization into organisms representing two trophic levels. Environ Sci Technol 46:9646–9652.
31. Cheng HG, Duan XL, Liu SQ, Lin CY, Shao X (2013) Thallium, arsenic, and mercury contamination of soil near the world's largest and longest-operating tungsten mine. Polish J Environ Studies 22:301–305.
32. Lin C, He MC, Zhou YX, Guo W, Yang ZF (2008) Distribution and contamination assessment of heavy metals in sediment of the Second Songhua River, China. Environ Monit Assess 137:329–342.
33. Tessier A, Campbell PGC, Bisson M (1979) Sequential extraction procedure for the speciation of particulate trace metals. Anal Chem 51:844–851.
34. Wei FS, Chen JS, Wu YY, Zheng CJ, Jiang DZ (1990) Background contents of elements in China soils. Publishing House of Chinese Environmental Sciences, Beijing.
35. Bowen HJM (1979) Environmental chemistry of the elements. Academic Press, London.
36. Lin CY, Negev I, Eshel G, Banin A (2008) In Situ Accumulation of Copper, Chromium, Nickel, and Zinc in Soils Used for Long-term Waste Water Reclamation. J Environ Qual 37:1477–1487.
37. Bednar AJ, Boyd RE, Jones WT, McGrath CJ, Johnson DR, et al. (2009) Investigations of tungsten mobility in soil using column tests. Chemosphere 75:1049–1056.

38. Eriksson JE (2001) Concentrations of 61 trace elements in sewage sludge, farmyard manure, mineral fertilizers, precipitation, and in soil and crops. Swedish EPA Rep 5159.
39. Fu MH, Tabatabai MA (1988) Tungsten contents of soils, plants and sewage sludges in Iowa. J Environ Qual 17:146–148.
40. USEPA (2011) Exposure Factors Handbook: 2011 Edition. EPA/600/R-090/052F, www.epa.gov.

16

New Method for Automatic Body Length Measurement of the Collembolan, *Folsomia candida* Willem 1902 (Insecta: Collembola)

Oxána Bánszegi[1], András Kosztolányi[2], Gábor Bakonyi[3], Borbála Szabó[4], Miklós Dombos[1]*

1 Institute for Soil Sciences and Agricultural Chemistry, Centre for Agricultural Research, Hungarian Academy of Sciences, Budapest, Hungary, **2** MTA-DE "Lendület" Behavioural Ecology Research Group, Department of Evolutionary Zoology and Human Biology, University of Debrecen, Debrecen, Hungary, **3** Szent István University, Faculty of Agricultural and Environmental Science, Gödöllő, Hungary, **4** Szent István University, Faculty of Veterinary Science, Budapest, Hungary

Abstract

The collembolan, *Folsomia candida,* is widely used in soil ecotoxicology. In recent years, growth rate of collembolans has become as frequently used endpoint as reproduction rate in ecotoxicological studies. However, measuring collembolan body sizes to estimate growth rate is a complicated and time-consuming task. Here we present a new image analysis method, which facilitates and accelerates the body length measurement of the collembolan *Folsomia candida*. The new software package, called CollScope, consists of three elements: 1) an imaging device; 2) photographing software; 3) an ImageJ macro for image processing, measurement and data analysis. We give a complete description of the operation of the software, the image analyzing process and describe its accuracy and reliability. The software with a detailed usage manual is attached as Supplementary Material. We report a case study to demonstrate that the automated measurement of collembolan body sizes is highly correlated with the traditional manual measurements (estimated measuring accuracy 0.05 mm). Furthermore, we performed a dose-response ecotoxicity test using cadmium-sulfate by using CollScope as well as classical methods for size measurement. Size data measured by CollScope or manually did not differ significantly. Furthermore the new software package decreased time consumption of the measurements to 42% when tested on 35 animals. Consequently, methodological investigations performed in this study should be regarded as a recommendation for any other routine dose-response study where body growth is an endpoint.

Editor: Peter Shaw, Roehampton University, United Kingdom

Funding: This study was supported financially by MEDAPHON LIFE+ project (LIFE 08ENV/H/000292). The funders had no role in study design, data collection and analysis, decision to publish, or preparation of the manuscript.

Competing Interests: The authors have declared that no competing interests exist.

* E-mail: dombos.miklos@gmail.com

Introduction

Over the past three decades the interest in ecotoxicological effects of new chemicals on soil organisms has been continuous. Several ecotoxicological tests have become widely used and are now commercially available. Soil ecotoxicology investigates the effects of chemicals on soil organisms in the environment [1]. There are three standardized ISO tests and one OECD guideline to test chemicals on the collembolan *Folsomia candida* Willem 1902 [2–5]. These tests aim at assessing toxicity expressed in terms of effects on selected endpoints like survival, reproduction or behavior based on dose-response relations [1]. On collembolans, the reproduction inhibition test is the most widely used [3,6].

The protocol of the reproduction inhibition test characterizes the effects of a given toxic agent on the collembolan reproduction rate. This laboratory method is based on the direct visual identification of the number of offspring at the end of the test. The number of offspring may rise up to several thousand, therefore several automatized methods have been developed for this task [7–10].

Growth rate is a primary life-history characteristic, therefore it has also been suggested as another endpoint for toxicity tests. Reduced growth indicates the reduced fitness of the individual [11–13]. First, collembolans cannot reproduce until they gain a minimum weight [14]. Second, positive relationship was found between growth retardation and reduction in reproduction, probably due to lowered metabolic efficiency [15,16]. The importance of growth-studies have been demonstrated repeatedly as they were able to detect reduction in growth caused by several different toxic agents [16,17], including metals [15,18–21]. Fountain & Hopkin [13] presented a test system by which the effects of xenobiotics on growth were noticeable in as young as 7-day old collembolans. Even though the growth test is slightly less sensitive to most of the chemicals compared to the reproduction test, it requires much fewer animals and is more rapid [13,16,22].

However, manual measuring of the collembolan's length is time-consuming. For this process researchers initially used an ocular micrometer with magnification [23,24]. Later, the most common method to measure a collembolan's length was the methodology by Folker-Hansen *et al.* [17]. They introduced digital image processing equipment, which consisted of a stereo microscope mounted with a video camera. Pictures of the animals were displayed on the monitor while measurements were made manually using a digital pointing device. In the past few years digital cameras have become relatively cheap and free image analyzing software (e.g. ImageJ) can be used to measure the length of objects on the picture, so this method has become widely used. However, this method still requires a considerable amount of manual work in the laboratory. Collembola must be photographed first and then manually measured one by one on the images. According to the protocols, body length of adults is measured from the end of the posterior abdominal segment to the anterior margin of the head between the antennae, as described by Folker-Hansen *et al.* [17]. Since collembolans are soft bodied insects, they can be flattened or curved and some of their segments can be telescoped [25], the measurable length of the live animals may vary and as a consequence it is necessary to repeat both the taking of pictures and the measuring of individuals.

Our aim was to develop a digital imaging device and an image analysis software application linked to the procedure in order to measure the length of the living *F. candida* individuals automatically, which, at the same time, meets the accuracy demands of the body length measurement in ecotoxicological tests. We present here a new tool which can be used for constant monitoring of the collembolan growth throughout the exposure period in ecotoxicological experiments. We also report the accuracy of the method and its applicability in a standard ecotoxicological experiment.

Materials and Methods

Animals

F. candida belongs to Isotomidae family and is an eudaphic, non-pigmented, eyeless springtail with a parthenogenetic reproduction pattern.

Folsomia candida Willem 1902 (Collembola) was chosen as a model animal for our study because this species is widely used in soil ecotoxicology. We obtained a standard breed from the soil ecological laboratory of the Institute for Soil Sciences and Agricultural Chemistry, Centre for Agricultural Research, Hungarian Academy of Science. Collembolans were kept in 8 cm tall boxes with a base of 12×7 cm in constant darkness. 0.5 cm thick layer of plaster of Paris mixed with charcoal was added to the bottom of the boxes. The animals were fed 2 mg baker's yeast once a week. The culturing boxes were opened once a week for airing and moistening with distilled water.

Automatic Image Processing

Our software package consists of three elements: 1) the imaging device comprised of a digital camera with a lens and a light source; 2) software written in Delphi for automatizing the photographing; 3) an ImageJ macro for image processing, measurement and the data analysis. The whole method is called CollScope.

1) Imaging-CollScope Imaging Device

The imaging device consists of a 3.3 MP color CCD camera (TCC-3.3ICE-N, ICX412AQ Sony CCD), a Tamron Mega-pixel Machine Vision CCTV lens (23FM25SP Focal length 25 mm, Aperture 1.4), a PMMA sample holder with a view field of 48 mm×37 mm, and uniform white backlight illumination. The resolution of the instrument can be characterized as one pixel corresponding to 24.6 micrometers * 24.6 micrometers. The microscope was enclosed in anodized aluminum housing to avoid the influence of external light (Figure 1). Sample holder position as well as lighting of the device needed to be completely stable during photographing; any changes resulted in false imaging and analysis.

Figure 1. The CollScope imaging device. The device consisting of a 3.3 MP color CCD camera, a Tamron Mega-pixel Machine Vision CCTV lens, a sample holder with a view field of 48 mm×37 mm (bottom right) and uniform white backlight illumination.

2) Photographing the Animals–"CollScope Photographing" Software

The insects were removed from their boxes with an insect vacuum than were inserted into the PMMA plate, which was placed into the drawer of the microscope. Each plate contained one living individual of *F. candida*. All measurements were carried out in room temperature between 21–25°C. For this capturing process we developed a Delphi software, which simplified the image capturing and file storing processes. The software and its instructions can be found in Supplement S1. The software automatically captures ten consecutive images of each collembolan individual over a 30 second period in 3 second intervals and stores the images in JPEG format.

3) Analysis of Images–"CollScope Image Analysis" Software Application

In the image analysis process a major challenge was to distinguish the individuals from other, often overlapping particles like sand, excrement, vegetal particles etc. mixed in the field of view. In our method individuals are identified based on their movements during visualizing. In general, everything that changed position during the 30 seconds of image capture was treated as a live individual and everything else as background or noise that had to be excluded from the evaluation. During the analyses a pixel-extraction method was conducted resulting in patches corresponding to the moving animal. For this image analysis we wrote a plug-in for the open source image processing package ImageJ (version 1.48s; http://rsbweb.nih.gov/ij). Details of the plug-in and the plug-in can be found in Supplement S2. In short, the plug-in operates with differences of images. First, the difference between the current and previous image and the difference between the current and the next image is calculated, and then the two resulting difference images are used to create a final secondary difference image. Thus, out of the series of ten images, eight secondary difference images are created. On the secondary difference images, the displaced patches are in red, while the unchanged parts are in uniform gray. That is, if the animal moved, then it is a red patch on the image. Furthermore, the shots from the previous and next images are also showed in dark grey (Figure 2). These secondary difference images can then be thresholded and evaluated using the particle analysis tool of the image processing package ImageJ. Only particles within the 0.01 mm^2–10 mm^2 range are included in the image analysis. Although this plug-in calculates several parameters concerning the shapes of the patches, we used only the "Major" (M) to estimate the body length of the animals. This parameter corresponds to the major axis of the ellipse fitted to each patch of the picture. The result is exported into a simple text file. The file contains all data obtained from the eight secondary difference images. The table contains data calibrated in mm units; each row corresponding to one patch (see Supplement S2 for how to calibrate the system). Since there are eight secondary difference images for each animal, on average eight data rows can be obtained for each individual. However, there will be more patches and more M data if animals do not move sufficiently and therefore their patches overlap and their patches will be split, or there can be fewer, in case the animals do not move at all between images. If the animals do not move at all during the photographing, then no secondary difference images would be produced. However, since collembolans could be forced to move by air blowing, in our case this apparent drawback of the methodology did not cause serious difficulty.

4) Data Analysis–"CollScope Data Analysis" Software Application

Determining the accuracy and precision of the method. The raw M data stored in the text file required additional data processing, since they may contain errors occurring for two reasons: (a) inappropriate photographing or (b) biased length estimation from the patch.

(a) If the animal does not move sufficiently between two photographs, then only a part of its full size will appear on the secondary difference image. This means that the corresponding M parameter values obtained from these smaller spots will be lower, and the distribution of M values will be skewed. Furthermore, animals that are shot in an inappropriate position e.g. bending in a crescent or having a contracted shape will result also in a lower M value.

(b) As the plug-in fits an ellipse to the patches that takes into account the protruding appendages as well, we will always get a higher value for the larger diameter of this ellipse (i.e. the parameter M) compared to the body length measured between the head and abdomen manually.

First, errors caused by inappropriate photographing are corrected by data filtering. Since animals rarely stretched, the larger M parameter values are the more precise ones. Therefore the filtering process is based on the maximum value of the obtained M figures. We cannot use simply the maximum itself since collembolans are soft bodied and weakly sclerotized animals that do not have a stable body length. Therefore, we have to work with averaged data regardless whether an automatic or manual method is used. In this filtering process we determined a critical value below which we ignored the M data.

We analyzed images of 50 collembolans. Collembolans were randomly chosen from different rearing boxes; however, we tried to measure individuals of differing sizes. We analyzed the images using the plug-in, and then classified manually the original images whether the animal appeared in a regular or irregular position. We determined the critical value for the photographs of each individual as the lowest M value that corresponded to an animal in a regular position (i.e. in a position that would be regarded as appropriate in case of manual measurement). We divided this critical value by the highest M value of the individual. This ratio gives the critical ratio below which data are discarded and excluded from further analysis.

Second, the extent of the bias coming from the ellipse fitting was determined. In the sample of another 50 individuals, the length of the animals was manually measured on the first picture considered acceptable from the series of 10 pictures using the classic method (manual measuring of the animal's body length on the screen from the end of the posterior abdominal segment to the anterior margin of the head between the antennae). The manual measurement was compared to the body length estimated by the M value of the corresponding secondary difference image received after the CollScope data analysis procedure, as described above. The relationship between the two variables was determined by linear regression, which gave a calibration equation. The estimated body length of an individual obtained by our method is the average of the filtered and calibrated M values. This part of the analysis also included in the ImageJ macro file.

Third, we tested the accuracy of our method using yet another 50 animals. In this test we measured the body length of animals both manually and automatically and compared the data coming from these two measurement procedures. For the classic manual measurements the first three pictures were used where the animal was in regular position and the measured body lengths were averaged. For the ScollScope method all 10 images were used.

Application of the Device in an Ecotoxicological Test

A standard growth test on *F. candida* following Crouau and Moïa [16] was conducted, except that they used artificial soil, while in our experiment the animals were kept on plaster of Paris mixed with charcoal. Briefly, the effect of cadmium-sulphate ($CdSO_4$) on *F. candida* growth was studied using synchronized animals. On day 0, 10–12-day old juvenile collembolans were introduced individually into plastic Petri dishes of 9 cm in diameter containing plaster of Paris mixed with charcoal on the bottom. Each dish was opened once a week for airing and moistening with distilled water. Collembolans were fed with baker's yeast (about 2 mg per week). We used a control, and three nominal cadmium concentrations as follows: 0; 76.5; 153; 306 mg/kg dry soil $CdSO_4$ (Sigma-Aldrich, purity ≥99%). Plaster of Paris was moistened with $CdSO_4$ solution such that about 100% moisture content was achieved in the Petri dish. Each concentration had 15 replications resulting in 60 Petri dishes in all. Body length was chosen as an endpoint of the experiment. Measurements were taken on day 0, 7 and 14. Collembolans were measured manually as well as applying the new CollScope procedure at each sampling time.

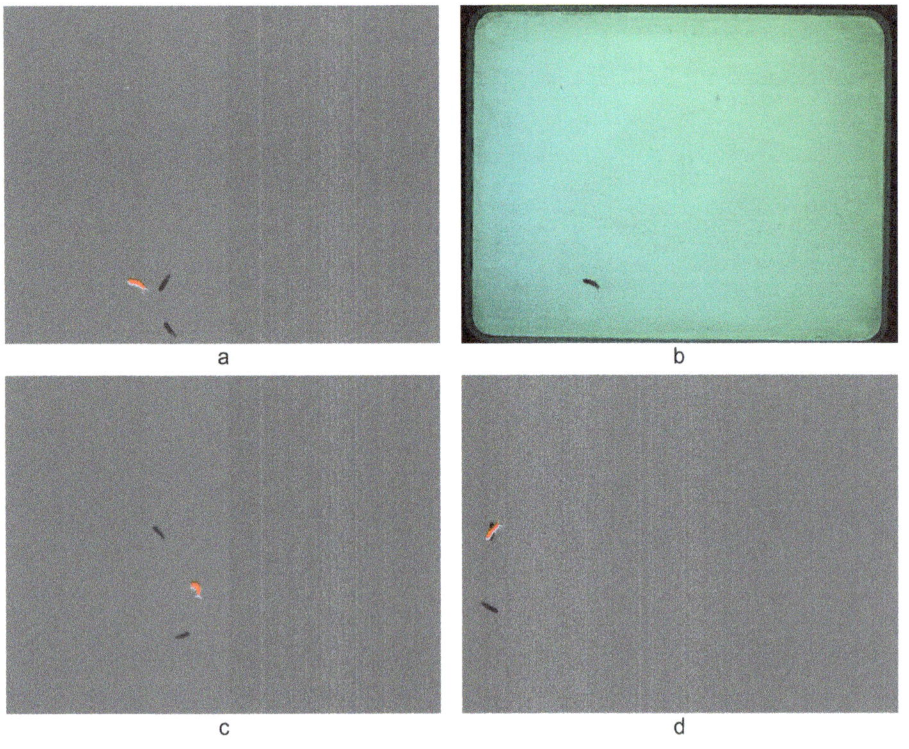

Figure 2. Secondary difference images produced by the CollScope image analysis software application. In the resulting pictures individuals appear in three representations: red patches correspond to the animal's spot involved in further calculations, while dark grey patches show its previous and next positions (a). Original photo (b), Collembolan bending in a crescent on the image or split during the image processing (c and d).

We had repeated measurements from each individual, however, because of dropouts (6.7%, 6.7%, 13.3% and 53.3% of the individuals died during the experiment in the control, 76.5, 153 and 306 mg/kg cadmium concentration groups, respectively), the number of measurements varied between individuals. Therefore, we analyzed the data by linear mixed-models (LMMs) using the nlme package in R to account for the non-independence of repeated measurements. The effect of measurement method and cadmium treatment on growth over time was investigated in random intercept LMMs with individual identity as grouping factor in the models to control for the repeated measurements. In the models, the response variable was body length, whereas the fixed variables were the measurement method as factor (manual or CollScope), treatment as factor (control, 76.5, 153 and 306 mg/kg cadmium concentration) and day of measurement as factor (0, 7, 14 day). We also included treatment×day interaction term to investigate whether the growth over time was similar in the treatment groups. Models were fitted with maximum likelihood method and the effect of fixed variables was tested by likelihood ratio tests (LRTs). As there was a strong interaction between treatment and day of measurement (see results), we further analyzed the effect of treatment for each measurement day separately by one-way ANOVAs. Because the measurements of Manual and CollScope method was not different (see results), in the ANOVAs the measurements of CollScope method were used.

Results

In the first study, from the 50 collembolans 408 M data were produced by Collscope Image analysis instead of the expected 400. On average, we obtained 8.16±1.54 (mean ± SD) M data for an individual. 285 out of the 408 pictures were evaluated as appropriate shoots, that is, the animals' movements were sufficient and animals were pictured in a straight body position on these images. On average, for each animal 5.7±1.36 (mean ± SD) M data were classified as appropriate. The mean critical ratio obtained was 0.939±0.027 (mean ± SD from the 50 animals). That is M data less than 93.9% of the highest M value obtained from image analysis of the eight secondary difference images has to be omitted from the determination of body length. The 93.9% threshold value was built into the CollScope data analysis.

The regression of the manually measured body lengths of the second set of 50 individuals to the Major value gave the calibration equation: body length = 0.930×Major −0.070 ($R^2 = 0.98$). The negative intercept indicates that the ellipse fitting slightly overestimated the true body length, whereas the slope smaller than one indicates that the bias is size dependent (larger individuals were overestimated more, Figure 3). This calibration equation was built into the ImageJ macro file. The result of the third study shows that the matching between the classic manual measurement and our newly developed method was significantly correlated (Pearson correlation: $r = 0.99$; $n = 50$; $p<0.001$) (Figure 4). In the ecotoxicological test no collembolan mortality was observed in the control group on day 7 and it was 6.7% at the end of the experiment. Consequently, experimental conditions can be regarded as appropriate during the experimental period.

Body length measurements did not differ between the two methods over the experiment (Figure 5, LRT, $\chi2 = 1.36$, $df = 1$, $p = 0.244$). There was a very strong interaction between treatment and day of measurement (Figure 5, LRT, $\chi2 = 396.84$, $df = 6$, $p< 0.001$): there was no difference between treatment groups at the beginning of the experiment (day 0: ANOVA, $F_{3,56} = 0.29$,

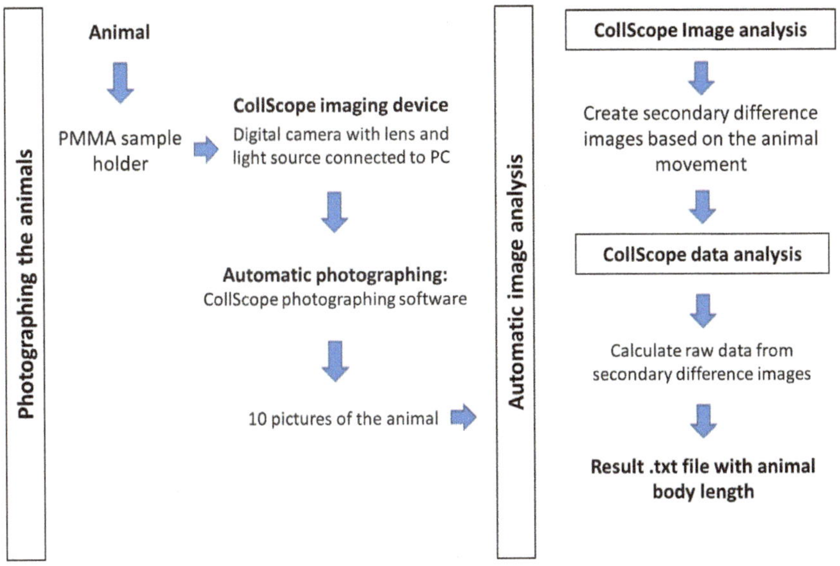

Figure 3. Flow chart of the CollScope method.

p = 0.833), whereas body lengths were significantly different on days 7 and 14 (ANOVAs, day 7:, $F_{3,47} = 58.27$, $p<0.001$; day 14: ANOVA, $F_{3,44} = 59.95$, $p<0.001$) (Figure 6).

Discussion

In the first study we developed a data filtering process and a calibration equation to estimate the body length of *F. candida*. Optimally, 400 M data were to be obtained for the 50 collembolans (8×50), however, 408 M data were produced. These extra 8 data must be because either some individuals did not move sufficiently during the photographing therefore their secondary images showed two or more patches (increase in the number of M data), or the animals did not move at all between three images and their secondary images contained no patches (decrease in the number of M data). Since on average we obtained out of the 8 M data generated for one collembolan 5.7 M data appropriate for analysis, we can conclude that the 3-second time interval for photographing was usually sufficient for the animals to completely change their position, which is essential for proper image analysis.

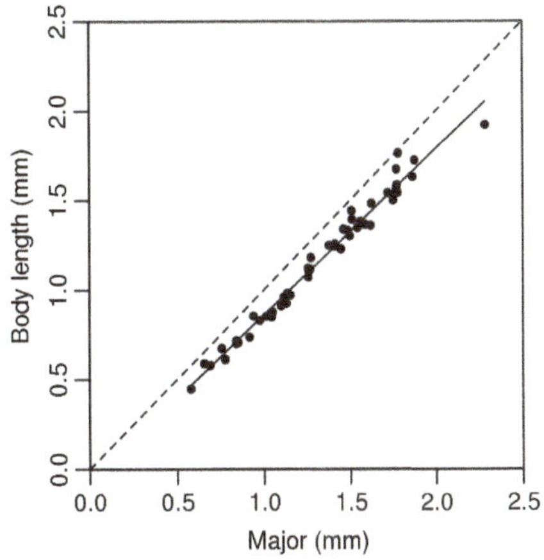

Figure 4. Relationship between the M values obtained from the image analysis and body length measured manually. The broken line corresponds to y = x. The continuous line shows the prediction used in the calculations (n = 50 individuals).

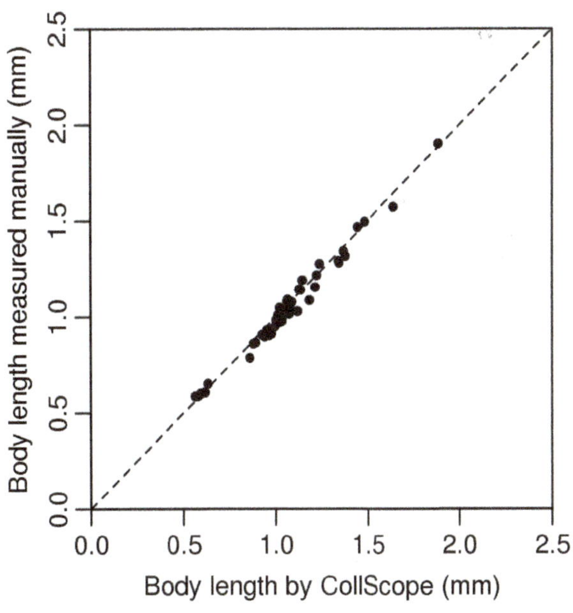

Figure 5. Accuracy of the CollScope method. Relationship between the body length estimated by CollScope and measured manually (n = 50 individuals). The broken line corresponds to y = x.

In the third study, we showed that the body length values measured with the new image analysis method and those of the classical manual method were highly correlated. The results show that our newly developed digital imaging device with the image analysis software can be used for the accurate measurement of living collembolan body length.

Measuring the growth of individuals as an endpoint in ecotoxicological studies is increasingly widespread. Improvement in the measurement of the body length of the animals both in accuracy and

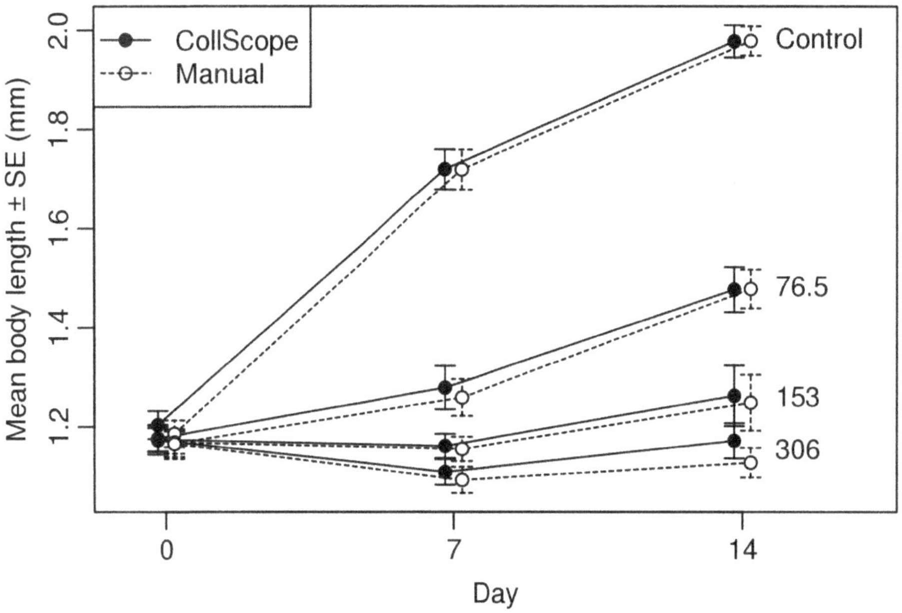

Figure 6. Body length of *F. candida* (mean ± SE) exposed to different $CdSO_4$ concentrations at sampling days 0, 7 and 14. Open circles correspond to manual measurements, closed circles to CollScope measurements.

speed has long been needed [16]. As presented here, using the CollScope software package the same degree of accuracy can be reached as with the classic manual measurement. However, it is less time-consuming, since the photographing method, selection of appropriate images and the measurement itself are each automatized.

By using the new CollScope software package we were able to reduce time period of the measurement to 42% (n = 35 individuals). In our test the minimum period of time necessary for the apparatus to take photos was 30 seconds, which can be modified according to the number of pictures set in the CollScope Shooting Software. If the animals are more or less mobile the time intervals between two pictures can be adjusted, as well. In addition to that the phase of putting the animals from the culture pot to the plate of the CollScope digital microscope and vice versa remained as the only time consuming and human involved part that lead to a method less sensitive to human error.

Although the growth of *Folsomia candida* has a lower sensitivity in toxicity tests than reproduction [20,21], it has become a complementary variable [13,16]. In a standard ecotoxicological study the endpoints are mortality and/or reproduction. A major disadvantage of reproduction tests is that the effects of the toxic agent will only be seen at the end of the exposure period, which is usually 28 days in the case of *F. candida* [3]. In contrast, some metal effects on body growth can be observed as early as 7 days after the beginning of the experiment [13]. Our results confirm the previous findings, that cadmium-sulphate reduces growth by day 7. And this result is consistent with the results found by Crommentuijn et al. [26].

Reproduction an important biomarker for the level of toxicity of a compound, however, in some cases, growth may be more sensitive and, therefore, more useful to provide an early warning system [17]. Van Straalen *et al.* [27] found in female *Orchesella cincta* that cadmium reduces the growth of individuals without affecting their reproduction. Toxic agents can influence an animal's fitness in several ways, some of the effects are more quickly measurable others are not. Hence, it is suggested to always consider a number of endpoint variables. Additionally, the decrease of the density in a population can occur not just via the lack of the production of juveniles, but via the impaired growth of the juveniles, which delays egg production [14,18,23,24].

We also demonstrated the application of the method in a standard ecotoxicology test that confirmed that the CollScope software package can be used in these types of experiments. Body length measurement has been performed on the same individual with both methods and the results showed no significant differences between the methods. This new method can be used for constant monitoring of Collembolan growth throughout the exposure period in ecotoxicological experiments.

Our methodology for body length measurements of *F. candida* by automatized processes provides a less labor-intensive way to carry out ecotoxicity tests and further ecological investigations. This methodology can be set up by end-users using a digital camera mounted on a stereo microscope; or by our professionals. Software can be downloaded for free. In this way, these ecotoxicological and ecological tests become time-effective.

Acknowledgments

Thanks to Tibor Hambek (managing director, Deákdelta Ltd.) and Tibor Virág (managing director, Kokabura Ltd.) for developing the equipment and the software. We are indebted to Paul Henning Krogh for his advice at the start of the project. We thank Robyn Hudson for suggestions regarding the manuscript.

Author Contributions

Conceived and designed the experiments: OB AK GB BS MD. Performed the experiments: OB AK GB BS MD. Analyzed the data: OB AK GB BS MD. Contributed reagents/materials/analysis tools: OB AK GB BS MD. Wrote the paper: OB AK GB BS MD.

References

1. van Gestel CAM (2012) Soil ecotoxicology: state of the art and future directions. ZooKeys 176: 275–296.
2. OECD (2009) OECD Guidelines for the testing of Chemicals 232. Collembolan reproduction test in soil. 232. Paris: Organisation for Economic Co-operation and Development.
3. ISO (1999) Soil quality–Inhibition of reproduction of Collembola (Folsomia candida) by soil pollutants. ISO 11267. Geneva: International Standardization Organization.
4. ISO (2006) Soil quality–Sampling of soil invertebrates – Part 2: Sampling and extraction of micro-arthropods (Collembola and Acarina). ISO 23611-2: 2006. Geneva: International Standardization Organization.
5. ISO (2011) Soil quality–Avoidance test for determining the quality of soils and effects of chemicals on behaviour–Part 2: Test with collembolans (Folsomia candida). ISO 17512-2: 2011. Geneva: International Standardization Organization.
6. Filser J, Wiegmann S, Schröder B (2013) Collembola in ecotoxicology–Any news or just boring routine? Appl Soil Ecol. Available: http://dx.doi.org/10.1016/j.apsoil.2013.07.007.
7. Caridade CMR, Marcal ARS, Mendonca T, Natal-da-Luz T, Sousa JP (2011) Automatic counting the number of Collembola in digital images. 4th International Congress on Image and Signal Processing: IEEE. 1837–1841.
8. Mallard F, Le Bourlot V, Tully T (2013) An Automated Image Analysis System to Measure and Count Organisms in Laboratory Microcosms. PloS One 8: e64387.
9. Marçal ARS, Caridade CMR (2006) A system for automatic counting the number of collembola individuals on petri disk images. In: Campilho A, Kamel M, editors. Image Analysis and Recognition: Springer. 814–822.
10. Krogh PH, Johansen K, Holmstrup M (1998) Automatic counting of collembolans for laboratory experiments. Appl Soil Ecol 7: 201–205.
11. Scheu S, Simmerling F (2004) Growth and reproduction of fungal feeding Collembola as affected by fungal species, melanin and mixed diets. Oecologia 139: 347–353.
12. Hopkin SP (1997) Biology of the Springtails (Insecta: Collembola). New York: Oxford University Press. 334 p.
13. Fountain MT, Hopkin SP (2001) Continuous Monitoring of Folsomia candida (Insecta: Collembola) in a Metal Exposure Test. Ecotoxicol Environ Saf 48: 275–286.
14. Crommentuijn T, Brils J, VanStraalen NM (1993) Influence of Cadmium on Life-History Characteristics of Folsomia candida (Willem) in an Artificial Soil Substrate. Ecotoxicol Environ Saf 26: 216–227.
15. Smit CE, van Gestel CAM (1997) Influence of Temperature on the Regulation and Toxicity of Zinc in Folsomia candida (Collembola). Ecotoxicol Environ Saf 37: 213–222.
16. Crouau Y, Moïa C (2006) The relative sensitivity of growth and reproduction in the springtail, Folsomia candida, exposed to xenobiotics in the laboratory: An indicator of soil toxicity. Ecotoxicol Environ Saf 64: 115–121.
17. Folker-Hansen P, Krogh PH, Holmstrup M (1996) Effect of dimethoate on body growth of representatives of the soil living mesofauna. Ecotoxicol Environ Saf 33: 207–216.
18. Scott-Fordsmand JJ, Krogh PH, Hopkin SP (1999) Toxicity of Nickel to a Soil-Dwelling Springtail, Folsomia fimetaria (Collembola: Isotomidae). Ecotoxicol Environ Saf 43: 57–61.
19. Nursita AI, Singh B, Lees E (2005) The effects of cadmium, copper, lead, and zinc on the growth and reproduction of Proisotoma minuta Tullberg (Collembola). Ecotoxicol Environ Saf 60: 306–314.
20. Bur T, Crouau Y, Bianco A, Gandois L, Probst A (2012) Toxicity of Pb and of Pb/Cd combination on the springtail Folsomia candida in natural soils: Reproduction, growth and bioaccumulation as indicators. Sci Total Environ 414: 187–197.
21. Bur T, Probst A, Bianco A, Gandois L, Crouau Y (2010) Determining cadmium critical concentrations in natural soils by assessing Collembola mortality, reproduction and growth. Ecotoxicol Environ Saf 73: 415–422.
22. Xu J, Ke X, Krogh PH, Wang Y, Luo Y, et al. (2009) Evaluation of growth and reproduction as indicators of soil metal toxicity to the Collembolan, Sinella curviseta. Insect Sci 16: 57–63.
23. Bengtsson G, Gunnarsson T, Rundgren S (1983) Growth changes caused by metal uptake in a population of Onychiurus armatus (Collembola) feeding on metal polluted fungi. Oikos 40: 216–225.
24. Tranvik L, Bengtsson G, Rundgren S (1993) Relative abundance and resistance traits of two Collembola species under metal stress. J Appl Ecol 30: 43–52.
25. Staempfli C, Tarradellas J, Becker-van Slooten K (2007) Effects of dinoseb on energy reserves in the soil arthropod Folsomia candida. Ecotoxicol Environ Saf 68: 263–271.
26. Crommentuijn T, Doodeman CJAM, Doornekamp A, Van Gestel CAM (1997) Life-table study with the springtail Folsomia candida (Willem) exposed to cadmium, chlorpyrifos and triphenyltin hydroxide. In: van Straalen NM, Løkke H, editors. Ecological risk assessment of contaminants in soil. London: Springer US. 275–291.
27. van Straalen NM, Schobben JHM, de Goede RGM (1989) Population consequences of cadmium toxicity in soil microarthropods. Ecotoxicol Environ Saf 17: 190–204.

17

Investment in Seed Physical Defence Is Associated with Species' Light Requirement for Regeneration and Seed Persistence: Evidence from *Macaranga* Species in Borneo

Pimonrat Tiansawat[1*¤], Adam S. Davis[2], Mark A. Berhow[3], Paul-Camilo Zalamea[4], James W. Dalling[1,4]

1 Department of Plant Biology, University of Illinois, Urbana, Illinois, United States of America, **2** Global Change and Photosynthesis Research Unit, United Department of Agricultural – Agricultural Research Service, Urbana, Illinois, United States of America, **3** National Center of Agricultural Utilization Research, United Department of Agricultural – Agricultural Research Service, Peoria, Illinois, United States of America, **4** Smithsonian Tropical Research Institute, Ancon, Republic of Panama

Abstract

The seed stage is often critical in determining the regeneration success of plants. Seeds must survive an array of seed predators and pathogens and germinate under conditions favourable for seedling establishment. To maximise recruitment success plants protect seeds using a diverse set of chemical and physical defences. However, the relationship between these defence classes, and their association with other life history traits, is not well understood. Data on seed coat thickness and fracture resistance, and the abundance and diversity of potential defensive compounds were collected for 10 tree species of *Macaranga* from Borneo. The data were used to test whether there is a trade-off in physical versus chemical defence investment, and to determine how investment varies with seed mass, and light requirement for regeneration. Across species there was no correlation between seed coat thickness and abundance of potential defensive compounds, indicating the absence of a direct trade-off between defence classes. While chemical defences were not correlated to other traits, physical defences were positively correlated with light requirement for regeneration. For a subset of five *Macaranga* species we evaluated the relative investment in chemical and physical defence to seed persistence in the soil, measured as the time to half initial seed viability (seed half-life). Half-life was negatively related to the ratio of potential defensive compound abundance to seed coat thickness, suggesting that species with long persistence invested in physical defence more than stored chemical defences. These results indicate that investment in seed defences are associated with species' light requirements for regeneration, rather than scaling positively with seed mass. Furthermore, chemical defences, although highly variable among species, do not appear to be critical to long term persistence of *Macaranga* seeds, and may be important in defending seeds from natural enemies distinct from those found in the soil.

Editor: Jin Chen, Key Laboratory of Tropical Forest Ecology, Xishuangbanna Tropical Botanical Garden, Chinese Academy of Sciences, China

Funding: The funding for this work was provided by the Center of Tropical Forest Science (2010 Research Grant) to PT, the Royal Thai Government to PT and The National Science Foundation Grant 1120205 to JWD and ASD. The funders had no role in study design, data collection and analysis, decision to publish, or preparation of the manuscript.

Competing Interests: The authors have declared that no competing interests exist.

* E-mail: tiansawat@yahoo.co.th

¤ Current address: Department of Biology, Chiang Mai University, Chiang Mai, Thailand

Introduction

Natural enemies strongly influence plant population growth, plant traits, and the local and regional distribution of plant species [1], [2], [3]. Many of these effects are most strongly exerted at the seed stage [4], [5]. Since the seeds of most plant species need to contain sufficient nutrients to support seedling establishment, they represent rich resources that are consumed by a wide range of animals [6], [7], [8] while also providing a substrate for microbial infection in the soil [9], [10], [11]. As a consequence, maternal resources are allocated not only to provisioning seeds, but also to the protection of seeds from predators and pathogens [12], [13].

Seed defences are often sub-divided into traits that provide physical or chemical protection. These protective traits are usually associated with seed-enclosing structures (e.g., pericarp and testa; [14], [15]). Physical defences are provided by the thickness and fracture resistance of enclosing structures that protect against penetration by insect predators [16], or seed rupture when passed through the jaws and digestive tracts of dispersers [14]. However, while physical defences can provide effective mechanical protection against vertebrates and invertebrates) [17], [18] seed-protecting structures may still be consumed by predators and seeds may be susceptible to pathogen infection [19]. Therefore, to discourage predators and pathogens seeds may also develop other means of protection based on chemical defences.

Plants contain many classes of secondary compounds that potentially serve as chemical defences against predators and pathogens [20], [21], [22]. In seeds, several classes of compounds such as phenolics, and alkaloids have been reported to be to be effective in protecting against insects [23], molluscs [24], vertebrates [25], and fungal and bacterial pathogens [19], [20], [26]. In *Macaranga beccariana*, potential stored defensive compounds in a class of dihydroxyphenols such as gentisic acid, protocatechuic acid, and gamma-resorcylic acid have been found, especially in the seed coat (Tiansawat, unpublished data). The presence of potential defensive compounds in a hard seed coat suggests that *Macaranga* species invest their resources in both physical and chemical protection of their seeds.

Many theories have been developed to explain the distribution of defences across plant structures and among plant species on the basis that defences are costly to produce [27], [28], [29]. Consequently, the allocation of resources to defences is hypothesized to be subjected to trade-offs such that high investment in one defence type should reduce investment in others with similar effectiveness [30], [31], [32]. However, most studies that have explored trade-offs in plant defences have focused on the leaves of seedlings and adult plants (e.g., [30], [31], [32], [33], and it remains unclear whether a trade-off exists between physical and chemical defensive traits in seeds.

In addition to undergoing trade-offs, allocation to seed defences may also be habitat specific. In general, seedlings and adults of light-demanding species are characterized by a low investment in defensive traits relative to shade-tolerant species [2], [27], [34]. However, it is unclear whether defence allocation patterns post-germination also extend to seeds. Light-demanding species often form persistent soil seed banks that require seeds to survive long exposure to post-dispersal seed predators and pathogens [35], [36]. Therefore, prolonged seed persistence may instead be associated with higher investment in defence.

The relative investment in physical and chemical defences associated with prolonged seed persistence remains unclear. Previous work with 80 native British species and six neotropical pioneer species has shown that longer persistence in the soil is associated with increased investment in chemical protection [37], [38]. In contrast, the most persistent temperate weed species allocate relatively more resources to physical defence than those species with short persistent seed banks [39], [40]. These contrasting results suggest complex relationships may exist linking seed size, persistence and seed chemical and physical defences, and may also vary among ecosystems and phylogenetic clades.

In this study, three questions were asked regarding the distribution of seed defences among species: First, does a trade-off exist among species between the physical defences and the quantity of potential defensive compounds? Second, does seed mass variation underlie interspecific differences in seed physical and chemical defences? We hypothesized physical defences should become more effective as seed mass increases, because the absolute thickness of seed protecting structures will scale positively with seed mass even if investment in physical defence is unchanged. Furthermore, if defence investment is subject to trade-offs, then larger-seeded species will be less dependent on chemical defences than smaller-seeded ones.

Third, does light requirement for regeneration influence the distribution of seed physical and chemical defences among species? We hypothesized that light-demanding species that form persistent seed banks will have greater investment in both physical and chemical defences than shade-tolerant species that germinate soon after dispersal. Alternatively, if physical damage is a primary cause of seed loss, then the half-life of seeds in the soil and the ratio of chemical to physical defences will be negatively correlated, indicating species that persist longer in the soil will invest more in physical defence relative to chemical defence.

Materials and Methods

Study species and site

In ever-wet paleotropical forests, the genus *Macaranga* serves as a unique system for studying the relationships among seed defence, habitat requirements and seed persistence. For the 10 Macaranga species included in this study seed mass ranged from 1.72 to 64.09 mg, and in habitat preference from shade-tolerant shrubs to light-demanding trees. Mature fruits of 10 *Macaranga* species (Table 1) were collected between July 2009 – September 2011 from primary and secondary forests of Lambir Hills National Park (hereafter Lambir), Sarawak, Malaysia (NW Borneo) (4°12' N, 114°02' E) (Permit number NCCD.907.4.4(V)-97 and NCCD.907.4.4(Jld.VI)-84). The mean temperature is 26°C. Lambir receives annual rainfall ranging between 2100 and 3300 mm [41] with all months averaging >100 mm [41]. Lambir has no clear seasonality; however, unpredictable short dry spells (<1 month) often occur throughout the year [42].

Light requirement for regeneration and seed collection

The light requirement of species was represented by the crown illumination index (CI index), [43]. CI index scores were obtained from published literature [44], and were based on a visual assessment of the crown exposure to light. Species were classified on a scale of 1–5 from very shade-tolerant to highly light-demanding. Since light requirements may change during ontogeny [45], only the CI index of *Macaranga* individuals <4 cm diameter were used, and were obtained by averaging the value of CI index across the first two columns of Table 3 in [44].

For seed collection, mature fruits of 10 *Macaranga* species (Table 1) were collected between July 2009 and September 2011 from primary and secondary forests of Lambir. All fruits were collected directly from the crowns of at least two maternal sources (except one parent for fruits of the rare species *M. lamellata*, *M. umbrosa* and *M. hypoleuca*). The numbers of seed sources and seeds used in the study were relatively small. This is a consequence of low fruit availability during non-mast years, and further constraints on seed production that are a consequence of resource limitation to relatively large-seeded trees that grow beneath a forest canopy.

Macaranga seeds are enclosed with oil-rich arils. The presence of aril is recognized as a reward for seed dispersal agents. The soft nutrient-rich tissues are likely to be decomposed rapidly and would not persist long enough to affect persistence in the soil. Therefore, in this study examinations of physical and chemical defences, and seed persistence in the soil excluded the aril. Seeds (Figure 1) were cleaned the same day of collection to remove arils and to exclude non-viable seeds that floated in water. To measure seed dry mass, two replicates of 10 seeds from each maternal source were randomly sampled and were oven-dried at 65°C for three days and reweighed. For each species, seeds from different maternal sources were pooled, stored in a refrigerator, and shipped by air courier in airtight plastic bags to the University of Illinois for physical and chemical defence analyses.

Measurement of seed physical defence

Quantitative differences in seed physical protection among species were represented by a measurement of seed coat thickness. In *Macaranga*, seed coat thickness was strongly correlated with seed fracture resistance, i.e. the minimum force required to initiate seed rupture (Appendix S1, Figure S1). Therefore, the seed coat thickness served as a good estimate of seed physical strength. Measurements of seed coat thickness were made using a dissecting microscope (ZEISS SteREO Discovery V8, Carl Zeiss Microscopy, Germany). Ten seeds of each species, (except *M. lamellata* and *M. umbrosa* for which five seeds were measured) were randomly selected and cut in half. The half-seeds were photographed under the dissecting microscope and seed coat thickness measured at three random points using graphic tools of ZEN 2011 imaging software (Carl Zeiss Microscopy, Germany). The mean of seed coat thickness was obtained for each species.

Table 1. Summary of seed shape and surface sculpturing, species crown illumination index (CI index), and physical and chemical defence determined for 10 *Macaranga* species.

Species	Seed shape and sculpturing*	CI**	Seed dry mass (mg)		Seed coat thickness (μm)		Total mass-standardized peak area (10^8 mV × min/g sample)	Maternal source
Macaranga			Mean	SE†	Mean	SE‡		
beccariana	S	3.20	12.19	1.70	319.9	4.8	3.47	3
hypoleuca	S	3.20	19.58	.	375.8	5.6	6.69	1
umbrosa[1]	O	2.00	60.47	.	126.0	1.2	0.61	1
lamellata	O	1.80	64.09	.	130.8	2.9	3.06	1
bancaca[2]	O	3.05	18.73	0.61	218.3	4.7	19.60	3
trachyphylla	O	3.05	20.41	0.14	234.3	24.6	14.92	2
hullettii	O	2.05	24.72	1.57	178.0	3.1	13.72	3
havilandii	O	2.75	33.22	0.66	171.3	1.8	22.11	2
gigantea	L	3.50	17.60	1.81	410.0	3.3	3.45	3
winkleri	V	3.70	1.72	0.13	94.7	2.9	0.47	3

*Seed shape and surface sculpturing from Davies 2001 [63] and Whitmore 2008 [64].
O: ovoid or subtriangular-ovoid with shallow grooves, S: spheroidal or slightly flattened with large shallow round pits, L: lenticular with shallow grooves, V: ovoid shape with coarsely verrucose (peg-like structure) surface (Figure 1).
**Mean crown light index (CI) of the seedling and sapling stages (diameter size class of 0–4 cm) from Davies et al. (1998) [44]. The index, ranging from 1 to 5, scores the source (lateral and vertical light) and relative amount of light available for individual crowns [43].
†Standard error among seed sources. Missing value indicates one maternal tree contributing the seeds.
‡Standard error among seeds of pooled from maternal sources or seeds of one maternal source.
[1]Previously reported as *Macaranga kingii* in Davies et al. 1998 and Davies and Ashton 1999.
[2]Previously reported as *Macaranga triloba* in Davies et al. 1998 and Davies and Ashton 1999.

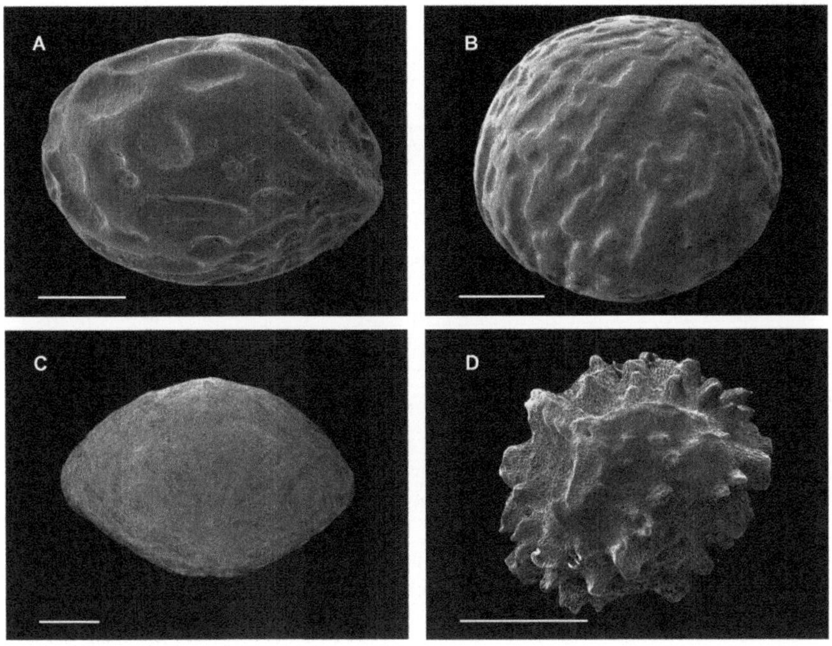

Figure 1. Four types of seed shape and sculpturing of Bornean *Macaranga* species (scale bar: 1 mm) – (a) *M. hypolueca*: spheroidal or slightly flattened with large shallow round pits (S), (b) *M. bancanca*: ovoid or subtriangular-ovoid with shallow grooves (O), (c) *M. gigantea*: lenticular with shallow grooves (L), and (d) *M. winkleri*: ovoid shape with coarsely verrucose surface (V).

Measurement of seed chemical defence

Rather than targeting a specific class of compounds, seed chemical defence investment was characterized as the quantity of potential defensive organic compounds in methanolic extracts of seed coat and may include phenols, flavonoids and alkaloids. Seeds were pooled from different maternal sources and the seed coats and seed contents separated by gently breaking seeds open in a mortar and pestle. The seed coats were then ground to a fine homogenate using a Wiley Mini Mill (model 3383-L10 Thomas Scientific, USA). For each species, three replicates of 0.1 g of ground seed coat were extracted in methanol (Appendix S2). The triplicates of methanol supernatant were then analysed for compounds using high performance liquid chromatography (HPLC; Appendix S3).

In this study, potential defensive compounds were not identified but were instead distinguished according to their retention times. An absorbance peak (at 280 nm), shown in a chromatogram, was used as a relative abundance measure for a series of unknown peaks. To compare compound abundance across samples, the area of each peak in each sample was first standardized to the mass of the seed coat sample.

$$Mass - standardized\ peak\ area = \frac{\frac{Raw\ peak\ area}{Injection\ volume} \times Total\ volume\ of\ extract}{Sample\ mass}$$

where peak area was measured in mV×min, injection volume was 25 μL, extract volume was 1.5 mL, and sample mass was measured in g. Mean peak areas were calculated from the three replicates per species. The mass-standardized peak area of all potential defensive compounds present in each species was then summed. In *Macaranga* the number of absorbance peaks and the mass-standardized peak area of potential defensive compounds were strongly and positively associated (Figure S2). Therefore, only the mass-standardized peak area of compounds was used as a measurement of the abundance of potential defensive compounds [46].

Seed persistence in the soil

A burial experiment to determine the loss rate of seed viability over time was conducted at Lambir from July 2009 to November 2011. Due to the low availability seeds, this experiment was limited to five species: *M. bancana*, *M. beccariana*, *M. gigantea*, *M. trachyphylla* and *M. winkleri*. There were two maternal sources for *M. beccariana* and *M. winkleri*, and one maternal source each for *M. bancana*, *M. gigantea* and *M. trachyphylla*. For initial seed germination, at least two replicates of 30 seeds per maternal source were sown in Petri dishes lined with two layers of sterile cotton gauze pad (Dynarex Corporation, New York, USA) for eight weeks. The cotton gauze pad was saturated with 5 ml of tap water and the dishes were covered with lids. Dishes were checked daily, and water was added every 2–3 days to resaturate the gauze pad.

Buried seeds of each maternal source were placed in individual fibreglass netting (0.5 mm mesh size) bags together with 10 g of homogenized forest soil. Mesh bags retained the seeds but were permeable to fungi and small invertebrates [36]. Each bag of the same maternal source contained the same number of seeds; however, the number of seeds per bag varied among maternal sources depending on seed availability (15 to 30 seeds), and the number of seed bags per maternal sources ranged from 27 to 42 bags. The seed bags were buried 5 cm below the soil surface and 30 cm apart from one another in a common garden beneath the forest canopy. The burial site was in a shade environment (average R: FR of 0.14±0.04) of primary Dipterocarp forest canopy of Lambir at 102 m above the sea level. Daily air temperature ranged from 24.0 to 28.6°C (StowAway TidbiT temperature data loggers, Onset Computer Corporation, MA, USA).

Seed bags were recovered 1, 2, 3, 6, 12, 19, 22 and 28 months after burial. For some maternal sources with few seed bags, recovery periods of seed bags varied, but spanned the 28 months. Four to eight seed bags of each maternal source were retrieved at each time point. After recovery, the seed bag content, which was comprised of soil and seeds, was placed individually in a Petri dish lined with a piece of moist filter paper (No. 1 Whatman filter paper, GE Healthcare, New Jersey, USA). The Petri dishes were covered with lids and placed on the shelves in a growing house with 30% full sun (average R: FR 1.12). Averages of day and night time air temperature in the growing house were 31.0 and 23.5°C, respectively; the average diel temperature is 26.7°C. The seed bag contents were sprayed with water every three days to maintain moisture. Germination was scored over five weeks as visible radicle or hypocotyl emergence.

Data Analysis

Seed mass, light requirement for regeneration and seed defence

Unless stated, data were analysed using R version 2.15.3 [47]. To examine trait relationships across species, the pairwise Pearson correlation matrix was computed for seed dry mass, seed coat thickness, total mass-standardized peak area of potential defensive compounds, and crown illumination index (CI index). Phylogenetic analyses were also conducted on trait comparisons. The results accounting for phylogeny were qualitatively similar to the results of the cross-species analyses (Table S1 in Appendix S4).

Ordination of potential defensive compound composition and light requirement

Principal coordinate analysis (PCO) (package *ecodist* version 1.2.7) [48] was used to position species in space according to differences in potential defensive compounds present in seed coats. The PCO is a metric multidimensional scaling method based on projection, which uses spectral decomposition to approximate a dissimilarity matrix of the distances between a set of points in a few dimensions [49]. The PCO is useful for biological data with many zeroes in the data set. Bray-Curtis index was used to calculate the dissimilarity matrix. The number of dimensions was fixed to two.

Seed persistence in the soil

In our study, no dormancy was observed for any species when seeds were exposed to suitable environmental conditions. Therefore, seed germinability reflects seed viability. The relationship between seed germination percent (y) and time (t) was described as a negative exponential function [39], [50]. The exponential decay function was in the form of:

$$y = ae^{-kt}$$

where a represents initial percent germination and k is the exponential decay constant. The time over which viability of seeds was reduced to half of initial viability, was used as a comparison index of seed persistence in the soil. Half-lives of seeds in the soil were estimated under the assumption of a constant rate of viability loss over time (t). Seed half-life in the soil ($t_{0.5}$) for each species was calculated as $t_{0.5} = ln(0.5)/-k$.

To explore the relationship between seed persistence and seed defence, the ratio of total mass-standardized peak area of potential defensive compounds to seed coat thickness was used. This ratio provides a measure of the relative amount of investment to chemical and physical defence [39]. The $\log_e$-transformed ratio was regressed against $\log_e$-transformed seed half-life using a linear regression (*lm*).

Results

The trade-off between seed physical and chemical defence

There was no significant correlation between seed physical and chemical defence either before or after controlling for phylogenetic relationships (Table 2, Figure 2). Species with both relatively thin (<150 μm) and thick (>300 μm) seed coats had relatively low quantities of potential defensive compounds compared to species with intermediate seed coat thickness (170–220 μm) (Figure 2A).

Seed physical defence, light requirement for regeneration, and seed mass

The light requirement for regeneration, represented by the CI index of seedling and sapling stages, ranged from 1.80 (shade-tolerant) to 3.70 (strongly light-demanding) across the 10 *Macaranga* species (Table 1). The CI index was strongly negatively correlated with seed mass (Table 2), indicating that more light-demanding species produced smaller seeds. There was also a significant negative relationship between seed coat thickness and seed mass, where *M. winkleri*, the only *Macaranga* species with a sculpted seed coat, was a significant outlier (Figure 3).

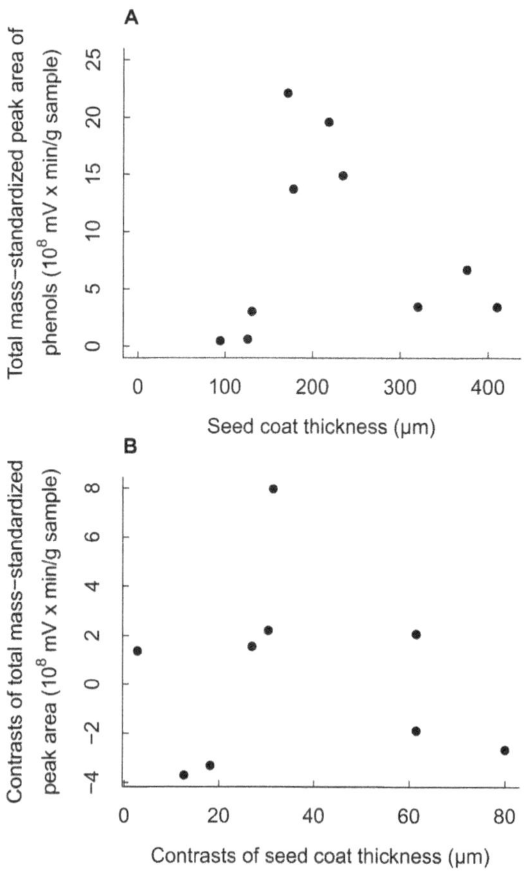

Figure 2. The cross-species analysis (a) and the analysis of phylogenetically independent contrasts (b) showed no linear relationship between seed coat thickness and total peak area of potential defensive compounds.

Table 2. Pearson product-moment correlation matrix between all paring seed attributes of 10 *Macaranga* species.

	Crown illumination index*	Seed dry mass*	Seed coat thickness*
Seed dry mass	**−0.87**		
Seed coat thickness	0.50	−0.41	
Total mass-standardized peak area	−0.04	−0.13	−0.04

* The coefficients with significance $P<0.001$ are indicated in boldface.

Because seed mass and CI index were strongly correlated (Table 2), their association with thickness of seed coat was examined using partial correlation. When controlling for CI index, seed coat thickness was uncorrelated with seed mass (Partial correlation coefficient $= -0.04$, $n=9$, $P=0.92$). When controlling for seed mass, and excluding the outlier *M. winkleri*, seed coat thickness was positively associated with CI index (Partial correlation coefficient $=0.66$, $n=9$, $P=0.03$), indicating that shade-tolerant species had relatively thinner seed coats than light-demanding species.

Seed chemical defence of Macaranga species

A total of 23 distinct peaks were recorded likely representing 23 compounds in the 10 *Macaranga* species. The lowest peak diversity was in *M. winkleri* with one compound peak, while *M. trachyphylla* and *M. havilandii* had the highest diversity with 20 distinguishable peaks. Total mass-standardized peak area of potential defensive compounds was not associated with CI index or seed mass (Table 2). The principal coordinate analysis (PCO) used to position species according to similarity in potential defensive composition revealed three distinct species groups (Figure 4): (1) *M. beccariana*, *M. gigantea*, *M. hypoleuca* and *M. lamellata*, (2) *M. bancana*, *M. trachyphylla*, *M. havilandii* and *M. hullettii*, and (3) *M. winkleri* and *M. umbrosa*. These groups were not associated with seed mass, or light requirement for regeneration.

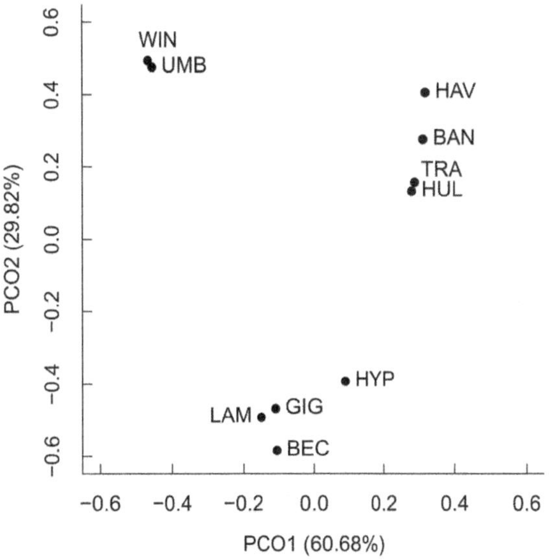

Figure 4. Principal coordinate analysis (PCO) plot showing similarity of potential defensive compound composition of 10 *Macaranga* species - *M. bannacana* (BAN), *M. becciariana* (BEC), *M. gigantea* (GIG), *M. havilandii* (HAV), *M. hullettii* (HUL), *M. hypoleuca* (HYP), *M. lamellata* (LAM), *M. trachyphylla* (TRA), *M. umbrosa* (UMB), and *M. winkleri* (WIN). The two axes explained 90.50% of the viability in the original dissimilarity matrix.

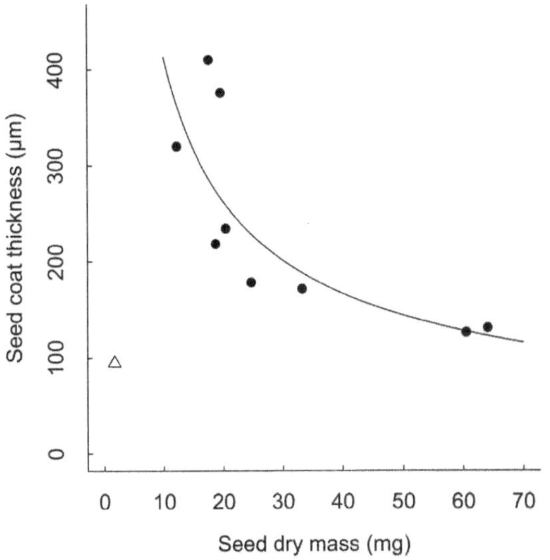

Figure 3. The relationship between seed mass and seed coat thickness for 10 *Macaranga* species. The outlier *M. winkleri* data (△) was excluded from the model ($\log(Y)=7.5328-0.6563\times\log(X)$, $n=9$, $R^2=0.7322$, $F_{(1,7)}=19.14$, $P=0.003$).

Seed persistence, seed mass, light requirement for regeneration and defence

Data on persistence (half-life) (Figure S3) of seeds in the soil were available for five *Macaranga* species: *M. bancana* (0.36 year), *M. trachyphylla* (0.48 year), *M. beccariana* (0.77 year), *M. winkleri* (1.26 years), and *M. gigantea* (1.78 years). When controlling for seed mass, the half-life of seeds in the soil was associated with CI index (Partial correlation coefficient $=0.94$, $n=5$, $P<0.0001$), but the half-life of seeds was unrelated to seed mass when CI index was controlled (Partial correlation coefficient $=0.02$, $n=5$, $P=0.98$).

Comparing physical and chemical defences, the ratio of total mass-standardized peak area of potential defensive compounds to seed coat thickness (chemical: physical defence) decreased as seed half-life in the soil increased (Figure 5). Linear regression with the PICs of log-seed half-life and log-ratio of total mass-standardized peak area also showed a negative relationship ($Y=-1.78X$, $R^2=0.84$, $F_{(1,\ 3)}=16.07$, $P=0.03$). Thus, the relative investment in chemical defence to physical defence decreased as seed persistence increased.

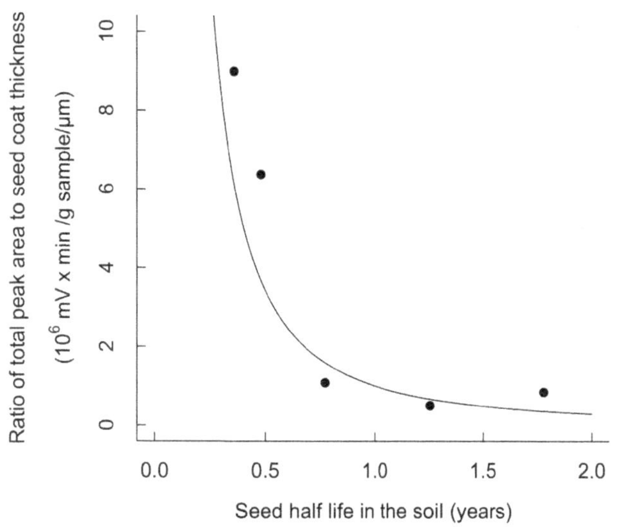

Figure 5. The ratio of seed chemical (total mass-standardized peak area of potential defensive compounds) to physical defence (seed coat thickness) of five *Macaranga* species decreased as seed half-life in the soil increased ($\log(Y) = -1.7732 \times \log(X)$, $n=5$, $R^2=0.8269$, $F_{(1,3)}=14.33$, $P=0.032$).

Discussion

This study tested hypotheses for how seed physical and chemical defences, seed mass, light requirement for regeneration and seed persistence are related. We found that variation in defence investment was primarily determined by species habitat requirements. Species that occupy more open environments, and that have longer-lived seed banks invested more in physical defences than more shade-tolerant species. Seed chemical defences, characterized as soluble polar molecules extracted from the seed coat, also varied markedly among species in both their diversity and abundance, and overall, appear to be a more important component of the defensive investment of more shade-tolerant *Macaranga* species.

Trade-off between seed physical and chemical defence

In general, studies on the allocation of physical and chemical defence in seeds are scarce. A single study, comparing six species of *Lithocarpus* (Fagaceae) with seeds dispersed by scatter-hoarding rodents has reported a trade-off between these defence classes [25]. No direct trade-off was observed in this study (Figure 2); however, the proportional allocation to defensive traits may be contingent on the suite of dispersers and predators that are associated with seeds, and may therefore be more apparent in comparisons among species that have similar capacities to persist in the soil.

The prediction of trade-offs among defensive traits also assumes that resource allocation rules apply at the individual seed level. However, allocation to defence in seeds is a component of the larger investment in seed production [51], [52]. Therefore, both seed size and seed defence investment potentially trade-off with the number of seeds produced.

The efficacy of chemical and physical defences also needs to be evaluated in the context of seed germination. Physical defences protect mature, ungerminated seeds; however, seed coats must soften and crack during seed germination, and seeds may then rely on existing chemical defences or on inducible mechanisms that transform stored precursors to defensive compounds [53], [54]. Future studies are needed to better understand temporal changes in physical and chemical attributes of defence after dispersal and the interactive effect of pathogens and environmental stresses.

Light requirement for regeneration, seed mass and seed defence

There was large variation in the investment in seed defences among species, which correlated with seed and plant regeneration traits. For seed physical defence, light requirement for regeneration rather than seed mass was the principle determinant of defence traits. Therefore, the results did not support the hypothesis that larger seeds are more reliant on physical defences. In contrast, the results supported the hypothesis that exposure to natural enemies in the soil determines investment in physical defence among species. Seeds of species that occupy more shaded habitats had lower levels of physical protection. Among the study species, shade-tolerant *Macaranga* species in this dataset showed no evidence of delayed germination and therefore may not require the physical defences that permit prolonged persistence in the soil after seed dispersal.

Variation in seed coat thickness may also be related to seed physiology. The thin seed coat of shade-tolerant species may also be explained in part by low desiccation risk in shade habitats. While seed coats play a role in defence, they also serve as a boundary to protect the embryo against fluctuations in humidity and temperature [19]. Shade habitats are moister than large gaps and open sites in tropical forests [55]. Therefore, seeds that germinate in drier high light environments may exhibit thicker seed enclosing structures to regulate water uptake [19], [56] and to prevent rapid desiccation [57]. In contrast, selection for thick seed covers may be low, especially in ever-wet forests, where seeds are not usually exposed to prolonged dry periods.

Seed chemical defence

The diversity of potential defensive compounds ranged widely, with 1–20 compounds present in the seed coat of the 10 *Macaranga* species. Only one unidentified compound with molecular weight of 330.1 was common to all species. Species grouped according to their similarity in chemical defences did not show associations with either light requirement for regeneration or seed mass. Other factors, such as soil specialization and phylogenetic relatedness among *Macaranga* species, may be correlated to the diversity, quantity and types of potential defensive compounds present in seed coat.

Seed persistence and seed defence

The results of this study supported the hypothesis that species with seeds that persist longer in the soil rely primarily on seed physical rather than chemical defence. These results were also in agreement with a previous study of six temperate weed species, where Davis et al. (2008) [39] explored the relationship between seed half-life and the relative importance of seed chemical defences, represented by the concentration of *ortho*-dihydroxyphenols, to physical defence, represented by seed coat thickness. They found that relative investment in phenolic defences declined with increasing half-life in the soil. Similarity in tropical tree species and temperate herbaceous species suggest that a reliance on physical defence for long persistence may be universal across different plant groups.

In this study, longer persistent *Macaranga* species had a lower ratio of total abundance of potential defensive compounds (total mass-standardized peak area) relative to seed coat thickness. When the number of potential defensive compounds among species and seed persistence in the soil were considered, the species with longer persistence also produced fewer potential defensive compounds. For the longer persistent species, it might be more resource efficient to produce relatively few potential defensive compounds with high defensive capacity. Alternatively, the few potential defensive compounds produced in species with high persistence may play little role in plant defence and species may rely on ongoing co-evolution between seeds and specialist microbes in the soil as a microbial defence [58].

Exceptional structure and seed defence of *Macaranga winkleri*

The external surface of enclosing structures of seeds of *Macaranga* species was smooth or only shallowly pitted with one exception, *M. winkleri*. Seed coats of *M. winkleri* had a verrucose surface (Figure 1D). Although many other taxa have similarly sculpted surfaces, the ecological significance of seed coat ornamentation is not well understood. There are, however, two suggested hypotheses that relate to abiotic interactions [59]. First, rough seed coverings could be advantageous in anchoring the seed to the soil (Werker 1997 referred to by [60]). Second, surface sculpturing may control temperature under conditions of high insolation [59]. Since seeds cannot regulate temperature through transpiration, roughness may increase seed surface area, which increases energy exchange with surrounding cooler air.

In addition to abiotic interactions, sculpturing of the seed surface may also play a role in seed microbial defence. The verrucose surface appears not to contribute to seed physical strength, given that *M. winkleri* had the lowest seed fracture resistance. However, the extra surface area may facilitate colonization of beneficial microbes around seeds that may prevent pathogen infection [58], [61]. Among the *Macaranga* species in this study, *M. winkleri* produced the smallest light-dependent seeds [62]. *M. winkleri* seeds were the least physically and chemically protected with low seed strength and only one soluble compound present. However, *M. winkleri* seeds had a half-life of 1.26 years and are abundant in the soil seed bank (Pimonrat Tiansawat, unpublished data). The persistence of *M. winkleri* and evidence of its poor physical and chemical protection may suggest other means of seed defence. Clearly, more work is needed to understand the ecological significance of seed surface characteristics that may influence species regeneration.

In conclusion, for the 10 *Macaranga* species studied here, the lack of correlation between physical defence and chemical defence indicated the absence of a direct trade-off between defence classes. Nevertheless, variation in physical and chemical defence relating to species traits was found among species. Light requirement for regeneration was unrelated to the quantity of potential defensive compounds present in the seed coat. However, light requirement for regeneration was positively correlated with seed coat thickness, when excluding the outlier *M. winkleri*, indicating thinner seed coats in more shade tolerant species. Short exposure to natural enemies after dispersal and physiological benefits may select for thinner seed coats in shade-tolerant species. Finally, this study indicates that seeds with prolonged persistence in the soil relied primarily on seed physical rather than chemical defence.

Supporting Information

Figure S1 The linear regression through the origin of seed coat thickness (µm) and seed fracture resistance (Newton: N). Seed coat thickness and seed fracture resistance were strongly positively related ($Y = 0.1717X$, $n = 10$, $R^2 = 0.9667$, $F_{(1,\ 9)} = 206.9$, $P < 0.001$).

Figure S2 The linear regression through the origin of the number and total mass-standardized peak area of potential defensive compounds. The number of potential defensive compounds was positively related with total mass-standardized peak area of potential defensive compounds ($Y = 0.893X$, $n = 10$, $R^2 = 0.9598$, $F_{(1,\ 9)} = 214.7$, $P<0.001$). The result corrected for phylogeny was qualitatively similar ($Y = 0.986X$, $n = 9$, $R^2 = 0.71$, $F_{(1,8)} = 19.62$, $P = 0.003$).

Figure S3 Exponential decay function fitted to percent seed germination after burial over time of five *Macaranga* species. For species with two maternal sources in the study, species' seed half-life is the average from the two maternal trees.

Appendix S1 Fracture resistance measurement and the correlation to seed coat thickness.

Appendix S2 Extraction of soluble seed phenolic compounds.

Appendix S3 Analytical method in high performance liquid chromatography (HPLC).

Appendix S4 Phylogenetic analyses and results of phylogenetic analyses. Table S1, Pearson partial correlation matrix of seed physical defence (seed coat thickness), chemical defence (total mass-standardized peak area), crown illumination index (CI index), and seed mass of nine *Macaranga* species. An outlier, *M. winkleri*, was excluded. The correlation coefficients were not statistically significant at $P<0.05$.

Acknowledgments

We acknowledge the Sarawak Forest Department and the Sarawak Forestry Cooperation for providing permission to conduct the field work at Lambir Hills National Park (Permit number NCCD.907.4.4(V)-97 and NCCD.907.4.4(Jld.VI)-84). The advice of Stuart Davies was a great help in identifying *Macaranga*. We thank Azam Bin Amat, Juari Mathius, Ah Chin, and Douglas for assistance in the field. We extend our gratitude to Christina Czekala, Dan Arrecis, Kristen Woodley, and Phil Jarboe, in helping with chemical extraction, and Carolina Sarmiento and Jorge Ceballos in editing seed SEM photos. We also thank Carol Augspurger, Uromi M Goodale and an anonymous reviewer for helpful comments on the manuscript.

Author Contributions

Conceived and designed the experiments: PT JWD. Performed the experiments: PT MAB PCZ. Analyzed the data: PT ASD. Contributed reagents/materials/analysis tools: PT ASD MAB JWD. Wrote the paper: PT ASD MAB PCZ JWD.

References

1. Bryant JP, Chapin FS III, Klein DR (1983) Carbon/Nutrient balance of boreal plants in relation to vertebrate herbivory. Oikos 40: 357–368.
2. Augspurger CK, Kelly CK (1984) Pathogen mortality of tropical tree seedlings: Experimental studies of the effects of dispersal distance, seedling density, and light conditions. Oecologia 61: 211–217.

3. Fine PVA, Mesones I, Coley PD (2004) Herbivores promote habitat specialization by trees in Amazonian forests. Science 305: 663–665.
4. Fenner M, Thompson K (2005) The ecology of seeds. Cambridge: Cambridge University Press. 250 p.
5. Fine PVA, Mesones I (2011) The role of natural enemies in the germination and establishment of *Pachira* (Malvaceae) trees in the Peruvian Amazon. Biotropica 43: 265–269.
6. Janzen DH (1971) Seed predation by animals. Annu Rev Ecol Syst 2: 465–492.
7. Zhang J, Drummond FA, Kiebman M, Hartke A (1997) Insect predation of seeds and plant population dynamics. Orono: Department of Applied Ecology and Environmental Studies, University of Maine. 28 p.
8. Kestring D, Menezes LCCR, Tomaz CA, Lima GPP, Rossi MN (2009) Relationship among phenolic contents, seed predation, and physical seed traits in *Mimosa bimucronata* plants. J Plant Biol 52: 569–576.
9. Dalling JW, Davis AS, Schutte BJ, Elizabeth Arnold A (2011) Seed survival in soil: Interacting effects of predation, dormancy and the soil microbial community. J Ecol 99: 89–95.
10. Cazetta E, Schaefer HM, Galetti M (2008) Does attraction to frugivores or defense against pathogens shape fruit pulp composition? Oecologia 155: 277–286.
11. Gallery RE (2007) Seed banks, seed mortality, and the role of fungal communities in neotropical forests. PhD Thesis. University of Illinois, Urbana-Champaign
12. Grubb PJ, Metcalfe DJ, Grubb EAA, Jones GD (1998) Nitrogen-richness and protection of seeds in Australian tropical rainforest: A test of plant defence theory. Oikos 82: 467–482.
13. Moles AT, Warton DI, Westoby M (2003) Do small-seeded species have higher survival through seed predation than large-seeded species? Ecology 84: 3148–3161.
14. Boesewinkel FD, Bouman F (1995) The seed: Structure and function. In: Kigel J, Galili G, editors. Seed Development and Germination.New York: Marcel Dekker. pp. 1–24.
15. Kuprewicz EK, Garca-Robledo C (2010) Mammal and insect predation of chemically and structurally defended *Mucuna holtonii* (Fabaceae) seeds in a Costa Rican rain forest. J Trop Ecol 26: 263–269.
16. Souza AJ, Santos PO, Pinto MST, Wermelinger TT, Ribeiro ES, et al. (2011) Natural seed coats provide protection against penetration by *Callosobruchus maculatus* (Coleoptera: Bruchidae) larvae. Crop Prot 30: 651–657.
17. Bodmer RE (1991) Strategies of seed dispersal and seed predation in Amazonian ungulates. Biotropica 23: 255–261.
18. Rodgerson L (1998) Mechanical defense in seeds adapted for ant dispersal. Ecology 79: 1669–1677.
19. Mohamed-Yasseen Y, Barringer SA, Splittstoesser WE, Costanza S (1994) The role of seed coats in seed viability. Bot Rev 60: 426–439.
20. Heil M, Delsinne T, Hilpert A, Schürkens S, Andary C, et al. (2002) Reduced chemical defence in ant-plants? A critical re-evaluation of a widely accepted hypothesis. OIKOS 99: 457–468.
21. Phrommart S, Sutthivaiyakit P, Chimnoi N, Ruchirawat S, Sutthivaiyakit S (2005) Constituents of the leaves of *Macaranga tanarius*. J Nat Prod 68: 927–930.
22. Yoder BJ (2005) Isolation and structure elucidation of cytotoxic natural products from the rainforests of Madagascar and Suriname. PhD Thesis. Virginia Polytechnic Institute and State University.
23. Cardinal-Aucoin M, Bauce E, Albert PJ (2009) Preingestive detection of tannins by *Choristoneura fumiferana* (Lepidoptera: Tortricidae). Ann Entomol Soc Am 102: 717–726.
24. Singh A, Singh DK, Misra TN, Agarwal RA (1996) Molluscicides of plant origin. Biol Agric Hortic 13: 205–252.
25. Chen X, Cannon CH, Conklin-Brittan NL (2012) Evidence for a trade-off strategy in stone oak (*Lithocarpus*) seeds between physical and chemical defense highlights fiber as an important antifeedant. PLOS ONE DOI: 10.1371/journal.pone.0032890.
26. Abdel-Monaim MF (2011) Role of riboflavin and thiamine in induced resistance against charcoal rot disease of soybean. Afr J Biotechnol 10: 10842–10855.
27. Coley PD (1983) Herbivory and defensive characteristics of tree species in a lowland tropical forest. Ecol Monogr 53: 209–234.
28. Stock WD, Pate JS, Rasins E (1991) Seed developmental patterns in *Banksia attenuata* R. Br. and B. laricina C. Gardner in relation to mechanical defence costs. New Phytol 117: 109–114.
29. Strauss SY, Rudgers JA, Lau JA, Irwin RE (2002) Direct and ecological costs of resistance to herbivory. Trends Ecol Evol17: 278–285.
30. Twigg LE, Socha LV (1996) Physical versus chemical defence mechanisms in toxic *Gastrolobium*. Oecologia 108: 21–28.
31. Hanley ME, Lamont BB (2002) Relationships between physical and chemical attributes of congeneric seedlings: How important is seedling defence? Funct Ecol 16: 216–222.
32. Read J, Sanson GD, Caldwell E, Clissold FJ, Chatain A, et al. (2009) Correlations between leaf toughness and phenolics among species in contrasting environments of Australia and New Caledonia. Ann Bot 103: 757–767.
33. Moles AT, Peco B, Wallis IR, Foley WJ, Poore AGB, et al. (2013) Correlations between physical and chemical defences in plants: Tradeoffs, syndromes, or just many different ways to skin a herbivorous cat? New Phytol 198: 252–263.
34. O'Hanlon-Manners DL, Kotanen PM (2004) Evidence that fungal pathogens inhibit recruitment of a shade-intolerant tree, white birch (*Betula papyrifera*), in understory habitats. Oecologia 140: 650–653.
35. Schafer M, Kotanen PM (2004) Impacts of naturally-occurring soil fungi on seeds of meadow plants. Plant Ecol 175: 19–35.
36. Gallery RE, Moore DJP, Dalling JW (2010) Interspecific variation in susceptibility to fungal pathogens in seeds of 10 tree species in the neotropical genus *Cecropia*. J Ecol 98: 147–155.
37. Hendry GAF, Thompson K, Moss CJ, Edwards E, Thorpe PC (1994) Seed persistence: A correlation between seed longevity in the soil and *ortho*-dihydroxyphenol concentration. Funct Ecol 8: 658–664.
38. Veldman JW, Greg Murray K, Hull AL, Mauricio Garcia-C J, Mungall WS, et al. (2007) Chemical defense and the persistence of pioneer plant seeds in the soil of a tropical cloud forest. Biotropica 39: 87–93.
39. Davis AS, Schutte BJ, Iannuzzi J, Renner KA (2008) Chemical and physical defense of weed seeds in relation to soil seedbank persistence. Weed Sci 56: 676–684.
40. Gardarin A, Durr C, Mannino MR, Busset H, Colbach N (2010) Seed mortality in the soil is related to seed coat thickness. Seed Sci Res 20: 243–256.
41. Kumagai T, Yoshifuji N, Tanaka N, Suzuki M, Kume T (2009) Comparison of soil moisture dynamics between a tropical rain forest and a tropical seasonal forest in Southeast Asia: Impact of seasonal and year-to-year variations in rainfall. Water Resour Res 45: W04413.
42. Kumagai T, Saitoh TM, Sato Y, Takahashi H, Manfroi OJ, et al. (2005) Annual water balance and seasonality of evapotranspiration in a Bornean tropical rainforest. Agric For Meterol 128: 81–92.
43. Davies SJ, Ashton PS (1999) Phenology and fecundity in 11 sympatric pioneer species of Macaranga (Euphorbiaceae) in Borneo. Am J Bot 86: 1786–1795.
44. Davies SJ, Palmiotto PA, Ashton PS, Lee HS, Lafrankie JV (1998) Comparative ecology of 11 sympatric species of *Macaranga* in Borneo: Tree distribution in relation to horizontal and vertical resource heterogeneity. J Ecol 86: 662–673.
45. Goodale UM, Ashton MS, Berlyn GP, Gregoire TG, Singhakumara BMP, et al. (2012) Disturbance and tropical pioneer species: Patterns of association across life history stages. For Ecol Manage 277: 54–56.
46. Xu Y, Gong F, Dixon SJ, Brereton RG, Soini HA, et al. (2007) Application of dissimilarity indices, principal coordinates analysis, and rank tests to peak tables in metabolomics of the gas chromatography/mass spectrometry of human sweat. Anal Chem 79: 5633–5641.
47. R Development Core Team (2013) R: A language and environment for statistical computing. R Foundation for Statistical Computing, Vienna, Austria2.15.3.
48. Goslee SC, Urban DL (2007) The ecodist package for dissimilarity-based analysis of ecological data. J Stat Softw 22: 1–19.
49. Gower JC (2005) Principal coordinates analysis. In: Anonymous Encyclopedia of Biostatistics: John Wiley and Sons, Ltd.
50. Conn JS, Beattie KL, Blanchard A (2006) Seed viability and dormancy of 17 weed species after 19.7 years of burial in Alaska. Weed Sci 54: 464–470.
51. Galen C, Weger HG (1986) Re-evaluating the significance of correlations between seed number and size: Evidence from a natural population of the lily, *Clintonia borealis*. Am J Bot 73: 346–352.
52. Venable DL (1992) Size-number trade-offs and the variation of seed size with plant resource status. Am Nat 140: 287–304.
53. Ding H, Lamb RJ, Ames N (2000) Inducible production of phenolic acids in wheat and antibiotic resistance to *Sitodiplosis mosellana*. J Chem Ecol 26: 969–985.
54. Roitto M, Markkola A, Julkunen-Tiitto R, Sarjala T, Rautio P, et al. (2003) Defoliation-induced responses in peroxidases, phenolics, and polyamines in scots pine (*Pinus sylvestris* L.) needles. J Chem Ecol 29: 1905–1918.
55. Garwood NC (1983) Seed germination in a seasonal tropical forest in Panama: A community study. Ecol Monogr 53: 159–181.
56. Koizumi M, Kikuchi K, Isobe S, Ishida N, Naito S, et al. (2008) Role of seed coat in imbibing soybean seeds observed by micro-magnetic resonance imaging. Ann Bot 102: 343–352.
57. Daws MI, Garwood NC, Pritchard HW (2005) Traits of recalcitrant seeds in a semi-deciduous tropical forest in Panamá: Some ecological implications. Funct Ecol 19: 874–885.
58. Dalling JW, Swaine MD, Garwood NC (1998) Dispersal patterns and seed bank dynamics of pioneer trees in moist tropical forest. Ecology 79: 564–578.
59. Barthlott W (1981) Epidermal and seed surface characters of plants: Systematic applicability and some evolutionary aspects. Nord J Bot 1: 345–355.
60. Metzing D, Thiede J (2001) Testa sculpture in the genus *Frailea* (Cactaceae). Bot J Linn Soc 137: 65–70.
61. Johnson HB (1975) Plant pubescence: An ecological perspective. Bot Rev 41: 233–258.
62. Tiansawat P (2013) Ecological significance of seed traits in the genus *Macaranga*. PhD thesis, University of Illinois, Urbana-Champaign.
63. Davies SJ (2001) Systematics of *Macaranga* sects. Pachystemon and Pruinosae (Euphorbiaceae). Harv Pap Bot 6: 371–448.
64. Whitmore TC (2008) The Genus *Macaranga*: A prodromus. Kew: Royal Botanic Gardens 312 p.

18

Effect of Byproducts of Flue Gas Desulfurization on the Soluble Salts Composition and Chemical Properties of Sodic Soils

Jinman Wang[1,2]*, Zhongke Bai[1,2], Peiling Yang[3]

1 College of Land Science and Technology, China University of Geosciences, Beijing, People's Republic of China, 2 Key Laboratory of Land Consolidation and Rehabilitation, Ministry of Land and Resources, Beijing, People's Republic of China, 3 College of Hydraulic and Civil Engineering, China Agricultural University, Beijing, People's Republic of China

Abstract

The byproducts of flue gas desulfurization (BFGD) are a useful external source of Ca^{2+} for the reclamation of sodic soils because they are comparatively cheap, generally available and have high gypsum content. The ion solution composition of sodic soils also plays an important role in the reclamation process. The effect of BFGD on the soluble salts composition and chemical properties of sodic soils were studied in a soil column experiment. The experiment consisted of four treatments using two different sodic soils (sodic soil I and sodic soil II) and two BFGD rates. After the application of BFGD and leaching, the soil soluble salts were transformed from sodic salts containing Na_2CO_3 and $NaHCO_3$ to neutral salts containing NaCl and Na_2SO_4. The sodium adsorption ratio (SAR), pH and electrical conductivity (EC) decreased at all soil depths, and more significantly in the top soil depth. At a depth of 0–40 cm in both sodic soil I and sodic soil II, the SAR, EC and pH were less than 13, 4 dS m^{-1} and 8.5, respectively. The changes in the chemical properties of the sodic soils reflected the changes in the ion composition of soluble salts. Leaching played a key role in the reclamation process and the reclamation effect was positively associated with the amount of leaching. The soil salts did not accumulate in the top soil layer, but there was a slight increase in the middle and bottom soil depths. The results demonstrate that the reclamation of sodic soils using BFGD is promising.

Editor: Stephen J. Johnson, University of Kansas, United States of America

Funding: This research was supported by the Fundamental Research Funds for the Central Universities of China (2652012072), and the National Natural Science Foundation of China (50749032). The funders had no role in study design, data collection and analysis, decision to publish, or preparation of the manuscript.

Competing Interests: The authors have declared that no competing interests exist.

* E-mail: wangjinman2002@163.com

Introduction

Wet flue gas desulfurization (FGD) is the dominant technology used in the control of SO_2 emissions from coal-fired power plants. The major byproduct of the process is $CaSO_4$, or a mixture of $CaSO_3$ and $CaSO_4$ [1]. With the rapid development of the energy and power industries in China, the installed capacity of power plants with FGD devices, and therefore the amount of byproducts of flue gas desulfurization (BFGD), is expected to increase rapidly. By the end of 2010, the installed capacity of power plants in China with FGD devices was about 57.8 GW, and the annual production of BFGD was about 7.1 million tons. According to the National Development Program of China, the installed capacity of power plants with FGD devices will be 530 GW by 2020, with an annual production of 90 million tons of BFGD. As BFGD contain large amounts of moisture and ash, they can only be used as building gypsum after purification and dehydration, hence they are more expensive compared with natural gypsum produced in China. If the BFGD were to be directly disposed of rather than being treated and used, a vast area of land would be required. Such an approach would be a waste of valuable land resources and represent a potential threat of secondary pollution to the environment [2–4].

Significantly, there are large areas of sodic soils in China. Excessive exchangeable sodium (ES) has gradually been adsorbed in soil colloid, causing deterioration in the soil's physical properties and inducing the formation of sodic soils [5]. Sodic soils contain an excess of exchangeable Na^+ in soil colloids and have soluble carbonates in the form of $NaHCO_3$ and Na_2CO_3. The pH value, SAR and exchangeable sodium percentage (ESP) are greater than 8.5, 13 and 15, respectively, and the electrical conductivity of the saturated paste (EC_{sat}) is less than 4.0 dS m^{-1} [5–6]. Amendments for sodic soils are materials that directly or indirectly supply divalent cations (usually Ca^{2+}) to replace ES [3,7]. The dissolution-exchange reaction in soil colloids is represented as [5].

$$2NaX + CaSO_4 = CaX_2 + Na^+ + SO_4{}^{2-}, \qquad (1)$$

where X is the soil exchange phase. The exchange of Ca^{2+} for Na^+ in soil colloid results in the flocculation of soil particles, which restores the porous structure and high permeability of the soil. Because the main component of BFGD is $CaSO_4$, substantial evidence exists that these BFGD can, with proper use, be valuable for the reclamation of sodic soils [1,3,8–12].

The degree of ES absorbed by soil colloid depends on the soil soluble salts composition. ES is most easily absorbed into soil colloid when the soil solution contains large amounts of Na_2CO_3 and $NaHCO_3$. In soil containing neutral salts (such as NaCl and

Table 1. Physical and chemical properties of the tested soils.

Soil type	Soil depth (cm)	Bulk density ($g \cdot cm^{-3}$)	pH	EC ($dS \cdot m^{-1}$)	ES ($cmol \cdot kg^{-1}$)	CEC ($cmol \cdot kg^{-1}$)	ESP (%)
Sodic soil I	0–5	1.41	8.95	11.45	1.65	17.14	9.63
	5–20	1.46	9.32	5.18	3.72	17.86	20.83
	20–40	1.41	9.29	4.31	2.44	16.95	14.38
	40–60	1.47	8.96	2.52	2.55	16.84	15.12
	60–100	1.44	8.96	2.31	2.19	16.71	13.12
Sodic soil II	0–5	1.58	9.07	13.09	9.46	21.23	44.55
	5–20	1.52	8.91	4.52	6.07	21.63	28.06
	20–40	1.46	9.23	1.89	5.60	20.15	29.74
	40–60	1.45	9.63	1.26	5.40	20.34	26.53
	60–100	1.46	9.44	0.84	4.91	19.86	24.71

Na_2SO_4), Na^+ can only be absorbed by soil colloid and cause soil sodication if $Na^+/(Ca^{2+}+Mg^{2+}) \geq 4$ in the cationic soil solution [13]. Therefore, the ion solution composition of sodic soils plays an important role in the reclamation process. Although several studies have examined the changes in the physicochemical properties of sodic soils and crop yield after applying BFGD, less attention has been paid to the ion solution composition [3,6,10,12–13,14–16].

The objective of this study was to determine whether the application of BFGD causes changes in the soluble salts composition, SAR, EC and pH value of sodic soils at different depths; to analyze the relationship between the soil's chemical properties and soluble salts composition; and to elucidate the mechanism of sodic soil reclamation by BFGD.

Materials and Methods

Ethics Statements

This study was conducted with the permission of the Changsheng Experimental Station of the Baoyannur League Institute of Water Resources. No endangered or protected species were present in the field studies.

The Physical and Chemical Properties of the Soils Tested and the BFGD

The soils tested were sampled from the Changsheng Experimental Station of the Baoyannur League Institute of Water Resources in northwestern China (N 40°20′, E 108°31′). Through on-site investigation, two different sodic soils (sodic soil I and sodic soil II) were sampled using spade. The soil sampling depth was divided into five layers (0–5 cm, 5–20 cm, 20–40 cm, 40–60 cm and 60–100 cm). The soils in different layers were air-dried, crushed, mixed and passed through a 2-mm sieve before the soil column experiments. The physical and chemical properties of the tested soils are listed in Table 1. The sodication degree of sodic soil II was higher than that of sodic soil I.

The BFGD used in the test were from Huaneng Power International, Inc. The $CaSO_4$ content in the by-product was 89.8% and the moisture content was almost 10.1% (Table 2). The concentrations of pollution elements were far below the tolerance limits regulated by the control standards for pollutants in fly ash for agricultural use (GB8173-87) and the control standards for pollutants in sludge for agricultural use (GB4284-84) [17–18].

The BFGD Application Rate

The principle underlying the reclamation of sodic soils by BFGD is that the calcium contained in the BFGD replaces the sodium in soil colloids. The application rate of the BFGD was determined using a modified method based on gypsum application rates [11,15]. It was calculated according to the relative content of calcium in the BFGD and gypsum as follows:

$$R_{BFGD} = 1.11 R_G = 0.095 H \gamma (CEC)(E_{Nai} - E_{Naf}), \quad (2)$$

where R_{BFGD} is the BFGD application rate in t hm^{-2}; R_G is the gypsum application rate in t hm^{-2}; 1.11 is the modified coefficient, determined by comparing the content of $CaSO_4$ in gypsum and BFGD; H is the soil depth to be reclaimed in cm, 20 cm in this study; γ is the soil bulk density in g cm^{-3}, and the values for sodic soil I and sodic soil II at a depth of 0–20 cm are shown in Table 1;

Table 2. Component and properties of the BFGD.

pH	Density ($g \cdot cm^{-3}$)	Free water (%)	$CaSO_4 \cdot 2H_2O$ (%)		$CaSO_4 \cdot 1/2H_2O$ (%)		$CaCO_3$ (%)
5.90	1.02	10.10	89.8		0.20		5.55
Cd ($mg \cdot kg^{-1}$)	Cr ($mg \cdot kg^{-1}$)	As ($mg \cdot kg^{-1}$)	Se ($mg \cdot kg^{-1}$)	Ni ($mg \cdot kg^{-1}$)	Cu ($mg \cdot kg^{-1}$)	Hg ($mg \cdot kg^{-1}$)	Pb ($mg \cdot kg^{-1}$)
0.01	83.4	5.04	4.24	21.35	83.26	0.32	99.38
5[a]	250[a]	75[a]	15[a]	200[a]	250 [a]	5.0[b]	250[a]

Note: [a]Control Standards for Pollutants in Fly Ash for Agricultural Use (GB8173-87);
[b]Control Standards for Pollutants in Sludge for Agricultural Use (GB4284-84).

Table 3. Experimental design.

Experimental treatment	BFGD application rate ($g \cdot kg^{-1}$)	BFGD application rate ($t \cdot hm^{-2}$) ($g \cdot kg^{-1}$)application rate ($g \cdot kg^{-1}$)	Sodic soils types
T1	3.00	1.65	Sodic soil I
T2	3.60	1.98	Sodic soil I
T3	7.00	3.86	Sodic soil II
T4	8.40	4.63	Sodic soil II

CEC is the cation exchange capacity in cmol kg^{-1}, and the capacities of sodic soil I and sodic soil II at 0–20 cm soil depth are shown in Table 1; $E_{Nai}$is the initial exchange sodium fraction, which is 1; and $E_{Naf}$is the desired final exchange sodium fraction, which was 0.1 in this study.

Experimental Design

The experiment consisted of four treatments (Table 3), with two sodic soils (sodic soil I and sodic soil II) and two BFGD rates. The BFGD application rates for sodic soil I and sodic soil II were divided into two levels. The amount of BFGD applied at the first level was determined according to the theoretical calculation of Formula 1. Taking into account the incomplete dissolution of the BFGD and deep percolation when it was applied, the amount of BFGD applied should be appropriately increased. The amount of the second application was determined by the theoretical calculation multiplied by a factor of 1.2. The BFGD application rates for different treatments are shown in Table 3 based on the initial properties of the sodic soil tested. The soil columns of four treatment were each leached four times.

The columns tested were contained in unplasticized polyvinyl chloride (UPVC) plastic pipes, with an outside diameter of 16 cm, inside diameter of 15.2 cm and height of 100 cm. The bottom of the soil column was closed with a UPVC board with a 1-cm drainage hole. A 5-cm layer of quartz sand was placed in the bottom of each column as the filter layer. The columns were filled with treated soils according to the soil depth, and the volumetric soil moisture content was 10%. The weight of dry soil to be filled at different depths was determined according to soil bulk density. The soils were evenly mixed in a sealed plastic container with water that is 10% by volume of the soils to be filled and maintained two days, and the soils with 10% of volumetric water content were obtained. Soil bulk density and soil weight in different column layers are shown in Table 4. The soil was filled in 5-cm depth at each time and tamped to a desired height in respective layers. The surface soil was loosened before the next soil layer was filled. During the filling process, BFGD was applied to and mixed evenly with the soil at 0–5 cm depth and 5–20 cm soil depth in the amount specified in the experimental design (Table 3). The soil column leaching tests were then carried out.

Because the soil water content in each soil column was only 10% (volumetric water content), the leaching water volume of each soil column was 4500 ml the first time (250 mm when translated into the field water volume; i.e. 2500 $m^3 \cdot hm^{-2}$), and was 1360 ml the second, third and fourth time (75 mm when translated into the field water volume; i.e. 750 $m^3 \cdot hm^{-2}$). The leaching interval was 12 d. The leaching soils were hierarchically sampled at soil depths of 0–5 cm, 5–10 cm, 10–20 cm, 20–40 cm, 40–60 cm, 60–80 cm and 80–100 cm. Each treatment soil column was sampled using an auger. The measured parameters were soluble salt composition (cations: CO_3^{2-}, HCO_3^-, Cl^- and SO_4^{2-}; anions: Ca^{2+}, Mg^{2+}, $K^+$and Na^+), pH, EC, ES and CEC.

Analytical Methods and Statistical Analyses

The samples were air-dried and passed through a 1-mm sieve. The EC, pH, soluble anions and soluble cations were measured using 1:5 water extracts. The soluble cations were measured using an atomic absorption spectrophotometer, soluble anions were determined by anion chromatography, and exchangeable cations were determined in 1 M ammonium acetate (pH = 7) extract. Following this extraction and washing with 96% alcohol, the cation exchange capacity was determined by the removal of ammonium ions by distillation [19]. Na and K were determined by flame emission spectroscopy in the extract, and Ca and Mg were determined by atomic absorption spectrophotometer. Soil pH was determined using the glass electrode method.

The SAR of 1:5 soil to water extracts (SAR1:5) was calculated as [5]

Table 4. The bulk density and weight of soils filled in each soil depth.

Sodic soil I				Sodic soil II			
Soil depth (cm)	Bulk density ($g \cdot cm^{-3}$)	Soil Weight (kg)	Total weight (kg)	Soil depth (cm)	Bulk density ($g \cdot cm^{-3}$)	Soil Weight (kg)	Total weight (kg)
0–5	1.41	1.28	26.13	0–5	1.58	1.43	26.72
5–20	1.46	3.97		5–20	1.52	4.12	
20–40	1.41	5.13		20–40	1.46	5.28	
40–60	1.47	5.32		40–60	1.45	5.28	
60–100	1.44	10.43		60–100	1.46	10.61	

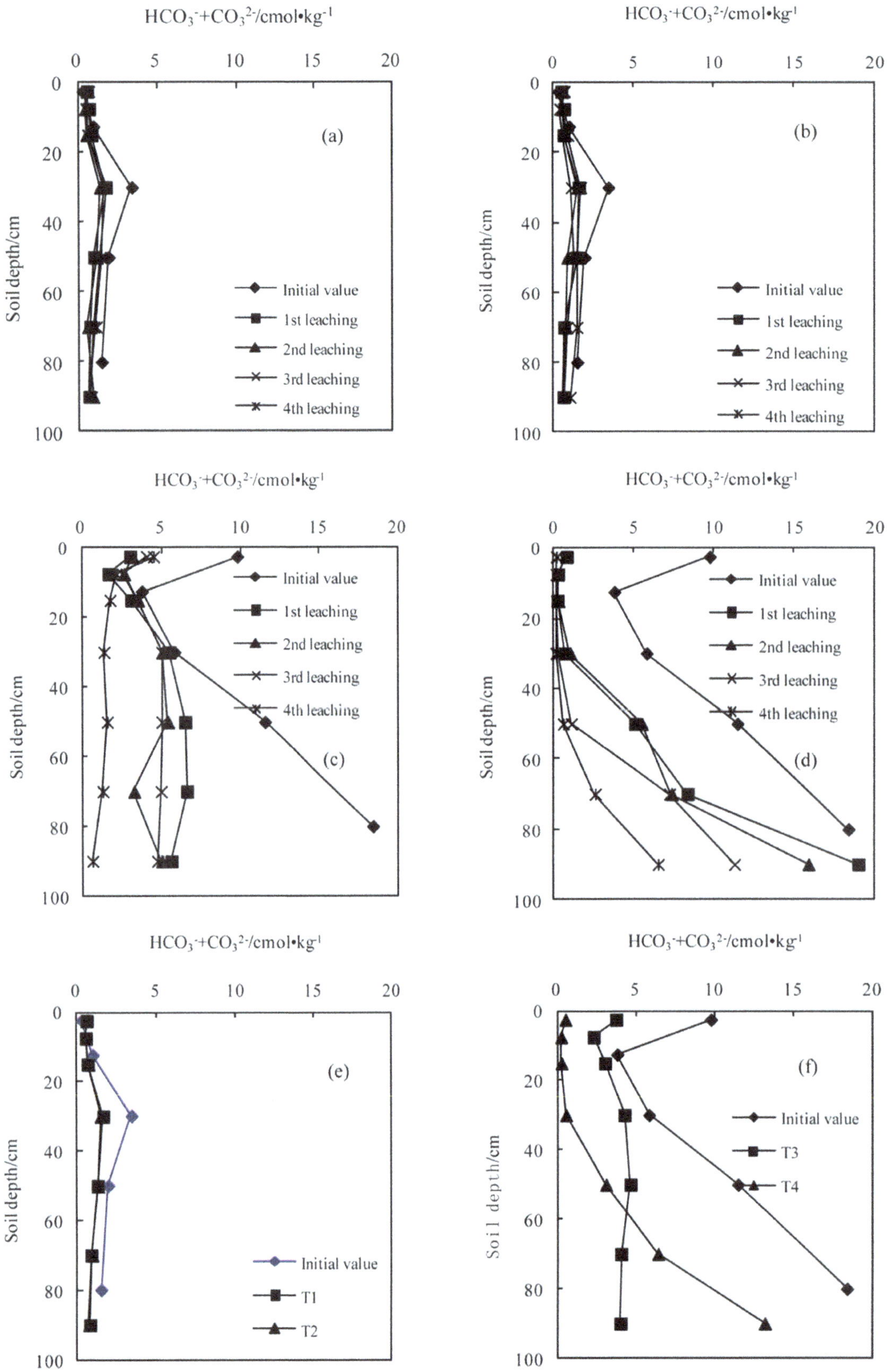

Figure 1. Variation in the concentration of carbonate and bicarbonate with successive leaching times (a.T1; b.T2; c.T3; d.T4); e. The average values in sodic soil I after leaching four times; f. The average value in sodic soil II after leaching four times).

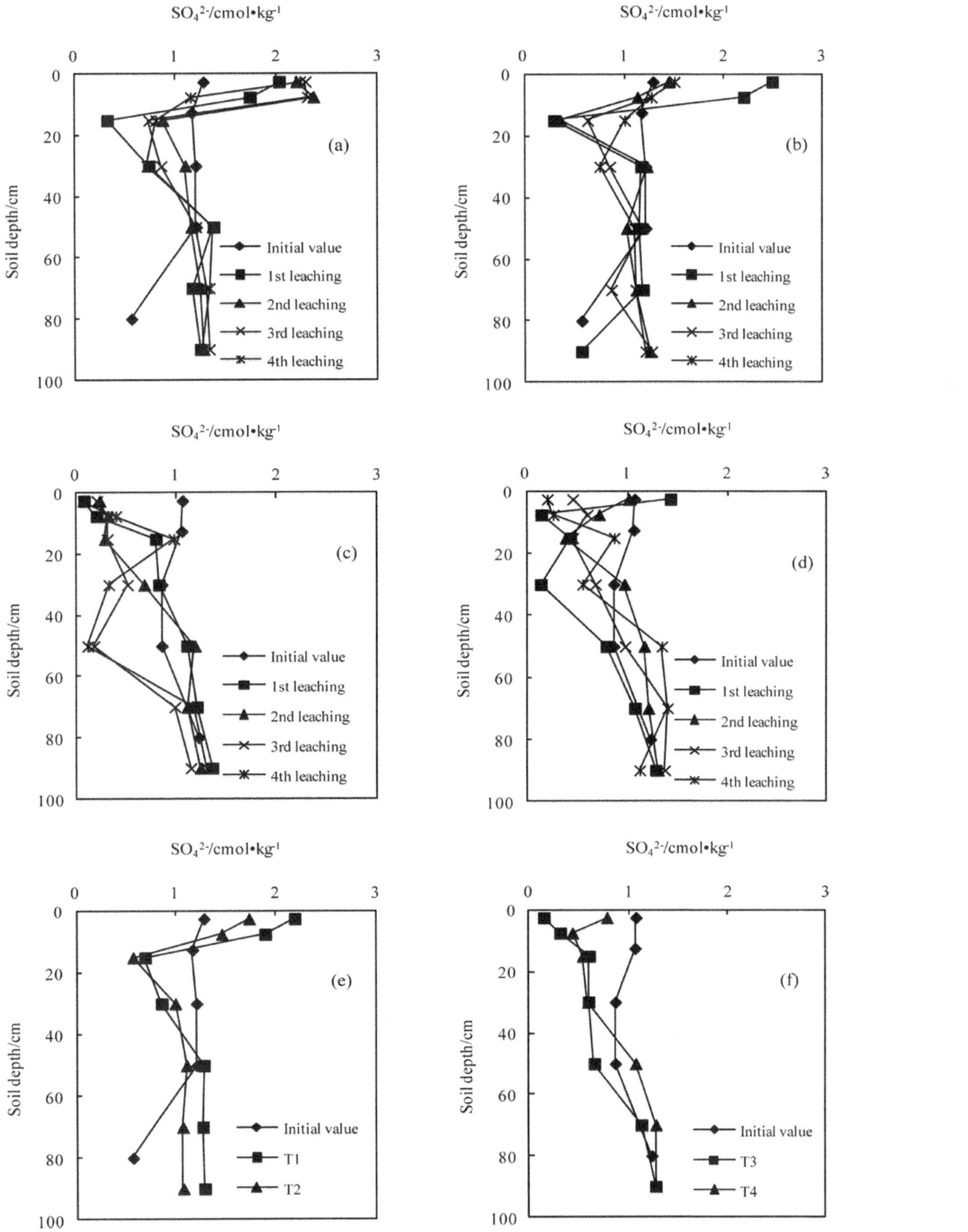

Figure 2. Variation in the concentration of sulfate with successive leaching times (a.T1; b.T2; c.T3; d.T4); e. The average values in sodic soil I after leaching four times; f. The average value in sodic soil II after leaching four times).

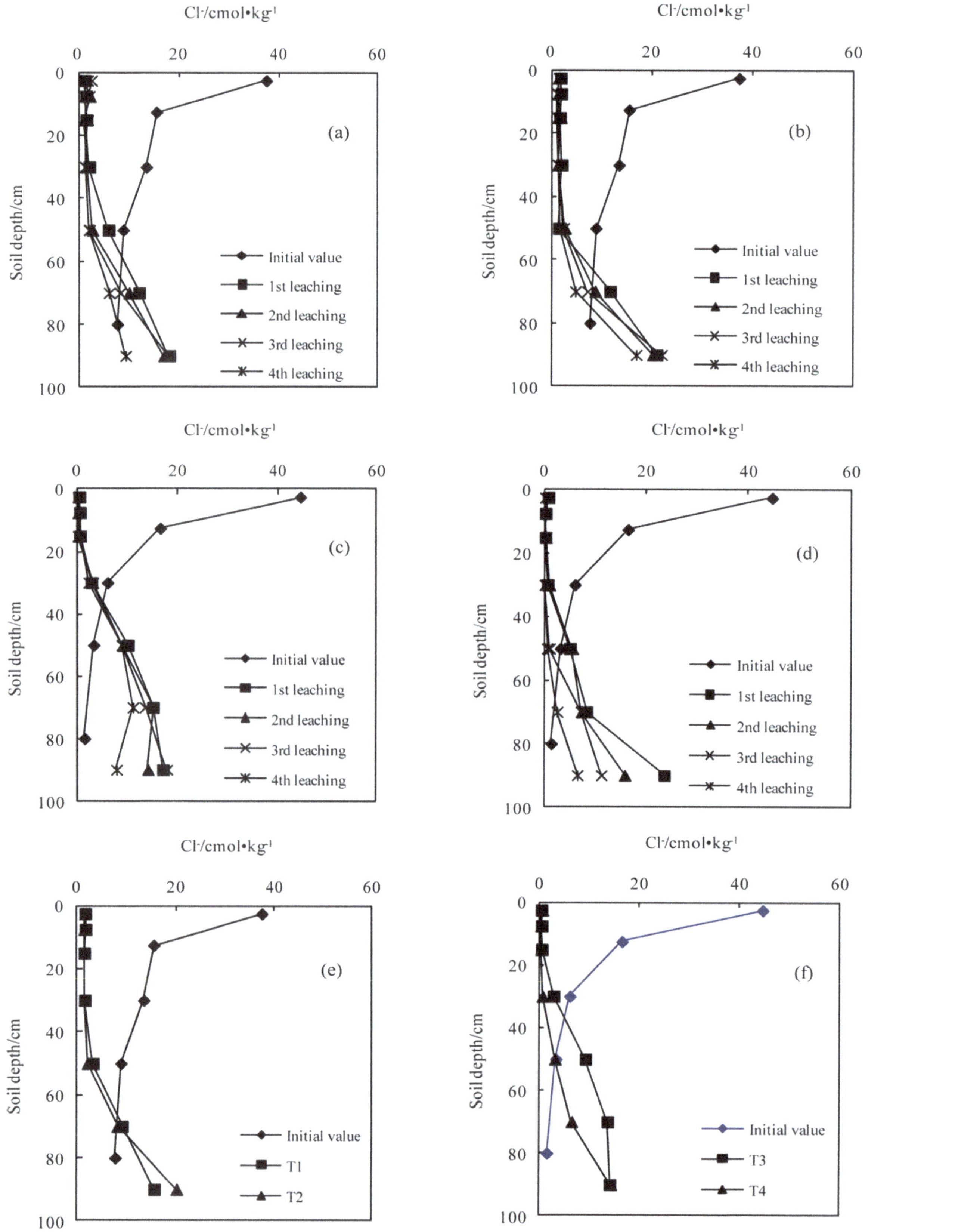

Figure 3. Variation in the concentration of chloride with successive leaching times (a.T1; b.T2; c.T3; d.T4); e. The average values in sodic soil I after leaching four times; f. The average value in sodic soil II after leaching four times).

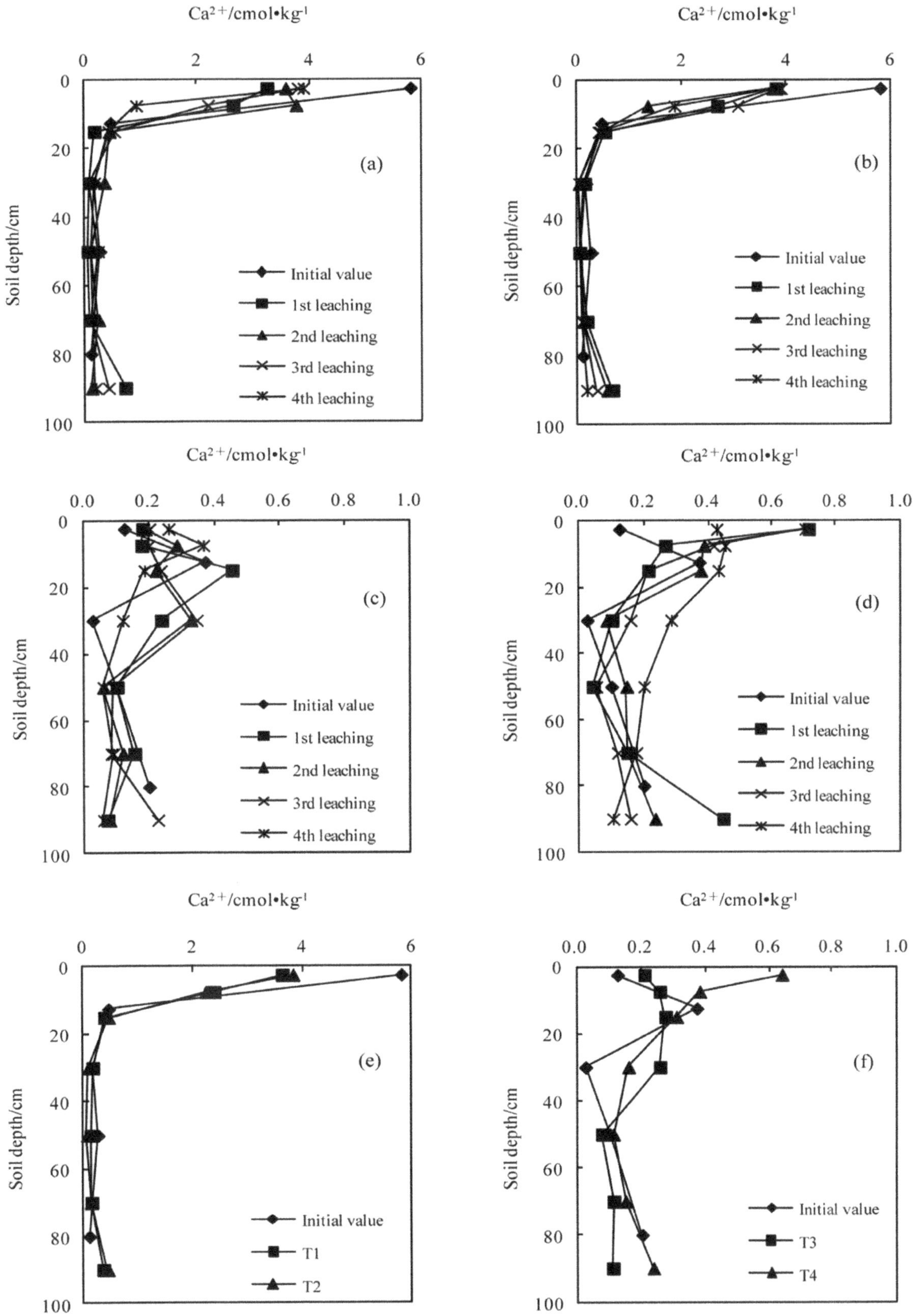

Figure 4. Variation in the concentration of calcium with successive leaching times (a.T1; b.T2; c.T3; d.T4); e. The average values in sodic soil I after leaching four times; f. The average value in sodic soil II after leaching four times).

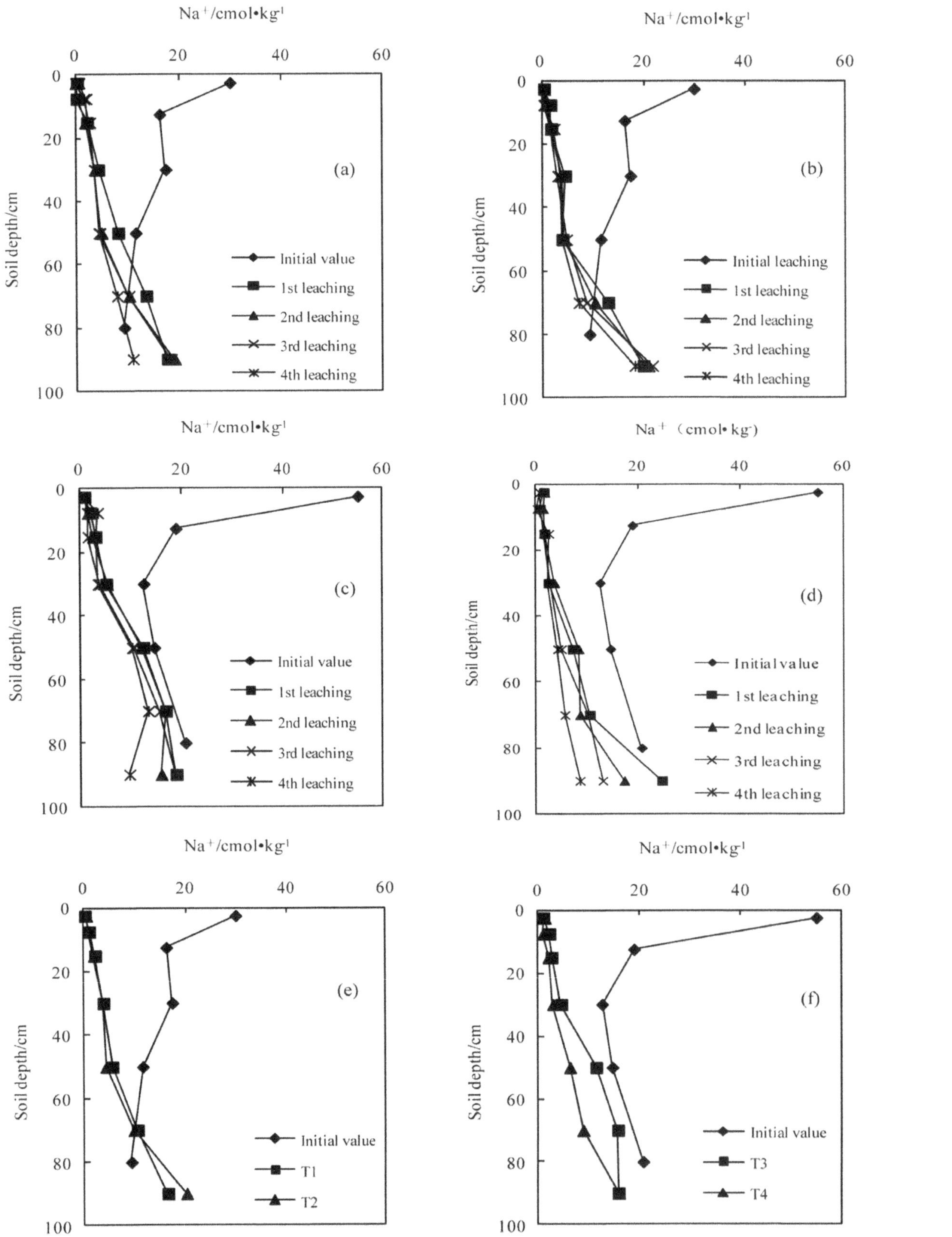

Figure 5. Variation in the concentration of sodium with successive leaching times (a.T1; b.T2; c.T3; d.T4); e. The average values in sodic soil I after leaching four times; f. The average value in sodic soil II after leaching four times).

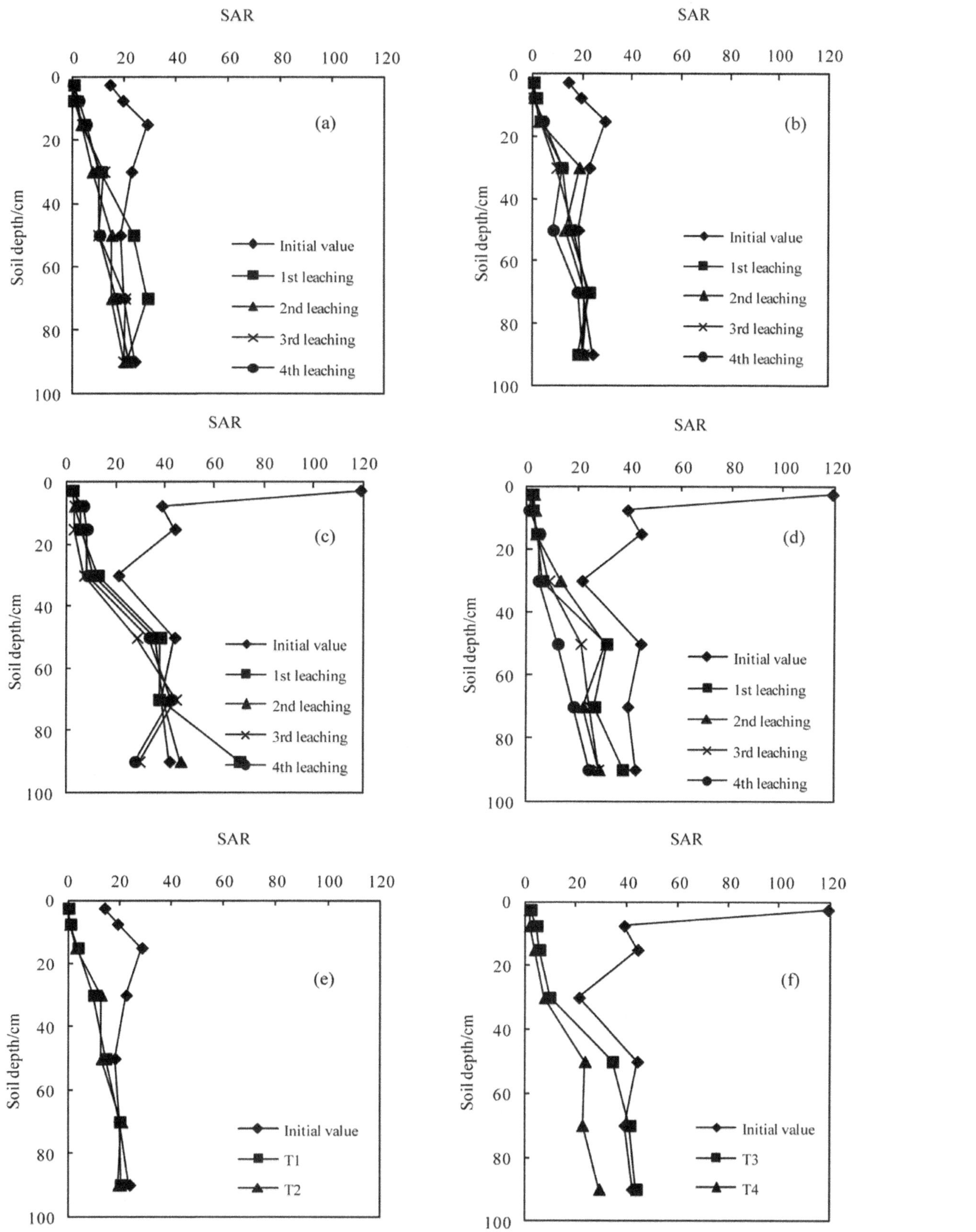

Figure 6. Variation in SAR with successive leaching times (a.T1; b.T2; c.T3; d.T4); e. The average values in sodic soil I after leaching four times; f. The average value in sodic soil II after leaching four times).

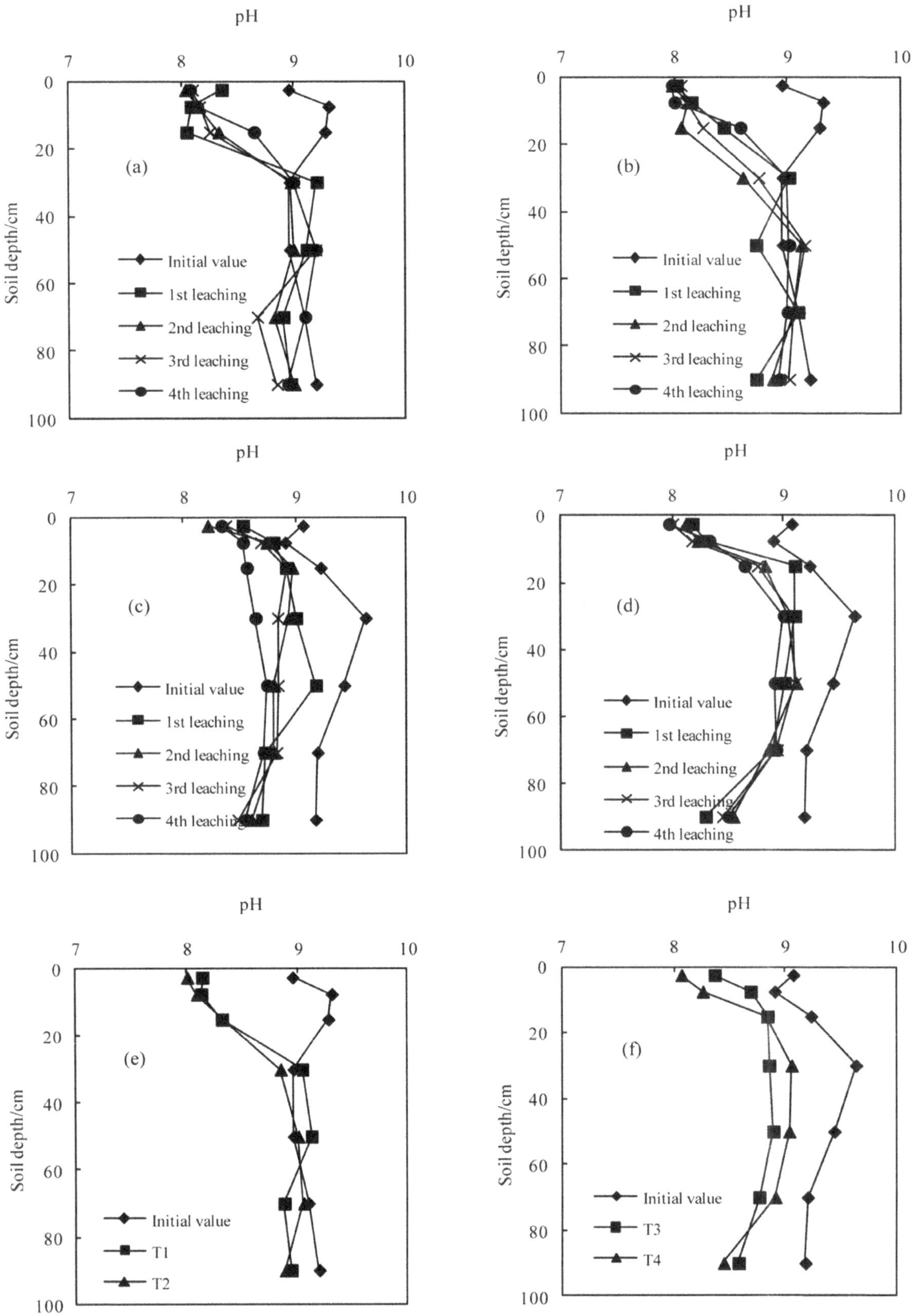

Figure 7. Variation in pH value with successive leaching times (a.T1; b.T2; c.T3; d.T4); e. The average values in sodic soil I after leaching four times; f. The average value in sodic soil II after leaching four times).

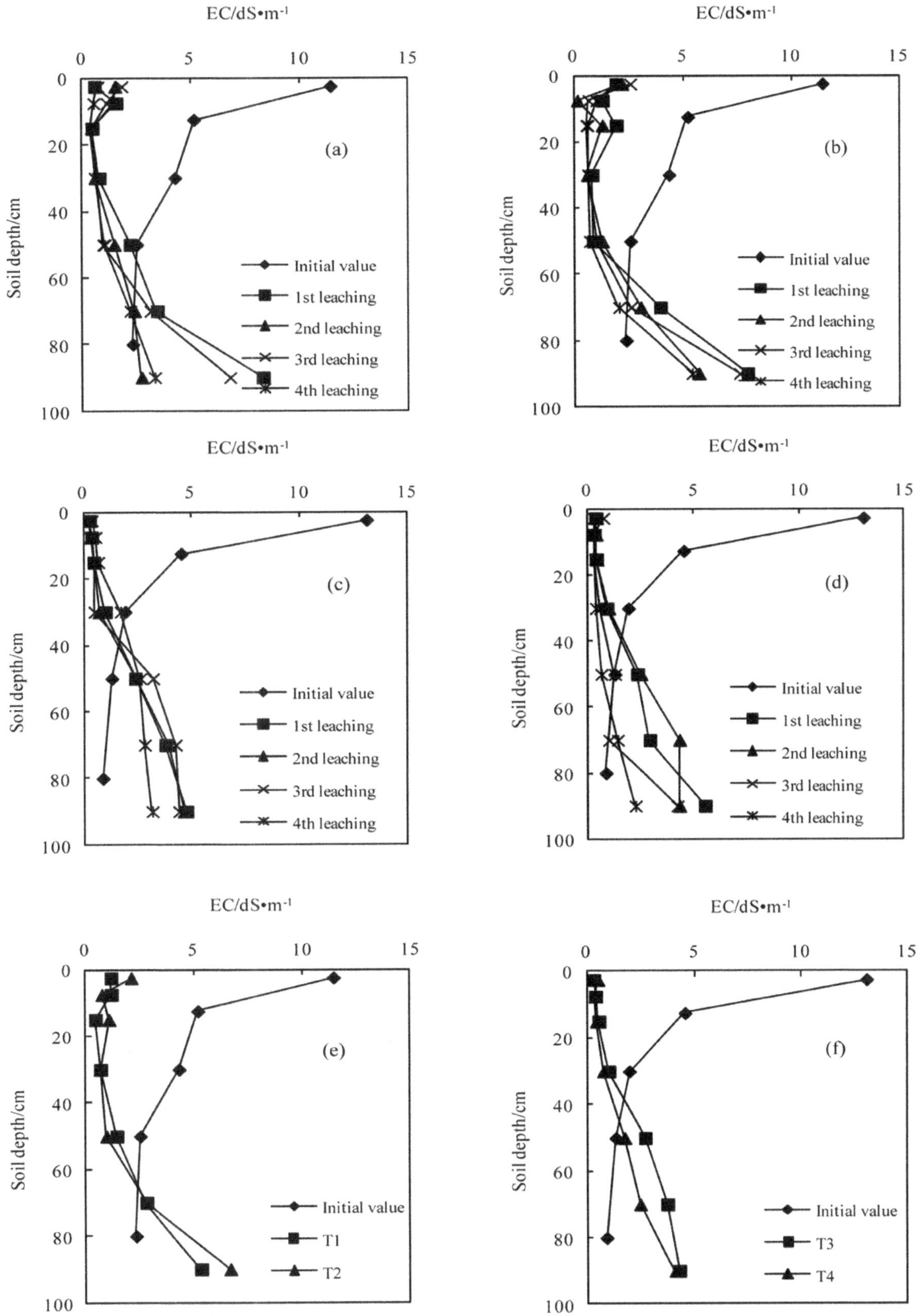

Figure 8. Variation in EC value with successive leaching times (a.T1; b.T2; c.T3; d.T4); e. The average values in sodic soil I after leaching four times; f. The average value in sodic soil II after leaching four times).

$$SAR_{1:5} = Na^+ / \sqrt{\frac{Ca^{2+} + Mg^{2+}}{2}}. \quad (3)$$

The physical and chemical items were measured twice, and the chemical analysis was replicated three times. The standard errors of the means of the three samples from each treatment were calculated. Descriptive statistics (mean, range, standard deviation (SD) and coefficient of variation (CV)) on soil parameters were analyzed with SAS9.1 (Statistical Analysis System; SAS Institute Inc.) and they were considered to be significantly different at $P<0.05$.

Results

Variations in the Soluble Ion Composition

After applying the BFGD to the sodic soils, the main component of BFGD, $CaSO_4$, was constantly dissolved under the leaching action of the water flow, producing a rapid chemical reaction between the $CaSO_4$ and the soluble salts in the soil solution. As the $CaSO_4$ exchanged with the sodium in the soil colloid, it changed the soluble salts composition of the soil solution. The variation in the soluble salts composition (anions: HCO_3^-+CO_3^{2-}, SO_4^{2-} and Cl^-; cations: Ca^{2+} and Na^+)of the soil solution are shown in Figs. 1–5.

The concentrations of HCO_3^- and CO_3^{2-}. The concentrations of HCO_3^- and CO_3^{2-} in sodic soil II were higher than those in sodic soil I at the initial stage. After application of the BFGD, the concentrations of HCO_3^- and CO_3^{2-} in the two sodic soils decreased with each leaching, and the concentrations in the surface soils (0–30 cm) were lower than those in the deep soils (30–100 cm).

The concentrations of SO_4^{2-}. During the initial leaching stage, the concentration of SO_4^{2-} at a depth of 0–20 cm in sodic soil I increased significantly, but there was no significant increase in sodic soil I. With each successive leaching, the concentration of SO_4^{2-} gradually reduced, and that in sodic soil II at a depth of 0–20 cm also began to decrease. The concentrations of SO_4^{2-} with a high BFGD application rate were higher than those with a low application rate at all soil depths in the two sodic soils.

The concentrations of Cl^-. The concentration of Cl^- gradually reduced from the top to the bottom soils during the initial stage. After application of the BFGD and leaching, the Cl^- concentrations in the top soils of the two sodic soils decreased. The Cl^- accumulated in the bottom soil layer and gradually increased from the top to the bottom. The Cl^- concentrations with a high BFGD application rate were lower than those with a low application rate at all soil depths in the two sodic soils.

The concentrations of Ca^{2+}. After application of the BFGD, the concentration of Ca^{2+} slightly increased in the top soil layer, but there was no obvious change in the other layers. The concentration of Ca^{2+} decreased significantly in the top soil layer with each subsequent leaching. The Ca^{2+} concentrations with a high BFGD application rate were higher than those with a low BFGD application rate at all soil depths in the two sodic soils.

The concentrations of Na^+. The concentration of Na^+ gradually reduced from the top to the bottom soils during the initial stage. After application of the BFGD and leaching, the concentration of Na^+in each soil layer decreased, with a more obvious decrease in the top soil layer, with a gradual increase from the top to the bottom. The Na^+ concentrations with a high BFGD application rate were lower than those with a low BFGD application rate at all soil depths in the two sodic soils.

Variations in the Properties of the Sodic Soils

Soil SAR. Under soil dispersion conditions, high SAR levels in the soil solution will cause severe soil clay dispersion and decrease hydraulic conductivity [20]. The variations in SAR following the application of BFGD are shown in Fig. 6. Leaching greatly reduced the SAR in the top soils, and to a much lesser extent in the bottom soils. The SAR was significantly reduced below the initial value at depths of 0–50 cm, 0–60 cm, 0–40 cm and 0–50 cm in T1, T2, T3 and T4 respectively, and the SAR was below 13 at depths of 0–50 cm, 0–50 cm, 0–30 cm and 0–30 cm respectively. For those soil depths at which the SAR was below the initial value, sodic soil I with a low ESP was greater than sodic soil II with a slightly higher ESP, and the treatments with high BFGD application rates were greater than those with low BFGD application rates. The extent of the SAR decrease was positively correlated with the amount of leaching. Under the same soil conditions, the reclamation effect improved with each leaching and the SAR showed an increasing trend from the top soil to the bottom soil. Under different treatments, the SAR of T2 was lower than that of T1 in sodic soil I and the SAR of T4 was lower than that of T3 in sodic soil II. The soil SAR was generally below the initial value in T4 at all depths, but only at a depth of 0–50 cm in T3. This indicates that the $Na^+$was constantly being swapped and moved to the bottom layer of soil, thus increasing the relative proportion of Na^+ in the bottom layer. Therefore, applying the BFGD and leaching reduced the Na^+ in the soil solution by hindering the diffusion of the soil particles. Hence, the physical structure of the soil was improved and the sodic soil was reclaimed.

Soil EC. The variations in soil EC after the application of BFGD are shown in Fig. 7. The soil EC gradually increased from the top to the bottom soils during the initial stage. The EC in the surface soil significantly reduced under different treatments with the application of the BFGD and leaching, but increased in the bottom soil. The EC significantly decreased at depths of 0–60 cm, 0–60 cm, 0–40 cm and 0–40 cm in T1, T2, T3 and T4 respectively, and was below 4 dS m^{-1} at depths of 0–70 cm, 0–70 cm, 0–50 cm and 0–50 cm respectively. Therefore, application of the BFGD resulted in consistent improvement in the physical and chemical properties of sodic soils, and leaching did not cause an accumulation of salt in the surface soil. However, leaching should be strengthened to ensure the rapid discharge of salt accumulated in the bottom soil, prevent soil salinization and consolidate the reclamation effect on sodic soils.

Soil pH. The variations in soil pH after the application of BFGD are shown in Fig. 8. The soil surface pH significantly reduced under different treatments with the application of the BFGD and leaching. The soil pH significantly decreased at depths of 0–40 cm, 0–40 cm, 0–20 cm and 0–20 cm in T1, T2, T3 and T4 respectively, and was below 8.5 at depths of 0–20 cm, 0–20 cm, 0–10 cm and 0–10 cm respectively. With different BFGD application rates, the pH value of T2 was less than that of T1 in sodic soil I at depths of 0–60 cm, but the value for T2 was higher than that for T1 at depths below 60 cm. The pH value of T4 was less than that of T3 in sodic soil II at depths of 0–20 cm, whereas the pH value of T4 was more than that of T3 in sodic soil II at depths of 20–100 cm. In the process of improving sodic soils, the Ca^{2+} concentration in the soil solution forms carbonate and precipitate due to excessive HCO_3^- concentration in the irrigation water, thereby increasing the relative content of Na^+in the soil solution [5,16,21]. The soil solution had a pH below 8.5, illustrating that the carbonate and bicarbonate had disappeared and the soluble salt had gradually made the transition from sodic salt to neutral salt, thus the sodic soils had been reclaimed.

Discussion

The Mechanism of Change in the Ionic Composition of Soluble Salts

The application of BFGD in sodic soils caused the main component of the BFGD, $CaSO_4$, to dissolve due to the water flow infiltration. The $CaSO_4$ comprised the dissolved part and the residual part, and the dissolved part reacted rapidly with the sodic salts (Na_2CO_3 and $NaHCO_3$); see Equations 4 and 5 [5,22–23].

$$Na_2CO_3 + CaSO_4 \rightarrow Na_2SO_4 + CaCO_3 \quad (4)$$

$$2NaHCO_3 + CaSO_4 \rightarrow Na_2SO_4 + Ca(HCO_3)_2 \quad (5)$$

The concentrations of HCO_3^- and CO_3^{2-} rapidly decreased with these reactions, thus also reducing the soil pH value. After the disappearance of the sodic salt, there was an exchange reaction between the BFGD and the exchangeable Na^+ in the soil colloids (see Equation 6), which generated the soluble salt sodium sulfate [3]. As the sodium sulfate leached into the deep soil layer, the sodic level of the soil continuously decreased.

$$[\text{Soil colloid}]_{Na}^{Na} + CaSO_4 \rightarrow [\text{Soil colloid}]\ Ca + Na_2SO_4 \quad (6)$$

The application of the BFGD should increase the concentration of SO_4^{2-} and Ca^{2+}due to the dissolution of $CaSO_4$. However, because the solubility of $CaSO_4$ is relatively low, only 2 $g \cdot l^{-1}$ (25°C), the dissolved Ca^{2+} reacted with the sodic salts in the soil solution and the exchangeable Na^+ in the soil colloids, thus consuming the Ca^{2+} in the soil solution. The chemical reaction between SO_4^{2-} and Na^+ generated soluble sodium sulfate, which was then leached into the deeper soil levels. Therefore, rather than increasing the concentrations of $Ca^{2+}$and SO_4^{2-}, the concentrations were actually lower than the initial values in some soil layers. This effect is predominantly due to the increased permeability of the sodic soil once the salt has been leached from it and the concentration of Ca^{2+} and SO_4^{2-} has decreased [3,6,16,24–26].

In sodic soil areas there is also a certain amount of salinization, in which the soils contain some salt and the NaCl content is often very high. NaCl has a salting effect on the BFGD, which improves the solubility of $CaSO_4$, increases the infiltration characteristics of sodic soils and accelerates their reclamation [5,26–30]. With the reclamation of sodic soils, NaCl was constantly leached from the soil, thus reducing the NaCl content in the surface soil, and accumulating it in the deeper soil layers. There was a decrease in the Cl^- content at depths of 0–60 cm and 0–40 cm for sodic soil I and sodic soil II respectively, and an increase in the Cl^- content at depths of 60–100 cm and 40–100 cm.

As a result, the soil soluble salts made the transition from sodic salts containing Na_2CO_3 and $NaHCO_3$ to neutral salts containing NaCl and Na_2SO_4. Leaching of the soluble salts from the soil resulted in continuous improvement in the sodic soils.

The Relationship between Soil Properties and Soluble Salts Composition

The BFGD also immediately increased the EC of the soil solution and increased the ionic strength, which leads to proton generation and a reduction in the soil pH value [31]. The higher BFGD application rate reduced the soil SAR significantly more efficiently than the lower BFGD application rate because the BFGD increased Ca^{2+} in the soil solution, which promoted the displacement of the adsorbed Na^+ and subsequent leaching. A decrease in SAR_e was also found in the control treatment. This was attributed to the "valence dilution" mechanism reported by Reeve and Bower [32]. In a soil-water system, there is equilibrium between the monovalent and divalent cations at the soil exchange site and in the solution. The equilibrium is altered by adding water to the system. This dilution of the soil solution favors the adsorption of divalent cations such as Ca^{2+} at the cost of monovalent cations such as Na^+. Furthermore, if the soil solution is diluted by a factor of ΔC while maintaining the same ionic ratios, the initial SAR decreases by the factor $(\Delta C)^{1/2}$, i.e.,$SAR_{final} = (\Delta)^{0.5} \times SAR_{initial}$. Because of the dilution, the $(\Delta C)^{1/2}$ is less than 1. Thus, the SAR_{final} is less than the $SAR_{initial}$.

The application of BFGD was also found to be effective in lowering the soil pH. The significant decrease in soil pH may be attributed to leaching of the exchangeable Na^+, which was replaced by Ca^{2+} from the BFGD, and the decreased concentrations of HCO_3^- and CO_3^{2-}. This pH decrease was partly associated with the degree of soil reclamation and especially with the reduction in soil sodicity [5,33]. The pH of sodic soils is a function of the activity of CO_3^{2-}+HCO_3^-, and partially due to CO_2 and ionic strength [34]. The presence of soluble Na+ increases the pH of calcareous soils by increasing the activity of carbonate and bicarbonate. Sodium carbonate (Na_2CO_3) and sodium bicarbonate ($NaHCO_3$) are relatively soluble, and their removal by leaching can substantially reduce pH. Thus, the decrease in pH was in part associated with the leaching of soluble salts [35–36].

Conclusions

The following conclusions can be drawn from our findings.

(1) The application of BFGD and leaching caused the soil soluble salts to make the transition from sodic salts containing Na_2CO_3 and $NaHCO_3$ to neutral salts containing NaCl and Na_2SO_4. The poor physical properties of sodic soils meant that long leaching times were needed to obtain good reclamation effects.

(2) The soil SAR, pH and EC decreased at all soil depths, although it was more significant in the top soil. At a depth of 0–40 cm, the soil SAR, EC and pH were reduced to less than 13, 4 dS m^{-1} and 8.5 respectively, in both sodic soils I and II.

(3) Leaching played a key role in the reclamation process, and the reclamation effect was positively correlated with the amount of leaching. The soil salts did not accumulate in the top soil layer, and there was a slight increase in the soil salts in the middle and bottom layers. Stronger leaching is thus required to ensure the salt accumulated in the bottom soil is discharged rapidly, thus preventing soil salinization and consolidating the reclamation effect.

(4) The use of BFGD is inefficient, and the reclamation effect was improved by increasing the BFGD application rate by 20% above the theoretically calculated base rate.

(5) The results of this research demonstrate that the reclamation of sodic soils using BFGD is promising.

Author Contributions

Conceived and designed the experiments: JW PY. Performed the experiments: JW. Analyzed the data: JW ZB. Contributed reagents/materials/analysis tools: JW ZB. Wrote the paper: JW.

References

1. Wang SJ, Chen CH, Xu XC, L YJ (2008) Amelioration of alkali soil using flue gas desulfurization byproducts: Productivity and environmental quality. Environmental Pollution 151: 200–204.
2. Carlson CL, Adriano DC (1993) Environmental impacts of coal combustion residues. Journal of Environmental Quality 22: 227–247.
3. Chun S, Nishiyama M, Matsumoto S (2001) Sodic soils reclaimed with by-product from flue gas desulphurization: corn production and soil quality. Environmental Pollution 114: 453–459.
4. Stuczynski TL, McCarty GW, Wright RJ (1998) Impact of coal combustion product amendments on soil quality: Mobilization of soil organic nitrogen. Soil Science 163: 953–959.
5. USDA (1954) Diagnosis and improvement of saline and alkali soils. USDA, U.S. Govt. Printing Office, Washington, DC.
6. Wang JM, Bai ZK, Yang PL (2012) Sodic Soil Properties and Sunflower Growth as Affected by Byproducts of Flue Gas Desulfurization. PLoS ONE 7(12): e52437.
7. Frenke H, Gerstle Z, Alperovitch N (1989) Exchange-induced dissolution of gypsum and the reclamation of sodic soils. Soil Science 40: 599–611.
8. Clark RB, Ritchey KD, Baligar VC (2001) Benefits and constraints for use of FGD products on agricultural land. Fuel 80: 821–828.
9. Sloan JJ, Dowdy RH, Dolan MS, Rehm GW (1999) Plant and soil responses to field-applied flue gas desulphurization residue? Fuel 78: 169–174.
10. Wang JM, Yang PL, Ren SM, Xiang GM (2005) Variation of chemical indices of alkaline soil ameliorated with desulphurization byproducts. Acta Pedologica Sinica (in Chinese) 42(1): 98–105.
11. Wang JM, Yang PL (2004) The effect on physical and chemical properties of saline and sodic soils reclaimed with byproduct from flue gas desulphurization, In *Land and Water Management Decision Tools and Practices Vol.II*, Huang GH. and S.Pereria L (eds). China Agriculture Press, Beijing: 1015–1021.
12. Sakai Y, Matsumoto S, Sadakata M (2004) Alkali soil reclamation with flue gas desulphurization gypsum in China and assessment of metal content in corn grains. Soil & Sediment Contamination 13(1): 65–80.
13. Ilyas M, Qureshi RH, Qadir MA (1997) Chemical changes in saline-sodic soils after gypsum application and cropping. Soil Technologies 10: 247–260.
14. Chen L, Ramsier C, Bigham J, Slater B, Kost D, et al (2009) Oxidation of FGD-$CaSO_3$ and effect on soil chemical properties when applied to the soil surface. Fuel 88: 1167–1172.
15. Oster JD, Frenkel H (1980) The chemistry of the reclamation of sodic soils with gypsum and lime. Soil Science Society of America Journal 44: 41–45.
16. Chi CM, Zhao CW, Sun XJ, Wang ZC (2012) Reclamation of saline-sodic soil properties and improvement of rice (Oriza sativa L.) growth and yield using desulfurized gypsum in the west of Songnen Plain, northeast China. Geoderma 187–188: 24–30.
17. Ministry of Agriculture and Fisheries (1987) Control Standards for Pollutants in Fly Ash for Agricultural Use (GB8173–87). Beijing: China Standards Press (in Chinese).
18. Ministry of Agriculture and Fisheries (1984) Control Standards for Pollutants in Sludge for Agricultural Use (GB4284–84). Beijing: China Standards Press (in Chinese).
19. Rhodes DR (1982) Cation exchange capacity. In *Methods of soil analysis Part 2, Chemical methods.* Page AL, Miller RH and Keeney DR (eds). American Society of Agronomy:Madison, Wisconsin, 149–165.
20. Chorom M, Rengasamy P, Murry RS (1994) Clay dispersion as influenced by pH and net particle of sodic soils. Australian Journal of Soil Research 32: 1152–1163.
21. Truman CC, Nuti RC, Truman LR, Dean JD (2010) Feasibility of using FGD gypsum to conserve water and reduce erosion from an agricultural soil in Georgia. Catena 81: 234–239.
22. Carter MR, Pearen JR (1989) Amelioration of a saline-sodic soil with low application of calcium and nitrogen amendments. Arid Soil Research 3: 1–9.
23. Sakai Y, Aihara T, Yamamoto T, Sadakata M (2010) Evaluation of soil properties and additional application of desulfurization gypsum in an area of no amelioration after an initial application in sodic soil of China. Kagaku Kogaku Ronbunshu 36 (4): 275–280.
24. Gharaibeh MA, Eltaif NI, Shra'ah SH (2010) Reclamation of a calcareous saline-sodic soil using phosphoric acid and by-product gypsum. Soil Use and Management 26(2): 141–148.
25. Minashina NG (2005) Concentration and composition of salts in the solution of a gypsum-containing soil as dependent on the soil water content. Eurasian Soil Science 38(7): 718–726.
26. Ritvo G, Speed FM, Neill WH, Dixon JB, Lawrence AL, et al (1999) Regression analysis of soil chemical composition for two shrimp farms in Texas. Journal of the World Aquaculture Society 30(1): 26–35.
27. Reading LP, Baumgartl T, Bristow KL, Lockington DA (2012) Hydraulic conductivity increases in a sodic clay soil in response to gypsum applications: impacts of bulk density and cation exchange. Soil Science 177(3): 165–171.
28. Singh K, Singh B, Singh RR (2012) Changes in physical-chemical, microbial and enzymatic activities during restoration of degraded sodic lands: Ecological suitability of mixed forest over plantation. Catena 96: 57–67.
29. Rhoton FE, McChesney DS (2011) Erodibility of a Sodic Soil Amended With Flue Gas Desulfurization Gypsum. Soil Science 176(4): 190–195.
30. Muhammad D, Khattak RA (2011) Wheat yield and chemical composition as influenced by integrated use of gypsum, pressmud and FYM in saline-sodic soil. Journal of the Chemical Society of Pakistan 33(1): 82–89.
31. Simon M, Garcia I (1999) Physico-chemical properties of the soil-saturation extracts: estimation from electrical conductivity. Geoderma 90 (1–2): 99–109.
32. Reeve RC, Bower CA (1960) Use of high-salt waters as a flocculant and source of divalent cations for reclaiming sodic soils. Soil Science 90, 139–144.
33. Pandey VC, Singh K, Singh B, Singh RP (2011) New approaches to enhance eco- restoration efficiency of degraded sodic lands: critical research needs and future prospects. Restoration Ecology 29: 322–325.
34. Mashhady AS, Rowell DL (1978) Soil alkalinity. I. Equilibria and alkalinity development. Journal of Soil Science 29: 65–75.
35. Katerji N, Van Hoorn JW, Hamdy A, Mastrorilli M, Mou Karzel E (1997) Osmotic adjustment of sugar beets in response to soil salinity and its influence on stomata conductance, growth and yield. Agricultural Water Management 34: 57–69.
36. Shannon MC (1997) Adaptation of plants to salinity. In *Advances in Soil Science.* Lal R and Steward B A (eds). Springer-Verlag, 75–120.

A Decision Analytic Approach to Exposure-Based Chemical Prioritization

Jade Mitchell[1], Nicolas Pabon[2], Zachary A. Collier[3], Peter P. Egeghy[4], Elaine Cohen-Hubal[4], Igor Linkov[3]*, Daniel A. Vallero[4]

1 Biosystems & Agricultural Engineering, Michigan State University, East Lansing, Michigan, United States of America, **2** Physics Department, Carnegie Mellon University, Pittsburgh, Pennsylvania, United States of America, **3** Environmental Laboratory, Engineer Research and Development Center, United States Army Corps of Engineers, Concord, Massachusetts, United States of America, **4** Office of Research and Development, United States Environmental Protection Agency, Research Triangle Park, North Carolina, United States of America

Abstract

The manufacture of novel synthetic chemicals has increased in volume and variety, but often the environmental and health risks are not fully understood in terms of toxicity and, in particular, exposure. While efforts to assess risks have generally been effective when sufficient data are available, the hazard and exposure data necessary to assess risks adequately are unavailable for the vast majority of chemicals in commerce. The US Environmental Protection Agency has initiated the ExpoCast Program to develop tools for rapid chemical evaluation based on potential for exposure. In this context, a model is presented in which chemicals are evaluated based on inherent chemical properties and behaviorally-based usage characteristics over the chemical's life cycle. These criteria are assessed and integrated within a decision analytic framework, facilitating rapid assessment and prioritization for future targeted testing and systems modeling. A case study outlines the prioritization process using 51 chemicals. The results show a preliminary relative ranking of chemicals based on exposure potential. The strength of this approach is the ability to integrate relevant statistical and mechanistic data with expert judgment, allowing for an initial tier assessment that can further inform targeted testing and risk management strategies.

Editor: Shyamal D. Peddada, National Institute of Environmental and Health Sciences, United States of America

Funding: The United States Environmental Protection Agency through its Office of Research and Development (funded and managed or partially funded and collaborated in) the research described here under an Interagency Agreement with the U.S. Army Corps of Engineers (DW-96-95807601-0). The funders had no role in study design, data collection and analysis, decision to publish, or preparation of the manuscript.

Competing Interests: The authors have declared that no competing interests exist.

* E-mail: Igor.Linkov@usace.army.mil

Introduction

Manufactured chemicals are widely used in products such as cosmetics, plastics, and electronics, and have applications in almost all industrial processes in sectors including energy, agriculture, and pharmaceuticals [1]. Increasing dependence on manufactured chemicals has not, however, been matched by an adequate increase in our understanding of the risks these may pose to the environment and human health [2]. Many chemicals in U.S. commerce today have unknown environmental fates and poorly understood potential for human exposure, including some of the most ubiquitous commercial chemicals, such as surfactants, fragrances, cleaning agents and pesticides [3,4]. In this context, exposure is the contact of a stressor (i.e., a chemical agent) with a receptor (i.e., a human or a human population) for a specific duration of time [5]. Because of the lack of resources and sufficient scientific information on toxicity [6] and exposure [3] for the assessment of all chemicals, efforts are typically, and rationally, devoted to assessing those chemicals believed to pose the greatest potential risks based on production volume and chemical properties.

Within the domain of human health risk assessment, toxicity is an indication and measurement of the severity of adverse health effects a chemical causes in relation to an exposure level (dose). We broadly define exposure to be the contact of a stressor with a receptor for a specific duration of time [5]. The stressors of interest are chemical agents that can potentially lead to an adverse impact and the receptors of interest are individuals or population of individuals. Exposure is complex and dynamic in nature due to its spatial and temporal characteristics. For this reason, exposure-based prioritization efforts focus on relative exposure potential as a means to evaluate and rank chemicals. While prioritization is in of itself a risk management strategy, other risk management decisions may follow to include the allocation of scarce resources to complete future risk assessments, collection of additional data or testing, and/or (bio) monitoring. Therefore, the resolution and precision of the data incorporated in these efforts may vary according to the overall objective of the prioritization.

The U.S. EPA Office of Chemical Safety and Pollution Prevention recently performed a chemical prioritization exercise to identify 83 "TSCA Work Plan Chemicals" [7] as candidates for risk assessment during the next few years. Broad stakeholder input was used to identify prioritization and screening criteria and data sources. Chemicals were evaluated based on their combined hazard, exposure potential, and persistence and bioaccumulation characteristics using a two-step process. In the first step, a set of data sources was used to identify 1,235 chemicals meeting one or more criteria suggesting concern, namely: known reproductive or developmental effects; persistent, bioaccumulative, and toxic

(PBT) properties; known carcinogenicity; and presence in children's products. Excluding those chemicals not regulated under TSCA and those with physical and chemical characteristics that do not generally present significant health hazards narrowed the number of chemicals down to 345 candidates. In the second step, a numerical algorithm was used to score each chemical based on three characteristics: hazard, exposure, and potential for persistence or bioaccumulation. Candidate chemicals that ranked highest on the basis of their total score were identified as work plan chemicals; those that could not be scored because of an absence of exposure or hazard data were identified as candidates for information gathering.

Using the methodology described above, EPA has been able to identify a priority set of chemicals for near-term assessment based on criteria widely accepted as warranting concern. The scoring algorithm is transparent and the data sources are well documented. Focusing on chemicals with documented evidence of concern (i.e. "data-rich") is reasonable in light of limited prototypes for *post hoc* screening and the paucity of available resources. However, this approach may not adequately address the need to make decisions about the thousands of chemicals in commerce and the hundreds of new chemicals introduced each year for which there is *little or no* information [1,3].

To support the development of novel rapid approaches for evaluating potential exposure of both existing and emerging chemicals, the EPA has initiated the ExpoCast research program [8]. This program is keenly interested in characterizing exposures across the chemical life cycle –manufacturing, transportation, product formulation, consumer product usage and finally disposal. EPA seeks to build on current chemical exposure models and knowledge to generate robust new protocols that better support chemical evaluation, risk assessment and risk management. Recent activities under this program have evaluated utility of available approaches for the purpose of rapidly prioritizing large numbers of chemicals on the basis of exposure [9,10].

A number of exposure models were recently comparatively evaluated through the EPA Expocast model challenge, where a set of approximately 50 data-rich chemicals of different classes were ranked by several different approaches [10]. The chemicals were chosen to include high interest chemicals with a range of properties. Each modeling approach was capable of analyzing a different number of chemicals from the full set because of varying input requirements. Key findings of the comparative analysis among the prioritization schemes indicated significant differences in chemical ranking as a result of several factors: (1) which processes the model described across the source to effects continuum [11]; (2) the exposure metric or surrogate metric used for prioritization and which statistic (i.e., median, upper bound or lower bound estimate); (3) whether the model inputs included actual, modeled or unit emissions; (4) which exposure pathways were considered (i.e., from aggregated sources or through a dominant pathway); and (5) which type of exposure scenarios were considered (i.e., direct or indirect, diffuse source or concentrated source, etc.) [10]. Only mechanistic models characterizing exposure associated with environmental sources could rapidly evaluate and rank potential exposure for the majority of chemicals. To a great extent, this was due to both the minimum data requirements and the availability of predictive tools (i.e., QSARs) to generate model inputs that could be used to describe fate and transport under steady state and equilibrium conditions. Of the other models evaluated in the EPA Expocast model challenge, those designed for evaluation of chemicals in specific exposure scenarios lacked data for chemical and scenario specific input parameters and were thereby inhibited in their ability to produce ordinal rankings for the 55 chemicals.

Arguably, one of the major limitations of the models evaluated, and perhaps one of the larger knowledge gaps in exposure-based chemical prioritization itself, involves complex social behaviors that determine how humans come in contact with manufactured chemicals, particularly those emanating from near field sources (e.g., residential and consumer products). Thus there is a pressing need for enhancing current approaches with tools and techniques developed for understanding human behaviors, such as human factors engineering and marketing research, to better define scenarios describing how products are used. Accurate use scenarios among population groups of interest are necessary to properly characterize the consumer use component of a chemical's life cycle.

Decision support tools borne out of the social sciences may also have a place in chemical prioritization. Multi-criteria Decision Analysis (MCDA), a rule-based method of classification for priority setting, is both a set of techniques and an approach for ranking alternatives [12,13]. MCDA is a promising approach for exposure-based prioritization because it is transparent and understandable, yet complex and rigorous enough to include scenario-based reasoning, stochastic processes and value of information analysis. Moreover, it is amenable to sparse data [14,15,16,17]. These characteristics complement some of the limitations of currently available statistical, mechanistic, or logic models, which provide useful frameworks for gathering relevant data but lack the social and policy context for risk-informed decision making. MCDA can merge a variety of types of exposure metrics from descriptions of physical chemical properties to the socioeconomic measures which characterize human activity, chemical use and contact to ultimately inform screening level risk estimates. Permitting structured integration of different types of information, MCDA methods provide a means for combining quantitative chemical property, production and use data with expert judgments and stakeholder preferences. MCDA assessment criteria can be adaptively weighted and modified in real time to evaluate both data-rich and data-limited chemicals.

Use of MCDA methods to support prioritization decision making under high uncertainty has been demonstrated many times including hazard identification and assessment. Risk management alternatives of industrial hazards or industrial consequences were relatively ranked using an MCDA approach by Paralikas and Lygeros [18]. The method recognizes that a single factor could not be used to define flammability and that different methods, tools, codes and legislation use varying sets of fire hazard properties as an example. Using the MCDA framework, the different decision criteria were successfully integrated using fuzzy logic to deal with linguistic variables and uncertainties allowing broad application for chemical hazard ranking decisions. In another example, life cycle assessment (LCA) was incorporated within a decision framework to prioritize future research and evaluate sensitivities to missing information in an assessment of processes for synthesizing single walled carbon nanotubes [14]. Engineered nanomaterials present uncertainties similar to chemicals in consumer products in terms of unknown environmental and human health across all life stages from formulation to disposal.

This paper demonstrates how analytical tools, such as LCA and MCDA, can offer a versatile and transparent approach to exposure-based prioritization utilizing results from several approaches evaluated in the EPA ExpoCast model challenge. The purpose of prioritization within this context is to focus resources on further evaluation of safety for chemicals with high potential for exposure and risk. A combination of exposure assessment model output with qualitative exposure criteria within such a decision framework has been recommended in the exposure-based waiving protocol within Europe's REACH Regulation [19] which shares some similar goals for human and environmental health protection.

Materials and Methods

We propose a decision analytic approach for exposure-based chemical prioritization to address the need for novel, rapid exposure potential screening protocols. In this approach, we build on current research and existing models by evaluating relevant chemical exposure criteria within a larger MCDA framework. We employ a two-part prioritization model that incorporates both properties of the chemical itself and properties of the chemical's life cycle (**Figure 1**).

The chemical property and life cycle property assessments are structured to analyze exposure-related information associated with specific chemical properties and distinct life cycle phases, respectively. Relevant chemical and life cycle properties are grouped into several criteria based upon the means by which each property contributes to the chemical's overall exposure potential (e.g., properties associated with a chemical's ability to bioaccumulate vs. those associated with its ability to be metabolized by the human body). Chemical and life cycle properties in each criterion are then further divided into various sub-criteria. The numerical values associated with these properties for a given chemical serve as inputs to the model. Input data can be obtained from a number of different sources, including existing databases, current literature and expert judgment. The criteria within this decision model were selected by reviewing those used in the models submitted to the ExpoCast model challenge, [10] and then structured into a hierarchical framework based on discussions with exposure science experts.

Within each sub-criterion, the constituent chemical or life cycle property is evaluated to determine its contribution to overall exposure potential. Input values for individual properties are compared against established numerical thresholds, which define distinct levels of risk that span the range of possible values for the given sub-criterion. Thresholds are used to score property values based on the indicated level of risk (e.g., a compound with a longer half-life may have higher potential for exposure than a compound with a shorter half-life, all other things being equal).

Following an MCDA approach, sub-criterion scores are then combined according to explicit decision rules to derive scores for their higher-level criterion. Chemical property and life cycle phase criterion scores are then combined to produce a Chemical Properties Exposure Score (CPES) and a Life Cycle Exposure Score (LCES) for each chemical. These scores reflect relative estimates of chemical exposure potential as indicated by available chemical property and life cycle property data, respectively. Exposure scores may then be integrated to derive aggregate measures of exposure potential, which can be used to compare and prioritize chemicals on a relative basis, or can remain separate and be plotted on a risk matrix for a more qualitative assessment.

Chemical property and life cycle phase criteria can be weighted within each assessment to reflect their relevance to the user's management objectives. Weights may indicate a specific focus of the assessment or reflect expert judgment of a criterion's predictive reliability or relative importance. Criterion weights can be adjusted to refine the scope of a particular assessment to a particular class of chemicals (e.g., pesticides), a particular exposure scenario (e.g., occupational exposure), or a particular exposure target (e.g., environmental contamination). When eliciting subjective weights, it is important to utilize best practices to avoid potential biases and inconsistencies [20,21]. Numerous elicitation techniques exist, including rank-based methods and swing-weight methods [13,21,22].

Chemical Properties Assessment

As seen in **Figure 1**, the Chemical Properties Assessment considers four main criteria to estimate potential risk for human exposure: bioaccumulation potential, persistence, ADME (**A**bsorption, **D**istribution, **M**etabolism, and **E**limination), and physical hazard potential. Each criterion constitutes a unique set of sub-criteria, which define the distinct chemical property data points that serve as inputs to the assessment. Observed chemical properties used to estimate exposure potential are defined by the specific sub-criteria under each of the four main criteria. Using thresholds established for each sub-criterion, individual data points are evaluated and assigned scores representing the potential for exposure indicated by the observed chemical property. Once these initial scores have been calculated, the highest within each set of sub-criteria is assigned as that criterion's exposure score.

When certain chemical-specific data are unavailable, as is often the case in this context, it may not be possible to assign scores to

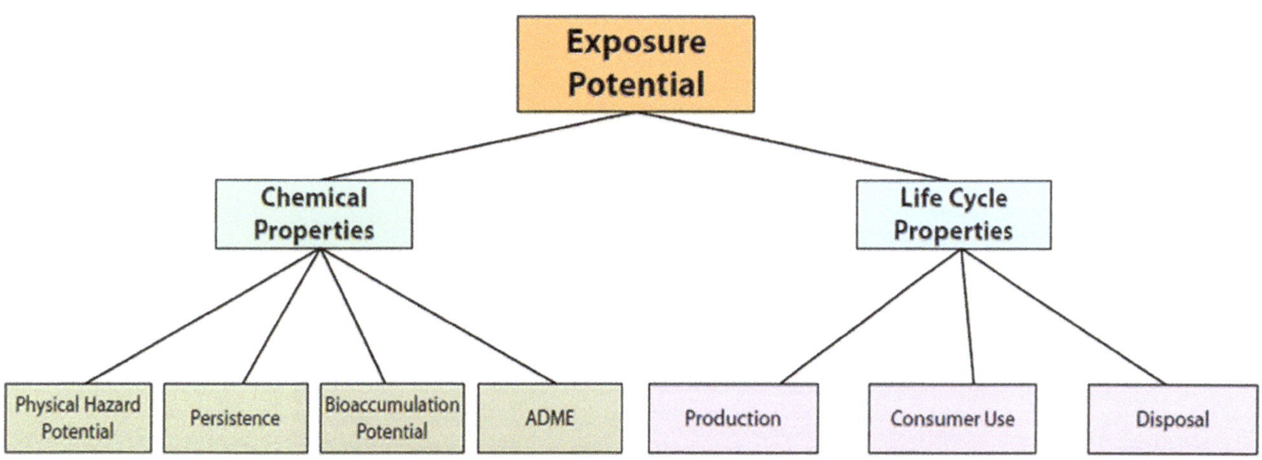

Figure 1. MCDA Framework for Exposure-Based Chemical Prioritization.

each sub-criterion. By defining each criterion's exposure score as the highest of its associated sub-criteria scores, we account for this possibility. By employing this approach, criterion scores can be assigned even in the presence of sparse data.

Each chemical's bioaccumulation, persistence, ADME, and physical hazard scores are combined with their associated weights. Weighted criteria exposure scores are then summed to produce initial chemical property exposure score for each chemical. Once this has been done for the set of chemicals being assessed, the initial chemical property exposure scores are normalized from 0 to 1 to produce relative rankings.

Bioaccumulation. Bioaccumulation is a process in which a chemical substance is absorbed by an organism via all routes of exposure in the natural environment, for example through dietary and ambient environmental sources, and increases in concentration over time [23]. Using three bioaccumulation-related sub-criteria, we evaluate surrogate chemical properties in order to predict the compound's ability to bioaccumulate.

Bioconcentration factor (BCF). A compound's BCF is a dimensionless number representing the relative concentration of the compound in organic tissues. In general, chemicals with relatively higher BCFs have greater potential for exposure, and thus are more likely to adversely impact human health and the environment. In this model, four distinct numerical thresholds were used to evaluate chemical BCF data. These thresholds are shown in **Table 1**, and were used to assign each chemical a BCF sub-criteria score from 1–4 based on the indicated level of bioaccumulation potential. Thresholds are based on previously published values employed by existing exposure assessment models: the EPA Design for the Environment Program [24], and the Clean Production Action's Green Screen for Safer Chemicals Initiative [25]. To address minor numerical discrepancies, the more conservative thresholds were chosen when values differed between models.

Log kow. A compound's K_{ow}, or octanol-water partition coefficient, describes its ability to transition between water and carbon-based media. Chemical compounds with relatively higher log K_{ow} are capable of greater movement within the environment; they are thus more adaptive and have higher potential for human exposure and absorption. In this model, four distinct numerical thresholds were used to evaluate chemical K_{ow} data. These thresholds are shown in **Table 1**, and were used to assign each chemical a log K_{ow} sub-criteria score from 1–4 based on the indicated level of bioaccumulation potential. Thresholds are based on previously published values employed by existing exposure assessment models: the EPA Design for the Environment Program [24], and the Clean Production Action's Green Screen for Safer Chemicals Initiative [25], with the more conservative threshold chosen when values differed between models.

Molecular weight. Previous studies have identified a significant correlation between a compound's molecular weight and its ability to bioaccumulate [26,27]. Results from these studies support the general conclusion that heavy molecules do not easily bioaccumulate, as their size hinders passage through lipid membranes. Lower weight chemicals thus possess a relatively greater potential for human exposure. These and similar findings have been used to inform chemical testing policy and legislation such as the OECD Chemical Substance Control Law (CSCL) in Japan [28] and the EPA Toxic Substances Control Act (TSCA) in the United States [29].

A single cut-off threshold is employed by our model to evaluate molecular weight data. Molecules 1000 amu or greater are given a bioaccumulation criteria score of 1, regardless of their other sub-criteria scores within the bioaccumulation category (BCF & log Kow). The 1000 amu cut-off follows TSCA premanufacture notification policy [29], and is based on current understanding that molecular weights in this range are generally better indicators of chemical bioaccumulation potential than other surrogate properties [26].

Persistence. Persistence corresponds to the length of time a chemical can exist in the environment before degrading or being transformed by natural processes [23]. Persistent chemicals are more likely to come into contact with humans compared to chemicals that degrade quickly in the environment. We consider the half-life in water, soil, sediment, and air for each chemical as surrogate indicators of persistence for the purpose of evaluating exposure potential.

The numerical thresholds used for evaluating chemical half-life data are shown below in **Table 1**. Thresholds were used to assign each chemical four distinct half-life sub-criteria scores from 1–4 based on the level of persistence indicated by each of the four half-lives (in water, soil, sediment, and air). Threshold values for water, soil, and sediment are based on previously published values employed by existing exposure assessment models: the EPA Design for the Environment Program [24], and the Clean Production Action's Green Screen for Safer Chemicals Initiative [25], using the more conservative thresholds. The threshold value for air follows science-based guidance for evaluating chemical long-range transport potential and overall persistence [30]. Chemicals with half-lives in air that are less than two days are assigned an associated sub-criteria score of 1 ("Low"), while those with half-lives in air greater than or equal to two days are assigned an score of 3 ("High").

Table 1. Thresholds for Bioaccumulation Potential and Environmental Persistence.

Score	1 (Low)	2 (Moderate)	3 (High)	4 (Very High)
Bioaccumulation				
BCF	<100	>100 to 1000	>1000 to 5000	>5000
Log Kow	<2	>2 to 3	>3 to 5	>5
Persistence				
Half Life in Water	<168 days	>168 to 960 days	>960 to 1440 days	>1440 days
Half Life in Soil	<384 days	>384 to 1440 days	>1440 to 4320 days	>4320 days
Half Life in Sediment	<384 days	>384 to 1440 days	>1440 to 4320 days	>4320 days
Half Life in Air	<2	n/a	>=2	n/a

ADME. Properties that describe a chemical's ability for absorption, distribution, metabolism, and excretion (ADME) are indicators of the potential for biologically relevant human exposure. Chemicals that can be easily absorbed by the body and that are resistive to metabolism or excretion pose a greater threat for extended exposure; therefore it is useful to focus on the entrance and exit of the chemicals within the context of the body. Though recent and current ADME-related research efforts have focused on establishing appropriate surrogate properties and developing predictive models, general consensus has not been reached regarding an accepted approach to ADME assessment for environmental chemicals [10]. Building on current research and existing models, a new ADME assessment protocol intended for screening-level exposure-based chemical prioritization was incorporated into the framework [10]. This method utilizes QikProp software Version 3.0 [31], a QSAR-based model to obtain surrogate chemical property values, which were then integrated to evaluate ADME properties along various sub-criteria briefly discussed below. All QikProp values are based on a 24-hour exposure period. Incidentally, QikProp is a three-dimensionally based structure method, so the SARs depend on the solvent accessible surface area. The properties calculated are dependent on the conformer adopted at the time of calculation and could be sensitive to molecular orientation. In addition, QikProp was designed exclusively to develop organic pharmaceutical compounds, so cannot be used for metals and inorganic compounds. Thus, if the analytics discussed herein are to be applied to metals and inorganic compounds, another QSAR system is needed.

Absorption. The chemical absorption assessment is based on two QikProp predictors which describe oral availability. The first descriptor represents a qualitative measure of oral absorption potential, and takes values of 1, 2, or 3 for low, medium, or high, respectively. The second descriptor represents a numerical probability of oral absorption on a 0 to 100% scale, with <25% and >80% designating low and high probability, respectively. These values were combined to derive an absorption score (1–3) for each chemical.

Distribution/Excretion: Distribution and excretion-related properties were combined into a single assessment. QikProp predicted octanol/water partition coefficients, serving as surrogates for half-life within the human body, were categorized into bins using subjective thresholds to derive a distribution/excretion score (1–4) for each chemical.

Metabolism. The assessment of metabolism was derived from the QikProp descriptor representing the number of expected possible metabolites for each chemical over a 24-hour period in the human body. These values were categorized based on the predicted half-life of each chemical in order to represent metabolism via natural degradation in the body. These values were combined to generate average metabolism scores (1–4) for each chemical.

Physical hazard potential. Highly flammable and reactive chemicals pose human and environmental threats that may not be considered in standard exposure or toxicity-based assessments. Though the properties that determine a given chemical's flammability and reactivity may be distinct from those that determine its environmental fate and transport, the threat of physical hazard is nonetheless directly related to the likelihood of exposure. The risk of physical hazards (e.g., combustion) is thus an exposure-related risk, and we assess each chemical's hazard-related properties in order to anticipate threats that may not be considered in other exposure or toxicity-based screenings. In accordance with existing National Fire Protection Association (NFPA) standards and classifications [32], flammability and reactivity were assigned scores of (1–4) using established NFPA thresholds.

Chemical Life Cycle Properties Assessment

Similarly to the assessment of chemical properties, we estimate potential for human exposure by assessing three main life cycle phases of manufactured chemicals: production, consumer use, and disposal. Each phase constitutes a unique subset of exposure-related criteria, which define the distinct life cycle characteristics that serve as inputs to the assessment.

The different criteria associated with each of the three life cycle phases designate the individual life cycle properties that will serve as indicators of a chemical's exposure potential during the relevant phase. All life cycle criteria are evaluated quantitatively, with higher values indicating higher potential for exposure. Instead of establishing thresholds for each sub-criteria as in the assessment of chemical properties, raw values are used but then normalized across the set of chemicals for each individual sub-criteria. This provides bounds for the range of values and assists in making comparative assessments.

Criteria scores are then calculated by summing the sub-criteria scores. Again, these scores are normalized across the set of chemicals to account for criteria containing more sub-criteria than others, and then multiplied by their weights to produce an initial Life Cycle Properties Exposure Score (LCES). Once initial LCESs have been calculated for all chemicals, we derive final LCESs by normalizing initial scores to the highest and lowest observed scores across all chemicals.

Production. Number of Potential Exposure Sources: Each chemical is evaluated to determine the possibility for human exposure during processes associated with production of the chemical. We consider one potential source (*occupational microenvironments*) defined as any workplace environment in which a release might occur during chemical manufacture and/or processing. Each chemical is assigned a score of either 0 or 1 based on whether the compound presents risk of exposure during production.

Projected average annual number of production sites. A chemical's exposure risk is increased if it is produced in many locations. Ubiquity classifications for each chemical were used to estimate the amount of chemical production sites [10]. Higher scores indicate increased potential for human exposure during chemical production: very widespread (5), widespread (4), moderate (3), localized (2), low (1).

Regional geometric mean production quantity (MQR). In addition to how widespread production is, estimates are made of the quantity produced. This is estimated using the Regional Geometric Mean Production Quantity (MQ_R), measured in units of kilotons per year. This is an estimated quantity, but production quantities could also be provided by industry.

Consumer Use

The assessment evaluates several sub-criteria relevant to the consumer use phase in the life cycle of manufactured chemicals. Based on the intended uses of each chemical, primary consumer class is defined as either strictly industrial, or industrial *and* individual. Chemicals used during industrial processes (e.g., monomers, solvents) and chemicals otherwise noted to have primarily industrial consumers were defined to have a strictly industrial consumer class. Chemicals used in agriculture (e.g., pesticides, insecticides, herbicides) or as food/cosmetic additives (e.g., preservatives, anti-microbials) were defined to have both industrial and individual consumers. Chemicals directly incorporated into consumer products during their production (e.g., plastics, coatings, fabrics, flame retardants) are also defined to have both industrial and individual consumers.

Number of potential exposure sources. Each chemical was evaluated to determine the possibility for human exposure during processes associated with both industrial and individual consumer uses of the chemical. Ten distinct potential sources associated with consumer exposure were considered (i.e., outdoor air, water, soil, biota, indoor air/dust, in-vehicle air, object contact, tap water, other water, food/beverages) by assigning each chemical a score from 0–10 based on possibility for exposure via each unique source during consumer use of the compound.

Projected average annual number of individual consumers. Chemicals defined as having industrial *and* individual consumer classes were assessed to determine their potential for exposure to individual consumers in non-industrial settings. Chemical ubiquity classifications were used to represent the relative size of each chemical's average, annual, individual consumer base. Chemicals defined as having strictly industrial consumer classes were assigned individual consumer scores of 0. Remaining chemicals were assigned scores from 1–5 based on their ubiquity, with higher scores indicating increased potential for individual consumer exposure during non-industrial use: very widespread (5), widespread (4), moderate (3), localized (2), low (1).

Projected average annual number of industrial consumers. To assess chemicals' potential for exposure to industrial consumers, we employ the ubiquity classification to estimate the average, annual size each chemical's industrial consumer base. As none of the chemicals assessed were defined as having a strictly individual (non-industrial) consumer base, all chemicals were assigned scores from 1–5 based on their ubiquity classification, with higher scores indicating increased potential for industrial consumer exposure during use of the chemical: very widespread (5), widespread (4), moderate (3), localized (2), low (1).

Projected average annual quantity consumed per individual/industrial consumer. The average annual quantity of each chemical consumed per consumer was predicted using the relative size of the chemical's total consumer base (including both individual and industrial consumers), and its MQ_R. Relative measures of consumption quantity per consumer (Q) were calculated by dividing each chemical's projected mean production volume by their total number of consumers, assuming chemicals with higher consumption quantities to have increased potential for consumer exposure. Projected annual quantities consumed per individual consumer were calculated using the same equation as that for industrial consumers:

$$Q = MQ_R/(n_{iIndividual} + n_{iIndustrial}) \quad (1)$$

where ($n_{iIndividual}+n_{iIndustrial}$) represents the chemical's total consumer base, or the number of individual consumers plus the number of industrial consumers.

Susceptible populations. To determine if there was a heightened exposure risk to susceptible populations (in this case, children), particular processes associated with individual consumer use of the chemical were evaluated. Nine distinct potential sources associated with exposure to children were considered (Outdoor Air, Water, Soil, Indoor Air/Dust, In-Vehicle Air, Object Contact, Tap Water, Other Water, and Food/Beverages), and each chemical was assigned a score from 0–9 based on possibility for exposure via each unique source.

Disposal

Number of potential exposure sources. Each chemical was evaluated to determine potential for human exposure resulting from disposal events. We consider four distinct disposal-related sources (Outdoor Air, Water, Soil, Biota), assigning each chemical a score from 0–4 based on potential for exposure via each unique source during and after disposal of the compound.

Projected average annual number of disposal events. Each chemical's total number of consumers was estimated to determine an annual number of associated chemical disposal events. Assuming that each chemical's industrial and individual consumers dispose of equal amounts of the compound, we define the projected number of disposal events as each chemical's total number consumers, and assign scores of 1–10, with higher scores representing greater potential for disposal-related human exposure.

Projected average annual quantity disposed. To account for assumed variations in the actual quantities disposed during industrial and individual consumer disposal events, we assume that 0.1% of the net production volume of each chemical is disposed of in order to evaluate disposal-related exposure potential. Note that the use of this unit value assumes that no chemical- or product-specific data were available. With larger disposal quantities indicating higher potential for post-disposal chemical exposure, we calculate relative disposal quantities of each chemical (Q_{DISP}) as:

$$Q_{DISP} = (.001) * MQ_R \quad (2)$$

Integrating Chemical Properties and Life Cycle Exposure Scores

Once assessments of chemical properties and life cycles have been performed on all chemicals, those chemicals lacking sufficient data to calculate either a chemical properties exposure score or life cycle exposure score are removed from the remainder of the prioritization. Though these chemical's available scores may indicate significant threat of exposure, they are excluded from the integration process as their scores can skew final exposure potential relationships. The remaining chemicals are renormalized as:

$$xES_{Final} = \frac{xES_{Initial} - xES_{Min}}{xES_{Max} - xES_{Min}} \quad (3)$$

where *xES* denotes the relevant exposure score (either chemical or life cycle). Next, the remaining chemicals' exposure scores (chemical property and life cycle property) are summed to produce aggregate exposure scores. These scores represent cumulative measures of exposure potential based on each chemical's distinct properties and characteristics of its projected life cycle. Aggregated exposure scores, which all lie in the range of 0–2, are used to numerically rank chemicals based on their potential for human exposure.

In addition to this quantitative integration, chemical property and life cycle scores can be visualized using a risk-reporting matrix (**Figure 2**) for a more qualitative assessment of aggregate chemical exposure potential.

In this method of integration, chemical property and life cycle exposure scores are converted from a scale of 0–1 to a scale of 0–5 by multiplying the initial score by a factor of five to place them within the 5×5 risk matrix, with each chemical's position representing a qualitative, cumulative measure of exposure potential based on both chemical and life cycle properties. Qualitative exposure potential thresholds (red, yellow, or green)

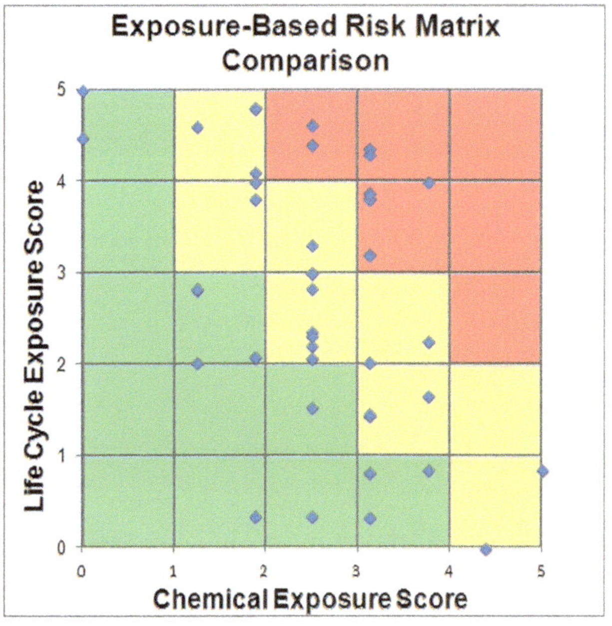

Figure 2. Example Chemical Exposure Potential Risk Matrix.

can be defined within the matrix to designate high, moderate, and low risk regions.

Case Study

Data Set

For the case study, a set of 51 chemicals was selected from those presented and evaluated in the model challenge (**Table 2**), representing a wide variety of chemical classifications (e.g., organics, metals, etc.). Sub-criteria scores for these chemicals were collected from numerous reports and online databases, and the sources for each sub-criterion are listed in **Table 3**. Case study data can be found in the online File S1.

Prioritization

First, the data for each chemical was compiled. It was found that some chemicals were difficult to assess due to a lack of readily available data. If a chemical did not have any sub-criteria scores for at least one of its criteria, that chemical was removed from the analysis process as having too little data for analysis. Nine of the 51 chemicals (largely metals) were removed for this reason.

Following the MCDA approach outlined above, each of the remaining test chemicals was assessed. Scores for each criterion were weighted by allocating equal weights (i.e., bioaccumulation, persistence, ADME, and physical hazards each weighted 25%; production, consumer use, disposal each weighted 33.33%). The final prioritization under this weighting distribution is shown in **Table 4**. The risk matrix comparison under this weighting distribution is shown in **Figure 3**.

Discussion

As stated above, one of the major limitations of currently available exposure models involves the inability to fully characterize the influence of complex social behaviors on resulting exposures or contact between humans and manufactured chemical across all life stages of the chemical. This is especially true for chemicals used in residential and consumer products, those arising from near field sources. A multi-criteria decision model was developed to combine typical physiochemical screening level data with measures to characterize human activities. As a proof of concept to show the utility of this approach, a case study was conducted on a small set of chemicals that were also analyzed using higher tiered statistical and mechanistic exposure models in a model challenge [10]. The models used in the model challenge considered different types of exposure scenarios including indirect exposures from diffuse environmental sources and direct, concentrated exposures from micro-environmental sources (i.e. from a personal care product or within a residence), though the latter had significant limitations in terms of necessary data to produce exposure estimates. Ranking results were obtained by three models and the comparative analysis is reported elsewhere [10]. Some agreement between ranking results was observed, but in general these models produced widely incongruous results across a number of different domains of information. Interestingly, some of the results using the MCDA model developed herein coincide with results from these more complex models. The majority of the chemicals (13 of 14) ranked in the top one-third of the list in **Table 4** (Rank 1–14), are also ranked in the top one-third of one of the models evaluated in the challenge. In general this agreement is with a "far-field" indirect diffuse source model which does not incorporate human activity at the micro-environmental level. Nonylphenol was the exception as it was ranked low by all other mechanistic models. Similarly, the bottom third of the ranked list in **Table 4** (Rank 28–42) shows high agreement with results from a model from the challenge. One model used characterized both far-field and near-field exposures and the other two were far-field models.

Because this analysis was conducted as a proof of concept, an exhaustive search for quality data and subsequent data validation was not conducted independently of the model challenge. However, the absence of the mechanistic relationships involved in the exposure models as well as the equal weighting scheme used in our example would lead to the assumption that the input drivers of the challenge models would be different than the input drivers of MCDA model. To fully explore this assumption and the utility of this methodology for larger scale research prioritization or policy guidance, the results of the case study underscore the need for quality data inputs. Only nine of the chemicals had to be excluded. These chemicals have properties that exclude them from the domain of applicability of the analytics, e.g. models, QSAR type, and other tools. As mentioned, metals and inorganic compounds are not characterized by the ADME models used in this study.

For the majority of compounds that fall within the domain of applicability, the MCDA approach is useful. As shown in **Table 4**, the majority of the chemicals used in plastics appear in the top half of the ranked list denoting highest exposure potential by highest aggregated exposure score. Plastics are broadly related to exposures that occur in all locations across the life-cycle of the chemicals. The chemicals in the bottom half of the ranked list (lower exposure potential) fit into a number of other of categories, but 11 of 21 are or were used as pesticides/herbicides, agriculturally, in homes or in public and commercial areas. The two pesticides/herbicides, Parathion and Methoxychlor, are ranked relatively low on the list in **Table 4**. Both chemicals were exclusively used in agriculture only, but have been previously banned or restricted by the EPA and do not have other uses like 1,2,3-trichlorobenzene, ethylene thiourea, and hexachlorobenzene which were also used exclusively in agriculture but are now used as a nonfood commercial additives. The remaining chemical in the agricultural only category is aldicarb. Aldicarb was restricted more

Table 2. Case Study Chemicals.

Chemical	CAS #	Chemical	CAS #
Formaldehyde	50000	Malathion	121755
DDT	50293	Perchloroethylene	127184
Parathion	56382	1-methoxy-4-(2-propen-1-yl)-benzene	140670
gamma-Hexachlorocyclohexane	58899	decaBDE	1163195
Carbaryl	63252	Trifluralin	1582098
Methoxychlor	72435	PFOS	1763231
Vinyl Chloride	75014	Atrazine	1912249
1,1,2,2-tetrachloroethane	79345	Lead	7439921
Tetrabromobisphenol A	79947	Manganese	7439965
Bisphenol-A	80057	Cadmium	7440439
p-tert-Pentylphenol	80466	Butylhydroxyanisole	8003245
Diethyl phthalate	84662	Perchlorate (Mg salt)	10034818
Di-n-butylphthalate	84742	Tris (l,3-dichloro-2-propyl) phosphate	13674878
1,2,3 Trichlorobenzene	87616	Methyl mercury	22967926
Pentachlorophenol	87865	Phenol, (l,l-dimethylethyl)-4-rnethoxy	25013165
2,4,5-Trichlorophenoxy acetic acid	93765	Nonylphenol	25154523
2,4-D	94757	Hexabromocyclododecane (HBCD)	25637994
Ethylene thiourea	96457	8-2 fluorotelomer acid	27854315
Methylparaben	99763	Aroclor_1260	35065271
Styrene	100425	Aroclor_1254	38380017
n-Hexane	110543	Vinclozolin	50471448
Tris (2-chloroethyl) phosphate	115968	Permethrin	52645531
Aldicarb	116063	Penta BDE	60348609
DEHP, Di(2-ethylhexyl)phthalate	117817	C10–C13 Chloroalkanes	85535848
Hexachlorobenzene	118741	octaBDE	207122165
Ethylparaben	120478		

recently in 2010 and will not be completely phased out until 2018, so exposure potential may be higher than the others in this category.

It should be noted that the nature of this analysis is to score chemicals in a comparative and relative manner, as opposed to assigning an absolute measure of exposure risk, which would not be practical or appropriate for a screening tool such as this. The relative assessment of chemical exposure potential is therefore dependent upon the set or sub-set of chemicals under consideration, and must be considered when designing the analysis and interpreting the results.

If a risk matrix is used for interpretation or communication of exposure potential results, it is important to note that a chemical with a high chemical property score and low life cycle property score (or vice versa) may be displayed has having a low exposure risk. When the risk matrix is used for score integration, however, these chemicals will appear on the boundaries of the matrix and can easily be identified as outliers that may warrant further assessment. Figure 3 shows the results of the case study on such a risk matrix. The risk matrix approach can be used to graphically visualize qualitative risk categories such as high, medium and low risk. The case study chemicals mostly fall within the same middle risk range of the matrix. Six chemicals fall into the higher exposure risk potential category and seven chemicals fall into the low exposure risk potential category based on the delineations shown in Figure 2. As a high tier screening, this type of representation may be useful for rapid visualization and categorization of large number of chemicals; however risk matrices should be used with caution when guiding risk management decisions [35].

Both the ranking and risk matrix approaches highlight the potential promise of multi-criteria decision analytic models for exposure-based prioritization, but further development beyond this effort is warranted. Given that the baseline weighting scenario – equal weights distributed among the chemical property and life cycle criteria – is likely an unrealistic one, a sensitivity analysis should be conducted to explore the effects of uncertainty in both the scoring of chemical parameters and the weighting schemes on the final chemical prioritization. This will help identify chemicals which are targets for further exposure assessment and data collection, ideally including better release characterization, proximal exposure assessment, and biomonitoring.

Finally, it is important to recognize that these results are strictly a measure of exposure potential and do not consider toxicological properties. Risk is a function of both hazard and exposure. The means by which organisms are exposed to stressors are complex; with many feedback loops (e.g., an outcome may itself become a stressor or modify other stressors). Risks related to chemical ingredients in products depend not only on the inherent properties of that chemical, but also the manner in which the chemical is formulated and used. Exposure potential therefore might be integrated with computational toxicology to paint a more

Table 3. Data Sources.

Criteria	Sub-Criteria	Data Sources
Chemical Properties		
ADME	Absorption (A)	QikProp software Version 3.0 [31]
ADME	Distribution/Excretion (D/E)	QikProp software Version 3.0 [31]
ADME	Metabolism (M)	QikProp software Version 3.0 [31]
Bioaccumulation	Bioconcentration Factor (BCF)	PBT Profiler [33]; Estimation Programs Interface Suite™ (EPI suite) [23]
Bioaccumulation	Log Kow	EPA Exposure-Based Prioritization Challenge [34]
Bioaccumulation	Molecular Weight	EPA Exposure-Based Prioritization Challenge [34]
Persistence	Half Life in Air	EPA Exposure-Based Prioritization Challenge [34]; Mitchell, et al. [10]
Persistence	Half Life in Water	EPA Exposure-Based Prioritization Challenge [34]; Mitchell, et al. [10]
Persistence	Half Life in Soil	EPA Exposure-Based Prioritization Challenge [34]; Mitchell, et al. [10]
Persistence	Half Life in Sediment	EPA Exposure-Based Prioritization Challenge [34]; Mitchell, et al. [10]
Physical Hazard	Flash Point (Flammability)	Material data safety sheets
Physical Hazard	Explosivity (Reactivity)	Material data safety sheets
Life Cycle Properties		
Production	Number of Potential Exposure Sources	EPA Exposure-Based Prioritization Challenge [34]
Production	Projected Avg. Annual Number of Production Sites	EPA Exposure-Based Prioritization Challenge [34]
Production	Regional Geometric Mean Production Quantity [MQR]	EPA Exposure-Based Prioritization Challenge [34]
Consumer Use	Number of Potential Exposure Sources	EPA Exposure-Based Prioritization Challenge [34]
Consumer Use	Projected Avg. Annual Number of Individual Consumers	EPA Exposure-Based Prioritization Challenge [34]
Consumer Use	Projected Avg. Annual Number of Industrial Consumers	EPA Exposure-Based Prioritization Challenge [34]
Consumer Use	Projected Avg. Annual Quantity Consumed Per Individual Consumer	EPA Exposure-Based Prioritization Challenge [34]
Consumer Use	Projected Avg. Annual Quantity Consumed Per Industrial Consumer	EPA Exposure-Based Prioritization Challenge [34]
Consumer Use	Susceptible Populations: Number of Potential Exposure Sources to Children	EPA Exposure-Based Prioritization Challenge [34]
Disposal	Number of Potential Exposure Sources	EPA Exposure-Based Prioritization Challenge [34]
Disposal	Projected Avg. Annual # of Disposal Events	EPA Exposure-Based Prioritization Challenge [34]
Disposal	Projected Avg. Annual Quantity Disposed	EPA Exposure-Based Prioritization Challenge [34]

complete picture of risk and to effectively prioritize the numerous chemicals in commerce.

Conclusions

In this paper, we have presented a decision analytic approach to exposure-based prioritization of manufactured chemicals. The proposed methodology allows for structured and transparent analysis of chemical exposure potential through integration of heterogeneous metrics used to evaluate exposure risk-related information associated with both chemical properties and life cycle phases. The model is scalable to assess as many chemicals as is necessary for the project scope, and the MCDA framework is able to accommodate varied inputs and exposure potential indicators, providing an adaptive and easy-to-use screening tool for rapid prioritization in the face of sparse data. In addition, the use of weighting in the model allows for specific user objectives, expert judgment, and data availability considerations to be explicitly implemented within the assessment.

The proposed approach builds on earlier models and current research relating to rapid evaluation of exposure potential. Specifically, it integrates the results of mechanistic and statistical approaches with semi-quantitative categorical data to describe exposure potential. In this paper, we attempt to address the need for high-level screening tools that (1) are capable of more detailed assessments than those provided by simpler predictive models (i.e., limited to persistence and bioaccumulation as indicators of exposure), and (2) have less intensive data requirements than more complex models, so as to remain efficient at the screening level.

It is important to note that work on this model is ongoing, and that the initial framework presented in this paper is primarily intended to illustrate the application of decision analytic methods to supplement existing exposure potential estimation techniques. Currently, our developmental efforts are focused on: (1) refining ADME assessment criteria and calculations; (2) identifying optimal surrogates for bioaccumulation potential; (3) implementing value of information (VOI) techniques to quantify data gaps and prioritize further research efforts; (4) improving normalization algorithms; and (5) developing a supplemental logic model for more specific exposure scenario evaluation. Additionally, we are working to develop formal means of considering expert judgment and empirical chemical exposure data within our assessments. In

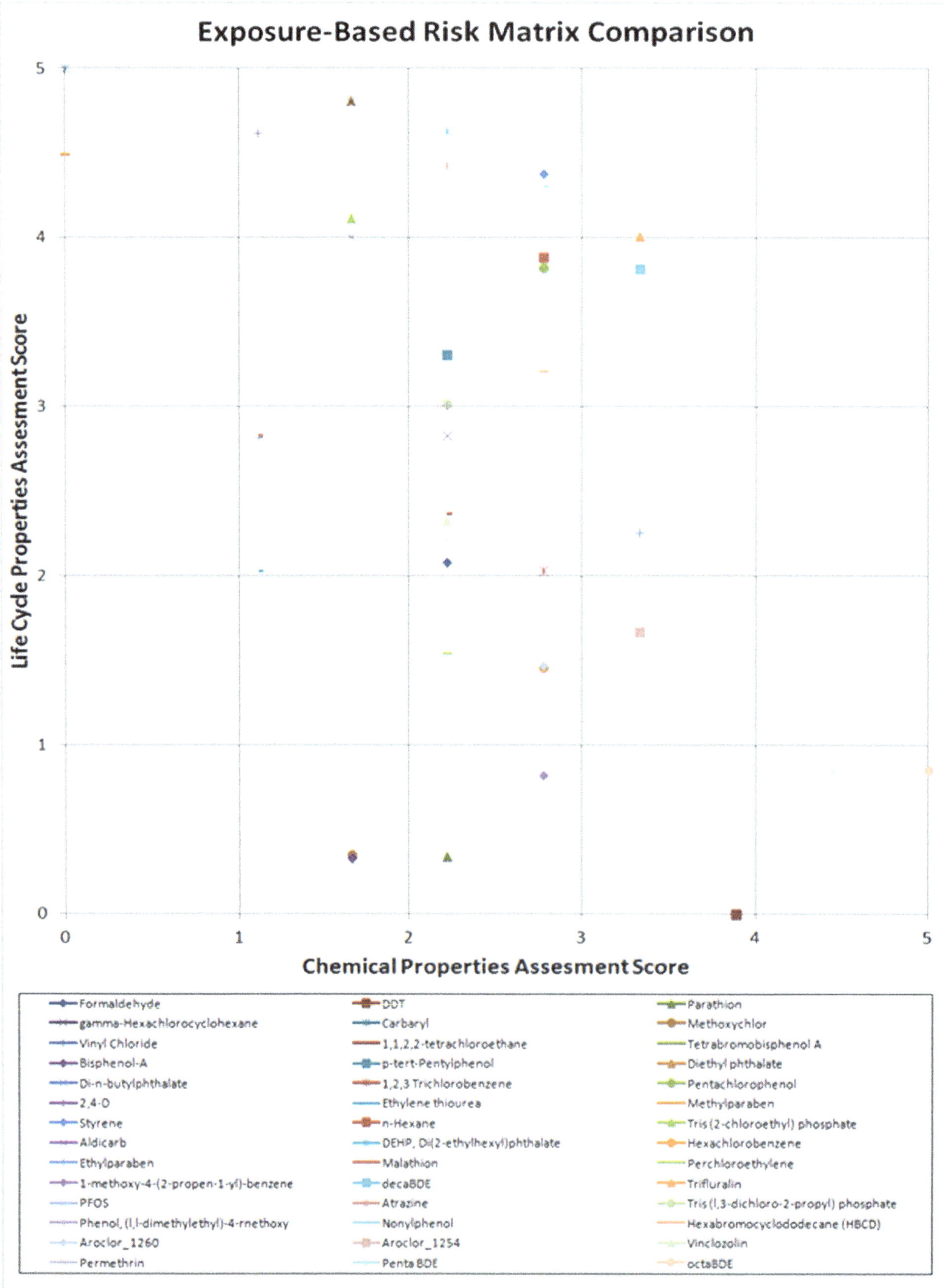

Figure 3. Risk Matrix Comparison of Exposure Potential with Even Weighting.

Table 4. Exposure Rankings with Even Weighting.

Rank	Chemical Name	CAS #	Chemical Property Score	Life Cycle Score	Aggregate Exposure Score
1	Trifluralin[3]	1582098	0.67	0.80	1.47
2	Styrene[2]	100425	0.56	0.87	1.43
3	decaBDE[4]	1163195	0.67	0.76	1.43
4	Nonylphenol[4]	25154523	0.56	0.86	1.42
5	DEHP, Di(2-ethylhexyl)phthalate[2]	117817	0.44	0.92	1.37
6	n-Hexane[4]	110543	0.56	0.78	1.33
7	Atrazine[3]	1912249	0.44	0.88	1.33
8	Tetrabromobisphenol A[2]	79947	0.56	0.76	1.32
9	Pentachlorophenol[3]	87865	0.56	0.76	1.32
10	Di-n-butylphthalate[2]	84742	0.33	0.96	1.29
11	Diethyl phthalate[2]	84662	0.33	0.96	1.29
12	Hexabromocyclododecane (HBCD)[2]	25637994	0.56	0.64	1.20
13	octaBDE[2]	207122165	1.00	0.17	1.17
14	Tris (2-chloroethyl) phosphate[2]	115968	0.33	0.82	1.15
15	2,4-D[3]	94757	0.22	0.92	1.15
16	Aldicarb[3]	116063	0.33	0.80	1.13
17	Vinyl Chloride[1]	75014	0.67	0.45	1.12
18	p-tert-Pentylphenol[3]	80466	0.44	0.66	1.11
19	Penta BDE[4]	60348609	0.89	0.17	1.06
20	Tris (l,3-dichloro-2-propyl) phosphate[2]	13674878	0.44	0.60	1.05
21	Phenol, (l,l-dimethylethyl)-4-rnethoxy[5]	25013165	0.44	0.60	1.05
22	gamma-Hexachlorocyclohexane[3]	58899	0.44	0.57	1.01
23	Carbaryl[3]	63252	0.00	1.00	1.00
24	Aroclor_1254[1]	38380017	0.67	0.33	1.00
25	1,2,3 Trichlorobenzene[3,4]	87616	0.56	0.41	0.96
26	1,1,2,2-tetrachloroethane[1]	79345	0.44	0.47	0.92
27	Vinclozolin[3]	50471448	0.44	0.46	0.91
28	Methylparaben[5]	99763	0.00	0.90	0.90
29	PFOS[4]	1763231	0.44	0.44	0.89
30	Formaldehyde[1,4]	50000	0.44	0.42	0.86
31	Aroclor_1260[1]	35065271	0.56	0.29	0.85
32	Hexachlorobenzene[3,4]	118741	0.56	0.29	0.85
33	Malathion[3]	121755	0.22	0.57	0.79
34	Ethylparaben[5]	120478	0.22	0.56	0.79
35	DDT[3]	50293	0.78	0.00	0.78
36	Perchloroethylene[1]	127184	0.44	0.31	0.75
37	Permethrin[3]	52645531	0.33	0.42	0.75
38	1-methoxy-4-(2-propen-1-yl)-benzene[5]	140670	0.56	0.16	0.72
39	Ethylene thiourea[3,4]	96457	0.22	0.41	0.63
40	Parathion[3]	56382	0.44	0.07	0.51
41	Methoxychlor[3]	72435	0.33	0.07	0.40
42	Bisphenol-A[2]	80057	0.33	0.07	0.40
	2,4,5-Trichlorophenoxy acetic acid[3]	93765	n/a	n/a	Insufficient Data
	Lead[4]	7439921	n/a	n/a	Insufficient Data
	Manganese[5]	7439965	n/a	n/a	Insufficient Data
	Cadmium[4]	7440439	n/a	n/a	Insufficient Data
	Butylhydroxyanisole[5]	8003245	n/a	n/a	Insufficient Data
	Perchlorate (Mg salt)[1]	10034818	n/a	n/a	Insufficient Data
	Methyl mercury[1]	22967926	n/a	n/a	Insufficient Data

Table 4. Cont.

Rank	Chemical Name	CAS #	Chemical Property Score	Life Cycle Score	Aggregate Exposure Score
	8-2 fluorotelomer acid[4]	27854315	**n/a**	**n/a**	Insufficient Data
	C10–C13 Chloroalkanes[4]	85535848	**n/a**	**n/a**	Insufficient Data

Key:
[1]Industrial/occupational additives and byproducts.
[2]Plastics.
[3]Pesticides and herbicides.
[4]Additives in commercial products.
[5]Additives in food and commercial products.

the future, we anticipate that the decision analytic approach will be able to provide decision makers with important and reliable information to support efficient, exposure-based prioritization of manufactured chemicals.

Acknowledgments

This material has been subjected to Environmental Protection Agency review and approved for publication. Mention of trade names or commercial products does not constitute endorsement or recommendation for use. Permission was granted by the US Army Corps of Engineers, Chief of Engineers to publish this material. The views and opinions expressed in this paper are those of the individual authors and not those of the US Army, or other sponsor organizations.

Author Contributions

Conceived and designed the experiments: JM PPE IL. Performed the experiments: JM NP ZAC. Analyzed the data: NP ZAC. Contributed reagents/materials/analysis tools: PPE ECH IL DAV. Wrote the paper: JM NP ZAC PPE ECH IL DAV.

References

1. Wilson MP, Schwarzman MR (2009) Toward a new U.S. chemicals policy: rebuilding the foundation to advance new science, green chemistry, and environmental health. Environ Health Perspect 117 (8): 1202–1209.
2. Dellarco V, Henry T, Sayre P, Seed J, Bradbury S (2010) Meeting the common needs of a more effective and efficient testing and assessment paradigm for chemical risk management. J Toxicol Environ Health B Crit Rev 13: 347–60.
3. Egeghy PP, Judson R, Gangwal S, Mosher S, Smith D, et al. (2012) The exposure data landscape for manufactured chemicals. Sci Total Environ, 414: 159–166.
4. Muir DCG, Howard PH (2006) Are there other persistent organic pollutants? A challenge for environmental chemists. Environ Sci Technol 40: 7157–7166.
5. Zartarian V, Bahadori T, McKone T (2005) Adoption of an official ISEA glossary. J Expo Anal Environ Epidemiol 15: 1–5.
6. Judson R, Riachrd A, Dix DJ, Houk K, Martin M, et al. (2009) The toxicity landscape for environmental chemicals. Environ Health Perspect 117: 685–695.
7. USEPA (2012) TSCA Work Plan Chemicals: Methods Document. Office of Pollution Prevention and Toxics. Washington, DC, USA. http://www.epa.gov/oppt/existingchemicals/pubs/wpmethods.pdf.
8. Cohen Hubal EA, Richard A, Aylward L, Edwards S, Gallagher J, et al. (2010) Advancing exposure characterization for chemical evaluation and risk assessment. J Toxicol Environ Health B Crit Rev 13: 299–313.
9. Egeghy PP, Vallero DA, Cohen Hubal, EA (2011) Exposure-based prioritization of chemicals for risk assessment. Environ Sci Pol 14 (8): 950–64.
10. Mitchell J, Arnot J, Jolliet O, Georgopolous P, Isukapalli S, et al. (2013) Comparison of modeling approaches to prioritize chemicals based on estimates of exposure and exposure potential. Sci Total Environ 458–460: 555–567.
11. USEPA (2009) A Conceptual Framework for U.S. EPA's National Exposure Research Laboratory. Office of Research and Development. Washington, DC, USA.
12. Shatkin J, Qian S (2005) Classification schemes for priority setting and decision making. In: Linkov I, Ramadan AB. Comparative risk assessment and environmental decision making. Dordrecht: Kluwer Academic Publishers. 213–243.
13. Linkov I, Moberg E (2011) Multi-criteria decision analysis: environmental applications and case studies. Boca Raton: CRC Press.
14. Canis L, Linkov I, Seager TP (2010) Application of stochastic multiattribute analysis to assessment of single walled carbon nanotube synthesis processes. Environ Sci Technol 44: 8704–8711.
15. Linkov I, Bates ME, Canis LJ, Seager TP, Keisler JM (2011) A decision-directed approach to prioritizing research into the impact of nanomaterials on the environment and human health. Nat Nanotechnol 6: 784–787.
16. Yokota F, Gray G, Hammitt JK, Thompson KM (2004) Tiered chemical testing: s value of information approach. Risk Anal 24 (6): 1625–1639.
17. Gabbert S, Weikard H-P (2010) A theory of chemicals regulation and testing. Nat Resour Forum 34: 155–164.
18. Paralikas AN, Lygeros AI (2005) A multi-criteria and fuzzy logic based methodology for the relative ranking of the fire hazard of chemical substances and installations. Process Saf Environ Prot 83 (2): 122–134.
19. Vermeire T, van de Bovenkamp M, Bruinen de Bruin Y, Delmaar C, van Engelen J, et al. (2010) Exposure-based waiving under REACH. Regul Toxicol Pharmacol 58: 408–420.
20. von Winterfeldt, D, Edwards W (1986) Decision analysis and behavioral research. Cambridge: Cambridge University Press.
21. Belton V, Stewart T (2002) Multiple criteria decision analysis: an integrated approach. Boston: Kluwer Academic Publishers.
22. Stillwell WG, Seaver DA, Edwards W (1981) A comparison of weight approximation techniques in multi-attribute utility decision making. Organ Behav Hum Perform 28 (1): 62–77.
23. USEPA (2011a) Estimation Programs Interface Suite™ for Microsoft® Windows, v 4.10. United States Environmental Protection Agency. Washington, DC, USA. http://www.epa.gov/opptintr/exposure/pubs/episuite.htm. Accessed 2012 Feb 14.
24. USEPA (2011b) Design for the Environment Program Alternative Assessment Criteria for Hazard Evaluation, v 2.0. Office of Pollution Prevention & Toxics, United States Environmental Protection Agency. Washington DC, USA. http://www.epa.gov/dfe/alternatives_assessment_criteria_for_hazard_eval.pdf.
25. Rossi M, Heine L (2009) The Green Screen for Safer Chemicals, v 1.0. North American Clean Production Action. Spring Brook, NY, USA. http://www.cleanproduction.org/library/cpa-fact%20grscreen_Jan09_final.pdf.
26. Dow Corning Corporation (1999) Polydimethylsiloxanes Do Not Bioaccumulate. Health Environment & Regulatory Affairs (HERA) Ref. no. 01-1128-01. Midland, MI, USA. http://www.dowcorning.com/content/publishedlit/01-1128-01.pdf.
27. Muller M, Nendza M (2007) Literature Study: Effects of Molecular Size and Lipid Solubility on Bioaccumulation Potential. Fraunhofer Institut Molekularbiologie and Angewandte Oekologie, FKZ 360 01 043. Dessau, Germany.http://www.reach-info.de/dokumente/bcf.pdf.
28. OECD. (1992) Test No. 301: Ready Biodegradability, OECD Guidelines for the Testing of Chemicals, Section 3, OECD Publishing. doi: 10.1787/9789264070349-en. http://search.oecd.org/officialdocuments/displaydocumentpdf/?cote = ocde/gd(95)43&doclanguage = en.
29. USEPA (1995) Premanufacture Notification Exemption; Revision of Exemption for Chemical Substances Manufactured in Small Quantities; Low Release and

Exposure Exemption; Final Rule. (60 FR 16336–16351).U.S. Environmental Protection Agency, Office of Pollution Prevention and Toxics. Washington, DC, USA. http://www.smallbiz-enviroweb.org/Resources/sbopubs/edocs/e04.pdf.

30. Scheringer M, Jones KC, Matthies M, Simonich S, van de Meent D (2009) Multimedia partitioning, overall persistence, and long-range transport potential in the context of POPs and PBT chemical sssessments. Integr Environ Assess Manag 5 (4): 557–76.
31. Schrödinger LLC (2007) QikProp 3.0 User Manual. http://www.schrodinger.com/supportdocs/18/17.
32. National Fire Protection Agency (NFPA) (2007) NFPA 704: Standard System for the Identification of the Hazards of Materials for Emergency Response, 2007 Edition.
33. Environmental Science Center (2007) PBT Profiler. U.S. Environmental Protection Agency, Environmental Science Center, Office of Chemical Safety and Pollution Prevention. Washington, DC, USA. <http://www.pbtprofiler.net/> Accessed 2013 Mar 18.
34. USEPA (2011c) Exposure-Based Chemical Prioritization Workshop II: Extending Capabilities for High Throughput Assessment. http://www.epa.gov/ncct/expocast/exposure_based_2.html. Accessed 2013 Mar 18.
35. Cox Jr., LA (2008) What's Wrong with Risk Matrices? Risk Anal 28 (2): 497–512.

20

Soil Particle Heterogeneity Affects the Growth of a Rhizomatous Wetland Plant

Lin Huang[⁹], Bi-Cheng Dong[⁹], Wei Xue, Yi-Ke Peng, Ming-Xiang Zhang, Fei-Hai Yu*

School of Nature Conservation, Beijing Forestry University, Beijing, China

Abstract

Soil is commonly composed of particles of different sizes, and soil particle size may greatly affect the growth of plants because it affects soil physical and chemical properties. However, no study has tested the effects of soil particle heterogeneity on the growth of clonal plants. We conducted a greenhouse experiment in which individual ramets of the wetland plant *Bolboschoenus planiculmis* were grown in three homogeneous soil treatments with uniformly sized quartz particles (small: 0.75 mm, medium: 1.5 mm, or large: 3 mm), one homogeneous treatment with an even mixture of large and medium particles, and two heterogeneous treatments consisting of 16 or 4 patches of large and medium particles. Biomass, ramet number, rhizome length and spacer length were significantly greater in the treatment with only medium particles than in the one with only large particles. Biomass, ramet number, rhizome length and tuber number in the patchy treatments were greater in patches of medium than of large particles; this difference was more pronounced when patches were small than when they were large. Soil particle size and soil particle heterogeneity can greatly affect the growth of clonal plants. Thus, studies to test the effects of soil heterogeneity on clonal plants should distinguish the effects of nutrient heterogeneity from those of particle heterogeneity.

Editor: Ben Bond-Lamberty, DOE Pacific Northwest National Laboratory, United States of America

Funding: This research was supported by the Forestry Commonwealth Project (201004078), the Fundamental Research Funds for the Central Universities (TD-JC-2013-1), the Program for New Century Excellent Talents in University (NECT-10-0234) and the National Science Foundation of China (31070371). The funders had no role in study design, data collection and analysis, decision to publish, or preparation of the manuscript.

Competing Interests: The authors have declared that no competing interests exist.

* E-mail: feihaiyu@bjfu.edu.cn

⁹ These authors contributed equally to this work.

Introduction

Spatial heterogeneity is an important feature of natural habitats [1–5], and plays major roles in shaping plant growth, species interactions, community structure and ecosystem functioning [6–15]. Heterogeneity occurs in various spatial scales [2,3,5], and the scale of heterogeneity is important because ecological processes that function at one spatial scale may not do so at other scales [16–21].

Clonal plants are very abundant in nature and often experience spatial heterogeneity in such a way that connected ramets of the same clone experience contrasting environmental conditions [22–24]. Many studies have addressed how spatial heterogeneity in essential resources such as light, soil water and mineral nutrients affects the growth of clonal plants, and these studies generally show that clonal plants can respond to heterogeneity in ways that increase their performance and likely fitness [7,19–20,25–29]. Moreover, clonal plants can respond differently to heterogeneity of soil and light at different spatial scales [18–21,30].

Particle size is an important feature of soil [31–34], and it can affect soil physical and chemical properties such as structure, oxygen concentration and water content [33,35–38]. Studies testing the effects of soil nutrient heterogeneity on the performance of clonal plants sometimes compare the growth in substrate with patches of fertile and infertile soil to the growth in a homogeneous mixture of the same amounts of fertile and infertile soil [6,20–21,39–44]. However, fertile and infertile soils generally differ in the size composition of particles, meaning that results of these studies might be partly due to effects of particle size rather than nutrient availability. To test the effects of soil nutrient heterogeneity, therefore, it would be good to create nutrient-rich and nutrient-poor patches with the same type of soil substrate but supplied with different amounts of nutrients [11,27,45–47]. On the other hand, it would be useful to test the effects of particle size separately. However, no published study appears to have tested the effects of soil particle heterogeneity on the growth of clonal plants.

To test the effects of soil particle heterogeneity on clonal growth, we conducted a greenhouse experiment in which we grew individual ramets of the rhizomatous, wetland plant *Bolboschoenus planiculmis* in three homogeneous soils each containing only one size class of quartz particles (large, medium or small), one homogeneous soil containing an even mixture of large and medium particles, and two patchy soils consisting of separate patches of large and medium particles and differing in patch scale (large and small patches). We asked the following questions. First, does soil particle size affect the growth (biomass, number of ramets, rhizome length and number of tubers) of *B. planiculmis*, as shown by comparing the treatments with only large, medium and small particles? Second, does patchiness or scale of patchiness of particle size affect the growth of *B. planiculmis* at the level of the whole clone, as shown by comparing the treatments with the homogeneously mixed and the two patchy large and medium particles? Based on effects of heterogeneity of other environmental factors on clonal growth, we expected that the growth in the homogeneously mixed treatment would be lowest and that the growth would differ between the two patchy treatments. Scaling theory predicts that organisms respond to spatial heterogeneity at scales related to their own size [16]. We therefore expected that *B. planiculmis* would show larger responses to soil particle heterogeneity at the scales closer to its spacer length (i.e. inter-ramet distance). Third, does patchiness or scale of patchiness of particle size affect the growth of *B. planiculmis* at the level of the patches? We expected that, in the patchy treatments, the growth of *B. planiculmis* would differ between patch types and that these differences would depend upon patch size.

Materials and Methods

Species and Sampling

Bolboschoenus planiculmis (F. Schmidt) T. V. Egorova (Cyperaceae), formerly called *Scirpus planiculmis* Fr. Schmidt, is a rhizomatous, perennial, herbaceous, wetland plant [48]. The species is native in China (e.g. in Beijing), and also occurs in India, Japan, Kazakhstan, Korea, Kyrgyzstan, Mongolia, Papua New Guinea, Russia, Tajikistan, southwestern Asia and Europe [48]. It attains heights of 0.6 to 1 m and forms solitary shoots from globose tubers. Shoots die back at the end of one growing season, but tubers can form new shoots and rhizomes that produce new tubers in the next growing season. Both tubers and rhizomes that connect them can survive for several years.

On 10 May 2011, more than 200 ramets, each consisting of a green shoot, a tuber at the base of the shoot and some roots, were collected from a natural population of *B. planiculmis* on the bank of the Beisha River in Beijing. Although the genetic background of these ramets is unknown, they were likely to originate from more than one genet because the longest distance between the sampled ramets was over 50 m. The sampling site did not belong to any farms or national parks and also did not involve any endangered or protected species, so we did not need any relevant permission for collecting plant samples. Each ramet was transplanted into a pot with sufficient nutrition to grow for 15 days. Then 48 ramets of similar size (10–15 cm high with 3–5 leaves) were chosen and used for the experiment.

Experimental Design

The experiment had three homogeneous treatments with a single type of particle size (termed "Small", "Medium" and "Large", respectively), two patchy treatments with different patch sizes ("Large patch" and "Small patch") and a mixture treatment ("Mixture"; Fig. 1). The substratum used in this experiment was pure quartz without any nutrients. In the Small, Medium and Large treatments, the particle sizes of the quartz were on average 0.75, 1.5 and 3.0 mm, respectively.

In a previous, eight-week, pilot experiment using the Small, Medium and Large treatments, it was found that *B. planiculmis* grew best in the Medium and worst in the Large treatment [49]. The medium and large particle sizes were, therefore, used in the Mixture and patchy treatments. In the Large patch treatment, each box (26 cm long×26 cm wide×13.0 cm deep) was divided equally into four parts (each 13 cm×13 cm), and two parts were filled with medium and two with large particles, in checkerboard fashion (Fig. 1). The Small patch treatment was similar except that each box was divided equally into 16 parts (each 6.5 cm×6.5 cm). In the Mixture treatment, each box was filled with a 1:1 (v: v) mixture of the medium and large particles (Fig. 1). There were no physical barriers between patches, and roots and rhizomes could grow freely between them. Each treatment had eight replicates (boxes) and there were a total of 48 boxes.

On 5 May 2011, the 48 ramets of *B. planiculmis* were randomly assigned to the six treatments and one ramet was planted in the center of each of the 48 boxes (Fig. 1). The boxes were closed at the bottom so that water and nutrients could not leak. We added 200 ml nutrient solution (1 g Peters Professional [Scotts Professional Limited, Nottingham, UK; N-P-K, 20-20-20] L^{-1} tap water) evenly to each box every week in the first eight weeks. The total amount of nutrients added corresponded to 14.201 g m^{-2} yr^{-1} of each of nitrogen, potassium and phosphorus. The concentration of the nutrient solution used was among the manufacturer's recommended application rate (Scotts Professional Limited, Nottingham, UK) and the nutrients are sufficient for the growth of the plants. Sufficient water was added to the boxes so that the soil in the boxes was flooded and there was always 1 to 3 cm deep water above the soil surface. These measures were to simulate typical wetland conditions experienced by *B. planiculmis* and also ensured that nutrients were evenly distributed in the boxes.

The experiment was conducted in a greenhouse at Forest Science Co., Ltd., of Beijing Forestry University, maintained at a mean temperature of 26.8°C and a mean relative humidity of 72% (measured by iButton DS 1923, Maxim Integrated Products, USA). The boxes were randomly positioned within an area of 18 m^2 and re-positioned systematically every two to three weeks to minimize effects of position.

Harvest and Measurements

The experiment lasted 16 weeks and was ended on 14 September 2011. One plant in each of the Small, Small patch and Large patch treatments died during the experiment, and these plants were excluded from harvest and subsequent analyses. In the homogeneous treatments (Small, Medium, Large and Mixture), the mother and offspring ramets were harvested separately. In the two heterogeneous treatments, the mother ramet and the offspring ramets in each type of patches were harvested separately. We counted number of ramets, number of non-sprouted tubers and measured rhizome length. These three variables are commonly used to measure the capacity of clonal growth, and tubers represent potential ramets. Roots, tubers, rhizomes, leaf blades and leaf sheaths were then separated, oven-dried at 70°C for at least 48 h, and weighed.

Data Analysis

We used one-way ANOVA to examine the effects of soil particle size on the growth (biomass, number of ramets, rhizome length and number of tubers), root-shoot ratio and spacer length (the distance between adjacent ramets) at the whole plant level. In these analyses, we included only the three homogeneous treatments with a single type of particle size (Small, Medium and Large). If a significant effect was detected, a Duncan test was used to compare individual means. We conducted one-way ANOVAs or non-parametric Kruskal-Wallis tests to examine the effects of spatial heterogeneity of soil particles on the growth, root-shoot ratio and spacer length at the whole plant level. In these analyses, three treatments (Small patch, Large patch and Mixture) were used.

We employed two-way ANOVA to examine the effects of particle size (large vs. medium) within patches and patchiness (Large patch vs. Small patch vs. the hypothesized "Null" treatment, see explanation below) on the growth of the plants at the patch level. Data from mother ramets were excluded because they were not located at any patches. The growth pattern

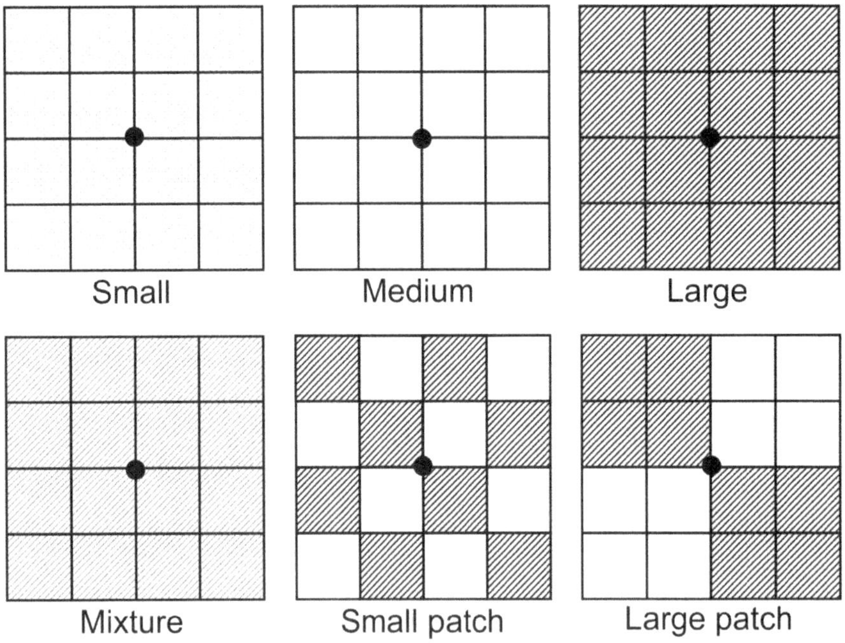

Figure 1. Schematic representation of the experimental design. The experiment had three homogeneous soil treatments with whole boxes filled with small (0.75 mm), medium (1.5 mm) or large (3 mm) quartz (coded as Small, Medium and Large, respectively), one homogeneous treatment with boxes filled with an even mixture of the same amount of large and medium quartz (coded as Mixture), and two heterogeneous treatments consisting of 16 or 4 patches of large and medium quartz (coded as Small patch and Large patch, respectively).

(distribution of biomass, ramets, rhizomes and tubers in the two types of soil particle patches) in the Null treatment represented the expected (null) growth pattern in soil without a real patch texture. For this treatment, half values of the corresponding growth variables in the treatment Large were used to represent the expected growth in the large particle patches, and half values of the corresponding growth variables in Medium were used to represent the expected growth in the medium particle patches. We used half values because in the Large patch and the Small patch treatment the plants in each type of patches occupied only half of the area of the whole box. The growth pattern in the Null treatment was used as a null model to show what the pattern was if the placement of biomass, ramets, rhizomes or tubers did not show selectivity (response). For the two heterogeneous treatments, if the growth pattern deviated significantly from the pattern of the Null treatment, as indicated by significant interaction effects of particle size and patchiness, then the plants showed significant selectivity in response to soil particle heterogeneity.

Before analysis, all data were checked for normality (by Shapiro-Wilk test) and homogeneity of variance (by Levene's test). When we tested the effects of particle heterogeneity on biomass and ramet number at the whole plant level, data could not meet the assumption of normality and/or homogeneity of variance even after transformation. Therefore, non-parametric tests were used for these analyses. Statistical analyses were conducted with SPSS 18.0 for Windows (SPSS, Chicago, IL, USA).

Results

Effects of Soil Particle Size at the Whole Plant Level

In the homogeneous treatments, biomass, number of ramets, rhizome length and spacer length of *B. planiculmis* were significantly greater in the Medium than in the Large treatment, but they did not differ between the Small and Medium treatments (Fig. 2A–C and F). Root-shoot ratio was higher in the Medium and Large treatments than in the Small treatments (Fig. 2E). Soil particle size did not affect number of tubers (Fig. 2D).

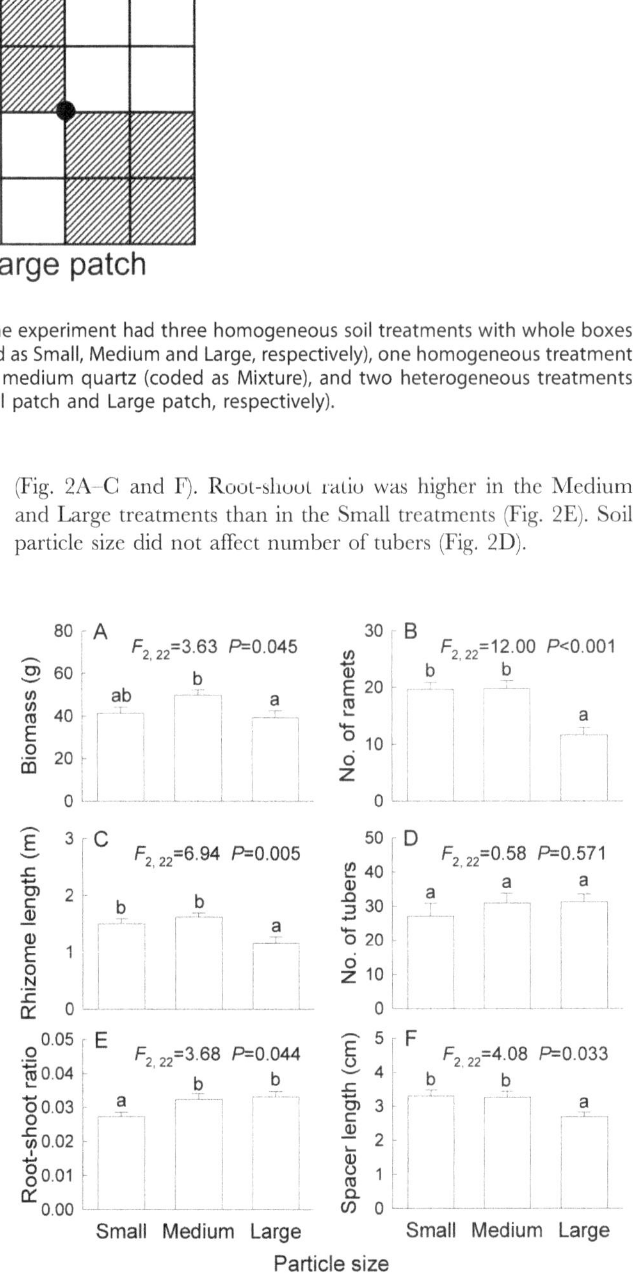

Figure 2. Effects of soil particle size on growth (A–D) and morphology (E, F) in the treatments with a single particle size. The treatment codes are described as in Fig. 1. Bars and vertical lines are means and SE. Bars sharing the same letters are not different at $P = 0.05$.

Effects of Soil Particle Heterogeneity at the Whole Plant Level

None of the growth or morphological measures of the whole plants differed significantly between the Mixture, Small patch and Large patch treatments ($\chi^2 = 2.47$–3.16, $P = 0.206$–0.290 for biomass and number of ramets; $F_{2, 21} = 0.501$–2.186, $P = 0.140$–0.614 for rhizome length, number of tubers, spacer length and root-shoot ratio). The result suggests that soil particle heterogeneity did not affect the growth or morphology at the whole plant level.

Effects of Soil Particle Heterogeneity at the Patch Level

At the patch level, particle size significantly affected biomass, number of ramets, rhizome length and number of tubers of *B. planiculmis* (Table 1). In general, biomass, number of ramets, rhizome length and number of tubers were greater in the medium than in the large particle patches (Fig. 3), and such effects were the largest in the Small patch and smallest in the Null treatment (Table 1: significant interaction effect of particle size and patchiness; Fig. 3).

Table 1. Effects of particle size within patches and patchiness on the growth at the patch level.

Effects	DF	Biomass	No. of ramets	Rhizome length	No. of tubers
Particle size	1,38	100.5***	70.0***	71.3***	35.5***
Patchiness	2,38	0.3^{ns}	0.3^{ns}	1.4^{ns}	1.4^{ns}
Interaction	2,38	23.3***	7.5**	14.0***	10.8***

F-values and degree of freedom (DF) of two-way ANOVAs are given. *Superscripts give P:* ***$P<0.001$, **$P<0.01$ and $^{ns}P\geq0.05$.

Discussion

Soil particle size significantly affected the growth of *B. planiculmis*: plants accumulated more ramets and biomass in soil composed of medium (diameter 1.5 mm) or small (0.75 mm) particles than of large (3.0 mm) particles. This result suggests that differences in soil particle size should be considered when we use different soil types to test the effects of soil nutrient availability. Soil particle size may greatly affect both chemical and physical properties of soil [33,36,50–51]. In the present study, the effects on soil chemical properties may be of little importance because sufficient nutrient solutions were supplied to each plant during the experiment. Previous studies have showed that changing soil physical properties such as soil mechanical resistance, infiltration capacity and adsorption capacity can significantly influence the growth and reproduction of plants [51–52]. For example, increasing soil mechanical resistance may slow down root growth and root expansion [52–56]. In the present study, each plant was supplied with the same, sufficient amount of nutrients and also grown in simulated wetland (flooded) conditions. Therefore, changing soil infiltration capacity or adsorption capacity caused by different soil particle sizes may also be of little importance. One likely explanation is that differences in soil mechanical resistance mediated by different soil particle sizes are responsible for the difference in the biomass production and clonal reproduction of *B. planiculmis*. Compared to soil composed of small or medium particles, soil composed of large particles is likely to have a higher

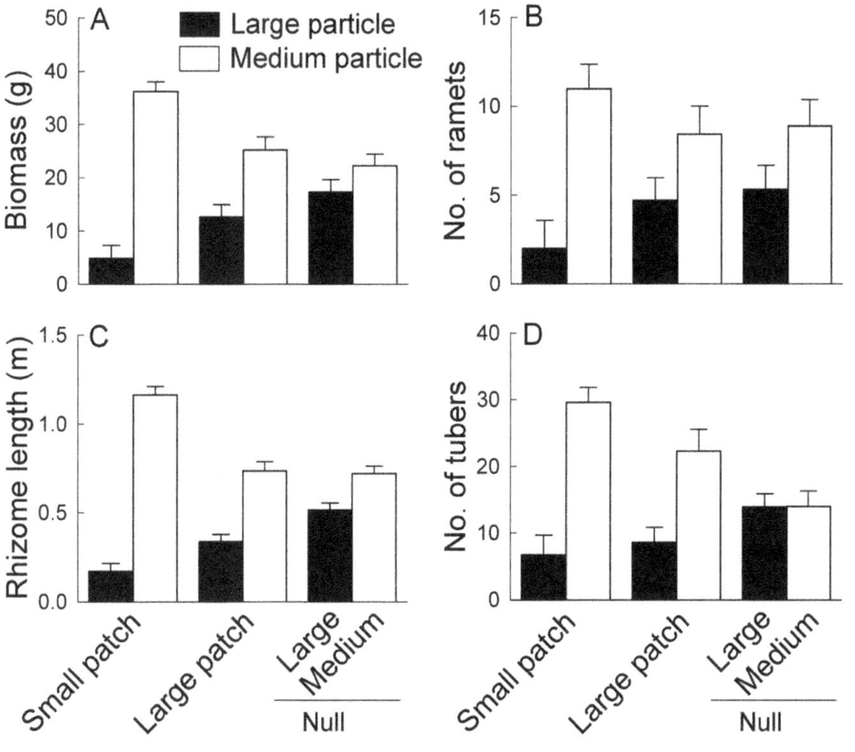

Figure 3. Effects of patchiness and particle size within patches on the growth within patch types. The treatment codes are described as in Fig. 1. In the "Null" treatment, half values of the corresponding variables in Large and Medium were used because in Large patch and Small patch the plants in each type of patches occupied only half of the area of the whole box. Bars and vertical lines are means and SE.

mechanical resistance, which significantly decreased the growth of *B. planiculmis*. Further studies are required to test the relationships between soil particle size and soil mechanical resistance.

Soil particle heterogeneity significantly affected the growth of *B. planiculmis* at the patch level. Compared to the homogeneous large particle treatment, *B. planiculmis* produced less biomass and tubers and shorter rhizomes in the comparable area in the large particle patches in the two heterogeneous treatments, and such effects were greater in the small patch than in the large patch treatment. Also, compared to the homogeneous medium particle treatment, *B. planiculmis* produced more biomass and ramets and longer rhizomes in the comparable area in the medium particle patches in the small patch treatment, but not in the large patch treatment. As a result, in the heterogeneous treatments, comparably more ramets, tubers, biomass and rhizomes were placed in the patches composed of medium particles, which are more favorable for the growth of *B. planiculmis* [49]. These responses are similar to the foraging responses of clonal plants when growing in resource-heterogeneous environments where clonal plants commonly position more ramets in favorable (resource-rich) patches and less in unfavorable (resource-poor) patches [21,24,41,57–58]. Such responses are thought to be adaptive since they potentially can increase resource uptake of the whole clone [57,59–62].

However, at the whole (clone) plant level, we found that soil particle heterogeneity did not affect the growth of *B. planiculmis*. In contrast, spatial heterogeneity consisting of patches of fertile and infertile soil could greatly increase the whole plant growth of some clonal plants such as *Glechoma hederacea* and *Alternanthera philoxeroides* [40–41,63]. Therefore, in these studies the effects of soil heterogeneity at the whole plant levels are very likely due to soil nutrient heterogeneity. When more ramets are located in nutrient-rich patches, both the efficiency and the total amount of nutrient uptake increase markedly [41]. This will result in a significant increase in nutrient uptake of the whole clone as well as its subsequent growth. In the present study, however, because differences in soil mechanical resistance might be the main mechanism underlying the effects of soil particle heterogeneity, the patch-level effects on the selectivity of where ramets are placed and their growth in the patches could not translate into the effects at the whole clone level.

The effects of soil particle heterogeneity were greater in the small patch than in the large patch treatment, suggesting that patch scale may have played an important role during the responses of *B. planiculmis* to soil particle heterogeneity. Previous studies have also shown that patch scale affected the responses of *Glechoma hederacea* [20–21] and *Colophospermum mopane* [16] to soil heterogeneity and *Duchesnea indica* to light heterogeneity [19]. In the present study, the smaller the patch scale was, the more the ramets were placed in the medium particle patches than in the large particle ones. The larger effects of soil particle heterogeneity at the smaller scale were likely because the mean inter-ramet distance (spacer length, 2.3–3.7 cm on average, Fig. 2F) of *B. planiculmis*, which partly determines the placement and density of ramets in a patch [57,60,64], was closer to the patch size (6.5 cm×6.5 cm) in the small patch treatment. This result agrees with scaling theory, which predicts that organisms respond to spatial heterogeneity at scales related to their own size [16,65]. Another possible interpretation is that in the small patch treatment plants had a higher probability of finding a favorable patch (i.e. consisting of medium particles) when such favorable patches were more spread out. However, this interpretation should be considered only when patch sizes are within a range that plants can sense them, such as the two patch sizes used in the present study. If patch size is set smaller than this range, plants will likely not be able to respond to spatial heterogeneity [65].

In conclusion, soil particle size and heterogeneity may greatly affect the growth of plants. Thus, when we address the effects of soil nutrient availability or the effects of soil nutrient heterogeneity, caution should be taken when different soil types are used to form low and high nutrient conditions, because different soil types may differ greatly in soil particle composition [31–32,66–67]. Further studies testing the effects of soil heterogeneity should distinguish the effects of soil nutrient heterogeneity from those of soil particle heterogeneity. The results also suggest that scale should always be considered when we address the effects of spatial heterogeneity [16,18–21].

Acknowledgments

We thank Prof. Peter Alpert and two anonymous reviewers for valuable comments on the manuscript, and Yu-Qiang Li, Qian Zhang, Xiong-Ying Wu and Ming-Lian Chen for assistance with the harvest of the experiment.

Author Contributions

Conceived and designed the experiments: BCD FHY. Performed the experiments: LH YKP WX. Analyzed the data: LH BCD FHY. Contributed reagents/materials/analysis tools: MXZ. Wrote the paper: LH BCD FHY.

References

1. Alpert P, Mooney HA (1996) Resource heterogeneity generated by shrubs and topography on coastal sand dunes. Vegetatio 122: 83–93.
2. Caldwell MM, Pearcy RW (1994) Exploitation of environmental heterogeneity by plants: ecophysiological processes above-and belowground. San Diego: Academic Press.
3. Jackson RB, Caldwell MM (1993) Geostatistical patterns of soil heterogeneity around individual perennial plants. Journal of Ecology 81: 683–692.
4. Li P-X, Wang N, He W-M, Krusi BO (2008) Fertile islands under *Artemisia ordosica* in inland dunes of northern China: Effects of habitats and plant developmental stages. Journal of Arid Environments 72: 953–963.
5. Chen YF, Yu F-H, Dong M (2002) Scale-dependent spatial heterogeneity of vegetation in Mu Us sandy land, a semi-arid area of China. Plant Ecology 162: 135–142.
6. Fransen B, de Kroon H, Berendse F (2001) Soil nutrient heterogeneity alters competition between two perennial grass species. Ecology 82: 2534–2546.
7. Day KJ, Hutchings MJ, John EA (2003) The effects of spatial pattern of nutrient supply on the early stages of growth in plant populations. Journal of Ecology 91: 305–315.
8. García-Palacios P, Maestre FT, Gallardo A (2011) Soil nutrient heterogeneity modulates ecosystem responses to changes in the identity and richness of plant functional groups. Journal of Ecology 99: 551–562.
9. Mommer L, Visser EJW, van Ruijven J, de Caluwe H, Pierik R, et al. (2011) Contrasting root behaviour in two grass species: a test of functionality in dynamic heterogeneous conditions. Plant and Soil 344: 347–360.
10. Maestre FT, Reynolds JF (2006) Spatial heterogeneity in soil nutrient supply modulates nutrient and biomass responses to multiple global change drivers in model grassland communities. Global Change Biology 12: 2431–2441.
11. Wijesinghe DK, John EA, Hutchings MJ (2005) Does pattern of soil resource heterogeneity determine plant community structure? An experimental investigation. Journal of Ecology 93: 99–112.
12. Maestre FT, Bradford MA, Reynolds JF (2005) Soil nutrient heterogeneity interacts with elevated CO_2 and nutrient availability to determine species and assemblage responses in a model grassland community. New Phytologist 168: 637–650.
13. Maestre FT, Reynolds JF (2007) Amount or pattern? Grassland responses to the heterogeneity and availability of two key resources. Ecology 88: 501–511.
14. Maestre FT, Reynolds JF (2006) Nutrient availability and atmospheric CO_2 partial pressure modulate the effects of nutrient heterogeneity on the size structure of populations in grassland species. Annals of Botany 98: 227–235.
15. García-Palacios P, Maestre FT, Bardgett RD, de Kroon H (2012) Plant responses to soil heterogeneity and global environmental change. Journal of Ecology 100: 1303–1314.

16. van der Waal C, de Kroon H, Heitkönig I, Skidmore AK, van Langevelde F, et al. (2011) Scale of nutrient patchiness mediates resource partitioning between trees and grasses in a semi-arid savanna. Journal of Ecology 99: 1124–1133.
17. Wiens JA (1989) Spatial scaling in ecology. Functional Ecology 3: 385–397.
18. Kume T, Sekiya N, Yano K (2006) Heterogeneity in spatial P-distribution and foraging capability by *Zea mays*: effects of patch size and barriers to restrict root proliferation within a patch. Annals of Botany 98: 1271–1277.
19. Wang P, Lei J-P, Li M-H, Yu F-H (2012) Spatial heterogeneity in light supply affects intraspecific competition of a stoloniferous clonal plant. PloS ONE 7: e39105.
20. Wijesinghe DK, Hutchings MJ (1997) The effects of spatial scale of environmental heterogeneity on the growth of a clonal plant: an experimental study with *Glechoma hederacea*. Journal of Ecology 85: 17–28.
21. Wijesinghe DK, Hutchings MJ (1999) The effects of environmental heterogeneity on the performance of *Glechoma hederacea*: the interactions between patch contrast and patch scale. Journal of Ecology 87: 860–872.
22. Kleijn D, van Groenendael JM (2001) The exploitation of heterogeneity by a clonal plant in habitats with contrasting productivity levels. Journal of Ecology 87: 873–884.
23. Klimes L, Klimesova J, Hendriks R, Van Groenendael J (1997) Clonal plant architecture: a comparative analysis of form and function. In: de Kroon H, van Groenendael J, editors. The ecology and evolution of clonal plants. Leiden: Backhuys Publishers. 1–29.
24. Roiloa SR, Retuerto R (2006) Small-scale heterogeneity in soil quality influences photosynthetic efficiency and habitat selection in a clonal plant. Annals of Botany 98: 1043–1052.
25. Alpert P (1999) Clonal integration in *Fragaria chiloensis* differs between populations: ramets from grassland are selfish. Oecologia 120: 69–76.
26. de Kroon H, Kreulen R, van Rheenen JWA, van Dijk A (1998) The interaction between water and nitrogen translocation in a rhizomatous sedge (*Carex flacca*). Oecologia 116: 38–49.
27. Mommer L, van Ruijven J, Jansen C, van de Steeg HM, de Kroon H (2012) Interactive effects of nutrient heterogeneity and competition: implications for root foraging theory? Functional Ecology 26: 66–73.
28. Roiloa SR, Retuerto R (2007) Responses of the clonal *Fragaria vesca* to microtopographic heterogeneity under different water and light conditions. Environmental and Experimental Botany 61: 1–9.
29. Zhang Y, Zhang Q, Luo P (2008) Photosynthetic response of *Fragaria orientalis* in different water contrast clonal integration. Ecological Research 24: 617–625.
30. Lamb EG, Haag JJ, Cahill JF (2004) Patch-background contrast and patch density have limited effects on root proliferation and plant performance in *Abutilon theophrasti*. Functional Ecology 18: 836–843.
31. Gee GW, Or D (2002) Methods of soil analysis: physical methods. Soil Science Society of America Journal 5: 255–278.
32. Shein EV (2009) The particle-size distribution in soils: problems of the methods of study, interpretation of the results, and classification. Eurasian soil science 42: 284–291.
33. Wang D, Fu B-J, Chen L-D, Zhao W-W, Wang Y-F (2007) Fractal analysis on soil particle size distributions under different land-use types: a case study in the loess hilly areas of the Loess Plateau, China. Acta Ecologica Sinica 27: 3081–3089.
34. Zhou X-F, Zhao R, Li Y-Y, Chen X-Y (2009) Effects of land use types on particle size distribution of reclaimed alluvial soil of the Yangtze Estuary. Acta Ecologica Sinica 29: 5544–5551.
35. Kettler T, Doran JW, Gilbert T (2001) Simplified method for soil particle-size determination to accompany soil-quality analyses. Soil Science Society of America Journal 65: 849–852.
36. Medinski TV, Mills AJ, Fey MV (2009) Relationships between soil particle size fractions and infiltrability. South African Journal of Plant and Soil 26: 147–156.
37. Montero E (2005) Rényi dimensions analysis of soil particle-size distributions. Ecological Modelling 182: 305–315.
38. Rawls WJ, Brakensiek DL, Saxton KE (1982) Estimation of soil water properties. Trans Asae 25: 1316–1320.
39. Poor A, Hershock C, Rosella K, Goldberg D (2005) Do physiological integration and soil heterogeneity influence the clonal growth and foraging of *Schoenoplectus pungens*? Plant Ecology 181: 45–56.
40. Zhou J, Dong B-C, Alpert P, Li H-L, Zhang M-X, et al. (2012) Effects of soil nutrient heterogeneity on intraspecific competition in the invasive, clonal plant *Alternanthera philoxeroides*. Annals of Botany 109: 813–818.
41. Birch CPD, Hutchings MJ (1994) Exploitation of patchily distributed soil resources by the clonal herb *Glechoma hederacea*. Journal of Ecology 82: 653–664.
42. Cahill JF, McNickle GG, Haag JJ, Lamb EG, Nyanumba SM, et al. (2010) Plants integrate information about nutrients and neighbors. Science 328: 1657.
43. McNickle GG, Cahill JF (2009) Plant root growth and the marginal value theorem. Proceedings of the National Academy of Sciences 106: 4747–4751.
44. Janecek S, Janeckova P, Leps J (2004) Influence of soil heterogeneity and competition on growth features of three meadow species. Flora 199: 3–11.
45. Gross KL, Peters A, Pregitzer KS (1993) Fine root growth and demographic responses to nutrient patches in four old-field plant species. Oecologia 95: 61–64.
46. Reynolds HL, Mittelbach GG, Darcy-Hall TL, Houseman GR, Gross KL (2007) No effect of varying soil resource heterogeneity on plant species richness in a low fertility grassland. Journal of Ecology 95: 723–733.
47. Day KJ, Hutchings MJ, John EA (2003) The effects of spatial pattern of nutrient supply on yield, structure and mortality in plant populations. Journal of Ecology 91: 541–553.
48. Wu Z-Y (2010) Flora of China, V 23. Beijing: Science Press. 180 p.
49. Huang L (2011) Effects of soil particle on the growth of the clonal plant *Scirpus planiculmis* [Dissertation]: Beijing Forestry University.
50. Bronick CJ, Lal R (2005) Soil structure and management: a review. Geoderma 124: 3–22.
51. Mando A, Ouattara B, Somado AE, Wopereis MCS, Stroosnijder L, et al. (2005) Long-term effects of fallow, tillage and manure application on soil organic matter and nitrogen fractions and on sorghum yield under Sudano-Sahelian conditions. Soil Use and Management 21: 25–31.
52. Tracy SR, Black CR, Roberts JA, Mooney SJ (2011) Soil compaction: a review of past and present techniques for investigating effects on root growth. Journal of the Science of Food and Agriculture 91: 1528–1537.
53. Bengough AG, McKenzie BM, Hallett PD, Valentine TA (2011) Root elongation, water stress, and mechanical impedance: a review of limiting stresses and beneficial root tip traits. Journal of Experimental Botany 62: 59–68.
54. Bengough AG, Mullins CE (1990) Mechanical impedance to root growth: a review of experimental techniques and root growth responses. Journal of Soil Science 41: 341–358.
55. Dexter AR (1987) Mechanics of root growth. Plant and Soil 98: 303–312.
56. Ehlers W, Köpke U, Hesse F, Böhm W (1983) Penetration resistance and root growth of oats in tilled and untilled loess soil. Soil and Tillage Research 3: 261–275.
57. de Kroon H, Hutchings MJ (1995) Morphological plasticity in clonal plants: the foraging concept reconsidered. Journal of Ecology 83: 143–152.
58. Evans JP, Cain ML (1995) A spatially explicit test of foraging behavior in a clonal plant. Ecology 76: 1147–1155.
59. Alpert P, Stuefer JF (1997) Division of labour in clonal plants. In: de Kroon H, van Groenendael J, editors. The ecology and evolution of clonal plants. Leiden: Backbuys Publishers. 137–154.
60. Hutchings MJ, de Kroon H (1994) Foraging in plants: the role of morphological plasticity in resource acquisition. Advances in Ecological Research 25: 159–238.
61. Hutchings MJ, Wijesinghe DK (1997) Patchy habitats, division of labour and growth dividends in clonal plants. Trends in Ecology & Evolution 12: 390–394.
62. de Kroon H, Visser EJ, Huber H, Mommer L, Hutchings MJ (2009) A modular concept of plant foraging behaviour: the interplay between local responses and systemic control. Plant, Cell & Environment 32: 704–712.
63. Hutchings M, Wijesinghe D (2008) Performance of a clonal species in patchy environments: effects of environmental context on yield at local and whole-plant scales. Evolutionary Ecology 22: 313–324.
64. Dong M (1995) Morphological responses to local light conditions in clonal herbs from contrasting habitats, and their modification due to physiological integration. Oecologia 101: 282–288.
65. Hutchings MJ, John EA, Wijesinghe DK (2003) Toward understanding the consequences of soil heterogeneity for plant populations and communities. Ecology 84: 2322–2334.
66. Keller T, Håkansson I (2010) Estimation of reference bulk density from soil particle size distribution and soil organic matter content. Geoderma 154: 398–406.
67. Miranda JGV, Montero E, Alves MC, Paz González A, Vidal Vázquez E (2006) Multifractal characterization of saprolite particle-size distributions after topsoil removal. Geoderma 134: 373–385.

21

Urban Community Gardeners' Knowledge and Perceptions of Soil Contaminant Risks

Brent F. Kim[1,3], Melissa N. Poulsen[2,4], Jared D. Margulies[1,3,5], Katie L. Dix[6], Anne M. Palmer[1,3,7], Keeve E. Nachman[1,3,8]*

1 Johns Hopkins Center for a Livable Future, Baltimore, Maryland, United States of America, **2** CLF-Lerner Fellow, Johns Hopkins Center for a Livable Future, Baltimore, Maryland, United States of America, **3** Department of Environmental Health Sciences, Johns Hopkins Bloomberg School of Public Health, Baltimore, Maryland, United States of America, **4** Department of International Health, Johns Hopkins Bloomberg School of Public Health, Baltimore, Maryland, United States of America, **5** Department of Geography and Environmental Systems, University of Maryland, Baltimore, Maryland, United States of America, **6** Community Greening Resource Network, Parks & People Foundation, Baltimore, Maryland, United States of America, **7** Department of Health, Behavior and Society, Johns Hopkins Bloomberg School of Public Health, Baltimore, Maryland, United States of America, **8** Department of Health Policy and Management, Johns Hopkins Bloomberg School of Public Health, Baltimore, Maryland, United States of America

Abstract

Although urban community gardening can offer health, social, environmental, and economic benefits, these benefits must be weighed against the potential health risks stemming from exposure to contaminants such as heavy metals and organic chemicals that may be present in urban soils. Individuals who garden at or eat food grown in contaminated urban garden sites may be at risk of exposure to such contaminants. Gardeners may be unaware of these risks and how to manage them. We used a mixed quantitative/qualitative research approach to characterize urban community gardeners' knowledge and perceptions of risks related to soil contaminant exposure. We conducted surveys with 70 gardeners from 15 community gardens in Baltimore, Maryland, and semi-structured interviews with 18 key informants knowledgeable about community gardening and soil contamination in Baltimore. We identified a range of factors, challenges, and needs related to Baltimore community gardeners' perceptions of risk related to soil contamination, including low levels of concern and inconsistent levels of knowledge about heavy metal and organic chemical contaminants, barriers to investigating a garden site's history and conducting soil tests, limited knowledge of best practices for reducing exposure, and a need for clear and concise information on how best to prevent and manage soil contamination. Key informants discussed various strategies for developing and disseminating educational materials to gardeners. For some challenges, such as barriers to conducting site history and soil tests, some informants recommended city-wide interventions that bypass the need for gardener knowledge altogether.

Editor: Sunghun Park, Kansas State University, United States of America

Funding: This work was funded by a Faculty-Community grant from the Johns Hopkins Urban Health Institute (http://www.urbanhealth.jhu.edu/). The Johns Hopkins Center for a Livable Future is supported by a grant from the GRACE Communications Foundation (but did not provide funding specific to this project). The funders had no role in study design, data collection and analysis, decision to publish, or preparation of the manuscript.

Competing Interests: The authors have declared that no competing interests exist.

* E-mail: knachman@jhsph.edu

Background

Urban community gardens—gardens tended by multiple households in an urban neighborhood—may offer a range of benefits. Studies have observed associations between community gardening and health [1–8], social [6,9], and economic benefits [6,7,10], and gardening in general has been associated with cardiovascular [11,12] and mental [13–15] health benefits. Historically, backyard and community gardens have made substantial contributions to the food supply; World War II "Victory Gardens" have been credited with providing an estimated 40% of the U.S. vegetable supply [16]. In urban settings, community gardens—and urban green spaces in general—may confer an additional set of social benefits [17–20] and ecosystem services [21,22]. Urban green spaces also provide educational opportunities for urban residents, for whom parks and gardens may be their primary source of experience, knowledge, and valuation of nature.

Gardening in urban settings may also present health risks, including those stemming from exposure to contaminants such as heavy metals, organic chemicals, and asbestos that may be present in urban soils. Urban soils are often close to pollution sources, such as industrial areas and heavily trafficked roads. As a result, many soil contaminants have been found at higher concentrations with increasing proximity to urban centers [23]. In Baltimore, Maryland, prior soil analyses (Table S1 in File S1) have revealed high concentrations of lead at some sites [24–28], reflecting the city's long history of industrial activity, incinerators, and vehicular traffic, and raising concerns about lead exposure [24]. Table S2 in File S1 summarizes some of the more common urban soil contaminants, their sources, and health effects associated with exposure.

Gardeners can be exposed to contaminants by inadvertently ingesting soil, inhaling soil particles, or via dermal contact. Soil ingestion is of particular concern among children, who may ingest larger quantities of soil than adults (e.g., by putting their hands in their mouths), absorb higher levels of certain contaminants into their bloodstream [29], and are generally more sensitive to their effects. People who consume produce grown in contaminated environments also risk ingesting soil particles on the surfaces of plants [30]. Contaminants in soils including lead [31], cadmium [32], and arsenic [33] may accumulate in the tissues of vegetables grown in contaminated soils, posing another potential route of ingestion.

Urban gardeners may be unaware of how to manage these risks. Harms et al. [34,35] surveyed 121 urban farmers and gardeners from Kansas, Indiana, and Washington, most of whom indicated they do not have sufficient knowledge of how to minimize health risks associated with gardening in contaminated environments and want more information on soil testing and best management practices. Gardeners may also benefit from information on soil remediation, i.e. removing, destroying, detoxifying, immobilizing or containing soil contaminants [36].

The purpose of our study is to characterize urban community gardeners' knowledge of risks associated with contaminated garden soils, their perceptions of these risks, their knowledge of how to assess and reduce these risks, the sources they draw upon for information on soil contamination, and the information and training needs they have related to soil contamination.

Methods

To characterize urban community gardeners' knowledge and perceptions of soil contamination risks, we conducted surveys among urban community gardeners and semi-structured interviews with key informants in the gardening community.

Gardener surveys

We conducted brief verbal surveys in-person or by phone with Baltimore community gardeners. The survey included questions regarding demographics, garden site history, and knowledge, perceptions, and practices related to soil contamination. To be eligible to participate, gardeners had to be at least 18 years of age and have been gardening at their current site for at least 6 months.

We partnered with the Community Greening Resource Network (CGRN) to identify gardens from which to recruit survey participants. CGRN is Baltimore's gardening support network and maintains a registry of community gardens in the Baltimore metropolitan area. We randomly selected 30 gardens from the CGRN registry of 70 food-producing community gardens, contacting leaders at the selected gardens to identify opportunities to survey gardeners. After experiencing difficulty reaching some garden leaders, we included additional gardens – recommended to us by representatives in the gardening community or identified through personal contacts – whose leaders were willing to help us arrange surveys.

As an incentive for participating gardeners, we collected soil samples from represented community gardens, sent the samples for analysis at the U.S. Department of Agriculture (USDA) soils lab, shared the results with garden leaders, and offered guidance in interpreting the results.

Key informant interviews

We conducted semi-structured interviews with 18 purposively selected informants knowledgeable about community gardening and soil contamination in Baltimore City: representatives from City government urban agriculture-related programs (4), federal agency employees (2), a representative of a Baltimore community gardening organization (1), agricultural extension employees (2), Master Gardeners - trained volunteers who advise and educate the public on gardening (2), community garden leaders (4), and urban farmers (3). We distinguish farming from gardening by the intent to produce goods for sale.

Interviews focused on informants' perceptions of community gardeners' concerns about soil contamination, barriers to soil testing, and information needs related to soil contamination. When applicable, informants were also asked relevant questions about their roles and perspectives related to their employment in city, state, or federal agencies.

To identify major themes in the qualitative data, three members of the research team first developed a codebook that was organized by axial codes and sub-codes. Two researchers coded each transcript using Atlas/ti (v7); when discrepancies arose, we included all quotes assigned to a particular code by either researcher. We then generated reports of the text assigned to each code, writing reflective memos and pulling out illustrative quotes for each theme.

Ethical considerations

The Johns Hopkins School of Public Health Institutional Review Board (IRB) approved this study. Study participants provided verbal informed consent prior to participating in surveys and interviews. Oral consent was deemed adequate by the IRB, eliminating the need to record identifying information in study documents. An IRB-approved oral consent script was read by trained investigators to study participants. A dated questionnaire served as a record that the oral consent process had been completed.

Results

Gardener demographics

Seventy gardeners, representing 15 community gardens from a range of socioeconomic census tracts, responded to our survey. Most were female (66%), lived within a quarter-mile of their garden plots (76%), and had been at their current gardens for less than four years (76%). The median age of surveyed gardeners was 45. See Figure 1 for additional gardener demographic information.

Gardening practices and harvest use

Most (73%) surveyed gardeners indicated they avoid using commercial pesticides and fertilizers, relying instead on practices such as composting and mulching to promote soil fertility and suppress pests. Others (19%) reported using one or more commercial fertilizers and/or insecticides, several of which are allowed for use under USDA organic standards. Interview informants indicated that Baltimore's community gardeners generally garden without chemical inputs.

Among surveyed gardeners, 86% grew produce for home consumption, while the other 14% grew produce primarily for soup kitchens and other charitable uses. Among gardeners who grew produce for home consumption, almost half (45%) supplied over 60% of their family's produce intake from their community garden during the growing season.

General knowledge and concerns about contaminants

To assess their knowledge about chemical contaminants, surveyed gardeners were asked to list the soil contaminants they are aware of (Table 1). Most (66%) gardeners mentioned lead, and to a lesser extent, other trace elements (19%) and some types of organic chemicals (36%).

Gardeners were also asked to list any health concerns they have as community gardeners (Table 2); half (51%) cited soil contaminants as among their concerns. When asked to express

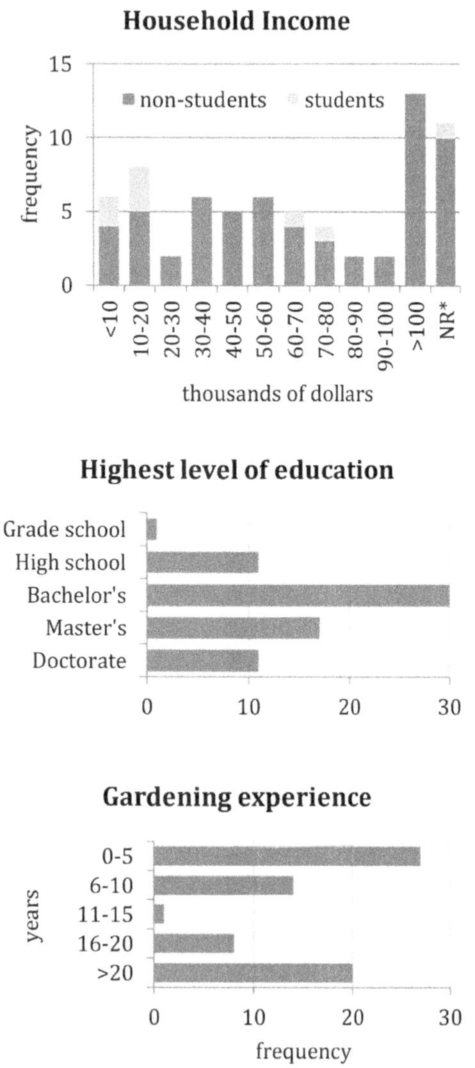

Figure 1. Additional gardener demographic information. * NR = No response.

their overall levels of concern about contaminants in their gardens, the average response was 2.3 on a scale of one to five, with five being the most concerned. There was no apparent association between levels of concern and whether gardeners thought their soil had been tested, or what they thought the test results indicated (Table 3).

When asked to list the ways in which one might come into contact with contaminants, most (70%) surveyed gardeners mentioned ingestion (e.g., "eating crops"). They did not specifically mention incidental ingestion, e.g., accidentally swallowing small amounts of soil while gardening. Other responses included dermal contact with (63%), and inhalation of (39%) contaminants.

Through interviews it became clear that lead is the contaminant of greatest concern among informants and, based on informants' perceptions, also the most common contaminant concern among gardeners. Informants expressed particular concerns about the vulnerability of children to contaminants—and specifically to lead—as compared to adults; among gardens where children may be present, some informants emphasized the heightened importance of testing their soil and making sure children do not ingest it. Informants were also concerned about other contaminants such as

Table 1. Open-ended responses to questions about soil contaminant knowledge: "What soil contaminants are you aware of that urban gardeners should be concerned about in general?" and "As a community gardener, do you have any concerns about hazards to your health?"

Response	%
Heavy metals and other trace elements	71
Non-specific	20
Lead	66
Arsenic	11
Mercury	4
Chromium	4
Cadmium	3
Copper	1
Organic chemicals	36
Petrochemicals (e.g., fuel, oil)	19
Pesticides	13
Persistent organic pollutants	7
Automotive fluids	6
Chemicals (non-specific)	16
Biological hazards	11
Human excreta	7
Animal excreta	6
Building materials (e.g., asbestos, asphalt, roofing tar)	11
Foreign objects (e.g., trash, needles)	21
Other	9
No response	10

trash, drug paraphernalia, and animal feces, as well as potential contaminants in fill dirt, compost, and water. Informants also expressed concerns about chemical inputs, such as pesticides, and indicated that gardeners may view their use as more harmful to

Table 2. Open-ended responses to "As a community gardener, do you have any concerns about hazards to your health?"

Response	%
Soil contaminants	51
Non-specific	20
Organic chemicals (e.g., pesticides)	16
Heavy metals and other trace elements	11
Trash	11
Discarded needles	9
Human or animal excreta	4
Crime	6
Animal pests	4
Injury	3
Air quality	3
No concerns	44

Table 3. Gardeners' levels of concern about contaminants in their garden, by perceived soil test status.

"Has your soil been tested for contaminants?"	"Did the results reveal a problem with contamination?"	Number of respondents	Average level of concern (1–5)* among respondents
No	NA	14	2.5
Unsure	NA	16	1.7
Yes	No	20	2.6
Yes	Unsure	9	2.0
Yes	Yes**	11	2.4

NA: Not applicable.
*On a scale of one to five, with five being the most concerned.
**8 of these 11 respondents indicated they discontinued growing food crops in contaminated areas; two indicated the soil was remediated; one was unsure whether corrective action was taken.

health than contaminants directly in soil. Table S3 in File S1 illustrates the range of concerns noted in these discussions.

Most informants did not express a high level of concern about soil contaminants. Frequently, informants indicated that their concerns about soil contamination were alleviated by the use of raised beds or after seeing safe results come back from soil tests. Informants also made repeated comments about soil quality, suggesting issues of soil fertility may be more salient than contaminant concerns.

> *For me, the most important thing is that we have good soil structure here. … So, really my energies have been not around contaminations, but just building healthy soil so that we get the best vegetable production out of here. (Community garden leader)*

In discussing gardeners' levels of awareness or concern about soil contamination, informants' views varied widely. Several had little confidence that gardeners think about soil contamination as an issue or are aware that soil testing can and should be conducted. Informants working with municipal programs related to community gardening noted that few people ask about soil contamination when starting a garden. At the other end of the spectrum, some informants noted broad concern about soil contamination among community gardeners.

> *We talk about air. We talk about water. … Nobody talks about soil, and, essentially, it became very obvious that soil's probably the most contaminated thing that we have in our environment, particularly in urban areas. (MasterGardener)*

> *I think each year, the new gardeners ask do we have to be concerned about soil, and I say we tested it and everything was okay. (Community garden leader)*

> *And I've noticed that most gardens want raised beds, because they think there is a lead issue. (City government representative)*

Informants noted that knowledge and concerns about contaminants vary with different populations. Younger and more educated individuals, for example, were thought to have greater awareness of soil contaminant issues. A few informants noted that soil contamination was of less concern for gardeners and volunteers at urban farms, because they trusted that the appropriate steps had been taken to ensure the safety of the soil. One urban farmer noted that community members were concerned about soil contamination when the farm was first getting started, but once it became "established", these concerns disappeared.

Site history

One of the first steps in determining soil safety is learning how a particular site was used. This was a top concern of one City government representative, who worried that by testing soil without the knowledge of a site's history, gardeners may be unaware of potential contaminants:

> *If you do a test for lead and other heavy metals and you pat yourself on the back and you go on, are you missing the fact that there used to be a gas station on that site and there could be types of contaminants that you don't even know how to test for that could pose a risk?*

Most (73%) surveyed gardeners said they knew the site history of their gardens. Likewise, the community garden leaders and urban farmers we interviewed indicated they knew the past use of their garden and farm sites.

When asked how they learned about their site's history, most community garden leaders and urban farmers we interviewed indicated they spoke to residents in the surrounding area. One urban farmer said they would start by referencing Sanborn Maps, which are available through local libraries and depict historical land uses from 1867 to 1970. Surveyed gardeners reported obtaining information on site histories primarily from other gardeners (42%), from neighbors (26%) or based on their own observations (23%). A small proportion (7%) had obtained information from a government office, such as the Department of Planning.

When asked if site history is important information in determining if a site is suitable for gardening, nearly all (99%) surveyed gardeners agreed. In contrast, one City government representative suggested that most gardeners would not be interested in trying to uncover information on site history, noting, "I think, based on the people I've met and talked to, they just wanna grow something."

Informants also suggested that gardeners may lack the expertise necessary to conduct a site history. One City government representative noted that in an ideal world, assistance would be provided to new gardens to test the soil and "sit down with somebody and go over the site history in a way that's simple and doesn't take too long and is very clear." One federal agency employee indicated that expert guidance is the "crucial part" of a Phase 1 Environmental Site Assessment, in which an auditor reviews historical and other records, visits the site, and interviews previous landowners and other informants:

[T]he site history will get you … maybe 85% there.... [T]here's nothing about [a Phase 1 assessment], less judgment, that a person can't do at the library. But that is still missing that crucial part of the equation – the expertise piece.... Because you'll see maybe it was a paint factory. Well, will everybody think about, "Okay, well, how was a paint made at that time"? Well, it was linseed oil and the white lead was actually mixed in, and it was about at 40% lead.

Another federal agency employee described additional limitations of site histories, noting that "the use and dumping on backyards and gardens is so idiosyncratic that it's impossible to learn what happened even ten years before," and that conducting sporadic site histories on a per-garden basis may not reveal evidence of contaminants from other parts of the city (e.g., from nearby industries that have since shut down). For these reasons, this informant recommended a more comprehensive approach of conducting site histories and soil testing at a city-wide level.

Soil testing

One theme that emerged during informant interviews was the barriers that deter gardeners from testing their soil for contaminants.

Cost was perceived to be prohibitive, particularly in situations in which a gardener wanted to test for a contaminant outside the scope of the usual metals panel, such as asbestos. According to informants, these situations also required additional knowledge about what to test for and how to find a service that offered those tests.

Informants also suggested that gardeners might perceive the steps involved in taking and sending away soil samples to be cumbersome or too complicated. Some informants suggested the need for a local testing service or a government-funded public service for soil testing, although not all agreed that a local testing service would be worth the cost.

There should be an immediate way to get testing … because some people will not go through the process of sending something away. (Community garden leader)

I don't really see a huge a demand out there for this kind of information … it takes a whole lot to get the right people together to have a soil-testing lab here, in the city. And then, again, it might not even be used. (City government representative)

Other barriers mentioned by informants included fear of discovering contamination after having already made investments in the land, and the need to document the exact locations from where soil samples are taken within a garden (since contaminant levels may vary spatially across a garden).

A few informants also noted that once soil tests are conducted, gardeners might have difficulty interpreting the results. Our survey results also hinted at this – in responding to questions about tests for contaminants, several gardeners referenced nutrient levels and soil tilth, suggesting that some gardeners may conflate tests of soil fertility with tests for contaminant levels.

Several informants also perceived a lack of scientific consensus about what levels of contamination are considered safe.

[H]ow much lead is too much lead? I have read different numbers. In Canada, the safe level is different than in the United States, and I think that in Europe it's different. (City government representative)

[T]here's a lot of conflicting information from EPA and different universities about what an action level would be for total lead...(Agricultural extension employee)

Both federal agency employees – experts in contamination – also noted imperfections in the science in determining risk standards.

Reducing exposure to contaminants

Another aim of this study was to explore knowledge and practices related to reducing exposure when working in potentially contaminated gardens.

We asked surveyed gardeners how they thought one should approach working in contaminated environments, and what they would do if their soil was found to be contaminated; these results are summarized in Table 4.

Among interviewed informants, most cited building raised beds and filling them with clean, imported soil as the safest and easiest way to manage potential soil contamination. Among surveyed gardeners, using raised beds was a common practice. The majority (74%) reported growing at least some crops in raised beds, and 50% said they use raised beds exclusively. Some informants, however, alluded to concerns regarding limitations of raised beds. One federal agency employee noted the possibility that the soil used to fill raised beds may be contaminated, particularly if it was taken from a questionable source, and that plant roots may extend down into contaminated soils below the raised bed, potentially allowing contaminants to enter plant tissues. One urban farmer also suggested that people may be exposed to contaminants from soil not covered by the raised bed (e.g., if native soil is kicked up), and that gardeners who use raised beds might underestimate these risks and not test the underlying soil.

We also asked informants and surveyed gardeners about soil remediation. One City government representative suggested that gardeners may not necessarily know how to remediate the soil and that guidance is needed to provide direction. When surveyed gardeners were asked, for example, whether they thought planting sunflowers in contaminated soil would effectively remediate it, 9% incorrectly said yes and 51% were unsure.

Information sources

Another aim of this study was to understand where gardeners obtain information about soil contamination. Surveyed gardeners (see Table 5) and interviewed informants most commonly mentioned gardening support institutions, particularly the agricultural extension office and its Master Gardener program.

One theme that emerged from informant interviews was the need for a central repository where gardeners could access information about soil contamination. Most informants thought this should be offered through an organization that gardeners already associate with gardening information. Compared to the more formal services of the agricultural extension, the community-based CGRN was cited as having "the biggest direct communication with community gardeners in the city" and being "more accessible" than more "bureaucratic" organizations. One informant noted, however, that while it provides a valuable network for gardeners, within CGRN "a lot of misinformation gets shared." The agricultural extension was thought to be the traditional place where gardeners and farmers would think to access soil

Table 4. Open-ended responses to questions about reducing exposure in contaminated environments.

Response	%
Stop growing produce in contaminated areas, and/or stop eating produce grown in contaminated areas	50
Remove surface contaminants	26
Wash produce	26
Peel root crops	3
Remediate soil	26
Install a barrier over contaminated soil	9
Add soil amendments (e.g., compost or minerals)	9
Bioremediate, phytoremediate, and/or mycoremediate	9
Remove contaminated soil	9
Remediate (non-specific)	4
Grow in raised beds or containers	17
Only grow certain crops (e.g., not root vegetables)	13
Wear gloves	9
Wash hands	6
Apply mulch (e.g., to reduce splashing on crops)	3
Continue using the same methods	3
Seek out more information	29
Don't know	24

"What methods should one use to grow, harvest or handle produce grown in contaminated environments?" and "What would you do if you found out your soil was contaminated?"

contamination information, but informants noted it had "not historically had a real urban presence" and was not prepared to deal with issues common in cities, such as urban soil contamination. One City government representative saw the municipal government as best positioned to gather, hold, and disseminate information on soil contamination:

> *[G]overnment can play a really important role in doing the due diligence and gathering all of that stuff and then making it publicly available. … I think we have the institutional longevity to hold on to that [information]. … And we need to have this historical data so if that [soil] test has already been done you don't need to go do it again.*

Table 5. Open-ended responses to "Where do you get information on gardening practices?"

Response	%
Gardening support organizations	84
Extension office/Master Gardeners	37
CGRN	19
Other	4
Online	53
Other gardeners	44
Books/magazines	29
Friends/family	16

Information needs

Through informant interviews, we aimed to identify urban community gardeners' information needs related to soil contamination. Responses fell under four main topics: site history, soil testing, remediation, and minimizing exposure (see Table 6).

Regarding how best to present this information to gardeners, the overriding theme was that it must be concise, clearly organized, and use simple, straightforward language that speaks to people of varying educational levels. One City government representative noted that the information that is needed already exists, but should be combined "in a document with clear instructions that a layperson would feel comfortable using."

One community gardening organization representative also placed value in communicating the "fluid" nature of determining what levels of contaminants are considered safe, and that the "best way … [to approaching contaminant issues is] not black and white." Furthermore, informants emphasized the need to balance risk reduction messages related to soil contamination with the health, social, and environmental benefits of gardening, as well as "the values in gardening that are beyond measure."

> *[W]e don't want to create barriers. We want more people to be growing food. (City government representative)*

Disseminating information

Another consideration is how best to disseminate information about soil contamination to community gardeners. Informants had a broad range of suggestions, and several noted that a combination of dissemination strategies – including print, online, and face-to-face information – is needed in order to reach all types of gardeners.

Table 6. Community gardeners' information needs related to soil contamination, as reported by key informants.

Site history	How to find information about past uses of a plot of land
	Which contaminants to test for, given specific past land uses
	Geographic areas of the city where there are likely to be high levels of contamination
Soil testing	Importance of obtaining a soil test prior to gardening
	Which contaminants to test for
	Why to test for certain contaminants and not others
	Where to get soil testing done
	How much soil testing costs
	How to correctly take a soil sample for a soil test
Remediation	Best practices for remediating contaminated urban soils
Minimizing exposure	How to reduce exposure risks when gardening
	Contamination risks associated with imported materials such as compost or mulch

Most informants expressed a need for more interactive, face-to-face methods of dissemination, such as individual consultations, workshops, or a citywide gardening conference. The important role of Master Gardeners was noted, for, as stated by one City government representative, they are "already tapped into the lives of the gardeners out there in the world."

[N]o matter how hard we try we may never reach half the people who want to garden. Because they're not going to ask, they're not going to look, they're not going to read, they're not going to do a Web search. ... So only by person-to-person communication pushed out from the community master gardeners. (Federal agency employee)

Training needs

While we did not ask informants about their level of knowledge related to soil contamination, many spontaneously acknowledged confusion and a lack of understanding about the issue. One federal agency employee recommended that to help urban community gardeners make decisions about the safety of garden sites, training is needed to provide plant and soil experts with expertise on soil contamination.

[T]he only really education I've had about soil is what we had in our Master Gardener class. And to be honest, that was very cursory, tip of the iceberg, basic soil composition. And I don't really remember ... there being a lot of information shared about contamination. (Community garden leader)

Discussion

Through surveys with urban community gardeners and key informant interviews, we characterized urban community gardeners' knowledge and perceptions of soil contaminant risks, including their knowledge of how to assess and reduce these risks, sources of information on soil contamination, and information needs related to soil contamination.

Knowledge and concerns

Our results suggest that concern about soil contaminants among community gardeners in Baltimore is generally low, particularly among established gardens. Informants indicated this is likely because gardeners assume soil contamination has already been addressed through safe soil test results, remediation, or the use of raised beds. Concern may be warranted, however, since prior studies of Baltimore soils suggest that soil contaminant levels vary widely [24–26]—even within the same garden plot [25]—and at some sites lead levels greatly exceed EPA screening levels (Table S1 in File S1). Soil lead levels have also been shown to increase over time due to atmospheric deposition [37]. Informants called for extra precautions where children may be present in gardens, echoing evidence of children's vulnerability to soil contaminants [24,29].

Surveyed gardeners' knowledge and concerns related to soil contaminants largely focused around lead. Their awareness may have been informed by recent state and city interventions in Baltimore aimed at raising awareness of child lead poisoning [38,39]. Gardeners' concerns were in accordance with tests of Baltimore soils (Table S1 in File S1), which identified high levels of lead more often than other trace elements included in analyses. Gardeners demonstrated inconsistent awareness about other potential contaminants.

Among key informants, issues that affect gardeners' ability to cultivate plants often appeared to be more salient than contamination concerns. Additionally, our results suggest that Baltimore's gardeners are more concerned about chemicals added to the gardening environment than what contaminants may already be present in soil. Gardeners' concerns about pesticides reflect the results of prior surveys that found the primary reason consumers purchase organic produce is to reduce exposure to pesticides and other chemicals [40].

Our results identified areas where gardeners' knowledge and concerns may not be concomitant with the potential health risks associated with urban soil contaminants. Efforts to address discrepancies in gardeners' knowledge, however, must be carefully crafted so as to not elevate levels of concern beyond those at which they would discontinue gardening altogether. Informants also made the important point that risk reduction messages must be balanced with the health, social, and environmental benefits of gardening.

Site history

Our results suggest gardeners recognize the importance of knowing a garden site's prior uses. Several informants indicated, however, that gardeners may lack the motivation, information and expertise to determine accurately the prior use of their garden site, or to anticipate the contaminants that may be present as a result. City-wide documentation of site histories may be an effective means to alleviate these responsibilities from gardeners.

Soil testing

Our findings also revealed potential barriers to soil testing, including not knowing how to properly sample soil from a garden, where to send soil samples for testing, and which contaminants to test for under various circumstances, as well as the perception that testing is too expensive, complicated, or cumbersome. Even when soil tests are conducted, gardeners may have difficulty interpreting the results. Given surveyed gardeners' knowledge and concerns were largely centered around lead, they may interpret a negative test result for lead as a "clean bill of health" and neglect to consider the presence of other contaminants. These and other concerns speak to the potential value of providing gardeners with cost-assistance and guidance on testing soil and interpreting results. Such services could, for example, be included as part of a lease to adopt city-owned vacant lots.

Reducing exposure to contaminants

When surveyed gardeners were asked to list practices to reduce exposure in contaminated environments (Table 4), several best management practices [25,41,42] were notably absent from responses, including reducing soil ingestion among children (e.g., by washing their hands, and reducing hand-to-mouth contact), growing produce away from busy streets, railways, and older buildings, and removing shoes to avoid tracking contaminants into the home. Few respondents alluded to the use of mulch (3%) to reduce splashing on crops; or the removal of surface contaminants, e.g., by washing produce (26%) and peeling root crops (3%). Other gaps in practices included the application of soil amendments (9%) to dilute contaminants and/or reduce their mobility or bioavailability (e.g., applying phosphorus to reduce the bioavailability of lead [29]), though such amendments may also *increase* the mobility or bioavailability of certain contaminants [43]. Some gardeners were quick to acknowledge their limited knowledge on how to approach contaminated environments, and indicated they would seek out more information (29%) or take the conservative approach of not growing produce in and/or eating produce from contaminated areas (50%).

Contaminant concerns among gardeners and informants were often alleviated by the use of raised beds, which were viewed as an easy and effective solution to managing soil contamination. As some informants noted, raised beds do not remedy the presence of contaminated soil surrounding the bed, which may be kicked up during gardening activities. Clark and colleagues [37] raise particular concerns regarding children: based on a model specific to lead-contaminated gardens that considered incidental soil ingestion, inhalation of ambient air, and consumption of tap water and garden-grown produce, an estimated 72–91% of children's lead exposure is via incidental soil ingestion. Because raised beds only cover a small percentage of land, they offer limited protection against incidental ingestion via hand-to-mouth behavior among children playing in areas with contaminated soils. Raised bed soils may also become contaminated over time; lead levels measured in urban raised beds in Dorchester, Massachusetts were found to increase by roughly 185 parts per million over a four year period due to wind-transported fine grain soil [37]. Another study explored concerns related to the use of timbers in the construction of raised beds, detecting elevated levels of arsenic in garden plots framed by chromated copper arsenate treated lumber and elevated levels of polycyclic aromatic hydrocarbons in plots framed by railroad ties [44]. In addition to these concerns, as one informants noted, soil used to fill raised beds may be contaminated, and plant roots may extend into contaminated soils below the bed. While raised beds provide some protection against contaminant exposure, they are not a panacea, and recommendations for their use should be tempered with information about their limitations.

Information and training needs

Informants voiced a need for specific information related to the management of soil contamination and indicated that gardeners need a central place where they can access such information. They noted several potential organizations that could serve such a role. Informants suggested collaboration between such organizations and City government to develop and disseminate a single set of information could yield the greatest reach.

Informant interviews suggest the major challenges in providing such information are the complexity and uncertainty surrounding the issue. Thus, there may be a role for two levels of guidance: additional training for gardening experts, to help them better understand the issues around soil contamination and how to effectively communicate risks to community gardeners; and concise, straightforward messaging for gardeners.

Study limitations

Our sample population was small, and the median income bracket among surveyed gardeners ($50,000 – 60,000) and percentage with a bachelor's degree (83%) were high relative to the population of Baltimore City in 2007–2011 [45]. For these reasons, our study population may not be representative of the average Baltimore gardener. Our qualitative findings may also be unique to the Baltimore context; however, given the lack of research on this topic, we believe the results of this study can be used as a starting point to inform educational interventions for reducing soil contamination risks among gardeners in a variety of urban contexts.

Conclusions

Through this study, we identified a range of factors, challenges, and needs related to Baltimore community gardeners' perceptions of risk related to soil contamination, including low levels of concern and inconsistent levels of knowledge about heavy metal and organic chemical contaminants, barriers to investigating a garden site's history and conducting soil tests, limited knowledge of best practices for reducing exposure, and a need for clear and concise information on how best to prevent and manage soil contamination. Key informants discussed various strategies for developing and disseminating educational material to gardeners. For some challenges, such as barriers to conducting site history and soil tests, some informants recommended city-wide interventions that bypass the need for gardener knowledge altogether. In cases where public health messages about the risks from soil contaminants are implemented, informants stressed the importance of crafting messages in ways that do not dissuade gardeners from continuing to garden in urban environments. Given the health, social, environmental, and economic benefits associated with participating in and supporting urban green spaces, it is critical to protect the viability of urban community gardens while also ensuring a safe gardening environment.

Supporting Information

File S1 Supporting Information Tables. Table S1, Levels of selected heavy metals and other trace elements detected in Baltimore soils. Heavy metals panels do not detect for the range of contaminants that may be present in soil (e.g., organic chemicals), and should be paired with information on a site's prior use before proceeding with gardening activities. Reported levels may be capped at minimum and maximum detectable levels. **Table S2, Examples of potential urban soil contaminants, recognized sources, and health effects associated with exposure. Table S3, Examples of contaminant concerns expressed by key informants.**

Acknowledgments

The authors thank the Community Greening Resource Network for their partnership on the project and access to the Baltimore urban gardening community, and Rufus Chaney of the U.S. Department of Agriculture's Agricultural Research Service for analyzing soil samples from participating gardens.

Author Contributions

Conceived and designed the experiments: JM KD AP KN. Performed the experiments: BK MP JM KD AP KN. Analyzed the data: BK MP AP. Wrote the paper: BK MP JM KD AP KN. Acquired funding for the study: KN.

References

1. Alaimo K, Packnett E, Miles RA, Kruger DJ (2008) Fruit and vegetable intake among urban community gardeners. J Nutr Educ Behav 40: 94–101. doi:10.1016/j.jneb.2006.12.003.
2. Blair D, Giesecke CC, Sherman S (1991) A dietary, social and economic evaluation of the Philadelphia urban gardening project. J Nutr Educ 23: 161–167.
3. McCormack LA, Laska MN, Larson NI, Story M (2010) Review of the nutritional implications of farmers' markets and community gardens: a call for evaluation and research efforts. J Am Diet Assoc 110: 399–408. doi:10.1016/j.jada.2009.11.023.
4. Litt JS, Soobader MJ, Turbin MS, Hale JW, Buchenau M, et al. (2011) The influence of social involvement, neighborhood aesthetics, and community garden participation on fruit and vegetable consumption. Am J Public Health 101: 1466–1473. doi:10.2105/AJPH.2010.300111.
5. Armstrong D (2000) A survey of community gardens in upstate New York: implications for health promotion and community development. Health Place 6: 319–327.
6. Patel IC (1991) Gardening's socioeconomic impacts. J Ext 29.
7. Wakefield S, Yeudall F, Taron C, Reynolds J, Skinner A (2007) Growing urban health: community gardening in South-East Toronto. Health Promot Int 22: 92–101. doi:10.1093/heapro/dam001.
8. Twiss J, Dickinson J, Duma S, Kleinman T, Paulsen H, et al. (2003) Community gardens: lessons learned from California Healthy Cities and Communities. Am J Public Health 93: 1435–1438.
9. Kuo FE, Sullivan WC, Coley RL (1998) Fertile ground for community: inner-city neighborhood common spaces. Am J Community Psychol 26: 823–851.
10. Voicu I, Been V (2006) The effect of community gardens on neighboring property values. Real Estate Econ 36: 241–283.
11. Caspersen CJ, Bloemberg M, Saris WHM, Merritt RK, Kromhout D (1991) The prevalence of selected physical activities and their relation with coronary heart disease risk factors in elderly men: The Zutphen Study, 1985. Am J Epidemiol 133: 1078–1092
12. Magnus K, Matroos A, Strackee J (1979) Walking, cycling, or gardening, with or without seasonal interruption, in relation to acute coronary events. Am J Epidemiol 110: 724–733.
13. Kaplan R (1973) Some psychological benefits of gardening. Environ Behav 5: 145–162. doi:10.1177/001391657300500202.
14. Fabrigoule C, Letenneur L, Dartigues JF, Zarrouk M, Commenges D, et al. (1995) Social and leisure activities and risk of dementia: a prospective longitudinal study. J Am Geriatr Soc 43: 485–490.
15. Simons LA, Simons J, McCallum J, Friedlander Y (2006) Lifestyle factors and risk of dementia: Dubbo Study of the elderly. Med J Aust 184: 68–70.
16. Brown KH, Jameton AL (2000) Public health implications of urban agriculture. J Public Health Policy 21: 20–39.
17. Kuo FE (2001) Coping with poverty: impacts of environment and attention in the inner city. Environ Behav 33: 5–34. doi:10.1177/00139160121972846.
18. Kuo FE, Taylor AF (2004) A potential natural treatment for attention-deficit/hyperactivity disorder: evidence from a national study. Am J Public Health 94: 1580–1587.
19. Wells NM (2000) At home with nature: effects of "greenness" on children's cognitive functioning. Environ Behav 32: 775–795.
20. Kuo FE, Sullivan WC (2001) Environment and crime in the inner city: does vegetation reduce crime? Environ Behav 33: 343–367. doi:10.1177/0013916501333002.
21. Bolund P, Hunhammar S (1999) Ecosystem services in urban areas. Ecol Econ 29: 293–301. doi:10.1016/S0921-8009(99)00013-0.
22. Andersson E, Barthel S, Ahrné K (2007) Measuring social-ecological dynamics behind the generation of ecosystem services. Ecol Appl A Publ Ecol Soc Am 17: 1267–1278.
23. Meuser H (2010) Causes of soil contamination in the urban environment. In: Alloway BJ, Trevors JT, editors. Contaminated urban soils. Springer. pp. 29–94.
24. Mielke HW, Anderson JC, Berry KJ, Mielke PW, Chaney RL, et al. (1983) Lead concentrations in inner-city soils as a factor in the child lead problem. Am J Public Health 73: 1366–1369.
25. Chaney R, Sterrett SB, Mielke HW (1984) The potential for heavy metal exposure from urban gardens and soils. In: Preer JR, editor. Proc. symp. heavy metals in urban gardens. USDA ARS. pp. 37–84.
26. Pouyat R V, Yesilonis ID, Russell-Anelli J, Neerchal NK (2007) Soil chemical and physical properties that differentiate urban land-use and cover types. Soil Sci Soc Am J 71: 1010. doi:10.2136/sssaj2006.0164.
27. Schwarz K, Pickett STA, Lathrop RG, Weathers KC, Pouyat R V, et al. (2012) The effects of the urban built environment on the spatial distribution of lead in residential soils. Environ Pollut 163: 32–39. doi:10.1016/j.envpol.2011.12.003.
28. Yesilonis ID, Pouyat RV, Neerchal NK (2008) Spatial distribution of metals in soils in Baltimore, Maryland: role of native parent material, proximity to major roads, housing age and screening guidelines. Environ Pollut 156: 723–731. doi:10.1016/j.envpol.2008.06.010.
29. Hettiarachchi GM, Pierzynski GM (2004) Soil lead bioavailability and in situ remediation of lead-contaminated soils: a review. Environ Prog 23: 78–93. doi:10.1002/ep.10004.
30. Morgan R (2013) Soil, heavy metals, and human health. In: Brevik EC, Burgess LC, editors. Soils and human health. CRC Press. pp. 59–82.
31. Finster ME, Gray KA, Binns HJ (2004) Lead levels of edibles grown in contaminated residential soils: a field survey. Sci Total Environ 320: 245–257. doi:10.1016/j.scitotenv.2003.08.009.
32. Smolders E (2001) Cadmium uptake by plants. Int J Occup Med Environ Health 14: 177–183.
33. Ramirez-Andreotta MD, Brusseau ML, Artiola JF, Maier RM (2013) A greenhouse and field-based study to determine the accumulation of arsenic in common homegrown vegetables grown in mining-affected soils. Sci Total Environ 443: 299–306. doi:10.1016/j.scitotenv.2012.10.095.
34. Harms AMR (2011) Determining and meeting the educational needs of students and urban gardeners and farmers on urban soil quality and contamination topics.Master's Thesis, Kansas State University.
35. Harms AMR, Presley DR, Hettiarachchi GM, Thien SJ (2013) Assessing the educational needs of urban gardeners and farmers on the subject of soil contamination. J Ext 51.
36. Meuser H (2012) Soil remediation and rehabilitation: treatment of contaminated and disturbed land. Alloway BJ, Trevors JT, editors. Spinger. p. 128.
37. Clark HF, Hausladen DM, Brabander DJ (2008) Urban gardens: lead exposure, recontamination mechanisms, and implications for remediation design. Environ Res 107: 312–319. doi:10.1016/j.envres.2008.03.003.
38. Maryland Department of the Environment (2000) State, city unveil new lead awareness campaign. MD Environ 4.
39. City of Baltimore Health Department (2010) Data watch: Maryland Department of the Environment releases 2009 lead poisoning data. Available: http://baltimorehealth.org/press/2010-08-27-lead-report-pr.pdf. Accessed 01 January 2014.
40. Hughner S, McDonagh P, Prothero A, Shultz CJ, Stanton J (2007) Who are organic food consumers? A compilation and review of why people purchase organic food. J Consum Behav 6: 94–110. doi:10.1002/cb.
41. U.S. Environmental Protection Agency (2011) Brownfields and urban agriculture: interim guidelines for safe gardening practices. Available: http://www.epa.gov/brownfields/urbanag/pdf/bf_urban_ag.pdf. Accessed 2014 Jan 1..
42. Kessler R (2013) Urban gardening: managing the risks of contaminated soil. Environ Health Perspect 121. A327–A333.
43. Scheckel KG, Diamond GL, Burgess MF, Klotzbach JM, Maddaloni M, et al. (2013) Amending soils with phosphate as a means to mitigate soil lead hazard: a critical review of the state of the science. J Toxicol Env Heal B Crit Rev 16: 337–380. doi:10.1080/10937404.2013.825216.
44. Heiger-Bernays W, Fraser A, Burns V, Diskin K, Pierotti D, et al. (2009) Characterization and low-cost remediation of soils contaminated by timbers in community gardens. Int J Soil, Sediment Water 2: 1–19.
45. U.S. Census Bureau (2013) Baltimore City, Maryland. state & county quickfacts. Available: http://quickfacts.census.gov/qfd/states/24/24510.html. Accessed 2013 June 28.

22

Climatic Control on Plant and Soil $\delta^{13}C$ along an Altitudinal Transect of Lushan Mountain in Subtropical China: Characteristics and Interpretation of Soil Carbon Dynamics

Baoming Du[1], Chunjiang Liu[1,2*], Hongzhang Kang[1,2], Penghua Zhu[2], Shan Yin[1,2], Guangrong Shen[1,2], Jingli Hou[3], Hannu Ilvesniemi[4]

1 School of Agriculture and Biology and Research Center for Low-Carbon Agriculture, Shanghai Jiao Tong University, Shanghai, China, **2** Key Laboratory of Urban Agriculture (South), Ministry of Agriculture, People's Republic of China, Shanghai, China, **3** Instrumental Analysis Center of SJTU, Shanghai Jiao Tong University, Shanghai, China, **4** Finnish Forest Research Institute, Vantaa, Finland

Abstract

Decreasing temperature and increasing precipitation along altitude gradients are typical mountain climate in subtropical China. In such a climate regime, identifying the patterns of the C stable isotope composition ($\delta^{13}C$) in plants and soils and their relations to the context of climate change is essential. In this study, the patterns of $\delta^{13}C$ variation were investigated for tree leaves, litters, and soils in the natural secondary forests at four altitudes (219, 405, 780, and 1268 m a.s.l.) in Lushan Mountain, central subtropical China. For the dominant trees, both leaf and leaf-litter $\delta^{13}C$ decreased as altitude increased from low to high altitude, whereas surface soil $\delta^{13}C$ increased. The lower leaf $\delta^{13}C$ at high altitudes was associated with the high moisture-related discrimination, while the high soil $\delta^{13}C$ is attributed to the low temperature-induced decay. At each altitude, soil $\delta^{13}C$ became enriched with soil depth. Soil $\delta^{13}C$ increased with soil C concentrations and altitude, but decreased with soil depth. A negative relationship was also found between O-alkyl C and $\delta^{13}C$ in litter and soil, whereas a positive relationship was observed between aromatic C and $\delta^{13}C$. Lower temperature and higher moisture at high altitudes are the predominant control factors of $\delta^{13}C$ variation in plants and soils. These results help understand C dynamics in the context of global warming.

Editor: Shuijin Hu, North Carolina State University, United States of America

Funding: This work was supported by the State Key Basic Research and Development Plan of China (2011CB403201) and the CFERN&GENE Award Funds on ecological paper. The funders had no role in study design, data collection and analysis, decision to publish, or preparation of the manuscript.

Competing Interests: The authors have declared that no competing interests exist.

* E-mail: chjliu@sjtu.edu.cn

Introduction

In terrestrial plant ecosystems, ^{13}C enrichment is generally greater in the soils than in vegetation due to the different fractionation of ^{13}C and ^{12}C in the biogeochemical processes [1,2]. In the last three decades, the C isotope technique has been a useful approach for understanding the effects of water deficit on plants [3,4], dynamics of soil organic C (SOC) [5,6,7], and local climate history [8]. $\delta^{13}C$ has been used as an index to study ecosystem response to climate and as a surrogate variable in modeling C fluxes in terrestrial ecosystems [9,10].

At a site scale, soil $\delta^{13}C$ is closely associated with local vegetation and climate history, and varies with the soil depth owing to its longer-term fractionation in deeper soil [11]. The impact of vegetation on soil $\delta^{13}C$ could vary with plant species and tissues. For instance, C_3 (Calvin cycle photosynthetic pathway) plants have $\delta^{13}C$ values of −22‰ to −35‰ (average −26.5‰), compared to C_4 plants whose $\delta^{13}C$ values are in the range of −8‰ to −16‰ (average −12.5‰) [8,12]. This difference help determine the relative proportion each vegetation type in soil organic C [13,14]. Among plants tissues, $\delta^{13}C$ is typically enriched by 1‰ to 2‰ in celluose and hemicelluloses, but depleted in ^{13}C by 2‰ to 6‰ in lignin, in comparison with whole-plant material [6,15].

At a broader scale, soil ^{13}C varies with spatial gradients by altitude, latitude, longitude due to changes of vegetation and climate [16]. Bird et al. (1996) reported that the average $\delta^{13}C$ was −28.3±0.6‰ in low latitude (0°N to 20°N or S), −27.7±0.6‰ in mid latitude soils (20° to 40°), and −27.3±0.7‰ in high latitude soils (40° to 90°) [17]. The overall increase of SOC $\delta^{13}C$ from low to high-latitude forests was −1‰. In mountain areas, climate varies with altitude resulting in different vegetation, litter input into soils, and litter decomposition rate, and therefore soil $\delta^{13}C$. Wei and Jia (2009) reported that soil $\delta^{13}C$ first decreased and then increased as altitude increased from 1000 m to 3800 m in Mount Gongga, southwestern China [18]. Consequently, ecosystems in mountain areas are very sensitive to changes in climate [19,20].

Lushan Mountain is located in the middle-lower plain of the Yangtze River in central subtropical China, with altitude from 30 m a.s.l. to 1470 m a.s.l. [21]. Along with the change of elevation is the opposite trend of heat and water (OHW), i.e., temperature decreases and precipitation increases from low to high elevations. Correspondingly, vegetation changes from evergreen broadleaf forests at the foothill to deciduous forests at the top of mountain. The OHW is a common mountain climate in subtropical China and has unique community composition, litterfall decomposition ratio, soil C biochemical processes for exploring the variations of plant and soil $\delta^{13}C$ with altitude, particularly in relation to climate change.

In this study, we examined the variation of $\delta^{13}C$ in fresh plant leaves, litter, and semi-decomposed and soil humic substances in order to characterize soil carbon dynamics at a local or regional scale in subtropical forest ecosystems in Lushan Mountain, China. The specific objectives were to: 1) determine the patterns of $\delta^{13}C$ variation in dominant plant species and soil with altitude, 2) investigate how soil $\delta^{13}C$ varies with soil depth at different altitudes, and 3) examine the effects of climatic factors on plant and soil $\delta^{13}C$ along the elevation gradient in Lushan Mountain.

Materials and Methods

Study Area

Study area is located in the Lushan Nature Reserve (29°31′–29°41′N, 115°51′– 116°07′E), the south of Jiujiang City, Jiangxi Province, China (Fig. 1). Permission was granted by the Nature Reserve of Lushan Jiangxi. Lushan is an isolated mountain body situated in the center of the vast plain of the middle and lower reaches of the Yangtze River, covering an area of about 300 km^2 along a altitude range from 50 m to 1474 m. This area has a subtropical monsoon climate. The mean annual precipitation (MAP) ranges from 1308 mm to 2068 mm, and the mean annual temperature (MAT) from 17.1°C to 11.6°C [21].

Owing to the variations in geology, climate, and vegetation along elevation, dominant soil types change from Ferric alisols at low elevations to Haplic alisols at high elevations [21] according to the FAO soil texture classification. The corresponding vegetation types are evergreen forests dominated by several Fagaceae tree species including *Castanopsis sclerophylla*, *Castanopsis eyrei* and *Lithocarpus glaber*, and some evergreen woodland species and shrubs at low altitudes of approximately 50 m to 600 m, evergreen broadleaf forests and some deciduous trees are mid altitudes of 600 m and 1000 m, and *Lindera obtusiloba* forest consisting of *Cerasus serrulata*, *Castanea seguinii*, *Tilia breviradiata*, and a few shrubs at about 1200 m.s.l. (Table S1). Some *Cryptomeria japonica* plantations were established about 50 years ago at mid altitudes.

Sample Stands and Collection

Permission was granted by the Nature Reserve of Lushan Jiangxi (29°31′–29°41′N, 115°51′–116°07′E) to carry out our study. Sample stands were chosen at an irregular altitude interval owing to natural forest fragmentation in mountain. In total, four study sites were established at altitudes of 219 m, 405 m, 780 m, and 1268 m (Fig. 1, Table 1). These study sites were not associated with endangered or protected species.At each site, three plots (each 20×20 m) were randomly delineated. In each plot, four 4 m^2 subplots were randomly chosen for shrub layer and four 1 m^2 subplots for herbaceous layer. In each plot, 100 leaves were collected from dominant tree (Table S1). Sample were collected from L (Litter) and LF (Semi-decomposition litter) horizons and mineral soil layers at 0–10 cm, 10–20 cm, 20–30 cm, 30–40 cm, 40–50 cm, and 50–60 cm depths. Five soil cores were randomly collected within each plot using a 2 cm-diameter stainless steel borer and bulked to make one composite sample by soil depth.

Soil samples were air dried, ground, and passed through a 2 mm sieve to remove coarse living roots and gravel before being ground and passed through a 0.149 mm mesh sieve prior to chemical analysis. Leaf and litter samples were oven dried (65°C) for a week to constant weight and ground to fine powder using a Tecator sample mill (Subang, Shanghai, China) prior to the chemical and isotopic analyses.

Chemical Analysis

The C isotope ratio ($\delta^{13}C$) of leaf, litter, and soil samples was determined using an elemental analysis–stable isotope ratio mass spectrometer (VarioElIII/Isoprime, Elementar, Hanau, Germany) operated at the Instrumental Analysis Center of Shanghai Jiao Tong University (SJTU). The results are reported as parts per thousand (‰) deviations from the Vienna–Pee Dee Belemnite (PDB) standard (uncertainty of ±0.1‰ uncertainty), which The is expressed as follows:

$$\delta^{13}C = \left({}^{\delta_{sample}} / {}_{\delta_{standard}} - 1 \right) \times 10^3$$

Where δ_{sample} is the $^{13}C/^{12}C$ ratio of the samples and $\delta_{standard}$ is the $^{13}C/^{12}C$ ratio of the reference standard (PDB) [22].

To express the absolute variation of soil $\delta^{13}C$ enrichment relative to litter, we define an absolute enrichment factor F_A as

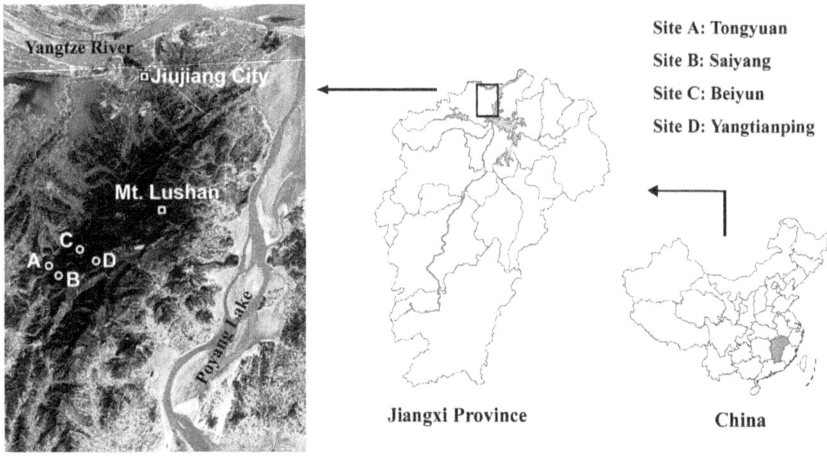

Figure 1. Location of the study area and the distribution of sample stands in Lushan Mountain, subtropical China.

Table 1. Features of climate and vegetation at different altitudes in Lushan Mountain.

Location	Altitude (ma.s.l.)	MAP[a] (mm)	MAT (°C)	TCM[b] (°C)	THM[c] (°C)	Growing season (Days)	Vegetation types[d]
Tongyuan	219	1429	16.2	3.8	28.5	262	EBF
Saiyang	405	1549	15.3	3.2	27.4	253	EBF
Beiyun	780	1794	13.6	1.9	25.1	234	EBMF
Yangtianping	1268	2112	11.3	0.3	22.2	209	DBF

[a]The climatic data from years 1971 to 2000 were obtained from the Lushan Meteorological Bureau.
[b]Temperature of the coldest month.
[c]Temperature of the hottest month.
[d]EBF represents for evergreen broadleaf forest; EBMF for evergreen broadleaf and needle-leaf mixed forest; and DBFfor deciduous broadleaf forest.

follows:

$$F_A = \delta^{13}C_{soili} - \delta^{13}C_{litter}$$

Where $\delta^{13}C_{soil\ i}$ is the $\delta^{13}C$ at *i*th soil layer and $\delta^{13}C_{litter}$is the $\delta^{13}C$ at the litter layer.

The rate of soil $\delta^{13}C$ enrichment varies with soil depth [1,2]. To express the relative enrichment of adjacent soil layers, we define the relative soil $\delta^{13}C$ enrichment factor F_R as follows:

$$F_R = \delta^{13}C_{soili} - \delta^{13}C_{soili-l}$$

Where $\delta^{13}C_{soil\ i}$ is the $\delta^{13}C$ at *i*th soil layer and $\delta^{13}C_{soil\ i-1}$ is the $\delta^{13}C$ at *i-1*th soil layer.

To explore the relationship between soil $\delta^{13}C$ and SOC concentration, as well as soil C functional groups, two soil variables were determined. The variation patterns in SOC concentration and soil C functional groups along the altitude gradient in Lushan Mountain will be presented in another paper. The SOC concentrations of soil samples from different depths were measured using dichromate oxidation method [23].

The chemical compositions of C in litter and soil layers at 0–10 cm, 30–40 cm, and 50–60 cm depths were analyzed with solid-state ^{13}C cross-polarization magic angle spinning–nuclear magnetic resonance (CP/MAS–NMR). The litter samples were dried to constant weight at 65°C and ground in a Wiley mill. The soil samples were pretreated with 10% (v/v) hydrofluoric acid (HF) before the NMR spectroscopy [24] to reduce Fe^{3+} and Mn^{2+} [25] and concentrate organic C for more accurate signal-to-noise ratio [24]. About 10 g of the ground sample was shaken with 50 ml HF for 2 h. After centrifugation (5,000 rpm) for 10 min, the supernatant was removed. The procedure was repeated five times. The remaining sediment was washed five times with 50 ml deionized water to remove residual HF before freeze drying.

The solid-state ^{13}C CP/MAS–NMR spectra of litter and soil samples were obtained at a frequency of 100.64 MHz using a Bruker AVANCEIII400 NMR spectrometer (BrukerBiospin, Rheinstetten, Germany) operated at 75.42 MHz for ^{13}C. The contact time was 1.5 ms, with 1 s recycle delay and the magnetic angle spinning rate was 5 kHz [24]. About 12,000 scans were collected for soil samples and 10,000 scans for litter samples [26]. The chemical shift regions 0–45 ppm, 45–110 ppm, 110–160 ppm, and 160–220 ppm were assigned to alkyl C, O-alkyl C, aromatic C, and carboxylic C, respectively [24,27]. The sources of organic carbon are: Alkyl C is derived from lipids, fatty acides and plant aliphatic polymers, O-alkyl C primarily from cellulose and hemicelluloses, as well as starch, proteins and carbohydrates, aromatic C from lignin and tannins, and carboxyl C from lipids, aliphatic esters, and amide carboxyls [28,29]. The signal intensities in the respective chemical-shift regions were expressed as a percentage of the area of the total spectra. The relative contents of different chemical structures were therefore calculated [26].

Statistical Analysis

Arithmetic means and standard deviation were calculated. A *t* test (i.e., least significant difference) was conducted to compare the means with a probability level of 0.05 for detecting significant differences. Linear regression analyses were used to examine the relationships between soil $\delta^{13}C$ and MAT, MAP, and SOC concentrations. All analyses were performed through SigmaPlot 10.0 (Systat Software, Richmond, CA, USA) and SAS V8.1 (SAS Institute Inc., Cary, North Carolina).

Results

Variations of Leaf, Litter, and Soil $\delta^{13}C$ with Altitude

The deciduous tree leaf $\delta^{13}C$ at 1268 m a.s.l. was –28.29‰, significantly lower than evergreen trees at lower altitudes (–27.65‰ to −26.98‰) (Fig. 2). A similar $\delta^{13}C$-altitude relationship was also evident in leaf litter, but not in semi-decomposed litter. Comparatively, fresh leaves had higher $\delta^{13}C$ than leaf litter or

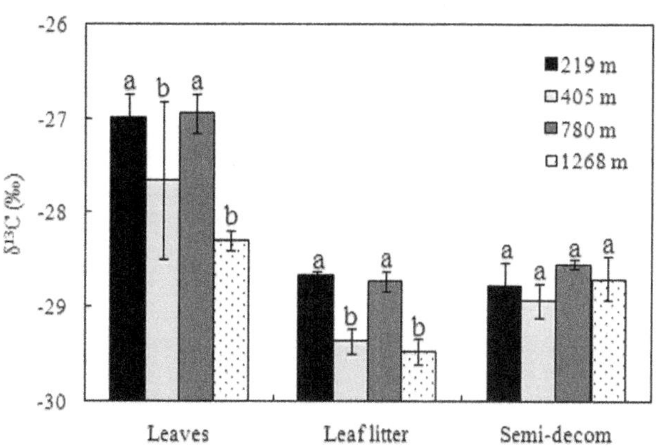

Figure 2. $\delta^{13}C$ values of leaves, leaf litter, and semi-decomposed litter in the natural secondary broadleaf stands at altitudes of 219, 405, 780, and 1268 m in the Lushan Mountain. The error bars are the standard errors (n = 3 for leaf litter and semi-decomposed litter, and n = 3 for 300 leaves). Different letters indicate significant differences among all classes at the different altitudes ($P<0.05$).

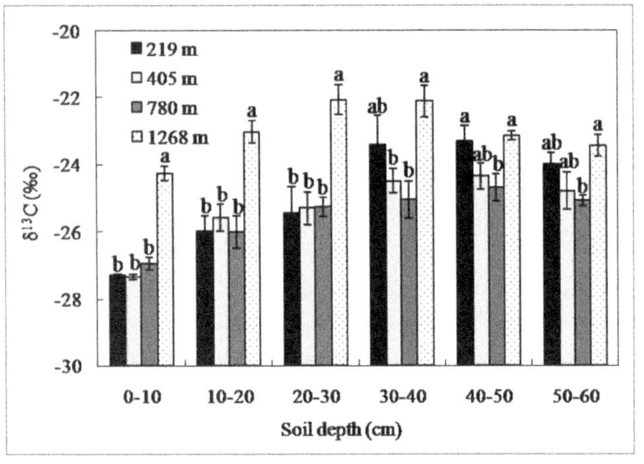

Figure 3. Soil $\delta^{13}C$ values of different layers in the natural secondary broadleaf stands at altitudes of 219, 405, 780, and 1268 m in the Lushan Mountain. The error bars represent standard errors means (n = 3). Different letters indicate significant differences among altitudes by soil depth ($P<0.05$).

semi-decomposed litter. For instance, $\delta^{13}C$ decreased from −26.98‰ in fresh leaves to −28.65‰ in leaf litter at 219 m a.s.l. (Fig. 2).

The surface layers of soils (0–10 cm, 10–20 cm, and 20–30 cm) at 1268 m was significantly higher than that at the other three lower altitudes. Ihe deeper soil layers (30–40 cm, 40–50 cm, and 50–60 cm), however, no significant difference by altitude occurred (Fig. 3).

Soil $\delta^{13}C$ Enrichment with Soil Depth

Soil $\delta^{13}C$ enrichment was greater with the increase of soil depth for all altitudes (Fig. S1). Absolute enrichment factor (F_A) generally increased from semi- decomposed layer, peaked at 30–40 cm, and remained stable at deeper soil (Fig. 4). The soil F_A at 1268 m were generally greater than that at the other three altitudes. For instance, the maximum F_A was 7.39‰ at 30–40 cm soil depth and 1268 m altitude, compared to the values of 3.66‰ to 5.23‰ at the same layer of three lower altitudes. The relative enrichment factor (F_R) was generally higher in the surface soil layers of 0–10 cm (1.49‰ to 4.43‰) and 10–20 cm (0.93‰ to 1.76‰) than in the deeper layers (Fig. 5).

Relationships of Soil $\delta^{13}C$ with SOC Concentration and Chemical Composition

Soil $\delta^{13}C$ increased (Fig. S1) with SOC concentration at each altitude, following a strong negative relationship ($p<0.01$) (Fig. 6). The relationship at the three sites of lower altitudes (219 m, 405 m and 780 m, all covered by evergreen forests) was different from that at the site of highest altitude (1268 m, covered by deciduous forests). Soil $\delta^{13}C$ was negatively correlated with O-alkyl C, but positively with aromatic C and carboxyl C (Fig. 7).

Relationships between Soil $\delta^{13}C$ and the Climatic Factors

In Lushan Mountain, temperature decreases with, whereas precipitation increases with the increase of elevation. The soil $\delta^{13}C$ decreased with MAT and increased with MAP in the three upper layers (0–10 cm, 10–20 cm, and 20–30 cm), whereas no clear trend existed for deeper layers (Fig. 8).

Discussion

Decoupled Patterns of Variations in Plant and Soil $\delta^{13}C$ along Elevation

In terrestrial ecosystems, plant functional types strongly affect site-level soil $\delta^{13}C$ through litter inputs [5,30,31]. For instance, Peri et al. (2012) reported that the soil $\delta^{13}C$ in *Nothofagus* forests was significantly associated with foliar $\delta^{13}C$, both of which decreased with precipitation [28]. In the present study, however, fresh leaf and leaf litter $\delta^{13}C$ decreased with, whereas soil $\delta^{13}C$ increased with altitude (Figs. 2 and 3), a pattern that cannot be explained alone with the increasing precipitation by altitude in Lushan Mountain.

Different from some previous studies [33,34], the tree leaf $\delta^{13}C$ was significantly lower at the highest altitude, likely due to the special climate regime in Lushan Mountain where precipitation increases with and temperature decreases with altitude (Fig. 8). According to a general notion about plant ^{13}C discrimination [35], plants at moist sites tend to have high stomatal conductance (close to maximum), low water use efficiency, and high intercellular CO_2 concentration. This results in increasing discrimination against $^{13}CO_2$ during photosynthesis leading to low $\delta^{13}C$ values, in comparison with arid sites. In Lushan Mountain, plants experience greater drought stress at lower altitude sites owing to low precipitation and high temperature, resulting in high tissues $\delta^{13}C$.

In the present study, soil $\delta^{13}C$ increased with altitude, consistent with the pattern found in previous studies [7,34,36,37,38]. For example, Townsend et al. (1995) reported an increase of soil $\delta^{13}C$ from −26.70 ‰ at 900 m a.s.l. to −25.90 ‰ at 1500 m a.s.l. in the island of Hawaii [32]. Similar results are also reported by Zimmermann et al. (2012) in a tropical forest in Peru where soil $\delta^{13}C$ values increased with elevation from −27.16‰ at 1700 m a.s.l. to −25.79‰ at 3030 m a.s.l. [38]. The major reason for the altitudinal variation of soil $\delta^{13}C$ in those studies is probably the influence of plant communities through the deposition of leaf litter, dead root material, and rhizodeposition [32]. In Lushan Mountain, however, increasing precipitation and decreasing temperature with altitude may have predominantly influence over soil $\delta^{13}C$.

Relative to the bulk leaf $\delta^{13}C$, sugars, starch, cellulose, protein, and organic aids are enriched, whereas lignin and lipids are depleted in $\delta^{13}C$ [6]. Therefore, organic matter with high concentrations of sugars, starch, and cellulose displays high $\delta^{13}C$ values. On the other hand, however, sugars, starch, cellulose, and protein, are more easily lost through litter decomposition than lignin [39,40]. This helps explain the high soil $\delta^{13}C$ of top soil layers (0–10 cm, 10–20 cm, and 20–30 cm) at the altitude of 1268 m (Fig. 3) where high moisture and low temperature not only reduces forest productivity and litter (organic matter) input to the soil, but also shows down decomposing activities of microbes.This may have led to accumulation of more less-decomposed organic matter in soils and therefore high soil $\delta^{13}C$ (accumulation of sugars, starch, cellulose, and protein).

Soil $\delta^{13}C$ Enrichment with Soil Depth by Altitude

In previous studies, enrichment factors (F_A in this study) were used to describe the variation in soil $\delta^{13}C$ enrichment relative to litter. However, the results of this study suggest that the relative enrichment factor (F_R) introduced in the present study better detect the difference of soil ^{13}C enrichment by depth than F_A used by previous studies (Figs. 4 and 5). For instance, F_R shows a more rapid change from semi-decomposed litter to soil layers (0–10 cm, 10–20 cm) than F_A (Fig. 5).

At all altitudes, soil $\delta^{13}C$ was enriched with soil depths from litter to O-layer and to mineral soil layers (Figs. 4 and S1). This

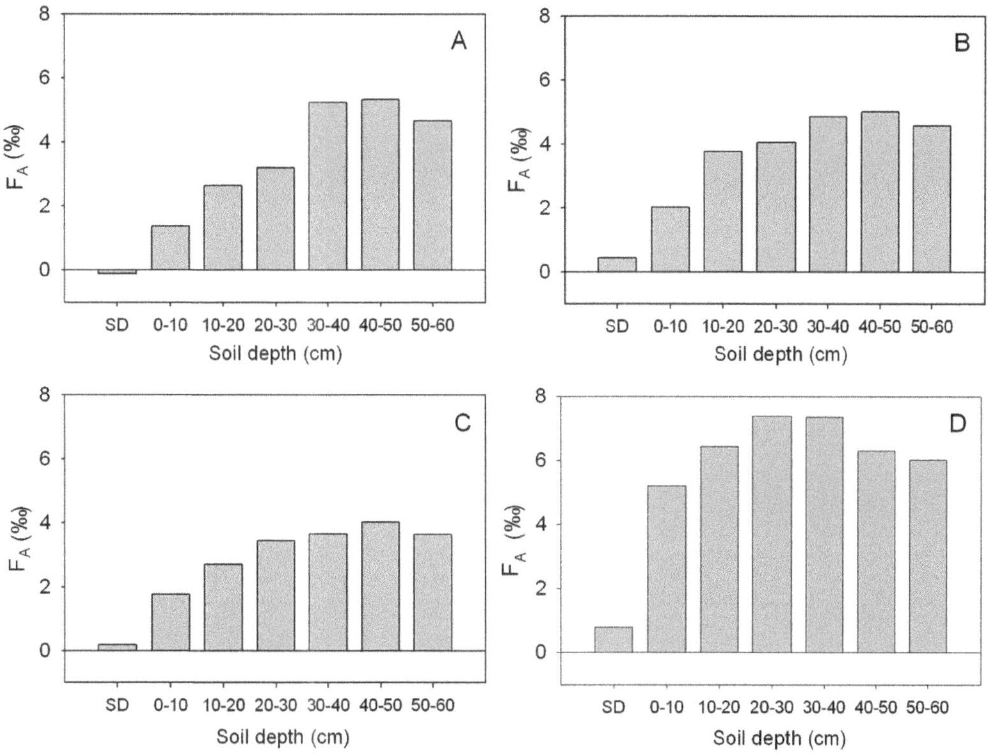

Figure 4. Absolute enrichment factor of soil δ^{13} (F_A) by soil depth at altitudes of 219 (A), 405 (B), 780 (C), and 1268 m (D) in the Lushan Mountain. In the figures, SD represents the semi-decomposed litter layer.

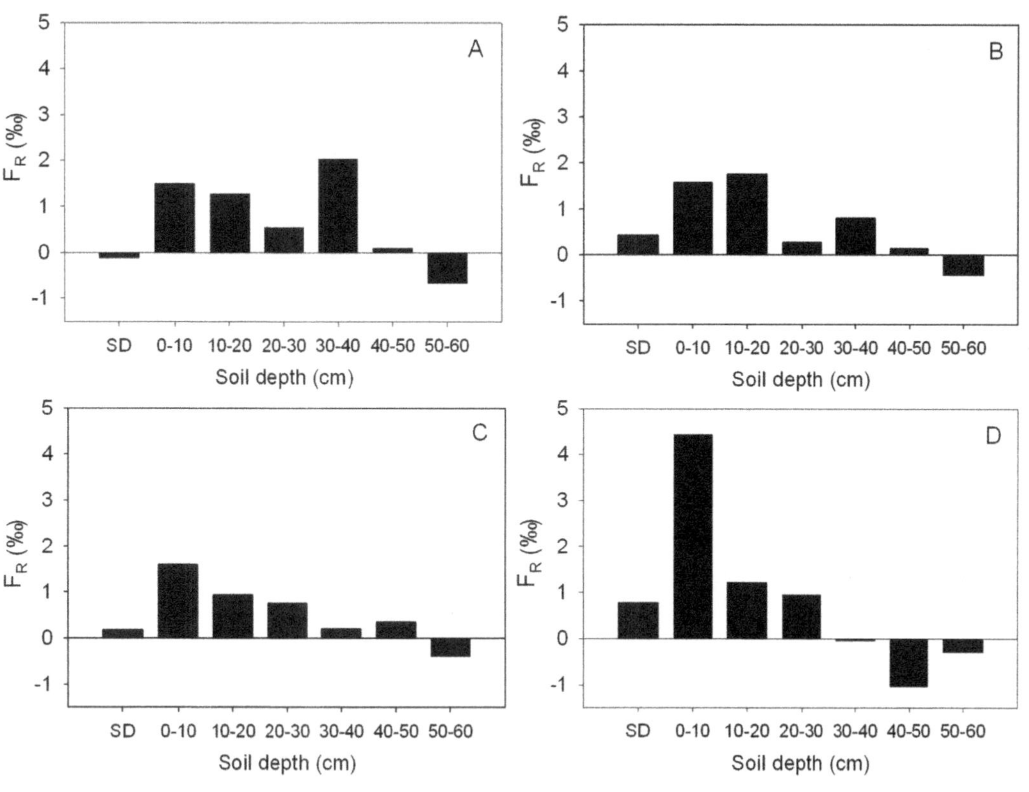

Figure 5. Relative enrichment factor of soil δ^{13} (F_R) by soil depth at altitudes of 219 (A), 405 (B), 780, (C), and 1268 m (D) in the Lushan Mountain.

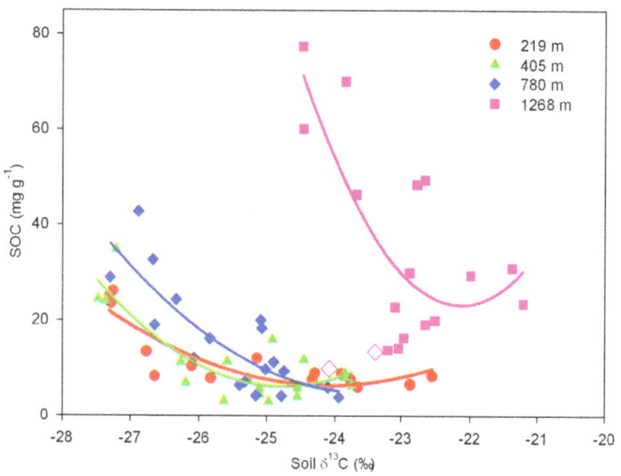

Figure 6. Relationships between SOC concentration (mg g^{-1}) and soil δ^{13}C by soil depth for four altitudes in the Lushan Mountain. The fitted models are: $y = 870.58 + 71.82x + 1.49x^2$, $r^2 = 0.76$, and $p = 0.0002$, for the 219 m site; $y = 1881.58 + 149.87x + 3.03x^2$, $r^2 = 0.72$, and $p < 0.0001$, for the 405 m site; $y = 1250.35 + 105.77x + 2.25x^2$, $r^2 = 0.68$, and $p = 0.0002$, for the 780 m site; and $y = 4265.44 + 383.58x + 8.67x^2$, $r^2 = 0.57$, and $p = 0.0041$, for the 1268 m site.

finding is consistent with the conclusion by previous studies [1,2,11,39]. In the present study, the absolute enrichment factor at 1268 m a.s.l. (F_A = 5‰ to 7.5‰) was greater than that at lower altitudes (F_A = 1.5‰ to 5.5‰). For example, The F_A of 0–10 cm at 1268 m a.s.l. was about 5‰ nearly twich of that of the same soil depth at all lower altitudes (Fig. S1). This result suggests that the climatic conditions (MAP = 2112 mm, MAT = 11.3°C) at 1268 m a.s.l. support a distinct fractionation compared to lower altitudes, particularly the sites at 219 m a.s.l. with MAP = 1429 and MAT = 16.2. Therefore, the climate regime lower temperature and higher moisture at high altitude strongly influences soil ^{13}C enrichment with soil depth [41,42].

Implications of Soil δ^{13}C in Ascertaining SOC Status

Soil C concentrations are strongly correlated with δ^{13}C in forest soil, following a negative relationship as demonstrated by previous studies [1,31,43]. In this study, soil δ^{13}C increased with soil depth, aromatic C and carbonyl C, but decreased with O-alkyl C (Figs. 3 and 7). The increase of ^{13}C and changes of SOC chemical composition with soil depth likely result from humification [44,45]. Microbial activities influence isotopic fractionation during SOC decomposition through differentiation use of substrate by different microbes [41,42,46] and isotopic effects on metabolic synthesis of secondary compounds (e.g., lipids, lignin, cellulose) [4,47]. However, lipids and lignin are degraded more slowly and tend to be ^{13}C depleted, whereas cellulose and carbohydrate degrade more rapidly and tend to be ^{13}C enriched [42,48]. Therefore, it is difficult to establish direct relationships between δ^{13}C and SOC chemical compositions with soil depth. The distinct high soil δ^{13}C at 1268 m a.s.l. is probably attributed to the accumulation of higher ^{13}C-based sugars, starch, cellulose, protein, and organic aids, resulting from slow litter decomposition in the surface soil layers (0–10 cm, 10–20 cm, and 20–30 cm) (Fig. 3). Therefore, the enrichment mechanisms of soil ^{13}C along the altitude gradient were different from those by soil depth in Lushan Mountain. The detail mechanisms need to be clarified in future research.

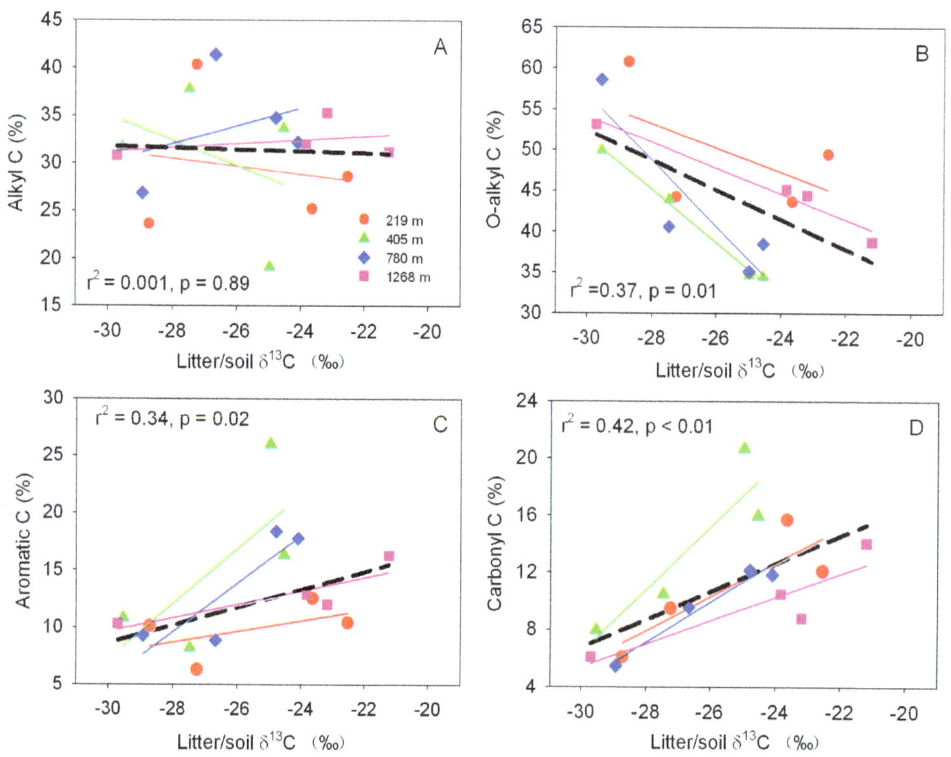

Figure 7. Relationships between soil δ^{13}C and Alkyl C (A), O-alkyl C (B), Aromatic C (C), and Carbonyl C (D) for the studied stands at the four latitudes in the Lushan Mountain. Dashed lines represent the general regression lines with all data, with significant level of $p < 0.05$, except for Alkyl C.

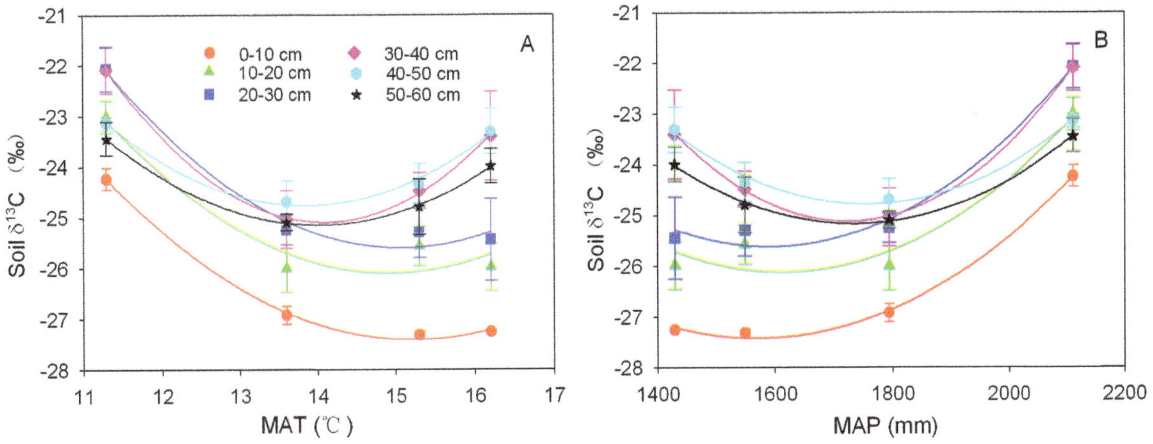

Figure 8. Variations of soil $\delta^{13}C$ with the mean annual temperature (MAT) (A) and mean annual precipitation (MAP) (B) by soil depth in the Lushan Mountain. The fitted model is $y = y_0 + ax + bx^2$. The regression lines wtih MAT have $r^2 = 0.998$ and $p = 0.0428$ for 0–10 cm layer, and $r^2 = 0.999$ and $p = 0.0289$ for 30–40 cm layer (A), and those with MAP have $r^2 = 0.998$ and $p = 0.0445$ for 0–10 cm layerand $r^2 = 0.999$ and $p = 0.0175$ for 30–40 cm layer (B).

Concluding Remarks

The patterns of plant and soil $\delta^{13}C$ variations and their relation to the climate regime along an altitudinal gradient were studied in Lushan Mountain of central subtropical China. The results indicated that tree leaf $\delta^{13}C$ decreased and soil $\delta^{13}C$ increased with altitude. The decoupled pattern of plant and soil $\delta^{13}C$ was due to the climate regime of decreasing temperature and increasing precipitation with altitude in the study area, which result in decreased litter decomposition at high-latitude sites. These results have important implications for understanding C dynamics of subtropical forest ecosystems in the context of global warming.

Acknowledgments

We thank Li Zhang, Shi Xu, and Jieli Wu for their assistance in the sample analysis, and Jinhao Qian, Qin Zou, Kongfan Qian and Ming Du for their assistance with fieldwork. Dr. Rongzhou Man at Ontario Forest Research Institute, Canada, is gratefully acknowledged for his constructive comments and language checking. Chemical analyses were conducted in the Instrumental Analysis Center of SJTU, and the Nature Reserve of Lushan Jiangxi.

Author Contributions

Conceived and designed the experiments: CL HK. Performed the experiments: BD HK PZ. Analyzed the data: BD JH. Contributed reagents/materials/analysis tools: SY GS. Wrote the paper: BD CL HI.

References

1. Garten CT, Cooper LW, Post IIIWM, Hanson PJ (2000) Climate controls on forest soil C isotope ratios in the southern Appalachian mountains. Ecology 81: 1108–1119.
2. Boström B, Comstedt D, Ekblad A (2007) Isotope fractionation and ^{13}C enrichment in soil profiles during the decomposition of soil organic matter. Oecologia 153: 89–98.
3. Gouveia A, Freitas H (2009) Modulation of leaf attributes and water use efficiency in *Quercus suber* along a rainfall gradient. Trees 23: 267–275.
4. Hartman G, Danin A (2010) Isotopic values of plants in relation to water availability in the Eastern Mediterranean region. Oecologia 162: 837–852.
5. Ehleringer JR, Buchmann N, Flanagan LB (2000) Carbon isotope ratios in belowground carbon cycle processes. Ecological Applications 10: 412–422.
6. Bowling DR, Pataki DE, Randerson JT (2008) Carbon isotopes in terrestrial ecosystem pools and CO_2 fluxes. New Phytologist 178: 24–40.
7. Chen P, Wang G, Han J, Liu X, Liu M (2010) $\delta^{13}C$ diference between plants and soil organic matter along the eastern slope of Mount Gongga. Chinese Science Bulletin 1: 55–62.
8. Kohn MJ (2010) Carbon isotope compositions of terrestrial C_3 plants as indicators of (paleo) ecology and (paleo) climate. Proc Natl Acad Sci U S A 107: 19691–19695.
9. Ito A (2003) A global-scale simulation of the CO_2 exchange between the atmosphere and the terrestrial biosphere with a mechanistic model including stable carbon isotopes, 1953–1999. Tellus B 55: 596–612.
10. Kaplan JO, Prentice IC, Buchmann N (2002) The stable carbon isotope composition of the terrestrial biosphere: Modeling at scales from the leaf to the globe. Global Biogeochemical Cycles 16: 1060.
11. Dzurec RS, Boutton TW, Caldwell MM, Smith BN (1985) Carbon isotope ratios of soil organic matter and their use in assessing community composition changes in Curlew Valley, Utah. Oecologia 66: 17–24.
12. Ambrose SH, Sikes NE (1991) Soil carbon isotope evidence for Holocene habitat change in the Kenya Rift Valley. Science 253: 1402–1405.
13. Balesdent J, Mariotti A, Guillet B (1987) Natural ^{13}C abundance as a tracer for studies of soil organic matter dynamics. Soil Biology and Biochemistry 19: 25–30.
14. Skjemstad JO, Krull ES, Swift RS, Szarvas S (2008) Mechanisms of protection of soil organic matter under pasture following clearing of rainforest on an Oxisol. Geoderma 143: 231–242.
15. Benner R, Fogel ML, Sprague EK, Hodson RE (1987) Depletion of ^{13}C in lignin and its implications for stable carbon isotope studies. Nature 329: 708–710.
16. Li J, Ziegler SE, Lane CS, Billings SA (2013) Legacies of native climate regime govern responses of boreal soil microbes to litter stoichiometry and temperature. Soil Biology and Biochemistry 66: 204–213.
17. Bird MI, Chivas AR, Head J (1996) A latitudinal gradient in carbon turnover times in forest soils. Nature 381: 143–146.
18. Wei K, Jia G (2009) Soil n-alkane $\delta^{13}C$ along a mountain slope as an integrator of altitude effect on plant species $\delta^{13}C$. Geophysical Research Letters 36: L11401.
19. Löffler J, Anschlag K, Baker B, Finch OD, Diekkrüger B, et al. (2011) Mountain ecosystem response to global change. Erdkunde 65: 189–213.
20. Gottfried M, Pauli H, Futschik A, Akhalkatsi M, Barančok P, et al. (2012) Continent-wide response of mountain vegetation to climate change. Nature Climate Change 2: 111–115.

21. Liu X, Wang L (2009) Scientific survey and study of biodiversity on the Lushan nature reserve in Jiangxi province. Science Press.
22. Tu CL, Liu CQ, Lu XH, Yuan J, Lang YC (2011) Sources of dissolved organic carbon in forest soils: evidences from the differences of organic carbon concentration and isotope composition studies. Environmental Earth Sciences 63: 723–730.
23. Kalembasa SJ, Jenkinson DS (1973) A comparative study of titrimetric and gravimetric methods for the determination of organic carbon in soil. Journal of the Science of Food and Agriculture 24: 1085–1094.
24. Wang H, Liu SR, Mo JM, Wang JX, Makeschin F, Wolff M (2010) Soil organic carbon stock and chemical composition in four plantations of indigenous tree species in subtropical China. Ecological Research 25: 1071–1079.
25. Schmidt M, Knicker H, Hatcher PG, Kögel-Knabner I (1997) Improvement of ^{13}C and ^{15}N CPMAS NMR spectra of bulk soils, particle size fractions and organic material by treatment with 10% hydrofluoric acid. European Journal of soil Science 48: 319–328.
26. Jien SH, Chen TH, Chiu CY (2011) Effects of afforestation on soil organic matter characteristics under subtropical forests with low elevation. Journal of Forest Research 16: 275–283.
27. Rumpel C, Kögel-Knabner I, Bruhn F (2002) Vertical distribution, age, and chemical composition of organic carbon in two forest soils of different pedogenesis.Organic Geochemistry 33: 1131–1142.
28. Baldock JA, Oades JM, Nelson PN, Skene TM, Golchin A, et al. (1997) Assessing the extent of decomposition of natural organic materials using solid-state ^{13}CNMR spectroscopy. Australian Journal of Soil Research 35: 1061–1083.
29. García M, Faz CA (2012) Soil organic matter stocks and quality at high altitude grasslands of Apolobamba, Bolivia. Catena 94: 26–35.
30. Natelhoffer KJ, Fry B (1988) Controls on natural nitrogen-15 and carbon-13 abundances in forest soil organic matter. Soil Science Society of America Journal 52: 1633–1640.
31. Balesdent J, Girardin C, Mariotti A (1993) Site-related ^{13}C of tree leaves and soil organic matter in a temperate forest. Ecology 74: 1713–1721.
32. Peri PL, Ladd B, Pepper DA, Bonser SP, Laffan SW, et al. (2012) Carbon ($\delta^{13}C$) and nitrogen ($\delta^{15}N$) stable isotope composition in plant and soil in Southern Patagonia's native forests. Global Change Biology 18: 311–321.
33. Körner C, Farquhar GD, Roksandic Z (1988) A global survey of carbon isotope discrimination in plants from high altitude. Oecologia 74: 623–632.
34. Bird MI, Haberle SG, Chivas AR (1994) Effect of altitude on the carbon-isotope composition of forest and grassland soils from Papua New Guinea. Global Biogeochemical Cycles 8: 13–22.
35. Farquhar GD, Ehleringer JR, Hubick KT (1989) Carbonisotope discriminationand photosynthesis. Annual Review of Plant Biology 40: 503–537.
36. Townsend AR, Vitousek PM, Trumbore SE (1995) Soil organic matter dynamics along gradients in temperature and land use on the island of Hawaii. Ecology 76: 721–733.
37. Powers JS, Schlesinger WH (2002) Geographic and vertical patterns of stable carbon isotopes in tropical rain forest soils of Costa Rica. Geoderma 109: 141–160.
38. Zimmermann M, Leifeld J, Conen F, Bird MI, Meir P (2012) Can composition and physical protection of soil organic matter explain soil respiration temperature sensitivity? Biogeochemistry 107: 1–14.
39. Melillo JM, Aber JD, Linkins AE, Ricca A, Fry B, et al. (1989) Carbon and nitrogen dynamics along the decay continuum: plant litter to soil organic matter. Plant and Soil 115: 189–198.
40. Coûteaux MM, Bottner P, Berg B (1995) Litter decomposition, climate and litter quality. Trends in Ecology and Evolution 10: 63–66.
41. Li J, Ziegler S, Lane CS, Billings SA (2012) Warming-enhanced preferential microbial mineralization of humified boreal forest soil organic matter: Interpretation of soil profiles along a climate transect using laboratory incubations. Journal of Geophysical Research: Biogeosciences 117: G02008.
42. Li J, Ziegler SE, Lane CS, Billings SA (2013) Legacies of native climate regime govern responses of boreal soil microbes to litter stoichiometry and temperature. Soil Biology and Biochemistry 66: 204–213.
43. Powers JS (2006) Spatial variation of soil organic carbon concentrations and stable isotopic composition in 1-ha plots of forest and pasture in Costa Rica: implications for the natural abundance technique. Biology and Fertility of Soils 42: 580–584.
44. Kramer MG, Sollins P, Sletten RS, Swart PK (2003) N isotope fractionation and measures of organic matter alteration during decomposition. Ecology 84: 2021–2025.
45. Quideau SA, Anderson MA, Graham RC, Chadwick OA, Trumbore SE (2000) Soil organic matter processes: characterization by ^{13}C NMR and ^{14}C measurements. Forest Ecology and Management 138: 19–27.
46. Ziegler SE, Billings SA, Lane CS, Li J, Fogel ML (2013) Warming alters routing of labile and slower-turnover carbon through distinct microbial groups in boreal forest organic soils. Soil Biology and Biochemistry 60: 23–32.
47. Cotrufo MF, Drake B, Ehleringer JR (2005) Palatability trials on hardwood leaf litter grown under elevated CO_2: a stable carbon isotope study. Soil Biology and Biochemistry 37: 1105–1112.
48. Ehleringer JR, Buchmann N, Flanagan LB (2000) Carbon isotope ratios in belowground carbon cycle processes. Ecological Applications 10: 412–422.

23

Effects of Soil Moisture on the Temperature Sensitivity of Soil Heterotrophic Respiration: A Laboratory Incubation Study

Weiping Zhou[1,2], Dafeng Hui[3], Weijun Shen[1]*

1 Key laboratory of Vegetation Restoration and Management of Degraded Ecosystems, South China Botanical Garden, Chinese Academy of Sciences, Guangzhou, China, **2** Graduate University of the Chinese Academy of Sciences, Beijing, PR China, **3** Department of Biological Sciences, Tennessee State University, Nashville, Tennessee, United States of America

Abstract

The temperature sensitivity (Q_{10}) of soil heterotrophic respiration (R_h) is an important ecological model parameter and may vary with temperature and moisture. While Q_{10} generally decreases with increasing temperature, the moisture effects on Q_{10} have been controversial. To address this, we conducted a 90-day laboratory incubation experiment using a subtropical forest soil with a full factorial combination of five moisture levels (20%, 40%, 60%, 80%, and 100% water holding capacity - WHC) and five temperature levels (10, 17, 24, 31, and 38°C). Under each moisture treatment, R_h was measured several times for each temperature treatment to derive Q_{10} based on the exponential relationships between R_h and temperature. Microbial biomass carbon (MBC), microbial community structure and soil nutrients were also measured several times to detect their potential contributions to the moisture-induced Q_{10} variation. We found that Q_{10} was significantly lower at lower moisture levels (60%, 40% and 20% WHC) than at higher moisture level (80% WHC) during the early stage of the incubation, but became significantly higher at 20%WHC than at 60% WHC and not significantly different from the other three moisture levels during the late stage of incubation. In contrast, soil R_h had the highest value at 60% WHC and the lowest at 20% WHC throughout the whole incubation period. Variations of Q_{10} were significantly associated with MBC during the early stages of incubation, but with the fungi-to-bacteria ratio during the later stages, suggesting that changes in microbial biomass and community structure are related to the moisture-induced Q_{10} changes. This study implies that global warming's impacts on soil CO_2 emission may depend upon soil moisture conditions. With the same temperature rise, wetter soils may emit more CO_2 into the atmosphere via heterotrophic respiration.

Editor: Shuijin Hu, North Carolina State University, United States of America

Funding: This study was financially supported by the Natural Science Foundation of China (31130011 and 31290222), and the Natural Science Foundation of Guangdong Province, China (S2012020011084). The funders had no role in study design, data collection and analysis, decision to publish, or preparation of the manuscript.

Competing Interests: Co-author DH is an academic editor of PLOS ONE. This does not alter our adherence to all the PLOS ONE policies on sharing data and materials.

* E-mail: shenweij@scbg.ac.cn

Introduction

Temperature sensitivity of soil respiration, usually termed as Q_{10}, is defined as the increase of soil respiration rate by a 10°C rise in temperature [1]. Q_{10} has been considered an important model parameter in predicting terrestrial ecosystem carbon cycle and feedback to climate warming [2]. In the past several decades, Q_{10} has been investigated extensively, particularly through field-observed soil respiration and environmental factor data [3,4]. It has been found that Q_{10} is not a constant of 2, but varies with vegetation and edaphic conditions such as temperature, moisture, and substrate availability [2]. As global temperature continues to rise [5], it is of paramount importance to understand how Q_{10} is influenced by these factors individually and interactively. Since under field conditions, effects of soil temperature and moisture on Q_{10} are often confounded with each other and with other factors, laboratory incubation has the advantage of deriving the primary and interactive effects of the environmental factors on Q_{10}.

Many studies have demonstrated that Q_{10} can be influenced by a variety of biological and environmental factors [1,6,7]. Soil temperature itself has been found to have a negative correlation with Q_{10}. For example, at lower temperature regions (e.g., tundra), Q_{10} tends to be higher than the estimates at warmer temperature regions (e.g., warm desert) [8]. A manipulated warming experiment also demonstrates that Q_{10} is significantly lower at high temperature treatments than at the low temperature control [1]. Thus, the temperature effects on Q_{10} have been generally consistent; i.e., Q_{10} decreases with increasing temperature. However, the effects of other factors such as soil moisture on Q_{10} have been less certain and deserve more research.

Soil moisture plays a critical role in soil respiration and may have a significant impact on Q_{10} [9–11]. The basic concepts and mechanisms of soil moisture on soil respiration have been discussed by many researchers [12–14]. The optimum soil moisture for soil respiration is frequently found at intermediate levels, above or below which soil respiration decreases [15]. At the optimum soil moisture, the macropore spaces are filled with adequate amounts of air and water which can facilitate the diffusion of both oxygen and soluble substrates [16]. In very wet soils oxygen limitation occurs, and in very dry soils the movement of soluble substrates via water films is restricted. Although the mechanistic understanding on the effects of soil moisture on R_h has been largely advanced, its influence on the Q_{10} of R_h is still inconclusive. For example, Wang et al. [17] reported that Q_{10} increased with soil moisture until reaching a threshold, and then declined in six temperate forests of China. Carlyle and Than [18] showed that soil moisture limited the Q_{10} of soil respiration beneath a *pinus radiata* stand in south-eastern Australia. But Reichstein et al. [19] found that Q_{10} was insensitive to the drying of a spruce forest soil. The inconsistency of soil moisture effects on Q_{10} is probably due to the confounding influences of different environmental factors under field conditions. One recent incubation study showed that soil moisture indeed influenced Q_{10} and the moisture-Q_{10} relationship differed between soils obtained at different topographic positions [20], but the underlying mechanisms remained unclear.

Effects of soil moisture on Q_{10} may be ascribed to changes in microbial biomass and community structure, and the physical and chemical properties of the soil [7,21]. Changes in soil moisture can affect the composition and function of soil microbial community due to differences in drought tolerance among taxonomic and functional groups of microorganisms [22]. For example, fungi can survive drought stress better than bacteria due to their ability to grow at lower matric potentials [23,24]. Soil moisture can also affect the quantity of soil microbial biomass carbon (MBC) and dissolved organic carbon (DOC) [25,26]. Despite a general understanding of the above processes, whether soil moisture effects on Q_{10} can be related to its influences on soil properties such as MBC, DOC, nutrient availability, and microbial community structure is still in active debate.

In this study, we investigated soil moisture effects on Q_{10} by incubating a subtropical forest soil under five temperature levels and five moisture levels over 90 days. Soil R_h and other properties such as MBC and DOC, nitrogen and phosphorous contents, and microbial community phospholipid fatty acids (PLFAs) were measured several times during the incubation period. Our objectives were first to analyze how changes in soil moisture influenced Q_{10}, and second to explore whether the moisture effects on Q_{10} could be related to its impacts on the soil microbial and chemical properties measured.

Materials and Methods

Ethics Statement

Soils were sampled from a study site that is maintained by the South China Botanical Garden, Chinese Academy of Sciences. All necessary permits were obtained for the described study. This study did not involve endangered or protected species.

Site Description

Incubation soils used in this study were collected from an evergreen broadleaved forest stand at the Heshan Hilly Land Interdisciplinary Experimental Station (22°34′N, 112°50′E) in Guangdong Province of China. The region has a subtropical humid monsoon climate with apparent dry and wet seasons. The wet season starts in April and ends in October, and the dry season begins in November and lasts through March of the following year. The mean annual precipitation and temperature are 1700 mm and 21.7°C, respectively. The forest stand is 29 years old and mainly dominated by native tree species (*Schima superba* and *Michelia macclurei*) with an average height of 15 m and an average diameter at breast height (DBH) of 30 cm. The soil is categorized as Oxisols based on the US Soil Classification System [27,28], with a bulk density of 1.4 g cm^{-3}, total organic carbon (TOC) of 2.80%, total nitrogen (TN) of 0.15%, and total phosphorous (TP) of 0.02% at the depth of 0–20 cm.

Incubation Experimental Design

In the field, four sampling areas, with a distance of at least 10 m between each, were selected to collect the incubation soils. In each area, five sampling sites (20×10 cm^2) were randomly selected and sampled to the depth of 20 cm. These five random samples were homogenized to form a composite sample. Before sampling, the uppermost layer of litter with visible un-decomposed materials was excluded. We had four composite samples as four experimental replicates, each one weighing about 50 kg in fresh weight. All soil samples were transported to the laboratory and passed through a 2 mm sieve with apparent plant roots and stones being removed.

To investigate soil moisture effects on the temperature sensitivity (Q_{10}) of soil R_h, we used five soil moisture levels: 20%, 40%, 60%, 80%, and 100% water holding capacity (WHC). For each moisture level, soils were incubated under five temperature levels: 10, 17, 24, 31, and 38°C. A full factorial combination of the two factors and five levels for each factor produced 25 experimental treatments. Each treatment had four replicates from the four composite samples. Each replicate further had 6 duplicates, with one duplicate for measuring R_h and the other 5 for destructive sampling. As a result, we had 600 incubation soil samples in total with 120 (= 5 moisture levels × 4 replicates × 6 duplicates) in each of the five static temperature incubators (RXZ-600B, Southeast Instrument Co., Ltd., Ningbo, China). The temperature and relative humidity deviations of the incubators are ±1.5°C and ±7%, respectively. Each air-dried incubation soil sample (equivalent to 50 g of oven-dried soil) was added to each triangle flask and its soil water content was adjusted to the corresponding soil moisture level by adding deionized water. The flasks were covered by rubber stoppers with small holes to reduce water loss via evaporation and maintain gas exchange. In order to maintain constant soil moisture levels, water loss was checked and corrected weekly by weighing each flask and adding water as necessary. At most, 1 ml of water (equivalent to 3.3% of changes in 100% WHC) was added every week to the flask at the temperature level of 38°C.

Measurements of Soil R_h, Microbial and Chemical Properties

The incubation experiment lasted for 90 days. Soil R_h rates were measured using the Li-6262 Infrared Gas Analyzer (Li-Cor Inc., Lincoln, NE) on days 1, 2, 3, 4, 6, 7, 13, 18, 27, 34, 41, 53, 62, 74, and 90. Electrical fans blowing air into each incubator for 30 min every four days were used to maintain an aerobic incubation environment. Before R_h measuring, each triangle flask was ventilated for 3 minutes to minimize gas accumulation in the headspace. After ventilation, another type of rubber stoppers with two plastic tubes for gas inlet and outlet was used to seal the flask and the tubes were connected to Li-6262 for measuring headspace CO_2 concentration. The CO_2 concentration in the headspace was recorded every second for 2 minutes and R_h rate was calculated using the linear portion of the response curve of CO_2 concentration versus time [29]. At each moisture level and each measurement day, Q_{10} was calculated by fitting an exponential function to the measured R_h against the 5 temperature levels:

$$R_h = ae^{bT} \tag{1}$$

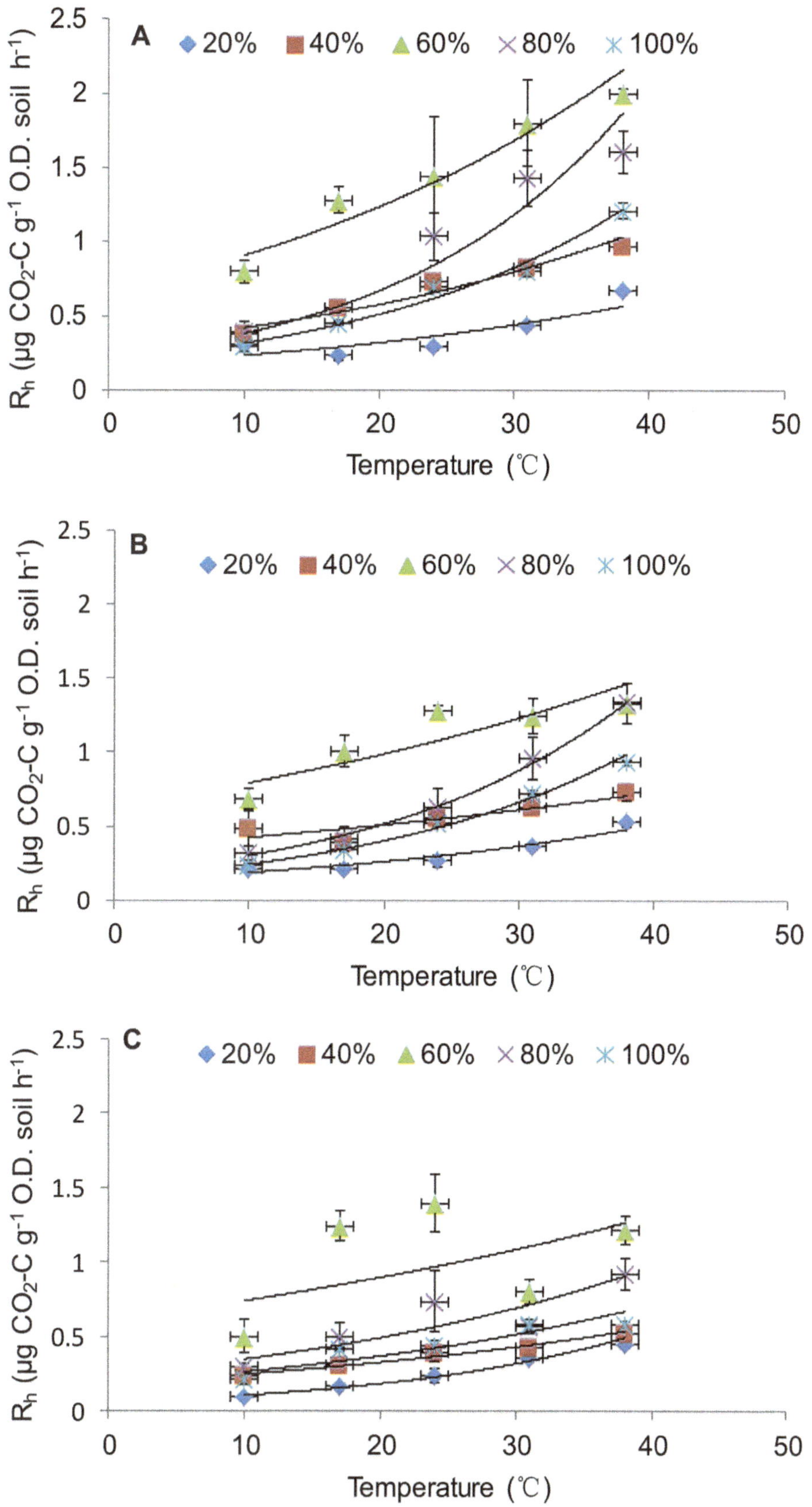

Figure 1. Responses of soil heterotrophic respiration (R_h) to changes in soil temperature after 7 (A), 30 (B), and 90 (C) days of incubation. Each data point is the mean of four replicates under each soil moisture treatment. Error bars represent standard errors (n = 4).

Table 1. Repeated-measures ANOVA for the temperature and time effects on R_h under different soil moisture treatments.

Moisture ([a] %WHC)	Temperature (°C)		Time (day)		Temperature×Time	
	$F_{(4, 15)}$	P	$F_{(14, 210)}$	P	$F_{(56, 210)}$	P
20%	36.96	<0.01	26.59	<0.01	5.46	<0.01
40%	35.77	<0.01	50.03	<0.01	3.88	<0.01
60%	10.90	<0.01	45.26	<0.01	6.63	<0.01
80%	12.24	<0.01	52.44	<0.01	6.47	<0.01
100%	139.15	<0.01	159.63	<0.01	17.60	<0.01

[a]%WHC: percent of water holding capacity.

where R_h is the measured soil heterotrophic respiration rate (μg g^{-1} oven dried soil h^{-1}), T is the incubation temperature (°C), a and b are the fitted model parameters. Q_{10} is calculated individually for each of the four replicates using the following equation:

$$Q_{10} = e^{10b} \tag{2}$$

As intrinsic Q_{10} is based on the kinetic reactions of molecular structure changes with temperature [2], the Q_{10} of this study is the apparent temperature sensitivity.

For measuring MBC, DOC, total organic carbon (TOC), total nitrogen (TN), total phosphorous (TP), inorganic N (NH_4^+ and NO_3^-), and PLFAs, three flasks (or replicates) of each treatment were harvested on days 7, 30, and 90. Part of the soil in each flask was collected and stored at −20°C for later analysis of PLFAs. The remaining soil was used to analyze microbial biomass and chemical properties. It is noted here that TOC, TN, and TP were only analyzed for the samples harvested on days 7 and 90.

Soil MBC was measured using the modified fumigation-extraction method [30]. MBC was calculated as the difference in extractable C concentrations between the fumigated and un-fumigated samples divided by a K_{EC} factor of 0.38 [30]. The extractable C concentrations of un-fumigated samples were the soil DOC [31]. Soil NH_4^+ and NO_3^- were measured by the method proposed in Dorich and Nelson [32]. Soil TOC was measured using the potassium bichromate-concentrated sulphuric acid heating method. Soil TN and TP were measured using the Kjeldahl resolution Auto Flow Injection method. Microbial community PLFAs were analyzed according to Bossio and Scow [33]. Concentrations of each PLFA were calculated relative to the 19:0 internal standard concentrations. 15:0, i15:0, a15:0, i16:0, 16:1w7c, 17:0, i17:0, a17:0, cy17:0, 18:1w7c, 19:0 cyclow8c were selected as bacterial biomarkers and 18:2w6, 9c were selected as fungal biomarkers [34,35].

Statistical Analysis

Repeated-measures ANOVA was employed to determine the effects of sampling time and soil moisture on Q_{10} and R_h on the 15 measuring days. One way ANOVA was used to analyze soil moisture effects on Q_{10} for days 7, 30, and 90; Tukey's HSD multiple comparison method was used to test Q_{10} differences among soil moisture levels. Regression analysis was used to derive the relationships between R_h and temperature, and between Q_{10} (or R_h) and incubation time. Pearson correlation analysis was applied to detect the potential contributions of soil microbial and chemical properties on the Q_{10} variations with moisture. All these statistical analyses were performed using SPSS software 16.0 (SPSS Inc., Chicago, IL).

Table 2. Regression equations of heterotrophic respiration (R_h) with temperature under different moisture treatments.

Soil moisture ([a] %WHC)	Day 7	Day 30	Day 90
20	[b] $R_h = 0.1684e^{0.0318T}$	$R_h = 0.1333e^{0.0335T}$	$R_h = 0.062e^{0.0546T}$
	[c] $R^2 = 0.72$*	$R^2 = 0.92$**	$R^2 = 0.98$**
40	$R_h = 0.3014e^{0.0324T}$	$R_h = 0.3526e^{0.0183T}$	$R_h = 0.1896e^{0.0273T}$
	$R^2 = 0.95$**	$R^2 = 0.72$*	$R^2 = 0.97$**
60	$R_h = 0.6639e^{0.031T}$	$R_h = 0.6382e^{0.0218T}$	$R_h = 0.6148e^{0.019T}$
	$R^2 = 0.92$**	$R^2 = 0.78$*	$R^2 = 0.25$
80	$R_h = 0.2136e^{0.057T}$	$R_h = 0.1774e^{0.0532T}$	$R_h = 0.2484e^{0.0342T}$
	$R^2 = 0.92$**	$R^2 = 0.99$**	$R^2 = 0.77$*
100	$R_h = 0.1926e^{0.0485T}$	$R_h = 0.1469e^{0.0501T}$	$R_h = 0.19e^{0.0333T}$
	$R^2 = 0.98$**	$R^2 = 0.99$**	$R^2 = 0.81$*

[a]%WHC: percent of water holding capacity.
[b]R_h represents soil heterotrophic respiration rate and T represents temperature.
[c]R^2 is the coefficient of determination; * and ** indicate significance at P≤0.05 and P≤0.01, respectively.

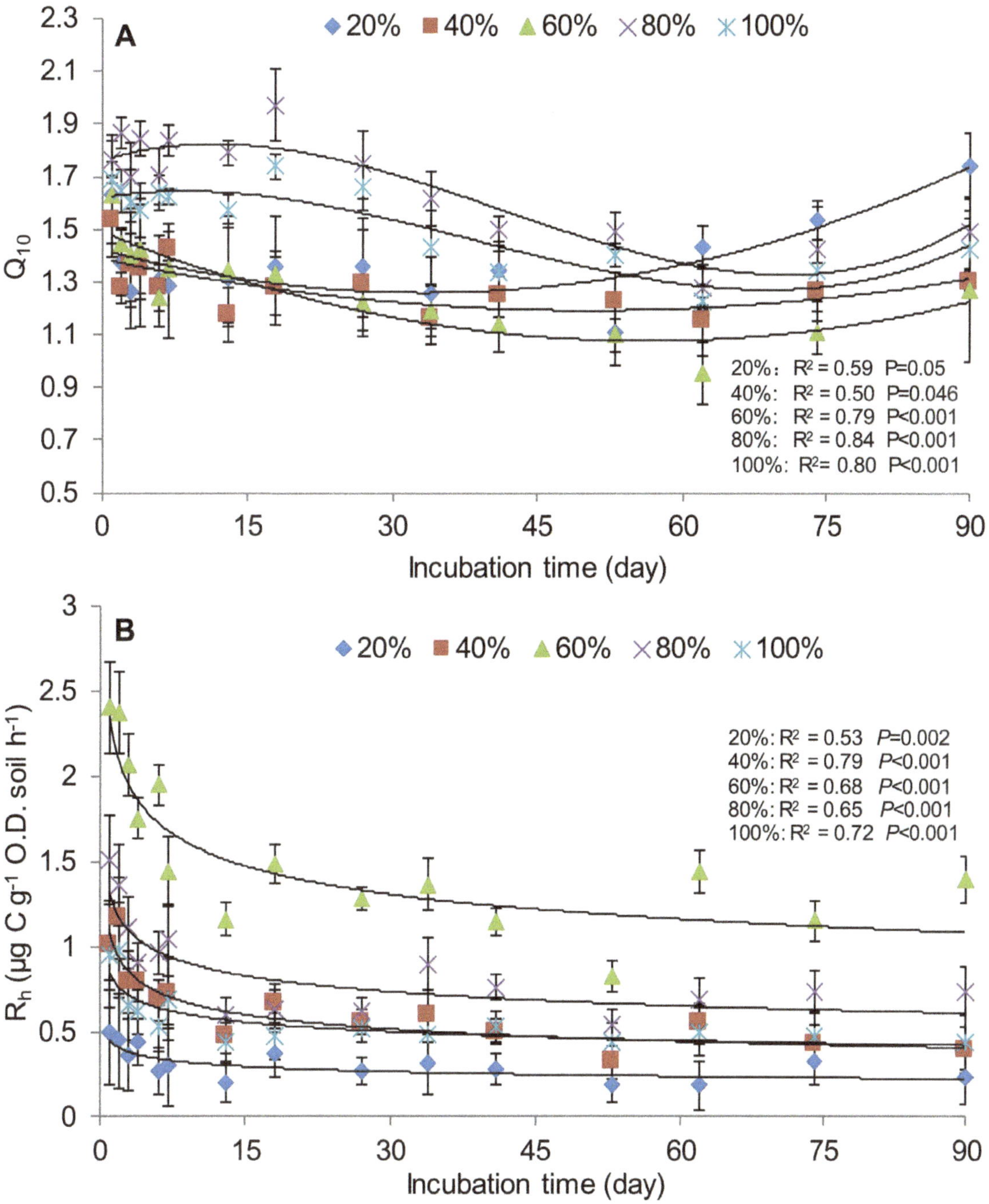

Figure 2. Changes of Q_{10} and R_h with incubation time at different soil moisture treatments. The Q_{10} regression functions are quadratic for 20%, 40% and 60% WHC and cubic for 80% and 100% WHC. Error bars (n = 4) represent standard deviations. R^2 is the coefficient of determination. P is the significance level.

Results

Relationships between Soil R_h and Temperature under Different Soil Moisture Treatments

Soil R_h varied significantly with temperature and incubation time under all the 5 moisture treatments (Table 1). R_h responses to temperature changes on days 7, 30, and 90 were displayed to represent the early, middle, and late incubation stages (Fig. 1). Throughout the whole incubation period, soil R_h was highest at 60% WHC, lowest at 20% WHC, and intermediate at the other moisture levels (Fig. 1). Among the three measurement days, soil R_h had the highest values on day 7, especially at 60% and 80% WHC, compared to those on days 30 and 90. Soil R_h declined with incubation time, declining in smaller magnitudes at 20% WHC and 100% WHC.

The temperature response of soil R_h could be well fitted using the exponential model for each soil moisture treatment and measurement day (Fig. 1). Model parameters for the three representative days are presented in Table 2. All models were significant with the coefficient of determination (R^2) ranging from

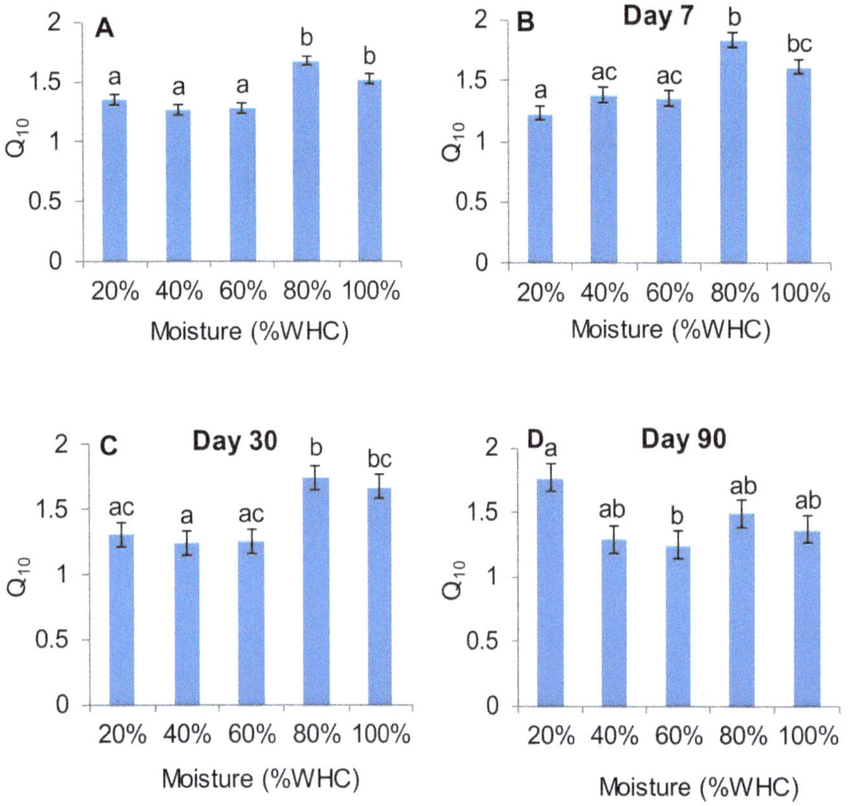

Figure 3. Comparisons of mean Q_{10} under different soil moisture treatments. The mean Q_{10} in panel (A) is calculated from 4 replicates across 15 measurement times (n = 60), and the mean Q_{10}s in panel (B), (C) and (D) are calculated from 4 replicates on day 7, day 30 and day 90 (n = 4), respectively. Error bars represent standard errors. Lower case letters represent significant difference at $P \leq 0.05$.

0.72 to 0.98, except for ($R^2 = 0.25$) at 60% WHC on day 90. Estimated basal soil respiration (i.e. parameter a) ranged from 0.06 to 0.66 and was higher at 60% WHC than at the other soil moisture treatments. The exponent (i.e. parameter b) ranged from 0.02 to 0.05 and was also lowest (0.019) at 60% WHC on day 90.

Changes of Q_{10} with Incubation Time under Different Soil Moisture Treatments

Under all the 5 moisture treatments, Q_{10} values ranged from 0.95 to 1.97 and varied markedly with incubation time. Q_{10} was higher in the beginning of the incubation, and declined with time to around day 60, then increased slightly to the end of the experiment (Fig. 2A). Among different soil moisture treatments, Q_{10} at 80% and 100% WHC was higher than at the other soil moisture treatments. Q_{10} at 60% WHC had the lowest value, especially after day 30. Q_{10} at 20% WHC was among the lowest in the beginning of the incubation, but increased with incubation time and had the highest values at the end of the experiment (Fig. 2A).

To better show the Q_{10} variation pattern with incubation time, polynomial regression models were used to fit Q_{10} with incubation time. Findings indicate that a quadratic regression model could fit Q_{10} well at 20%, 40% and 60% WHC with $R^2 \geq 0.50$ (Fig. 2A), while a cubic regression model should be applied at 80% and 100% WHC ($R^2 \geq 0.80$; Fig. 2A). Further inspection revealed the days on which the minimum and maximum Q_{10} appeared. At 20%, 40%, and 60% WHC, the lowest Q_{10} appeared on days 53, 62, and 62 with their values being 1.11, 1.15, and 0.95, respectively. At 80% and 100% WHC, Q_{10} showed the highest values of 1.94 and 1.74 on day 18, and the lowest values of 1.28 and 1.22 on day 62, respectively.

In contrast to the Q_{10} dynamics, R_h was always higher at 60% WHC, lower at 20% WHC and somewhere in between at the other three moisture levels (80%, 40%, and 100% WHC) (Fig. 2B). It is noted here that the R_h data shown in Fig. 2B are only those under the incubation temperature of 24°C, because under the other temperature treatments the general variation patterns of R_h with moisture and incubation time were similar. In our incubation experiment, the declination of R_h with incubation time could be best fitted using a power law decay function, with the coefficient of determination ranging from 0.53 to 0.78 (Fig. 2B).

Effects of Soil Moisture on Q_{10}

Repeated-measures ANOVA showed that Q_{10} was significantly influenced by soil moisture ($F_{(4,\ 15)} = 18.41$, $P<0.01$), incubation time ($F_{(14,\ 210)} = 15.41$, $P<0.01$) and the interaction of the two ($F_{(56,\ 210)} = 4.44$, $P<0.01$). Averaged over the 15 measurement times, Q_{10} was significantly lower at 60%, 40% and 20% WHC compared to those at 80% and 100% WHC (Fig. 3A). Q_{10} at 80% WHC had the highest value but was not significantly different from that at 100% WHC (Fig. 3A).

Since the interactive effect of soil moisture and incubation time was significant, we further compared Q_{10} values among soil moisture treatments on three typical measurement days (Fig. 3B–D). On days 7 and 30, Q_{10} at 80% WHC was not significantly different from that at 100% WHC but significantly higher than those at the three lower moisture levels, which is similar to the all-day average results shown in Fig. 3A. On day 90, Q_{10} at 20%

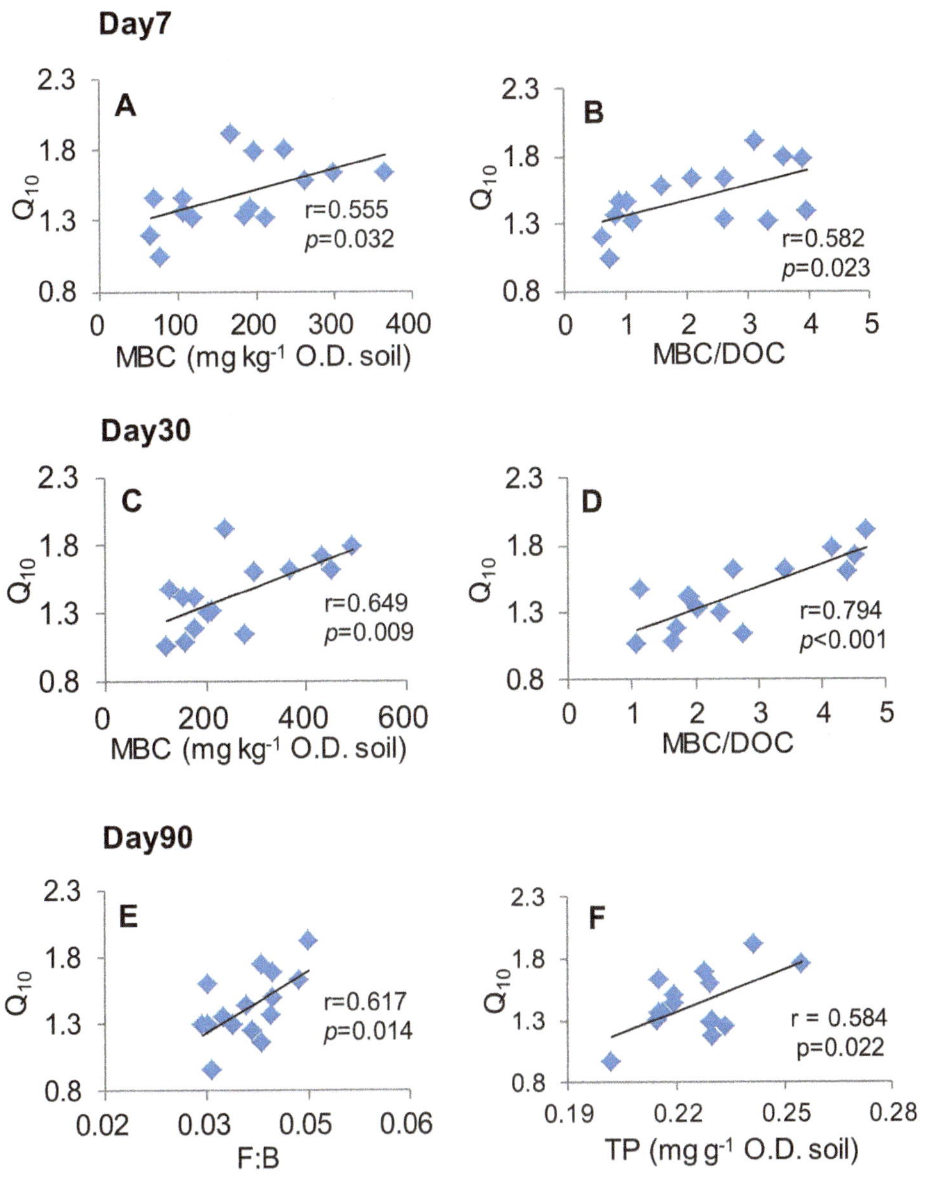

Figure 4. Correlations between Q_{10} and soil microbial and chemical properties. MBC: microbial biomass carbon, DOC: dissolved organic carbon, TP: total phosphorus, and F:B: ratio of fungi to bacteria. r is the correlation coefficient. P is the significance level.

WHC was significantly higher than at 60% WHC, but there was no significant difference among the other three soil moisture treatments.

Correlations between Q_{10} and Soil Properties

Pearson correlation analysis showed that on days 7 and 30, Q_{10} was positively correlated with MBC and the ratio of MBC to DOC (Fig. 4A–D). On day 90, no significant correlation was found between Q_{10} and MBC or MBC/DOC; instead, Q_{10} positively correlated with the F:B ratio (Fig. 4E) and TP (Fig. 4F). We did not find significant correlations between Q_{10} and other soil chemical properties (inorganic nitrogen, DOC, TOC, and TN), which was therefore not presented.

Discussion

Moisture Effects on R_h Responses to Temperature

Similar to many previous results [2,36,37], soil R_h increased with temperature exponentially in our study; however response curves varied among different soil moisture treatments (Fig. 1, Table 2). Some previous studies showed that soil respiration may be decoupled from temperature under certain soil moisture levels resulting in soil respiration that is unaffected by temperature under water stress [38]. For example, Yu et al. [39] found that temperature was the determinant factor and R_h increased with it exponentially only when soil moisture was not limited. However, our results showed that soil R_h could still increase with temperature, even at 20% WHC, though at a relatively slow rate (Table 2). The discrepancy between our results and theirs may be because the studied soils differ in chemical, physical and microbial properties.

Regardless of temperature variations, soil R_h at 60% WHC tended to be higher than at both lower and higher soil moisture treatments. This result is also consistent with some other studies that reported higher R_h at intermediate moisture content [40–42]. The decrease of soil R_h at lower soil moisture has been attributed to soluble substrate limitation, whereas at higher soil moisture level, especially at saturated soil moisture, R_h was mainly limited by oxygen [12,36]. The decrease of soil R_h over incubation time was probably caused by the depletion of labile substrate [43,44]. However, the persistently high R_h at 60% WHC during the 90-day incubation could be due to the persistently high microbial activity, because this was the optimal moisture level and high microbial activity might override the influence of substrate limitation.

Variations of Q_{10} with Incubation Time

The variation of Q_{10} with time has been found not to be uniform [45]. A laboratory incubation study found that Q_{10} increased with incubation time, which was ascribed to substrate quality change from labile to recalcitrant [46]. In some long-term warming experiments, Q_{10} was found to decline over time [1,45]. In our relatively short-term incubation experiment, we found that Q_{10} declined with incubation time initially but increased during later incubation stages, and quadratic or cubic regression models were fitted to quantify the changes of Q_{10} at different soil moisture treatments (Fig. 2A). Over the 90-day incubation period, mean Q_{10} values was mostly <2.0, which is lower than the conventional estimates (2.0–2.6) probably due to less confounding factors involved in our incubation experiment [47]. The changes of Q_{10} might be related to the changes of soil R_h, as many laboratory studies have shown that soil R_h decreases with incubation time [48]. The underlying mechanisms were ascribed to substrate depletion [43,44]: the longer the incubation time, the more time microbes had to consume the labile carbon, leaving less to remain in the soil. In the absence of labile carbon, microbial mediated soil R_h tends to have lower Q_{10} [49,50]. Similar variation patterns of Q_{10} with incubation time have been observed by Tuomi et al. [51] and Hamdi et al. [52], in which quadratic and cubic functions were also used to describe the relationships between Q_{10} and incubation time.

The increases of Q_{10} at the later stage might be related to soil substrate quality changes (Fig. 2A). As the labile carbon decreased, recalcitrant carbon could be decomposed. It has been previously reported that Q_{10} tends to be higher in this situation [53,54]. In this study, it was not clear what caused the higher Q_{10} at 20% WHC at the later stage, but NH_4^+-N and TP were also higher at 20% WHC, which may be related to the higher Q_{10}.

Moisture Effects on Q_{10}

We found that soil moisture had a significant effect on Q_{10}, which aligns with the findings from several previous studies [17,20,55]. Our results showed that at the intermediate soil moisture level (i.e. 60% WHC), Q_{10} was lower than at the other soil moisture levels. While there was no significant difference of Q_{10} among 60%, 40% and 20% WHC, Q_{10} at 60% WHC was significantly lower than at 80% and 100% WHC (Fig. 3). Previous studies have shown that drying can decrease Q_{10} of soil respiration and total ecosystem respiration [2,56], and this may be largely due to substrate limitation caused by the limited diffusion of solutes in thin soil water films [57,58].

We further tested which soil properties would influence Q_{10} at different incubation days and found that, at the early and middle incubation stages, Q_{10} had a significant positive correlation with MBC and the ratio of MBC to DOC (Fig. 4). The higher MBC and MBC to DOC ratio were particularly associated with higher soil moisture levels, under which labile substrate might be more available to microbes due to less water limitation. However, the Arrhenius equation shows that reactants with lower activation energies (i.e. more reactive and less recalcitrant) should have lower temperature sensitivity [2]. Our incubation results indicated that Q_{10} might not only be determined by substrate availability, but also by microbial properties such microbial biomass.

At the late stage of incubation, Q_{10} was significantly related to F:B and TP. The tight correlation of Q_{10} with F:B ratio was quite interesting. Both fungi and bacteria are important decomposers, but their structures and chemical compositions are very different. Fungi have hyphae that allow them to move, colonize and degrade surface litters, and fungal cell walls are the polymers of melanin and of chitin, much more resistant to degradation [59,60]. At the late stage, labile substrate diminishing may favor fungi communities which can degrade more recalcitrant substrate. As suggested by the carbon quality hypothesis [2,61], soils with more fungi or higher F:B ratio would have larger Q_{10}, as demonstrated here. Bradford et al. [62] also reported a shift in microbial community structure could alter the Q_{10} of R_h. The positive correlation between Q_{10} and TP suggested that P availability might also influence Q_{10}. For example, the Q_{10} value for the 20% WHC was higher than those for the other moisture levels (Fig. 2A) and TP was correspondingly higher, probably due to the lower rate of consumption by microbes at lower moisture levels. A field study also showed that summer drought caused a 22–64% reduction of microbial phosphorus [63], indicating lower microbial consumption of P under water stress. Furthermore, forest soils in subtropical China are often phosphorous limited [64]. The phosphorous saved by the lower rate of consumption might therefore contribute to the higher Q_{10} at 20%WHC during the late stage of the incubation.

Conclusions

By incubating a subtropical forest soil under five temperature levels and five moisture levels and measuring soil R_h and microbial and chemical properties throughout the incubation, we found that: 1) soil moisture significantly influenced Q_{10}, with Q_{10} being higher at higher soil moisture levels than at the lower moisture levels during the early stage of the incubation;2) soil heterotrophic respiration was highest at intermediate moisture and lowest when the soil was very dry; 3) Q_{10} mostly declined with incubation time and could be best described by quadratic or cubic functions; and 4) moisture-induced Q_{10} changes were associated with soil microbial biomass at the early stage of incubation, but to the ratio of fungi-to-bacteria at the late stage. These results imply that the response of soil R_h to future global warming may be shaped by changes in precipitation patterns. In dry conditions, global warming may stimulate less soil CO_2 emission, but in wet conditions, relatively more soil CO_2 may be emitted. Considering that more soil organic carbon has often been accumulated in the wet areas, with the same temperature rise high Q_{10} would mean more soil CO_2 emission to the atmosphere from these areas in the future.

Acknowledgments

We thank Ms. Chunqing Long for her assistance in field sampling and laboratory analysis, and Mr. Brett Seybert for his critical review and editing of the manuscript. Two anonymous reviewers provided insightful comments on the early versions of the manuscript.

Author Contributions

Conceived and designed the experiments: WS DH. Performed the experiments: WZ. Analyzed the data: WZ DH WS. Wrote the paper: WZ DH WS.

References

1. Luo Y, Wan S, Hui D, Wallace LL (2001) Acclimatization of soil respiration to warming in a tall grass prairie. Nature 413: 622–625.
2. Davidson EA, Janssens IA (2006) Temperature sensitivity of soil carbon decomposition and feedbacks to climate change. Nature 440: 165–173.
3. Davidson EA, Belk E, Boone RD (1998) Soil water content and temperature as independent or confounded factors controlling soil respiration in a temperate mixed hardwood forest. Global Change Biol 4: 217–227.
4. Jenkinson DS, Adams DE, Wild A (1991) Model estimates of CO_2 emissions from soil in response to global warming. Nature 351: 304–306.
5. Solomon S, Qin D, Manning M, Chen Z, Marquis M, et al. (2007) IPCC Climate Change 2007: The physical science basis. Contribution of working group I to the fourth assessment report of the intergovernmental panel on climate change, Cambridge University Press, Cambridge, United Kingdom and New York, NY, USA.
6. Cook FJ, Orchard VA (2008) Relationships between soil respiration and soil moisture. Soil Biol Biochem 40: 1013–1018.
7. Richardson J, Chatterjee A, Jenerette GD (2012) Optimum temperatures for soil respiration along a semi-arid elevation gradient in southern California. Soil Biol Biochem 46: 89–95.
8. Zhou T, Shi P, Hui D, Luo Y (2009) Spatial patterns in temperature sensitivity of soil respiration in China: Estimation with inverse modeling. Sci China Ser C 52: 982–989.
9. Raich JW, Schlesinger WH (1992) The global carbon dioxide flux in soil respiration and its relationship to vegetation and climate. Tellus B 44: 81–99.
10. Parton W, Silver WL, Burke IC, Grassens L, Harmon ME, et al. (2007) Global-scale similarities in nitrogen release patterns during long-term decomposition. Science 315: 361–364.
11. Hui D, Luo Y (2004) Evaluation of soil CO_2 production and transport in Duke Forest using a process-based modeling approach. Global Biogeochem Cycle 18: GB4029.
12. Orchard VA, Cook FJ (1983) Relationship between soil respiration and soil moisture. Soil Biol Biochem 15: 447–453.
13. Shen W, Jenerette GD, Hui D, Phillips RP, Ren H (2008) Effects of changing precipitation regimes on dryland soil respiration and C pool dynamics at rainfall event, seasonal and interannual scales. J Geophys Res-Biogeosci 113: G03024.
14. Shen W, Reynolds JF, Hui D (2009) Responses of dryland soil respiration and soil carbon pool size to abrupt vs. gradual and individual vs. combined changes in soil temperature, precipitation, and atmospheric [CO_2]: a simulation analysis. Global Change Biol 15: 2274–2294.
15. Oberbauer SF, Gillespie CT, Cheng W, Gebauer R, Serra AS, et al. (1992) Environmental effects on CO_2 efflux from riparian tundra in the northern foothills of the Brooks Range, Alaska, USA. Oecologia 92: 568–577.
16. Linn DM, Doran JW (1984) Aerobic and anaerobic microbial populations in no-till and plowed soils. Soil Sci Soc Amer J 48: 794–799.
17. Wang C, Yang J, Zhang Q (2006) Soil respiration in six temperate forests in China. Global Change Biol 12: 2103–2114.
18. Carlyle JC, Than UB (1988) Abiotic controls of soil respiration beneath an eighteen-year old *pinus* radiata stand in south-eastern Australia. J Eco 76: 654–662.
19. Reichstein M, Subke JA, Angeli AC, Tenhunen JD (2005) Does the temperature sensitivity of decomposition of soil organic matter depend upon water content, soil horizon, or incubation time? Global Change Biol 11: 1754–1767.
20. Craine JM, Gelderman TM (2011) Soil moisture controls on temperature sensitivity of soil organic carbon decomposition for a mesic grassland. Soil Biol Biochem 43: 455–457.
21. Moyano FE, Manzoni S, Chenu C (2013) Responses of soil heterotrophic respiration to moisture availability: An exploration of processes and models. Soil Biol Biochem 59: 72–85.
22. Gray SB, Classen AT, Kardol P, Yermakov Z, Michael Mille R (2011) Multiple climate change factors interact to alter soil microbial community structure in an old-field ecosystem. Soil Sci Soc Amer J 75: 2217–2226.
23. Schnürer J, Clarholm M, Boström S, Rosswall T (1986) Effects of moisture on soil microorganisms and nematodes: A field experiment. Microbial Ecol 12: 217–230.
24. Nakas JP, Klein DA (1979) Decomposition of microbial cell components in a semi-arid grassland soil. Appl Environ Microbiol 38: 454–460.
25. Christ MJ, David MB (1996a) Temperature and moisture effects on the production of dissolved organic carbon in a Spodosol. Soil Biol Biochem 28: 1191–1199.
26. Lundquist EJ, Jackson LE, Scow KM (1999) Wet–dry cycles affect dissolved organic carbon in two California agricultural soils. Soil Biol Biochem 31: 1031–1038.
27. Li Z, Peng S, Rae D, Zhou G (2001) Litter decomposition and nitrogen mineralization of soils in subtropical plantation forests of southern China, with special attention to comparisons between legumes and non-legumes. Plant Soil 229: 105–116.
28. Chen D, Zhang Y, Lin Y, Chen H, Fu S (2009) Stand level estimation of root respiration for two subtropical plantations based on *in situ* measurement of specific root respiration. Forest Ecol Manage 257: 2088–2097.
29. Yuste JC, Baldocchi DD, Gershenson A, Goldstein A, Misson L, et al. (2007) Microbial soil respiration and its dependency on carbon inputs, soil temperature and moisture. Global Change Biol 13: 2018–2035.
30. Vance ED, Brookes PC, Jenkinson DS (1987) An extraction method for measuring soil microbial biomass C. Soil Biol Biochem 19: 703–707.
31. Joergensen RG, Mueller T (1996) The fumigation-extraction method to estimate soil microbial biomass: Calibration of the k_{EC} value. Soil Biol Biochem 28: 33–37.
32. Dorich RA, Nelson DW (1983) Direct colorimetric measurement of ammonium in potassium chloride extracts of soils. Soil Sci Soc Amer J 47: 833–836.
33. Bossio DA, Scow KM (1998) Impacts of carbon and flooding on soil microbial communities: phospholipid fatty acid profiles and substrate utilization patterns. Microbial Ecol 35: 265–278.
34. Zak DR, Ringlberg DB, Pregitzer KS, Randlett DL, White DC, et al. (1996) Soil microbial communities beneath populus grandidentata grown under elevated atmospheric CO_2. Ecol Appl 1996: 257–262.
35. Frostegård A, Bååth E (1996) The use of phospholipid fatty acid analysis to estimate bacterial and fungal biomass in soil. Biol Fert Soils 22: 59–65.
36. Reichstein M, Rey A, Freibauer A, Tenhunen J, Valentini R, et al. (2003) Modeling temporal and large-scale spatial variability of soil respiration from soil water availability, temperature and vegetation productivity indices. Global Biogeochem Cycle 17: 1104–1118.
37. Deng Q, Hui D, Zhang D, Zhou G, Liu J, et al. (2012) Effects of precipitation increase on soil respiration: a three-year field experiment in subtropical forests in China. PLoS ONE 7: e41493.
38. Jassal RS, Black TA, Novak MD, Gaumont-Guay D, Nesic Z (2008) Effect of soil water stress on soil respiration and its temperature sensitivity in an 18-year-old temperate Douglas-fir stand. Global Change Biol 14: 1305–1318.
39. Yu X, Zha T, Pang Z, Wu B, Wang X, et al. (2011) Response of soil respiration to soil temperature and moisture in a 50-year-old oriental arborvitae plantation in China. PLoS ONE 6: e28397.
40. Suseela V, Conant RT, Wallenstein MD, Dukes JS (2012) Effects of soil moisture on the temperature sensitivity of heterotrophic respiration vary seasonally in an old-field climate change experiment. Global Change Biol 18: 336–348.
41. Xu L, Baldocchi DD, Tang J (2004) How soil moisture, rain pulses, and growth alter the response of ecosystem respiration to temperature. Global Biogeochem Cycle 18: GB4002.
42. Williams MA (2007) Response of microbial communities to water stress in irrigated and drought-prone tallgrass prairie soils. Soil Biol Biochem 39: 2750–2757.
43. Fang CM, Smith P, Moncrieff JB, Smith JU (2005) Similar response of labile and resistant soil organic matter pools to changes in temperature. Nature 433: 57–59.
44. Kirschbaum MUF (2006) The temperature dependence of organic-matter decomposition - still a topic of debate. Soil Biol Biochem 38: 2510–2518.
45. Billings SA, Ballantyne F (2013) How interactions between microbial resource demands, soil organic matter stoichiometry, and substrate reactivity determine the direction and magnitude of soil respiratory responses to warming. Global Change Biol 19: 90–102.
46. Wang G, Zhou Y, Xu X, Ruan H, Wang J (2013) Temperature sensitivity of soil organic carbon mineralization along an elevation gradient in the Wuyi Mountains, China. PLoS ONE 8: e53914.
47. Mahecha MD, Reichstein M, Carvalhais N, Lasslop G, Lange H, et al. (2010) Global convergence in the temperature sensitivity of respiration at ecosystem level. Science 329: 838–840.
48. Song M, Jiang J, Cao G, Xu X (2010) Effects of temperature, glucose and inorganic nitrogen inputs on carbon mineralization in a Tibetan alpine meadow soil. Eur J Soil Biol 46: 375–380.
49. Arevalo CBM, Bhatti JS, Chang SX, Jassal RS, Sidders D (2010) Soil respiration in four different land use systems in north central Alberta, Canada. J Geophys Res-Biogeosci 115: G01003.
50. Yuste Curiel J, Ma S, Baldocchi DD (2010) Plant-soil interactions and acclimation to temperature of microbial-mediated soil respiration may affect predictions of soil CO_2 efflux. Biogeochemistry 98: 127–138.
51. Tuomi M, Vanhala P, Karhu K, Fritze H, Liski J (2008) Heterotrophic soil respiration–Comparison of different models describing its temperature dependence. Ecol Model 211: 182–190.
52. Hamdi S, Moyano F, Sall S, Bernoux M, Chevallier T (2013) Synthesis analysis of the temperature sensitivity of soil respiration from laboratory studies in relation to incubation methods and soil conditions. Soil Biol Biochem 58: 115–126.

53. Conant RT, Steinweg JM, Haddix ML, Paul EA, Plante AF, et al. (2008) Experimental warming shows that decomposition temperature sensitivity increases with soil organic matter recalcitrance. Ecology 89: 2384–2391.
54. Craine JM, Fierer N, McLauchlan KK (2010) Widespread coupling between the rate and temperature sensitivity of organic matter decay. Nat Geosci 3: 854–857.
55. Moyano FE, Vasilyeva N, Bouckaert L, Cook F, Craine J, et al. (2012) The moisture response of soil heterotrophic respiration: interaction with soil properties. Biogeosciences 9: 1173–1182.
56. Reichstein M, Tenhunen JD, Roupsard O, Ourcival JM, Rambal S, et al. (2002) Ecosystem respiration in two Mediterranean evergreen Holm Oak forests: drought effects and decomposition dynamics. Funct Ecol 16: 27–39.
57. Rey A, Pegoraro E, Tedeschi V, De Parri I, Jarvis PG, et al. (2002) Annual variation in soil respiration and its components in a coppice oak forest in Central Italy. Global Change Biol 8: 851–866.
58. Curiel Yuste J, Janssens IA, Carrara A, Ceulemans R (2004) Annual Q_{10} of soil respiration reflects plant phenological patterns as well as temperature sensitivity. Global Change Biol 10: 161–169.
59. Holland EA, Coleman DC (1987) Litter placement effects on microbial and organic matter dynamics in an agroecosystem. Ecology 68: 425–433.
60. Guggenberger G, Frey SD, Six J, Paustian K, Elliott ET (1999) Bacterial and fungal cell-wall residues in conventional and no-tillage agroecosystems. Soil Sci Soc Amer J 63: 1188–1198.
61. Bosatta EÅ, Göran I. (1999) Soil organic matter quality interpreted thermodynamically. Soil Biol Biochem 31: 1889–1891.
62. Bradford MA, Davies CA, Frey SD, Maddox TR, Melillo JM, et al. (2008) Thermal adaptation of soil microbial respiration to elevated temperature. Eco Lett 11: 1316–1327.
63. He Z, Wu J, O'Donnell AG, Syers JK (1997) Seasonal responses in microbial biomass carbon, phosphorus and sulphur in soils under pasture. Biol and Fert Soils 24: 421–428.
64. Huang W, Liu J, Wang Y, Zhou G, Han T, et al. (2013) Increasing phosphorus limitation along three successional forests in southern China. Plant Soil 364: 181–191.

Long-Term Effect of Agricultural Reclamation on Soil Chemical Properties of a Coastal Saline Marsh in Bohai Rim, Northern China

Yidong Wang[1], Zhong-Liang Wang[1,2*], Xiaoping Feng[1,3], Changcheng Guo[1], Qing Chen[1]

1 Tianjin Key Laboratory of Water Resources and Environment, Tianjin Normal University, Tianjin, China, 2 State Key Laboratory of Environmental Geochemistry, Institute of Geochemistry, Chinese Academy of Sciences, Guiyang, China, 3 College of Urban and Environmental Sciences, Tianjin Normal University, Tianjin, China

Abstract

Over the past six decades, coastal wetlands in China have experienced rapid and extensive agricultural reclamation. In the context of saline conditions, long-term effect of cultivation after reclamation on soil chemical properties has not been well understood. We studied this issue using a case of approximately 60-years cultivation of a coastal saline marsh in Bohai Rim, northern China. The results showed that long-term reclamation significantly decreased soil organic carbon (SOC) (−42.2%) and total nitrogen (TN) (−25.8%) at surface layer (0–30 cm) as well as their stratification ratios (SRs) (0–5 cm:50–70 cm and 5–10 cm:50–70 cm). However, there was no significant change in total phosphorus (TP) as well as its SRs under cultivation. Cultivation markedly reduced ratios of SOC to TN, SOC to TP and TN to TP at surface layer (0–30 cm) and their SRs (0–5 cm:50–70 cm). After cultivation, electrical conductivity and salinity significantly decreased by 60.1% and 55.3% at 0–100 cm layer, respectively, suggesting a great desalinization. In contrast, soil pH at 20–70 cm horizons notably increased as an effect of reclamation. Cultivation also changed compositions of cations at 0–10 cm layer and anions at 5–100 cm layer, mainly decreasing the proportion of Na^+, Cl^- and SO_4^{2-}. Furthermore, cultivation significantly reduced the sodium adsorption ratio and exchangeable sodium percentage in plow-layer (0–20 cm) but not residual sodium carbonate, suggesting a reduction in sodium harm.

Editor: Jin-Song Zhang, Institute of Genetics and Developmental Biology, Chinese Academy of Sciences, China

Funding: This research was supported by the National Science & Technology Pillar Program (2012BAC07B02), the Program for New Century Excellent Talents in University (NCET-10-0954), the National Natural Science Foundation of China (31300381), and the Opening Fund of Tianjin Key Laboratory of Water Resources and Environment (YF11700102). The funders had no role in study design, data collection and analysis, decision to publish, or preparation of the manuscript.

Competing Interests: The authors have declared that no competing interests exist.

* E-mail: wangzhongliang@vip.skleg.cn

Introduction

Coastal wetlands provide essential ecosystem services to people and environment including flood protection, water supply and purification, food productivity, erosion control, wave attenuation, shoreline stabilization, wildlife habitat, biodiversity, climate regulation and amenity [1,2]. Over the past century, natural coastal wetlands all over the world have been rapidly shrunk due to intensive anthropogenic activities [3,4,5,6]. In China, approximately 51% (2.2×10^4 km^2) of coastal natural wetlands were lost or degraded since the 1950s, primarily due to agricultural reclamation [7]. Soil is one of the foundations of ecosystem services of wetland and cropland. Therefore, it is important to study the influence of agricultural cultivation after reclamation on soil properties of the coastal wetlands.

Some studies have evaluated the reclamation-induced changes of soil properties in coastal wetlands. Conversion of coastal wetlands to croplands has been reported to cause radical changes in soil chemical properties because of alterations of hydrology and agricultural activities. First, hydrologic alterations such as ditch drainage and diking led to increase in aeration and decrease in salinity [8,9,10,11]. Aeration accelerated soil organic matter decay [4,12,13] and affected substances' characteristics [5] and mobility through changing redox conditions [8,14]. Changes in salinity further influenced substances' cycle in soils [15,16]. Second, application of fertilizer affected nutrient contents such as carbon [17], nitrogen and phosphorus [14,18]. In addition, agricultural managements including planting, harvesting and tillage also influenced balances of substances inputs and outputs in soils [19,20,21]. Most of these studies were focusing on short-term scales (e.g. [5,8,11,13,14]). In contrast, only few studies were available at long-term scales [4,9,10].

Stratification of soil chemical properties with soil depth is common in many natural ecosystems [4,22]. Different soil horizons of coastal wetlands may have distinct responses to agricultural reclamation. The surface soil is the vital interface that receives intense impact from human disturbance. In support of this concept, chemical characteristics of the surface soil in coastal wetlands were demonstrated to be more sensitive to agricultural reclamation [4,18]. However, quantitative expression of the horizon-induced differences in response of coastal wetland to agricultural reclamation is rare. Stratification ratio (SR) is a quantitative indicator to represent changes of soil profile feature [22]. It is hypothesized that the SR can be used as an indicator of the responses of different soil horizons to human disturbance.

In Bohai Rim of northern China, coastal marsh geogenesises have been realized naturally by land-sea interactions since the middle-late Holocene. Since the founding of New China, natural coastal marshes have been largely reclaimed for agriculture to alleviate the pressure of the increasing population. For instance, approximately 55% of coastal natural wetlands in Tianjin were disappeared since the 1950s. However, it is unclear that how agricultural reclamation affects soil chemical properties and their SRs of the coastal saline marshes in Bohai Rim, northern China.

The objective of this study was to evaluate a long-term (approximately 60-years) impact of agricultural reclamation on soil chemical properties as well as their SR features of a coastal saline marsh in Bohai Rim, northern China.

Materials and Methods

Study site

This study was conducted at a coastal permanent marsh and a cropland cultivated from the marsh in Qilihai (39°18′ N, 117°29′ E, elevation 2 m) of Bohai Rim, northern China. No specific permits were required for the described field studies in the research site. The field studies did not involve endangered or protected species. The study site is flat and low-lying. The region was characterized by a warm temperate semi-humid monsoon climate. Annual mean air temperature was 11.2 °C; total precipitation amount was 500–600 mm year^{-1}; evapotranspiration was approximately 1900 mm year^{-1}; frost-free period was 180–194 d; and sunshine duration hours were 2600–2800 h. Rainfall occurred synchronously with the heat, with high values in summer (June–August).

The permanent marsh was formed from the land-sea interactions since middle-late Holocene. The permanent marsh was mainly covered by reed (*Phragmites australis*) and generally had a water level of 0–40 cm above surface recently. Croplands were cultivated from the *P. australis* marshes over approximately 60 years. During the first ~20 years (1950s–1960s), paddy was cultivated. Afterwards, the paddy lands converted to dry-lands mainly dominated by corn, cotton and sorghum in rotation. Organic (manure and straw) and chemical fertilizer applications during 1950s–1990s [23] were listed in Table 1. Since the new century, organic and chemical fertilizer applications [24] were presented in Table 2. The current water level of the dry-lands was about 80–100 cm below surface.

Soil sampling and chemical analysis

The *P. australis* marsh and the cropland were adjacent (about 500 m) and had similar altitudes. Soil samples were randomly collected using a soil drill (Eijkelkamp, Netherlands) at three locations in the permanent marsh and cropland in May 2012, respectively. In both marsh and cropland, the distance between any two sampling locations was more than 15 m. In each sampling location, two adjacent soil columns (0–100 cm) were collected and then mixed into one column according to 7 corresponding horizons: 0–10, 10–20, 20–30, 30–50, 50–70, and 70–100 cm.

Soil samples were brought to laboratory and air-dried naturally. The air-dried soil samples were sieved with a 2 mm mesh screen to remove the roots and coarse fraction before chemical analysis. Dilute HCl (0.5 mol L^{-1}) was used to remove soil inorganic carbon. Then deionized water was used to rinse off excess HCl and ensure the neutral pH of approximate 7. The contents of soil organic carbon (SOC) and total nitrogen (TN) were determined by an Elementar (Vario EL III, Elementar, Germany). The concentration of total phosphorus (TP) was determined by sulphuric and perchloric acids digestion and Mo-Sb Anti spectrophotometric method. Soil pH was measured with a pH meter (Star A 420C-01A, Thermo Orion, United States) using a soil:water ratio of 1:2.5 (g g^{-1}). Electrical conductivity (EC) was measured with a EC meter (Star A 420C-01A, Thermo Orion, United States) using a soil:water ratio of 1:5 (g g^{-1}). Soil major soluble cations (Na^+, Ca^{2+}, K^+ and Mg^{2+}) and anions (Cl^-, SO_4^{2-}, HCO_3^- and CO_3^{2-}) were measured using a soil:water ratio of 1:5 (g g^{-1}). Cation concentrations were determined by an Atomic Absorption Spectrophotometer (TAS-990, Beijing Purkinje General Instrument Co., Ltd., China). Concentrations of Cl^- and SO_4^{2-} were determined by an Ion Chromatography (ICS-2100, Dionex Corporation, Sunnyvale, California, United States). Concentrations of CO_3^{2-} and HCO_3^- were measured using neutralization titration method using phenolphthalein and methyl orange.

Data analysis

Sodium adsorption ratio (SAR), an easily measured property that gives information on the comparative concentrations of Na^+, Ca^{2+}, and Mg^{2+} in soil solutions, was calculated using equation (1) [25]. Exchangeable sodium percentage (ESP) was computed using equation (2) [26,27]. Residual sodium carbonate (RSC) was estimated using equation (3) [28]. Comparisons of mean SOC, TN, TP, pH, EC and salinity between the marsh and cropland were analyzed using one-way ANOVA of SPSS 13.0 (SPSS Inc., Chicago, Illinois, United States). Figures were drawn using Origin 8.0 (OriginLabs Corporation, Northampton, Massachusetts, United States) and CorelDraw 9 (Corel Corporation, Canada).

$$SAR = \frac{Na^+}{\sqrt{(Ca^{2+} + Mg^{2+})/2}} \quad (1)$$

$$ESP = 100 \times \frac{(-0.0126 + 0.01475SAR)}{1 + (-0.0126 + 0.01475SAR)} \quad (2)$$

Table 1. Organic and chemical fertilizer applications during 1950s–1990s (source from the reference [23]).

	Organic fertilizer (manure and straw)		Chemical fertilizer	
	Application rate (m^3 ha^{-1})	Proportion (%)	Application rate (kg ha^{-1})	Proportion (%)
1950s–1960s	30	93	53	7
1970s	40	61	600	39
1980s	20	40	690	60
1990s	20	29	1156	71

Table 2. Nitrogen and phosphorus inputs of organic and chemical fertilizer applications in 2007 (source from the reference [24]).

	Organic fertilizer (manure and straw)		Chemical fertilizer	
	Nitrogen (kg ha^{-1})	Phosphorus (kg ha^{-1})	Nitrogen (kg ha^{-1})	Phosphorus (kg ha^{-1})
2007	66	41	269	149

$$RSC = \left(CO_3^{2-} + HCO_3^{-}\right) - \left(Ca^{2+} + Mg^{2+}\right) \quad (3)$$

Results

Soil organic carbon, total nitrogen and phosphorus as well as their stratification ratios

Long-term agricultural reclamation significantly reduced the concentrations of SOC at surface soil (0–30 cm) but not lower soil layers (30–100 cm) (Fig. 1A). The weighted density of SOC at surface soil (0–30 cm) decreased by 10.1 g kg^{-1} (−42.2%). Similarly, the contents of TN of upper soil layers (0–50 cm) but not deeper soil layers (50–100 cm) were also evidently decreased under long-term agricultural cultivation (Fig. 1B). The weighted density of TN at surface soil (0–50 cm) reduced by 0.7 g kg^{-1} (−31.8%). In contrast, there were no significant effects of long-term agricultural reclamation on the contents of TP at the whole soil profile (0–100 cm) (Fig. 1C). The ratios of SOC/TN, SOC/TP and TN/TP were all notably decreased after agricultural reclamation at surface soil (0–30 cm, except for 10–20 cm of SOC/TN) but not lower soil layers (30–100 cm) (Fig. 2).

Stratification ratios of 0–5 cm:50–70 cm and 5–10 cm:50–70 cm of SOC were 12.3 and 3.7 for the marsh and were 3.4 and 3.0 for the cropland (Fig. 3A). Stratification ratios of 0–5 cm:50–70 cm and 5–10 cm:50–70 cm of TN were 4.0 and 1.9 for the marsh and were 1.5 and 1.4 for the cropland (Fig. 3A). Compared to the marsh, SRs of SOC and TN of the cropland were both remarkably dropped, suggesting greater decreases of SOC and TN in the upper soil layers (Fig. 3A and B). In contrast, there was no significant impact of reclamation on the SR of TP (Fig. 3C). The SRs (0–5 cm:50–70 cm) of ratios of SOC/TN, SOC/TP and TN/TP were all significantly decreased after agricultural reclamation (Fig. 4).

Soil EC, salinity and pH

Soil profile EC of the marsh ranged from 1026 to 2365 μs cm^{-1}, with a weighted average of 1488 μs cm^{-1}. Soil profile EC of the cropland varied from 392 to 854 μs cm^{-1}, with a weighted average of 594 μs cm^{-1} (Fig. 5A). Soil profile salinity of the marsh was from 3.2 to 6.9 g kg^{-1} dry weight, with a weighted average of 4.7 g kg^{-1} dry weight. Soil profile salinity of the cropland ranged from 1.3 to 2.5 g kg^{-1} dry weight, with a weighted average of 2.1 g kg^{-1} dry weight (Fig. 5B). Thus, after long-term reclamation, EC and salinity of the whole soil profile (0–100 cm) decreased by 60.1% and 55.3%, respectively. In support of these results, concentrations of water-extractable ions showed great declines after reclamation. In contrast, soil pH (20–70 cm) significantly increased after reclamation.

Compositions of cation and anion

Even though the cations of marsh and cropland were both dominated by Na^+, the proportion of Na^+ (0–10 cm) was significantly decreased after reclamation ($p<0.05$) (Fig. 6A). However, the proportion of Ca^{2+} (0–10 cm) and Mg^{2+}+K^+ (5–20 cm) were significantly increased after reclamation ($p<0.05$) (Fig. 6A). Contrasted to cation, the composition of anion changed greater at 5–100 cm after reclamation, with significant decreases

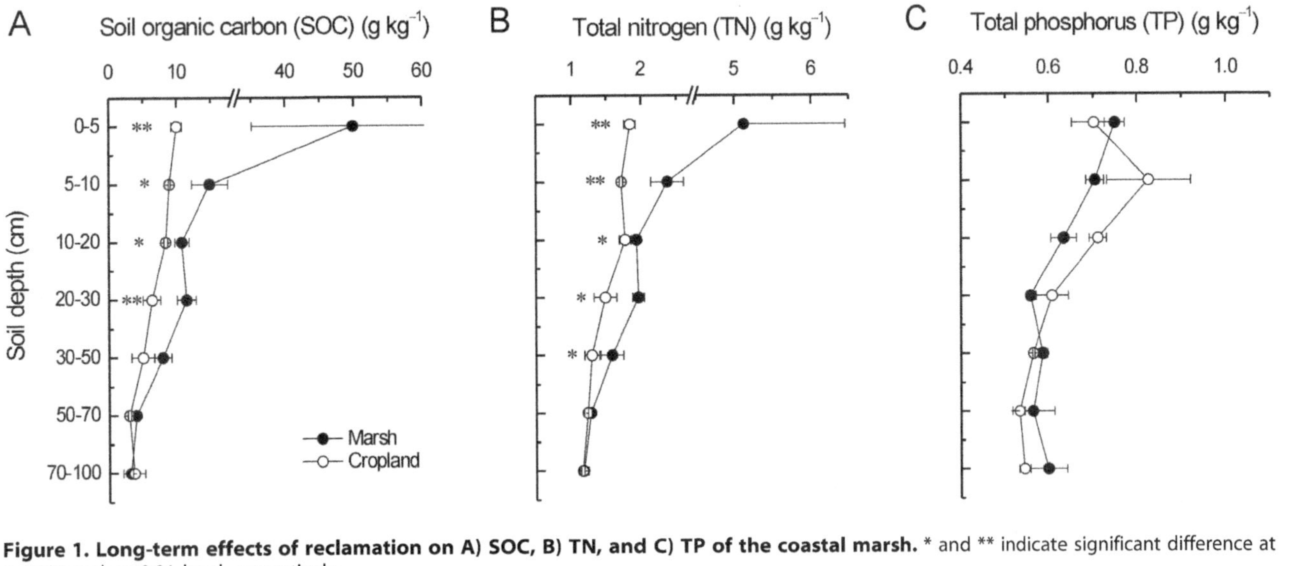

Figure 1. Long-term effects of reclamation on A) SOC, B) TN, and C) TP of the coastal marsh. * and ** indicate significant difference at $p<0.05$ and $p<0.01$ level, respectively.

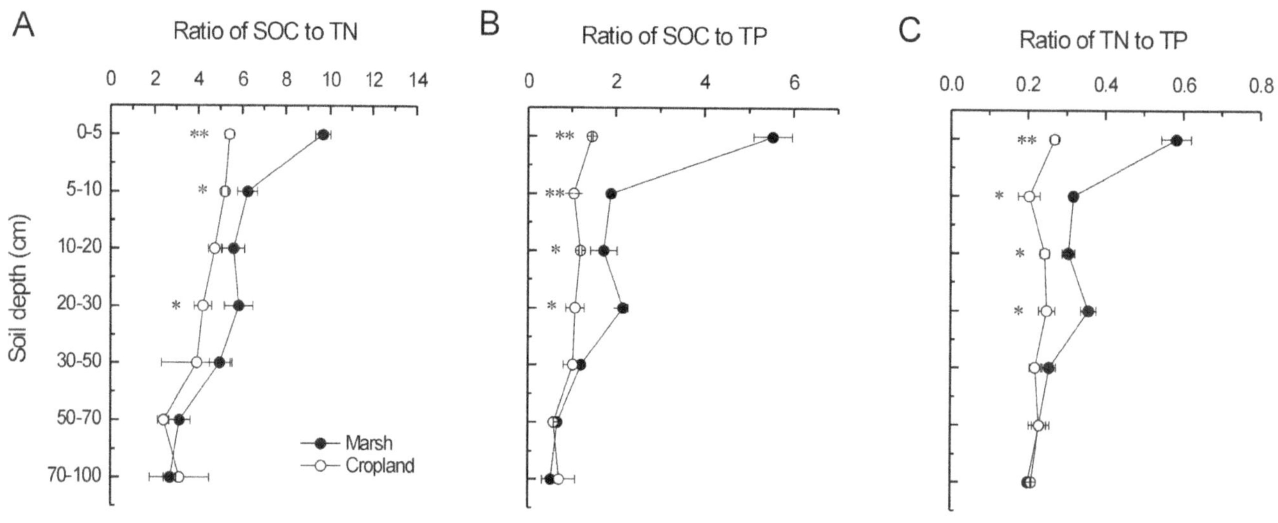

Figure 2. Long-term effects of reclamation on ratios of A) SOC/TN, B) SOC/TP, and C) TN/TP. * and ** represent significant difference at $p<0.05$ and $p<0.01$ level, respectively.

in Cl^- and SO_4^{2-} and increases in $HCO_3^-+CO_3^{2-}$ ($p<0.05$) (Fig. 6B).

Characteristics of alkalization

The SAR and ESP in plow-layer (0–20 cm) significantly reduced after reclamation ($p<0.05$) (Fig. 7A and B). This result suggested a reduction in soil alkalization and sodium harm. However, there was no significant difference in RSC under long-term cultivation (Fig. 7C).

Discussion

SOC, TN and TP as well as their stratification ratios

The soil storages of SOC, TN and TP are counterbalances of inputs (e.g. productivity and fertilizer) and outputs (e.g. loss of mineralization and harvest) during a certain period. The rates of inputs and outputs in soils have been widely recognized to be sensitive to environmental changes (e.g. [29,30,31]) and human disturbance (e.g. [19,20,32]). Agricultural reclamation of wetland substantially altered the soil environmental conditions, thus should change soil stocks of SOC, TN and TP through breaking the balance of inputs and outputs. In our study, even though many organic (manure and straw) and chemical fertilizers were applied for a long-term time (Table 1 and Table 2), SOC and TN decreased strongly in surface soil after reclamation (Fig. 1), suggesting that the outputs were much larger than the inputs. Our result was in agreement with the results in the Yangtze Estuary [10] and the Hangzhou Bay of China [14]. Four reasons may explain the above results. First, the mineralization and accumulation of soil organic matter were both regulated by hydrologic conditions [12,33]. Submerged soils of the marsh were under anaerobic conditions, which had lessened and incomplete decompositions of organic materials and decreased humifications of organic matter. In contrast, cropland soils lacked anaerobiosis and

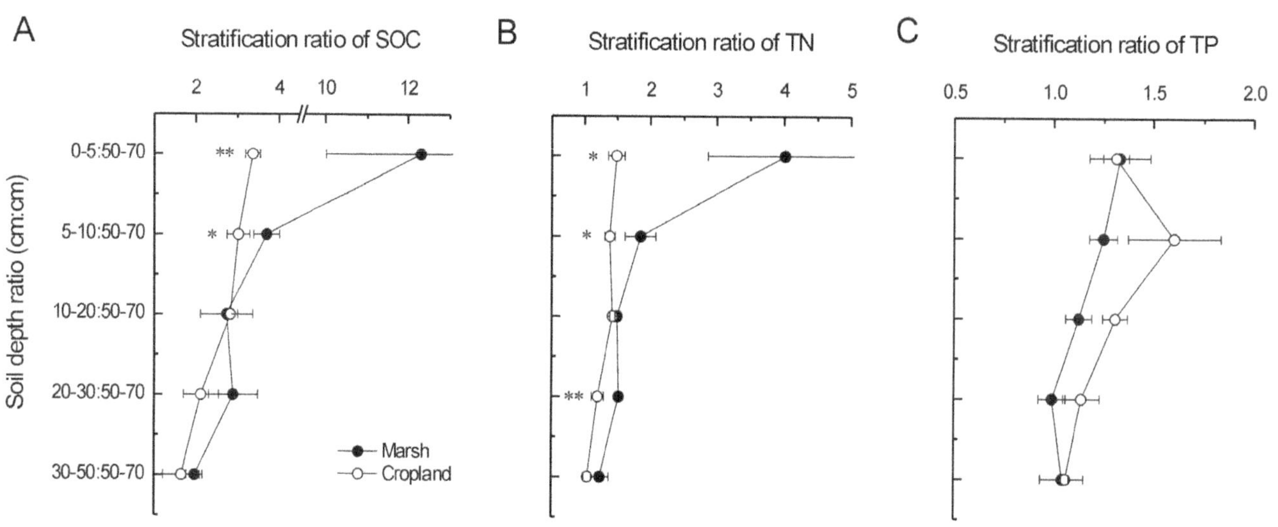

Figure 3. Long-term influences of reclamation on stratification ratios of A) SOC, B) TN, and C) TP. * and ** indicate significant difference at $p<0.05$ and $p<0.01$ level, respectively.

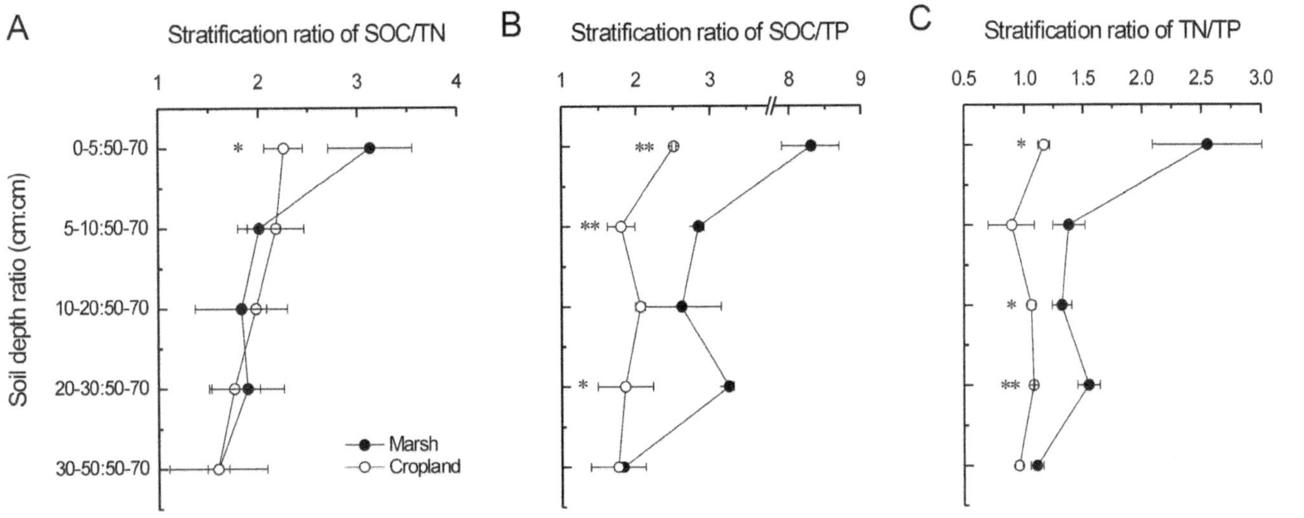

Figure 4. Long-term impacts of reclamation on stratification ratios of A) SOC/TN, B) SOC/TP, and C) TN/TP. * and ** indicate significant difference at $p<0.05$ and $p<0.01$ level, respectively.

had high rates of decomposition due to the aerobic conditions. Compared to cropland soils, there were preferential accumulations of organic matter in submerged soils [12,33]. Second, crop harvesting took lots of organic materials away from farmland ecosystem. Moreover, the net primary productivity of marshes is higher than cropland ecosystems [33]. In addition, decrease in soil salinity caused by agricultural reclamation might accelerate the mineralization of organic matters [16]. In our study, however, there was no significant change of SOC and TN in 50–100 cm soil under long-term reclamation (Fig. 1), indicating limited effects of reclamation on deeper soil horizons. This result was consistent with the results in the Sanjiang Plain [34], the Xingkai Lake [35] and the Hangzhou Bay of China [18]. Compared to SOC and TN, there was no significant variation in TP in whole soil profile after long-term reclamation (Fig. 1). However, Huang et al. [13] found significant decreases in TP caused by conversions of marshes to drylands. In marsh and cropland ecosystems, biogeochemical cycle of phosphorus is different from carbon and nitrogen. Phosphorus has poor mobility and is easy to be fixed in the soil. The input and output of phosphorus in the soil is mainly as fertilizer application and crop harvesting, respectively [36]. Thus, reclamation-induced impacts on TP may be explained by the balance of fertilizer application and crop harvesting.

Stratification ratio is a quantitative indicator to represent the responses of soil profile feature to anthropogenic disturbance [22], which was mainly applied in farmlands (e.g. [19,20]). However, the application of SR is not common in costal wetlands. Our results showed that SRs (0–5 cm:50–70 cm and 5–10 cm:50–70 cm) of SOC and TN decreased remarkably after long-term agricultural reclamation. In support of this result, chemical properties of the surface soil were reported to be more sensitive to reclamation than that of the deeper layers in two coastal wetlands [4,18]. These results are reasonable because the surface

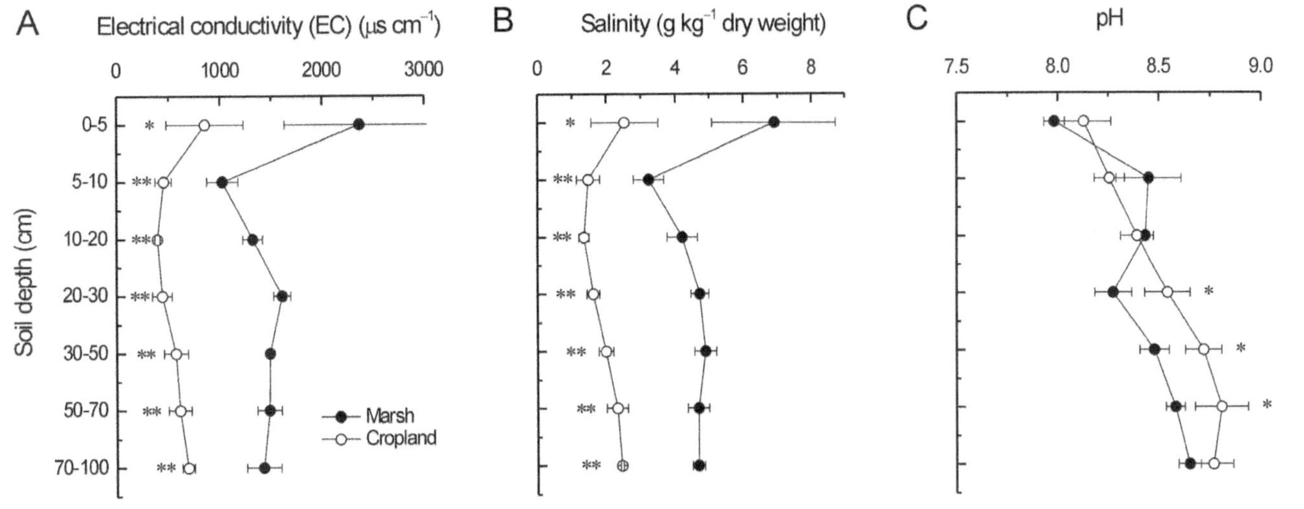

Figure 5. Long-term effects of reclamation on A) EC, B) salinity, and C) pH of coastal soils. * and ** indicate significant difference at $p<0.05$ and $p<0.01$ level, respectively.

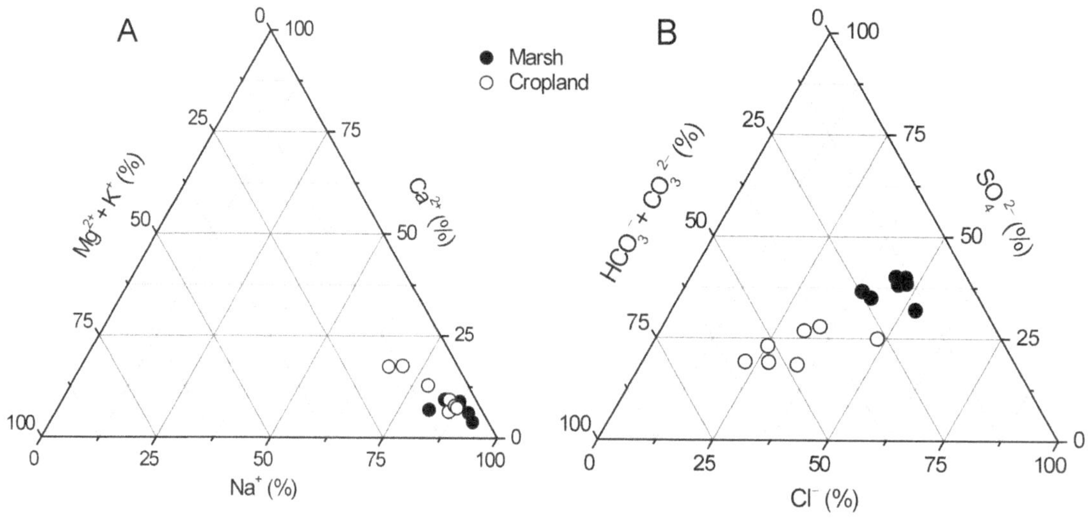

Figure 6. Long-term effects of agricultural cultivation on the compositions of A) cations and B) anions.

soils are the critical zones that receive intensive influences from anthropogenic activities.

Salinity, ion composition and alkalization

Our results showed that soil salinity and EC across the whole profile (0–100 cm) dropped significantly after agricultural reclamation (Fig. 5 A and B). This result was in agreement with the results in the Yellow River Delta [13], the Jiangsu province [11], the Yangtze Estuary [10], the Hangzhou Bay of China [14] and the Biscay Bay of Spain [9]. A reliable interpretation for this result is the effective leaching of the soil exchangeable ions induced by dewatering after reclamation. However, each of the major ions (anions and cations) in soil solutions has some unique properties which affect its production and mobility [37]. For example, the production of bicarbonate (HCO_3^-), one of the most common anions, is regulated by soil CO_2 pressure and pH. The sulfate (SO_4^{2-}) is involved in both biological and inorganic chemical reactions, whereas the chloride (Cl^-) is relatively uninvolved in these reactions and thus has high mobility [37]. The cations are held by the negatively charged clay and organic matter particles in the soil through electrostatic forces. Therefore, ion compositions were proposed to be sensitive to disturbance due to the specific mobility of each ion. In support of this concept, we found that the proportion of Na^+ (0–10 cm), Cl^- and SO_4^{2-} (5–100 cm) decreased significantly after reclamation. However, the proportion of Ca^{2+} (0–10 cm), Mg^{2+}+K^+ (5–20 cm) and HCO_3^-+CO_3^{2-} (5–100 cm) increased significantly after reclamation (Fig. 6). Knowing these reclamation-induced properties of the major ions, it is possible to protect and manage coastal saline soils of the wetlands and croplands as well as their ecosystem services. For example, reclamation-induced decreases in water table and soil salinity were beneficial to crop cultivation. High mobility of Na^+ further reduced the harm of salt ions after reclamation. Even though overlying water or saturated soil are basic properties of coastal wetlands, reclamation-induced hydrologic alterations such as ditch

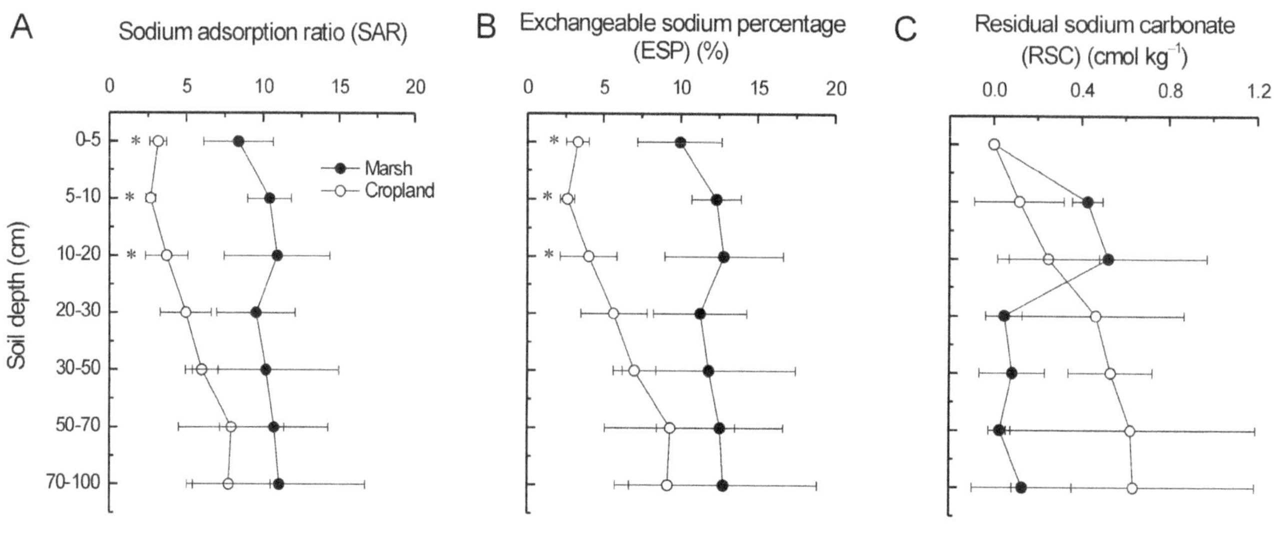

Figure 7. Long-term effects of reclamation on A) SAR, B) ESP, and C) RSC of the coastal marsh. * indicates significant difference at $p<0.05$.

drainage and diking result in decreases in salinity, which promotes plant activities and crop yields.

The SAR, ESP and RSC are three indicators of the degree of soil alkalization. The SAR of a soil extract takes into consideration that the adverse effect of Na^+ is moderated by the presence of Ca^{2+} and Mg^{2+}. When the SAR rises above 12 to 15, serious physical soil problems arise and plants have difficulty absorbing water [38]. In our study, all SARs of the marsh and cropland were below 12. Our results showed that agricultural reclamation significantly decreased the SAR and ESP but not RSC in the plow-layer (0–20 cm), suggesting a reduction in Na^+ harm (Fig. 7). These results were consistent with those in the Cape Cod, Massachusetts of USA [8] and the Hangzhou Bay of China [14]. Moreover, significant decreases of SAR and ESP in the upper soil horizons further supported the rapid desalinization.

Conclusions

In coastal marshes of northern China, long-term agricultural reclamation significantly reduced SOC and TN at surface soil (0–30 cm) as well as their SRs (0–5 cm:50–70 cm and 5–10 cm:50–70 cm). However, there was no significant change in TP as well as its SRs. Reclamation notably decreased ratios of SOC/TN, SOC/TP and TN/TP at surface soil (0–30 cm) as well as their SRs (0–5 cm:50–70 cm). A significant desalinization of whole soil profile occurred during cultivation. Soil pH (20–70 cm) markedly increased as an impact of reclamation. Compositions of cations (0–10 cm) and anions (5–100 cm) changed mainly as decreases in proportion of Na^+, Cl^- and SO_4^{2-}. Furthermore, reclamation significantly reduced the sodium adsorption ratio and exchangeable sodium percentage in plow-layer (0–20 cm) but not residual sodium carbonate, suggesting a reduction in sodium harm.

Acknowledgments

The authors thank Mr. Yuntao Shang for his help in laboratory analysis. The authors also appreciate Dr. Jin-Song Zhang and an anonymous reviewer for their helpful comments that improved this paper.

Author Contributions

Conceived and designed the experiments: ZW YW. Performed the experiments: YW XF CG QC. Analyzed the data: YW XF. Contributed reagents/materials/analysis tools: ZW YW. Wrote the paper: YW.

References

1. Costanza R, d'Arge R, de Groot R, Farber S, Grasso M, et al. (1997) The value of the world's ecosystem services and natural capital. Nature 387: 253–260.
2. Shepard CC, Crain CM, Beck MW (2011) The Protective Role of Coastal Marshes: A Systematic Review and Meta-analysis. PLoS ONE 6(11): e27374. doi:10.1371/journal.pone.0027374.
3. Cashin GE, Dorney JR, Richardson CJ (1992) Wetland alteration trends on the North Carolina coastal plain. Wetlands 12: 63–71.
4. Ellis S, Atherton JK (2003) Properties and development of soils on reclaimed alluvial sediments of the Humber estuary, eastern England. Catena 52: 129–147.
5. Santín C, de la Rosa JM, Knicker H, Otero XL, Álvarez MÁ, et al. (2009) Effects of reclamation and regeneration processes on organic matter from estuarine soils and sediments. Org Geochem 40: 931–941.
6. An S, Tian Z, Cai Y, Wen T, Xu D, et al. (2013) Wetlands of Northeast Asia and High Asia: an overview. Aquat Sci 75: 63–71.
7. An S, Li H, Guan B, Zhou C, Wang Z, et al. (2007) China's natural wetlands: past problems, current status, and future challenges. Ambio 36: 335–342.
8. Portnoy JW (1999) Salt marsh diking and restoration: biogeochemical implications of altered wetland hydrology. Environ Manage 24: 111–120.
9. Fernández S, Santín C, Marquínez J, Álvarez MA (2010) Saltmarsh soil evolution after land reclamation in Atlantic estuaries (Bay of Biscay, North coast of Spain). Geomorphology 114: 497–507.
10. Cui J, Liu C, Li Z, Wang L, Chen X, et al. (2012) Long-term changes in topsoil chemical properties under centuries of cultivation after reclamation of coastal wetlands in the Yangtze Estuary, China. Soil Till Res 123: 50–60.
11. Jin X, Huang J, Zhou Y (2012) Impact of coastal wetland cultivation on microbial biomass, ammonia-oxidizing bacteria, gross N transformation and N_2O and NO potential production. Biol Fertil Soils 48: 363–369.
12. Bridgham SD, Updegraff K, Pastor J (1998) Carbon, nitrogen, and phosphorus mineralization in northern wetlands. Ecology 79: 1545–1561.
13. Huang L, Bai J, Chen B, Zhang K, Huang C, et al. (2012) Two-decade wetland cultivation and its effects on soil properties in salt marshes in the Yellow River Delta, China. Ecol Inform 10: 49–55.
14. Iost S, Landgraf D, Makeschin F (2007) Chemical soil properties of reclaimed marsh soil from Zhejiang Province P.R. China. Geoderma 142: 245–250.
15. Weston NB, Dixon RE, Joye SB (2006) Ramifications of increased salinity in tidal freshwater sediments: geochemistry and microbial pathways of organic matter mineralization. J Geophys Res-Biogeoscience 111, G01009. doi:10.1029/2005JG000071.
16. Wong VNL, Greene RSB, Dalal RC, Murphy BW (2010) Soil carbon dynamics in saline and sodic soils: a review. Soil Use Manage 26: 2–11.
17. Lou Y, Wang J, Liang W (2011) Impacts of 22-year organic and inorganic N managements on soil organic C fractions in a maize field, northeast China. Catena 87: 386–390.
18. Wu M, Shao X, Hu F, Jiang K (2008) Effects of reclamation on soil nutrients distribution of coastal wetland in south Hangzhou Bay. Soils 40: 760–764. (in Chinese with English abstract)
19. Du Z, Ren T, Hu C (2010) Tillage and residue removal effects on soil carbon and nitrogen storage in the North China Plain. Soil Sci Soc Am J 74: 196–202.
20. Lou Y, Xu M, Chen X, He X, Zhao K (2012) Stratification of soil organic C, N and C:N ratio as affected by conservation tillage in two maize fields of China. Catena 95: 124–130.
21. Han G, Xing Q, Yu J, Luo Y, Li D, et al. (2013) Agricultural reclamation effects on ecosystem CO_2 exchange of a coastal wetland in the Yellow River Delta. Agr Ecosyst Environ doi:org/10.1016/j.agee.2013.09.012.
22. Franzluebbers AJ (2002) Soil organic matter stratification ratio as an indicator of soil quality. Soil Till Res 66: 95–106.
23. Liu H, Mou S (1999) Farmland fertilization conditions and countermeasures in Tianjin. Sci Technol Tianjin Agr For 2: 43–45. (in Chinese)
24. Zhu M (2011) Study on agricultural NPS loads of Haihe Basin and assessment on its environmental impact. Beijing: Doctoral Dissertation, Institute of Agricultural Economics and Development, Chinese Academy of Agricultural Sciences. (in Chinese with English abstract)
25. Robbins CW (1984) Sodium adsorption ratio-exchangeable sodium percentage relationships in a high potassium saline-sodic soil. Irrigation Sci 5: 173–179.
26. Oster JD, Sposito G (1980) The gapon coefficient and the exchangeable sodium percentage-sodium adsorption ratio relation. Soil Sci Soc Am J 44: 258–260.
27. Shi Y, Xin D (1983) Water-salt movement and comprehensive management of droughts, floods and saline-alkali in the Huang-Huai-Hai Plain. Hebei People's Publishing House, Shijiazhuang. (in Chinese)
28. Wan H, Yang J, Yu R (1983) Investigation on calculating method of soil alkalinity in the Huang-Huai-Hai Plain. Soils 23: 319–325. (in Chinese)
29. Wang Y, Wang Z-L, Wang H, Guo C, Bao W (2012) Rainfall pulse primarily drives litterfall respiration and its contribution to soil respiration in a young exotic pine plantation in subtropical China. Can J Forest Res 42: 657–666.
30. Wang Y, Wang H, Ma Z, Dai X, Wen X, et al. (2013) The litter layer acts as a moisture-induced bidirectional buffer for atmospheric methane uptake by soil of a subtropical pine plantation. Soil Biol Biochem 66: 45–50.
31. Wang Y, Wang H, Wang Z-L, Ma Z, Dai X, et al. (2014) Effect of litter layer on soil-atmosphere N_2O flux of a subtropical pine plantation in China. Atmos Environ 82: 106–112.
32. Macreadie PI, Hughes AR, Kimbro DL (2013) Loss of 'Blue Carbon' from Coastal Salt Marshes Following Habitat Disturbance. PLoS ONE 8(7): e69244. doi:10.1371/journal.pone.0069244.
33. Sahrawat KL (2003) Organic matter accumulation in submerged soils. Adv. Agron. 81: 169–201.
34. Wang L, Song C, Ge R, Song Y, Liu D (2009) Soil organic carbon storage under different land-use types in Sanjiang Plain. China Environ Sci 29: 656–660. (in Chinese with English abstract)
35. Huo L, Zou Y, Guo J, Lü X (2013) Effect of reclamation on the vertical distribution of SOC and retention of DOC. Environ Sci 34: 283–287. (in Chinese with English abstract)
36. Shen J, Yuan L, Zhang J, Li H, Bai Z, et al. (2011) Phosphorus dynamics: from soil to plant. Plant Physiol 156: 997–1005.
37. Johnson DW, Cole DW (1980) Anion mobility in soils: Relevance to nutrient transport from forest ecosystems. Environ Int 3: 79–90.
38. Munshower FF (1994) Practical handbook of disturbed land revegetation. Florida: Lewis Publishers.

25

Impacts of Edaphic Factors on Communities of Ammonia-Oxidizing Archaea, Ammonia-Oxidizing Bacteria and Nitrification in Tropical Soils

Vidya de Gannes[1], Gaius Eudoxie[1], William J. Hickey[2]*

1 Dept. Food Production, Faculty of Food and Agriculture, University of the West Indies, St. Augustine Campus, Republic of Trinidad and Tobago, **2** O.N. Allen Laboratory for Soil Microbiology, Dept. Soil Science, University of Wisconsin-Madison, Madison, Wisconsin, United States of America

Abstract

Nitrification is a key process in soil nitrogen (N) dynamics, but relatively little is known about it in tropical soils. In this study, we examined soils from Trinidad to determine the edaphic drivers affecting nitrification levels and community structure of ammonia-oxidizing bacteria (AOB) and ammonia-oxidizing archaea (AOA) in non-managed soils. The soils were naturally vegetated, ranged in texture from sands to clays and spanned pH 4 to 8. The AOA were detected by qPCR in all soils (*ca.* 10^5 to 10^6 copies archaeal *amoA* g^{-1} soil), but AOB levels were low and bacterial *amoA* was infrequently detected. AOA abundance showed a significant negative correlation ($p<0.001$) with levels of soil organic carbon, clay and ammonium, but was not correlated to pH. Structures of AOA and AOB communities, as determined by *amoA* terminal restriction fragment (TRF) analysis, differed significantly between soils ($p<0.001$). Variation in AOA TRF profiles was best explained by ammonium-N and either Kjeldahl N or total N ($p<0.001$) while variation in AOB TRF profiles was best explained by phosphorus, bulk density and iron ($p<0.01$). In clone libraries, phylotypes of archaeal *amoA* (predominantly *Nitrososphaera*) and bacterial *amoA* (predominanatly *Nitrosospira*) differed between soils, but variation was not correlated with pH. Nitrification potential was positively correlated with clay content and pH ($p<0.001$), but not to AOA or AOB abundance or community structure. Collectively, the study showed that AOA and AOB communities were affected by differing sets of edaphic factors, notably that soil N characteristics were significant for AOA, but not AOB, and that pH was not a major driver for either community. Thus, the effect of pH on nitrification appeared to mainly reflect impacts on AOA or AOB activity, rather than selection for AOA or AOB phylotypes differing in nitrifying capacity.

Editor: Christopher Quince, University of Glasgow, United Kingdom

Funding: These studies were supported by a grant (to Vidya de Gannes) from the Postgraduate Research Fund and the Department of Food Production, Faculty of Food and Agriculture University of the West Indies, St. Augustine Campus, Trinidad and Tobago, and an endowment from the O.N. Allen Professorship in Soil Microbiology (to William Hickey). The funders had no role in study design, data collection and analysis, decision to publish, or preparation of the manuscript.

Competing Interests: The authors have declared that no competing interests exist.

* E-mail: wjhickey@wisc.edu

Introduction

Nitrification is a key process in the nitrogen (N) cycle, as transformation of ammonium-N to nitrate-N can cause nitrate contamination of groundwater, and greenhouse gas production (*i.e.*, N_2O) directly and indirectly *via* denitrification. Thus, predicting the potential for N contamination of the environment necessitates a thorough understanding of the environmental factors affecting the prokaryotes mediating nitrification, a group that now includes ammonia-oxidizing archaea (AOA). The AOA are ubiquitous constituents of marine and terrestrial environments [1,2,3,4], and their discovery has changed the paradigm of aerobic nitrification mediated solely by ammonia-oxidizing bacteria (AOB). As such, for soils, there's now much interest in discerning the role AOA may play in nitrification relative to AOB, and predicting nitrification now requires an understanding of the environmental characteristics that may drive the structure and activity of both AOA and AOB communities (niche differentiation).

In soils, drivers of niche differentiation of AOA *vs.* AOB has centered largely on pH and, to a lesser extent, levels and forms of nitrogen. The former has been implicated to differentially shape AOA *vs.* AOB communities in strongly acidic soils, where AOA abundance increases, or remains unchanged, with decreasing pH, while that of AOB decreases [5,6,7,8]. However, in circum-neutral to alkaline soils, correlations of AOA or AOB abundance to pH have not been consistent [9,10]. Phylogenetic analyses have also implicated pH as a niche-shaping factor as some AOA phylotypes have been correlated with pH [11]. But, the strength of that correlation is uncertain in part because as the data set is skewed toward acidic environments, and it's often uncertain whether the correlations are attributed to pH alone and or other edaphic factors [5].

Potential differences between AOA and AOB in N relations have been evaluated by field and microcosm studies, in which large inputs of ammonium stimulated AOB, but not AOA [9]. In contrast, growth of AOA, but not AOB, was stimulated by addition of an organic N source [12]. These differences have been interpreted as reflecting a preference of AOA for low ammonium environments, which would be consistent with some physiological analyses of the few available AOA cultures [13,14]. But, field data establishing correlations of AOA to natural levels of ammonium or any other form of soil N are limited.

Soil is a highly complex microbial habitat and it's likely that AOA and AOB communities are affected by multiple edaphic factors [7,15,16]. Thus, studies have recently begun to focus on identification of multiple drivers of AOA and AOB community structure and function [17]. However, to date these studies have primarily focused on comparison of land use including soils with N inputs derived from management practices and/or varying in vegetation history [18,19,20,21,22]. Also, prior studies have spanned spatial scales sufficiently large as to include climatic variability as well [17,23].There has been relatively little focus specifically on effects of edaphic properties on AOA and AOB communities in soils with a common type of natural vegetation, and within a common climatic region. Furthermore, database surveys have revealed a strong biogeographical component to the distribution of AOA and AOB phylotypes [24,25]. But, the environmental characteristics affecting the biogeography of AOA and AOB are ill-defined, in part because the majority of studies on soil AOA and AOB communities have focused on soils from temperate regions, and relatively little on other biomes, such as the tropics. Studies of nitrifier communities in tropical soils could add much needed information on the biogeography of AOA and AOB. Moreover, a deeper understanding of nitrification in soils of the tropics is important as N cycling in these soils impacts food production for a large part of the world, and the dynamics of global climate change.

Thus, goals of the present study were to: 1) elucidate the structure of AOA and AOB communities in tropical soils from Trinidad, 2) identify edaphic drivers of variation in their composition and 3) determine the relation between nitrification levels, edaphic characteristics and AOA/AOB communities. The soils ranged from pH 4.8 to 8.2, encompassed a range of characteristics and were all sampled from non-managed locations under natural vegetation (primarily grasses) so that anthropogenic impacts were minimized. Soils were characterized for nitrification potential and a wide range of physico-chemical characteristics, while AOA and AOB communities were evaluated for abundance *via* quantitative PCR of bacterial *amoA* and archaeal *amoA*, and community structure assessed by *amoA* terminal restriction fragment analyses. Our hypotheses were: 1) soil pH would be the major factor affecting nitrifier communities, with acidic environments selecting for greater abundance of AOA and AOA phylotypes unique to low pH soils, and 2) nitrification levels would be correlated to pH as well as the abundance of AOA in acidic soils, and to abundance of AOB in alkaline soils.

Materials and Methods

Soil sampling, physical and chemical properties

Soils were collected from nine locations in Trinidad (Fig. 1, Table S1 in File S1). All soils were obtained from public land, for which no permissions were required for sampling, and no impacts on endangered species. At each location, three core samples (2.5 cm diameter × 10 cm length) were obtained and the top 0–10 cm were removed for analysis. Samples were composited, mixed, sieved (4.75 mm mesh size) and subsamples (20 g each) were taken for moisture content determination by weight loss after drying (105°C, 24 h) and for physical and chemical analyses (three subsamples per soil, 50 g each). Soil samples used for the nitrification potential assay (see below) were used immediately after field sampling. Soil pH was determined by the slurry method (1:5, w:v; sample:distilled deionized water) measured with an Eijkelkamp pH/mV/EC/Salinity/T/02m (Agrisearch Equipment ZG Giesbeek, the Netherlands). Total carbon (TC), organic carbon (OC) and inorganic carbon were determined by dry combustion using a LECO CNS-2000 analyzer. Total N (TN) was determined by combustion and by the Kjeldahl digestion method (TKN). Nitrate-N was measured in soil water extracts colorimetrically after reaction with phenoldisulphonic acid, and ammonium-N was determined by flow injection analysis. Major and minor trace elements were determined by inductively-coupled mass spectrometry. Cation exchange capacity (CEC) was determined by ammonium acetate extraction preceded by cation quantification *via* atomic absorption spectrometry.

Nitrification potential

Nitrification potential (NP) was determined using the shaken slurry procedure [26]. For each sample, 5 g soil was added to 100 mL assay solution (1 mM phosphate buffer adjusted to match pH of each soil, 800 μM ammonium) and then the flasks were loosely covered and incubated with mixing (150 rpm on a bench top orbital shaker) for 24 h, at 29°C, a temperature which approximated that of the soil. The experiment was conducted with 3 replicates per soil in a 9×3 randomized design. During incubation, a sterile pipet was used to remove 10 mL aliquots at 15 min, 2 h, 8 h and 24 h intervals. These samples were centrifuged (10,000× *g*, 10 min) and the supernatant were analyzed for nitrite and nitrate [27]. Nitrite and nitrate concentrations in the 15 min interval were used as baseline measurements for which nitrite and nitrate concentrations in all later time points were corrected.

DNA extraction, qPCR and *amoA* terminal restriction fragment analyses

The Power Soil DNA Isolation kit (MO BIO Laboratories Inc., West Carlsbad, CA) was used for DNA extraction and post-extraction clean-up was done with the Power Clean DNA Clean up kit (MO BIO Laboratories), with both procedures following the manufacturer's instructions. For each soil, triplicate 1 g soil sample extractions were done in batches of 0.25 g, giving 12 separate DNA extracts. These extracts were then pooled to give three samples (each the equivalent of one gram soil) and DNA concentrations were quantified fluorometrically by using a Quibit fluorometer (Invitrogen, Carlsbad, CA).

Each of the three replicate DNA extracts was analyzed by qPCR as described previously [1] using iQ SYBRgreen Supermix (BioRad Laboratories Inc., Hercules CA) and a MyiQ Real-Time qPCR detection system (Bio-Rad). Primer pairs used for qPCR of archaeal *amoA* were amoA_AF, amoA_AR [1], and bacterial *amoA* was amplified with primer pairs amoA_BF, amoA_BR [1]. The qPCR reactions for *amoA* quantification were done following the protocol of De Gannes et al. [1]. The qPCR analyses of archaeal 16S rRNA genes utilized primers 967F and 1060R [28] while qPCR of bacterial 16S rRNA genes was done with primers 338F and 518R [28]. Methods for qPCR of 16S rRNA genes followed those of Karlsson et al. [28]. All qPCR assays included a melting-curve protocol for analysis of primer specificity. Standard curves for qPCR were done with cloned segments of the gene of interest, with efficiencies of 91% to 100%, and r^2 values of 98% to 99.6%. Spiking experiments in which a dilution series of standard template was prepared in the sample matrix were done to verify the absence of PCR inhibitors.

Terminal restriction fragment (TRF) analyses utilized 6FAM-labelled forward primers for archael *amoA* and bacterial *amoA* and Phusion High Fidelity Master Mix with HF buffer (New England Biolabs, Ipswich, MA). Bacterial *amoA* was amplified using the primers and assay conditions of Horz et al. [29] while PCR of

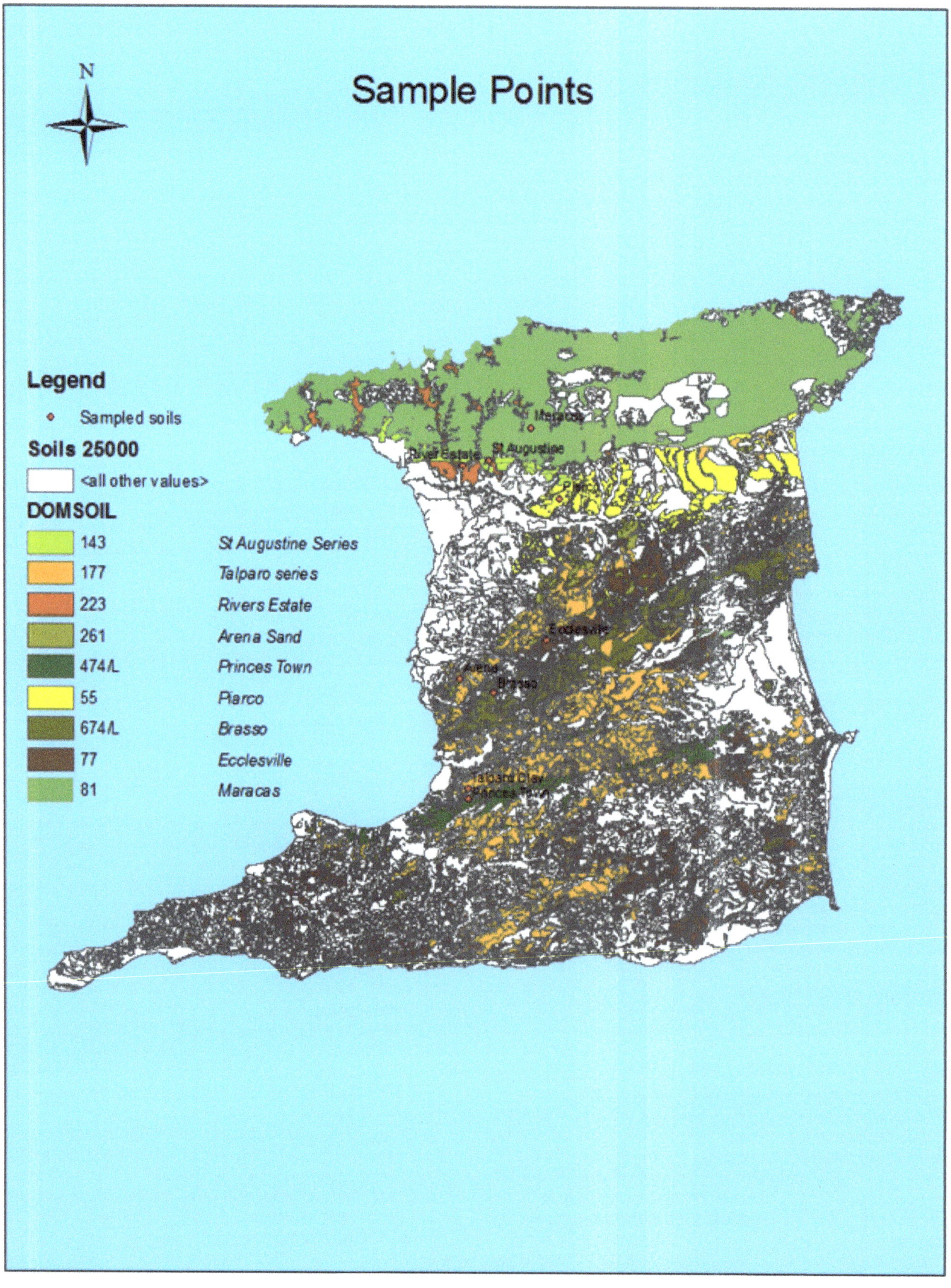

Figure 1. Map of Trinidad showing distribution of the soil series analyzed and the sampling locations.

archaeal *amoA* was done with the primers and conditions of Francis et al. [30]. Triplicate PCR were done for each soil by using an Eppendorf MasterCycler (Eppendorf, Hauppauge, NY), the PCR products were pooled, examined by electrophoresis, purified (QIAquick PCR Purification Kit; QIAGEN, Germantown, MD) and DNA concentrations determined fluorometrically as described above. Aliquots (*ca.* 500 ng) were then digested (3 h, 37°C) with either RsaI or MspI (New England Biolabs). Preliminary tests were done with these soils evaluating fragment generation by RsaI, MspI, HhaI and AluI; RsaI and MspI generated the greatest number of fragments, and were selected for use in the full study. DNA Fragments were precipitated by addition of ice cold 100% ethanol and overnight incubation at −20°C. Analysis of digests was done with an Applied Biosystems 3730xl DNA analyzer (Applied Biosystems, Foster City, CA). A preliminary test was done to determine DNA mass loads that were within the optimal range of quantification for the instrument, and then each sample was analyzed in triplicate at the optimized mass loading. This procedure was adopted from preliminary tests examining variability at each stage of the TRF workflow, which showed that the fragment analysis was the most significant source of error. Fragments of <50 fluorescence units were removed from the analysis, and baseline noise were eliminated by an interactive procedure similar to that of Rees et al. [31] that first excluded fragments <1% of the total peak area, followed by re-normalization and elimination of fragments constituting <5% of the total peak area. The replicate profiles were aligned and TRF profiles of averaged fragment abundance were developed for each sample.

All analyses of TRF data were done with the multivariate statistical software package Primer v. 6 (Primer-E Ltd, Plymouth, UK). A Bray-Curtis coefficient was used to generate similarity matrices, which were ordinated by non-metric multi-dimensional scaling (MDS) computed with 100 random restarts. Analysis of similarity (ANOSIM) was applied to assess the magnitude and statistical significance of dissimilarity between soils in TRF profiles. The BEST (Bioenvironmental Step) routine was applied to identify high rank correlations between the TRF similarity matrices and a matrix of all edaphic factors determined in this study. The BEST analysis was configured to include up to six variables in generating correlations, and run 999 permutations for testing correlation significance.

Construction of AOA and AOB clone libraries and bioinformatics

For all soils, aliquots of each of the replicate DNA extracts were pooled, and an aliquot of the pooled extract was used in PCR Amplification of archaeal *amo*A and bacterial *amoA* was done as described above for TRF analyses, except that the forward primers were not labelled. The PCR products were then purified using the QIAquick PCR Purification Kit (QIAGEN), cloned with the Zero Blunt TOPO system (Invitrogen), transformed into *E.coli* JM 109 competent cells (Promega, USA) and transformants were selected by plating on Luria Bertani medium containing ampicilin and X-gal (each at 100 μg mL^{-1}). Sixteen clones were selected from each soil for sequencing. Plasmid templates in these clones were amplified by using the TempliPhi system (GE HealthcareLife Sciences, Pittsburgh, PA), and then inserts sequenced by using an M13 primer and the BigDye Terminator Cycle Sequencing system (Applied Biosystems). Reactions were analyzed with an Applied Biosystems 3730xl DNA analyzer.

Raw sequence files were imported into Geneious Pro 5.4 (Biomatters Ltd., Auckland 1010, New Zealand) for manual curation and translation. Libraries were examined for potential chimeric sequences with UCHIME (Edgar et al. 2011; http://www.drive5.com/uchime/) and sequence identities were determined by searching GenBank with the BLAST-N web server (http://www.ncbi.nlm.nih.gov/). Sequences were aligned with MUSCLE (Edgar 2004; http://www.drive5.com/uchime/). Phylogenetic trees were constructed by the neighbour-joining method and the Jukes-Cantor distance model, and were created with Geneious Pro v. 5.4.5. The bootstrap re-sampling method was used with 1000 replicates and minimum bootstrap values of 50. Phylogenetic trees were illustrated with FigTree (v. 1.3.1, http://tree.bio.ed.ac.uk/software/figtree/). Distance matrices were created with PHYLIP v. 3.69 (http://evolution.genetics.washington.edu/phylip.htm) and assignment of operational taxonomic units (OTU) and rarefaction analyses were done with Mothur (http://www.mothur.org, [32]). Principle coordinate analyses (PCoA) were done with Fast Unifrac *via* webserver (http://bmf.colorado.edu/fastunifrac/).

Statistical analyses

Analysis of variance (ANOVA) was used to test for significant differences in nitrification potential between soils at the 95% confidence level with Genstat 13 (VSN International, Hemel Hempstead, HPI IESUK). Data from nitrification potential assays was examined for outliers by Grubbs' Test by using an online calculator (www.graphpad.com/quickcalcs/grubbs2). All other analyses were done by using Prism v. 5.4.5 (Graphpad Software, La Jolla, CA).

Sequence accession numbers

Representative sequences of each archaeal *amoA* and bacterial *amoA* OTU are deposited in Genbank under accessions KF888663-KF888698 for archaeal *amoA* and KF888699-KF888724 for bacterial *amoA*).

Results and Discussion

Soil physico-chemical characteristics and nitrification potential

All soils were sampled from non-managed locations vegetated primarily by grasses *viz*. Elephant grass (*Pennisetum purpureum*) and Fowl Foot grass (*Eleusine indica*) as the predominant species. The exception was the Arena sandy loam which was sampled from a preserve of seasonal evergreen forest, which was originally predominant in Trinidad. The soils developed from a variety of different parent materials (Table S1 in File S1), ranged in texture from sandy loam to clay, spanned pH 4.8 to 8.2 and varied in a number of other physico-chemical characteristics (Tables 1, 2). Five acidic soils (pH<6) spanned textural classes, two circum-neutral soils (pH 6.2, 6.6) were loams, and two alkaline soils (pH 7.8, 8.2) were clays.

The three clay soils were vertisols, a defining characteristic of which is a high content of the swelling phylosilicate, montmorillonite. The remainder of the soils either had non-swelling kaolinite as the main phylosilicate mineral or contained relatively little clay (Table 1). Clay content and pH were significantly correlated ($p = 0.0245$) while pH and TC levels were not. Total carbon levels ranged from 8.90 g kg^{-1} soil to 39.13 g kg^{-1} soil, but TC was not correlated with clay content ($p = 0.3317$). The CEC was strongly correlated with clay content ($p = 0.0003$), but not levels of soil OC ($p = 0.3438$). Total N content and TC contents were strongly correlated ($p<0.0001$), but neither parameter was correlated with levels of either ammonium-N or nitrate-N.

Nitrification potential (NP) differed significantly among soils (ANOVA $p<0.001$, Table 3) with the Arena sandy loam being lowest at 0.6 mg N kg^{-1} soil d^{-1} and Brasso being highest at

Table 1. Characteristics of soils used in this study[a].

Soil Name	Texture	pH	Clay	Sand	Silt	TOC	TC	TN	TKN	NH_4-N	NO_3-N	BD	CEC
			g/kg							mg/kg		Mg/m^3	cmol(+)/kg
Arena	Sandy Loam	5.4	4.57	74.00	21.43	6.1	8.90	0.65	0.7	16.4	2.1	1.78	0.3
Ecclesville	Silty Loam	4.8	21.81	34.2	43.99	26.2	3.91	3.46	3.2	46.9	56	1.54	8.8
Piarco	Silty Loam	4.8	17.71	32.00	50.28	13.6	17.87	1.56	1.5	20.5	6.65	1.58	3.1
Maracas	Loam	4.8	10.96	47.32	41.72	13.6	15.89	1.39	1.5	69.5	<0.01	1.56	0.9
River Estate	Loam	6.6	11.79	49.54	38.67	7.9	10.81	0.97	0.9	3.6	0.01	1.56	2.3
St. Augustine	Loam	6.2	19.76	39.16	41.08	24.5	24.75	1.58	1.5	16.4	35.6	1.52	5.6
Brasso	Clay	8.2	57.17	15.21	27.61	16.5	25.46	2.37	1.8	27.1	73.0	1.22	35.3
Princes Town	Clay	7.8	52.52	14.59	32.89	15.5	21.19	1.76	1.1	24.8	24.0	1.24	17.1
Talparo	Clay	5.8	46.71	17.32	35.97	21.3	29.81	2.69	2.3	52.0	46.1	1.28	29.1

[a]TOC, Total Organic Carbon; TC, Total Carbon; TN, Total Nitrogen; TKN, Total Kjeldahl Nitrogen; BD, Bulk Density; CEC, Cation Exchange Capacity.

14 mg N kg^{-1} soil d^{-1}). Rates were linear over the incubation period with $r^2 = 0.952$ to 0.998. Within the clays, there was a clear impact of pH on NP, as the rate in the pH 8.2 Brasso soil was nearly three times greater that in the pH 5.8 Talparo soil (Table 3). The high NP of the Brasso soil was a significant outlier (Grubbs Test $p<0.01$) from that of the other soils, and was thus excluded from the correlation analyses. The NP rates had a significant correlation ($p<0.0001$) with pH (Table 4) and with edaphic characteristics relating to soil texture/structure including: clay content, sand content, bulk density and cation exchange capacity (Table 4). The elemental composition also strongly correlated with NP, as only one element (P) of the 11 determined, lacked a signification relation to NP (Table 4). All of the carbon and nitrogen parameters measured were strongly correlated to NP, except total organic carbon and ammonium-N, which were not significant (Table 4). The significant positive correlation between NP and soil nitrate concentrations could indicate that variation in soil nitrate levels was at least in part attributable to varying rates of nitrification.

The positive correlation of NP rates with soil pH observed in the present study data was consistent with the majority of the literature [33,34,35,36]. But, our study also demonstrated that clay content had a significant factor modulating the pH effect. For example, the Arena sandy loam and Talparo clay were two acidic soils of roughly similar pH (5.4 and 5.8, respectively), yet NP in the montmorrillonic clay soil was *ca.* eight-fold greater than that of the sandy loam. Prior investigators determined that ammonia oxidation by AOB pure cultures was stimulated by expanding clays (montmorillonite) but not non-expanding clays, such as kaolinite [37,38]. Jiang et al. [39] demonstrated that amending a kaolinitic Oxisol with montmorillonite greatly increased NP as well as abundance of both AOA and AOB. For both pure culture and soil studies, the stimulatory effect of montmorrillonite was hypothesized to reflect the ability of this high-CEC mineral to enable localized exchange of ammonium and protons and thereby buffer acidity [37]. Our study builds on prior findings from laboratory experiments, and illustrated with soil field samples the potential impact on nitrification of naturally existing levels of expanding clays like montmorillonite.

Microbial community analysis by qPCR

The majority of the samples (93%) yielded amplification of archaeal *amoA*, which ranged in abundance from 300 copies (g soil)$^{-1}$ in the Maracas loam to 11830 copies g^{-1} soil in the Arena sandy loam (Fig. 2A). The abundance of archaeal *amoA* showed

Table 2. Soil Elemental Properties[a].

Soil Name	P	K	Ca	Mg	S	Zn	B	Mn	Fe	Cu	Al	Na
	g/kg					mg/kg						
Arena	0.03	0.1	0	0.04	0.1	2.77	<2	4.31	1046.9	<0.5	642.9	17.6
Eccelville	1	2.8	5	3.5	0.4	112.2	<2	777.73	24453.3	32.13	22058.6	125.5
Piarco	0.9	1.2	2	0.3	0.2	41.45	<2	48.55	4724.8	14.35	12731.3	160.1
Maracas	0.5	0.6	0	0.5	0.2	23.87	<2	392.56	26648	34.76	11909.5	42.7
River Estate	0.3	0.8	1	0.9	0.1	28.23	<2	127.27	10617.5	11.1	6599.8	110.2
St.Augustine	1	1.9	2	0.4	0.2	126.3	<2	333.66	20695.8	34.49	12277.3	245.4
Brasso	0.9	4.5	10	5	0.5	130	5.48	604.38	34142.7	34.37	46458.4	198.6
Princes Town	0.5	4.1	15	8.2	0.3	165	8.2	977.84	31128.9	44.68	54124	204.5
Talparo	1.9	4.6	9	4.2	0.5	158.7	4.96	624.97	33908.2	28.45	45810.3	207.2

[a]Abbreviations: P = Phosphorus; K = Potassium; Ca = Calcium; Mg = Magnesium; S = Sulphur; Zn = Zinc, B = Boron; Mn = Manganese; Fe = Iron, Cu = Copper; Al = Aluminium; Na = Sodium.

Table 3. Nitrification Potential[a].

Soil	pH	mg N kg^{-1} soil d^{-1}
Arena	5.4	0.7
Ecclesville	4.8	2.1
Piarco	4.8	0.6
Maracas	4.8	1.1
River Estate	6.6	2.4
St. Augustine	6.2	2.7
Brasso	8.2	14.1
Princes Town	7.8	6.6
Talparo	5.8	5.4

[a]N = NO_3-N+NO_2-N, $p<0.001$, LSD (5%) = 0.17, SEM = 0.06.

significant negative correlations with: ammonium concentrations ($r^2 = 0.2504$, $p = 0.0079$), OC levels ($r^2 = 0.2148$, $p = 0.0149$) and clay content ($r^2 = 0.1589$, $p = 0.0394$). Archaeal *amoA* abundance was also inversely related to total N and total carbon ($p = 0.0569$ and 0.0789, respectively). But, there was no significant correlation between archaeal *amoA* abundance and NP ($p = 0.4371$) or pH ($p = 0.8794$). In contrast to archaeal *amoA*, amplification of bacterial *amoA* occurred in only 33% of the samples (Table S2 in File S1). For most soils, amplification occurred in either just one of the three replicates or none of these (Table S2 in File S1). For example, the highest bacterial *amoA* abundance was 12,010 copies g^{-1} soil obtained from one of the three Piarco silty loam replicates, while the other two Piarco soil replicates did not yield bacterial *amoA* signals that were within the quantification limit (Table S2 in File S1). The low frequency of detection for bacterial *amoA* precluded assessing relationships between AOB abundance and soil properties. It should be noted that ammonia-oxidzers could exist that are not detected by currently available PCR primers.

Table 4. Correlations between nitrification potential and edaphic properties[a].

Property[b]	*p* value	r^2
pH	<0.0001	0.5929
clay	<0.0001	0.8888
sand	<0.0001	0.5756
silt	0.3230	0.0444
TOC	0.1161	0.1085
Total C	0.0031	0.3342
Total N	0.0897	0.1253
TKN	0.6111	0.0119
NH_4-N	0.7743	0.0038
NO_3-N	0.0172	0.2319
BD	<0.0001	0.8612
CEC	<0.0001	0.7205
P	0.1897	0.0822
K	<0.0001	0.7887
Ca	<0.0001	0.8573
Mg	<0.0001	0.7965
S	<0.0019	0.3623
Zn	<0.0001	0.7404
Mn	<0.0001	0.5508
Fe	0.0001	0.4576
Cu	0.0012	0.3850
Al	<0.0001	0.8560
Na	0.0006	0.4206

[a]Correlations exclude the Brasso soil;
[b]See Table 1 for abbreviations.

Genes for bacterial 16S rRNA and archaeal 16S rRNA were amplified from all replicates of all soils. Archaeal 16S rRNA gene numbers ranged from 9.24×10^3 copies g^{-1} soil in the Princes Town clay to 2.85×10^6 copies g^{-1} soil in the Piarco silty loam (Fig. 2B). There were no significant correlations with archaeal 16S rRNA gene abundance and any soil properties. Bacterial 16S rRNA copies were *ca.* 10^6-fold greater than archaea. Bacterial 16S rRNA gene numbers ranged from 8.1×10^8 copies g^{-1} soil in the Princes Town clay to 3.5×10^{11} copies g^{-1} soil in the Piarco silty loam (Fig. 2C), the same two soils that showed minimum and maximum archaeal 16S rRNA gene abundance. There were no significant correlations with bacterial 16S rRNA gene abundance and any soil property, and there was no correlation between NP

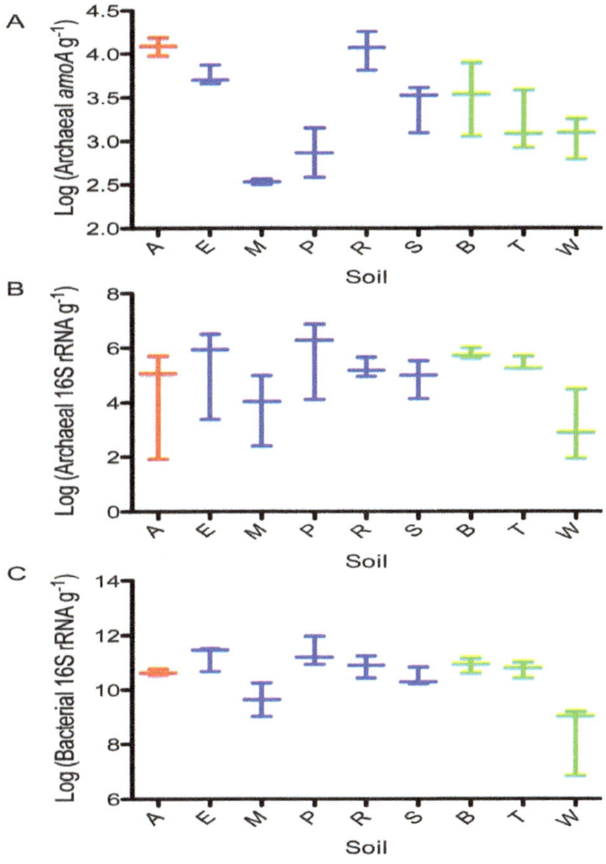

Figure 2. Box and whisker plots of gene abundance determined by qPCR. Panel A, Archaeal *amoA*; Panel B, Archaeal 16S rRNA; Panel C, Bacterial 16S rRNA. Symbol colors correspond to soil types: red, sandy loam; blue, silty loam; green, clay. Soil name abbreviations are: A, Arena; B, Brasso; E, Ecclesville; P, Piarco; R, River Estate; S, St. Augustine; T, Talparo; W, Princes Town.

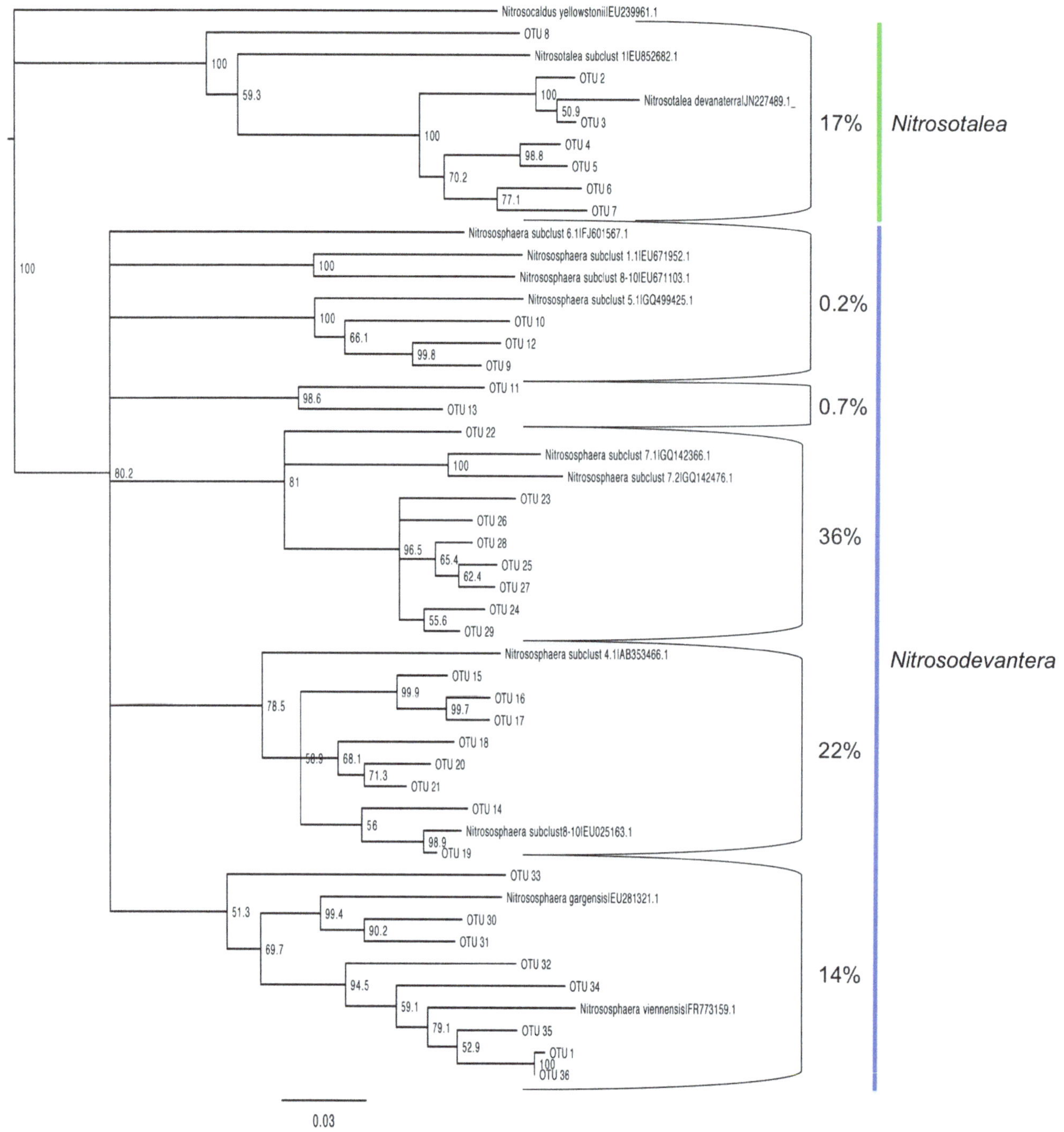

Figure 3. Neighbour joining consensus tree of archaeal *amoA* phylotypes. Colored branches indicate the *Nitrosotalea* cluster (green) and *Nitrososphaera* cluster (blue). Boot strap values are indicated at nodes. Cluster designations follow the classification of Pester et al. [25] based on nucleic acid alignment of archaeal *amoA*.

and the abundance of 16S rRNA genes for either archaea or bacteria.

The negative relation of AOA abundance with ammonium levels was in accordance with similar findings by prior investigations indicating a preference of soil AOA for low ammonium levels [18,21,40,41,42,51]. Since all soils in the present study were from non-managed locations, the soil N dynamics were not impacted by N inputs derived from agricultural practices (*e.g.*, mineral or organic N fertilizers, tillage, grazing). Therefore, differences in AOA abundance that arose from variation in ammonium levels resulted from N cycling processes intrinsic to the soils. Mineralization would be a major mechanism for ammonium input, and the present study thus supported the concept that a key factor favouring soil habitation by AOA *vs.* AOB is the predominance of mineralization as the N input pathway [12,43].

The negative correlation of AOA abundance to OC levels could have reflected the preference of AOA for low organic nutrient conditions [44,45]. But, soil organic matter affects a wide range of

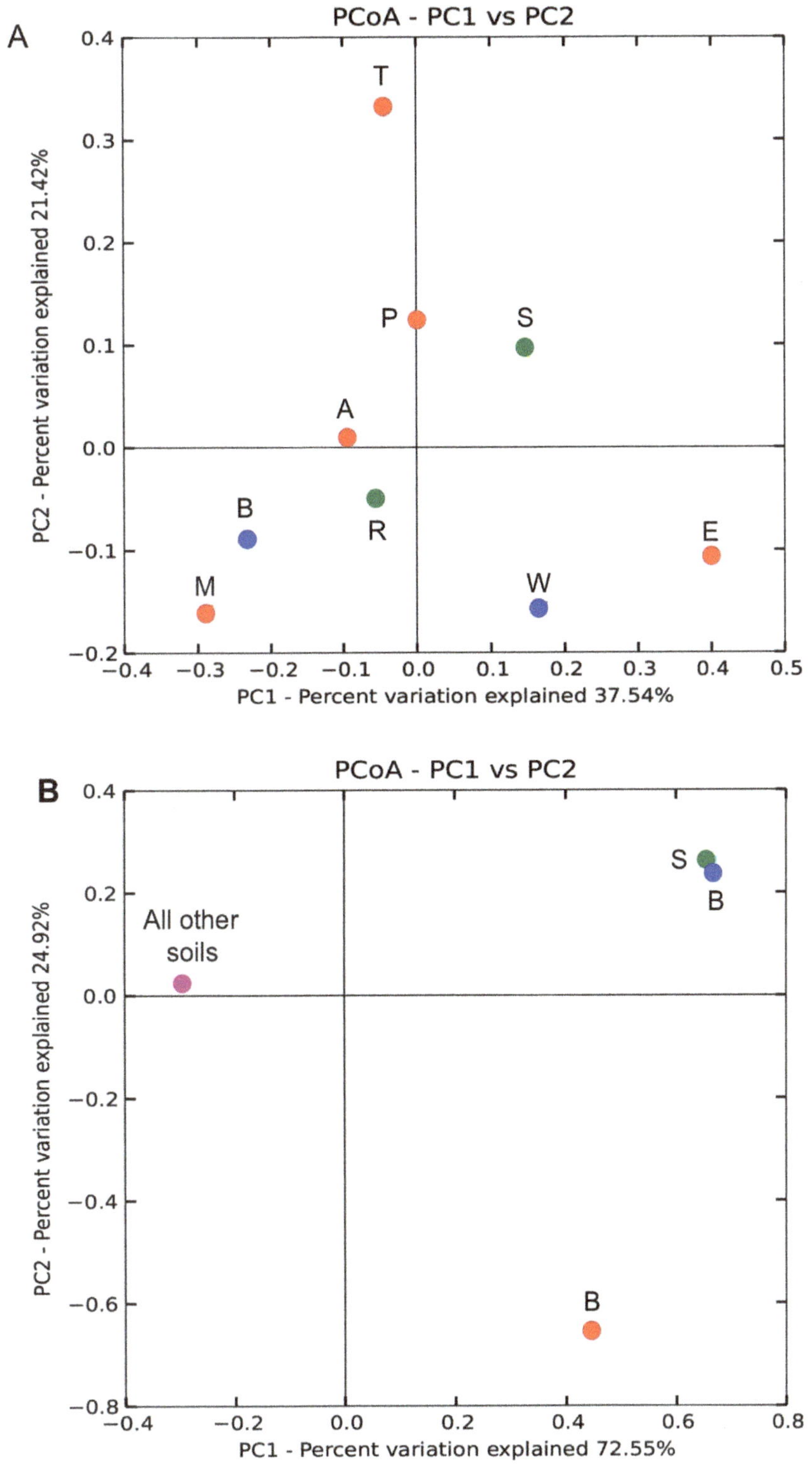

Figure 4. Principle coordinate plot of archaeal *amoA* clone libraries (Panel A) and bacterial *amoA* clone libraries (Panel B). Symbol colors correspond to soil pH: red, acidic (pH<6); green, circum-neutral (pH 6.2, 6.6), blue, alkaline (pH 7.8, 8.2), purple, multiple pH. Letters indicate soil names that are abbreviated as: A, Arena; B, Brasso; E, Ecclesville; P, Piarco; R, River Estate; S, St. Augustine; T, Talparo; W, Princes Town.

soil physico-chemical properties (*e.g.*, bulk density, water-holding capacity, aeration, *etc.*) that also influence microbial communities, and one or more of these broad soil properties cannot be ruled out as exerting effects on AOA. Similarly, clay minerals have many impacts on soil characteristics that affect microbial activity, which complicates interpretation of the negative correlation observed here of clay content with AOA abundance. In a survey of an organic farm, Wessen and colleagues [21] determined a negative

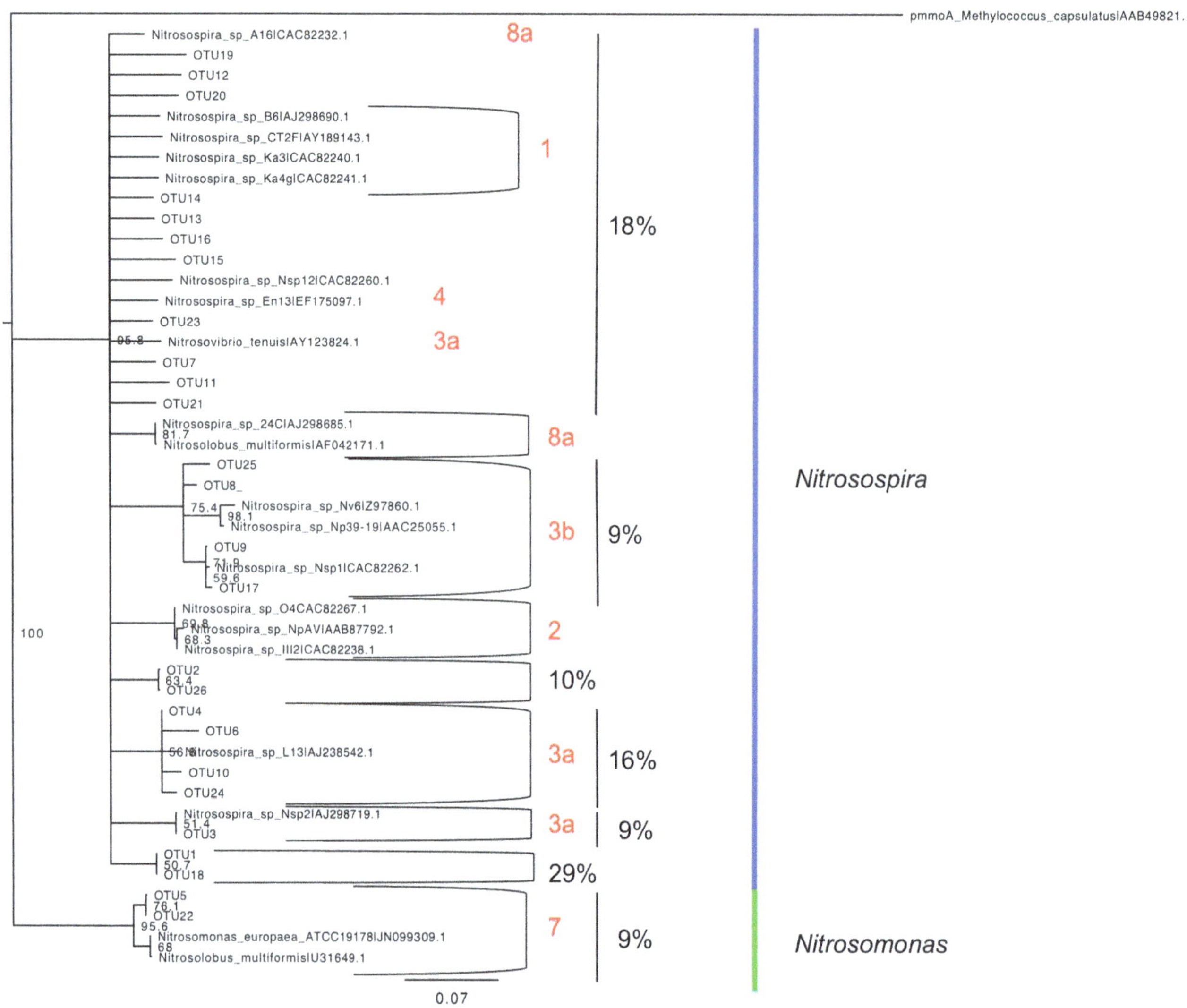

Figure 5. Neighbor joining consensus tree of bacterial AmoA phylotypes. Colored branches indicate the *Nitrosomonas* cluster (green) and *Nitrosospira* cluster (blue). Boot strap values are indicated at nodes. Values in red are cluster designations following the classification of Avrahami et al. [48] based on amino acid alignment of bacterial AmoA.

correlation of AOA with clay content, which they theorized to reflect increased binding of ammonium, implying reduced ammonium bioavailability. But, the potential for clays to impede AOA growth *via* ammonium binding is unclear, as even ammonium tightly bound between interlayers of expanding clays ("fixed ammonium") can be readily accessed by nitrifiers [46]. Also, as described above, expanding clays could have a stimulatory effect on nitrification. Thus, while a clay-associated inhibitory effect cannot be ruled out, we hypothesize that these minerals could exert a negative impact on AOA abundance by their effects on broader physico-chemical characteristics of soil (*e.g.*, bulk density, water-holding capacity, aeration, *etc.*).

The AOB populations in these tropical soils were low, and at levels not readily assayed by qPCR. Similarly low AOB populations have been reported by other investigators [7,43,47]. For example, Pett-Ridge and co-workers [47] were unable to amplify bacterial *amoA* from soils of a tropical forest and, in a survey of soils across Scotland, AOB were below qPCR detection limits in *ca.* 62% of the 184 sites sampled [17]. Thus, for soils having AOB populations below qPCR detection, AOA have been inferred to be the primary group driving nitrification [7,43,47]. By this reasoning, AOA are implicated as dominant drivers of nitrification in the tropical soils studied here. While AOA appeared to play a key role in nitrification, AOA numbers were not correlated with NP. A lack of correspondence between nitrification levels and nitrifier community density (AOB and/or AOA) has been reported by many other investigators [23,43,47], and the reasons for this disconnection are unknown. But, as mentioned above, it's possible that ammonia-oxidzers exist that are divergent from known betaproteobacterial AOB and thaumarchaeal AOA, and thus evade detection by PCR primers developed from currently available sequence data.

Analysis of community composition by *amoA* sequencing and TRF

Sequencing of archaeal *amoA* clone libraries generated 35 OTUs from a total of 121 sequences (Table S3 in File S1, Fig. S1). Most phylotypes (83%) were identified as *Nitrososphaera* and the

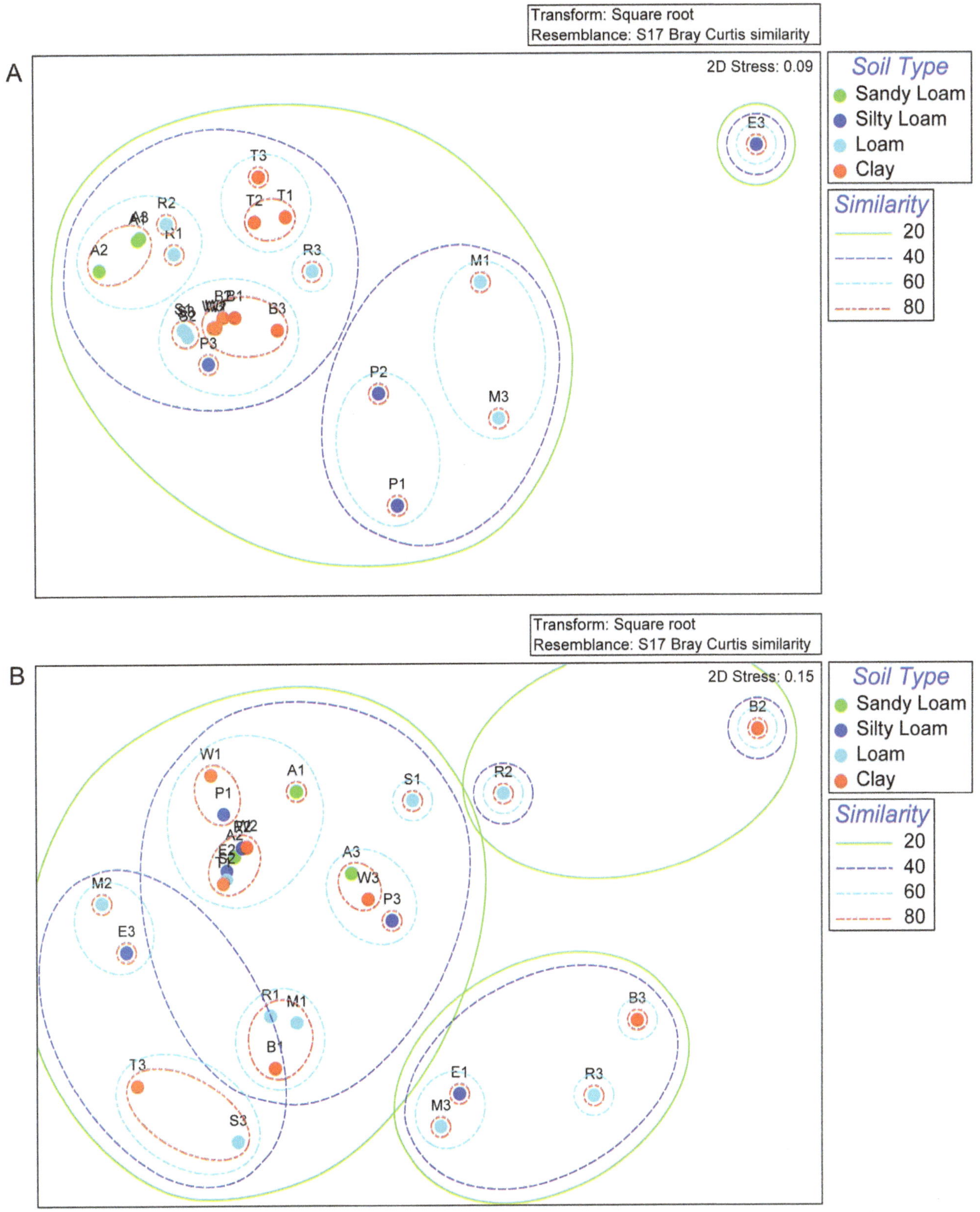

Figure 6. Non-metric multidimensional scaling plots of archaeal *amoA* terminal restriction fragment profiles generated by digestion with MspI (Panel A) or RsaI (Panel B). Symbol colors correspond to soil types, letters indicate soil names and values are the replicate number for the indicated soil. Soil name abbreviations: A, Arena; B, Brasso; E, Ecclesville; P, Piarco; R, River Estate; S, St. Augustine; T, Talparo; W, Princes Town.

Table 5. BEST variables explaining variation between soils in archaeal *amoA* RsaI terminal restriction fragment profiles[a].

No. Variables	Correlation	Variable						
		NH_4	TN	TKN	Zn	TC	pH	P
2	0.630	X		X				
2	0.624	X	X					
3	0.620	X		X	X			
2	0.619	X				X		
3	0.618	X	X		X			
3	0.618	X	X					
3	0.613	X	X				X	
5	0.612	X	X			X	X	X
4	0.611	X	X	X	X			
4	0.609	X		X		X	X	

[a]Variables included in correlation indicated by "X". Abbreviations: P, Phosphorus; Zn, Zinc; TC, Total Carbon; TN, Total Nitrogen; TKN, Total Kjeldahl Nitrogen.

remainder as *Nitrosotalea*. The largest group of sequences (36%) were associated with *Nitrososphaera* subclusters 7.1 and 7.2 (Fig. 3; Table S3 in File S1), and identified in all soils except the Princes Town clay (Table S3 in File S1). Sequences grouping wih *Nitrososphaera* subcluster 4 and *Nitrososphaera* subclusters 8–10 ranked second in abundance accounting for 22% of the sequences (Fig. 3; Table S3 in File S1). The *Nitrosotalea* phylotypes were identified in clone libraries from the St. Augustine, Maracas and River Estate loams, but were most abundant in the latter two (Table S3 in File S1). Analysis of the archaeal *amoA* phylotypes by PCoA showed no significant correlation with pH (Fig. 4A). Furthermore, while all *Nitrosotalea* phylotypes were identified from loams, there was no consistent soil characteristic that explained their more frequent detection in these soils.

Sequencing of bacterial *amoA* clone libraries generated 26 OTUs from a total of 128 sequences (Table S4 in File S1). Most OTU (91%) were *Nitrosospira* phylotypes with the remainder associated with the *Nitrosomonas* (Fig. 5). Of the *Nitrosospira* phylotypes, 27% were affiliated with Cluster 3a and Cluster 3b (Fig. 5), and the remainder did not group within AOB clusters defined by Avrahami and colleagues [48]. Among those not grouping with a defined cluster was the largest phylotype (OTU 1), which comprised 29% of the sequences and occurred in all soils except the St. Augustine loam and Talparo clay (Table S3 in File S1). Analysis of bacterial *amoA* phylotypes by PCoA showed no significant corrrelation of phylotypes with pH (Fig. 4B). While one alkaline soil and one cirum-neutral soil clustered apart from one acidic soil, the rest of the soils, representing a range of pH, did not segregate (Fig. 4B).

In TRF analysis of archaeal *amoA*, RsaI profiles were more effective than those of MspI in differentiating soils by MDS (Figs. 6A,B). The ANOSIM global R of the RsaI profiles was 0.752 ($p<0.001$) indicating significant dissimilarity between soils, whereas for the MspI profiles the ANOSIM global R was 0.058 and not significant. Comparison of the observed TRF phylotypes with those predicted from archaeal *amoA* sequences indicated that the majority of TRF mapped to the above-mentioned *Nitrososphaera* phylotypes. One *Nitrosotalea*-affiliated TRF (RsaI 315) was most abundant in the River Estate soil, which also had the highest proportion of *Nitrosotalea*-like archaeal *amoA* clones. For archaeal *amoA*, BEST analysis identified the combination of ammonium levels and total Kjeldahl N (TKN) as most important in explaining variation in the RsaI profiles (correlation 0.630, $p<0.001$, Table 5). Nitrogen-related variables were a dominant factor, as ammonium appeared in all ten of the significant combinations, and was accompanied by TKN and/or TN in all but one. Four other factors (Zn, TC, pH, P) were included in some explanations with these N variables, but none yielded a more significant correlation with TRF similarity patterns than did the combination of ammonium with either TKN or TN (Table 5).

In the TRF analysis of bacterial *amoA*, soil differentiation was more effective with MspI profiles than with those of RsaI (Figs. 7A,B). The former had an ANOSIM global R of 0.819 ($p<0.001$) indicating significant dissimilarity between groups. But, the RsaI ANOSIM Global R of 0.058 was not significant. From BEST analysis of the bacterial *amoA* MspI profiles, TRF phylotype variation across the soils was best explained by phosphorus (P), bulk density (BD) and Fe (correlation 0.648, $p<0.001$, Table 6). These variables were dominant factors explaining MspI TRF profile variation, as P occurred in all ten correlations, BD in nine and Fe in eight (Table 4). Aluminium was also prominent, and was included in five correlations (Table 6). Other significant factors were Cu, Na, clay, S, B, OC, and K. Notably, neither pH nor any N-related variable was included in a significant correlation.

Our analysis of archaeal *amoA* clone libraries was consistent with prior reports establishing *Nitrososphaera* as a cosmopolitan soil inhabitant [5,7,49]. But, our identification of *Nitrosotalea* phylotypes in soils of widely varying pH (and other edpahic factors) contrasted with prior studies, which have indicated *Nitrosotalea* as associated primarily with acidic soils [25]. Furthermore, PCoA analyses of archaeal *amoA* clone libraries did not support the hypothesis that pH was the primary factor in AOA phylotype selection, and contrasted with results from prior investigators [3,7,50]. For bacterial *amoA* clone libraries, predomination by *Nitrsosospira* was congruent with the findings of other investigators [23,48,49]. But, as with AOA, the data did not confirm the hypothesized effect of pH as a major driver of AOB community structure.

Collectively, results from the BEST and qPCR analyses indicated that communities of AOA and AOB were shaped by differing sets of edaphic factors. The study conducted by Wessen et al. [52] revealed a similar conclusion, although specific soil characteristics that differentially affected these communities were not identified. In the present study, soil N characteristics

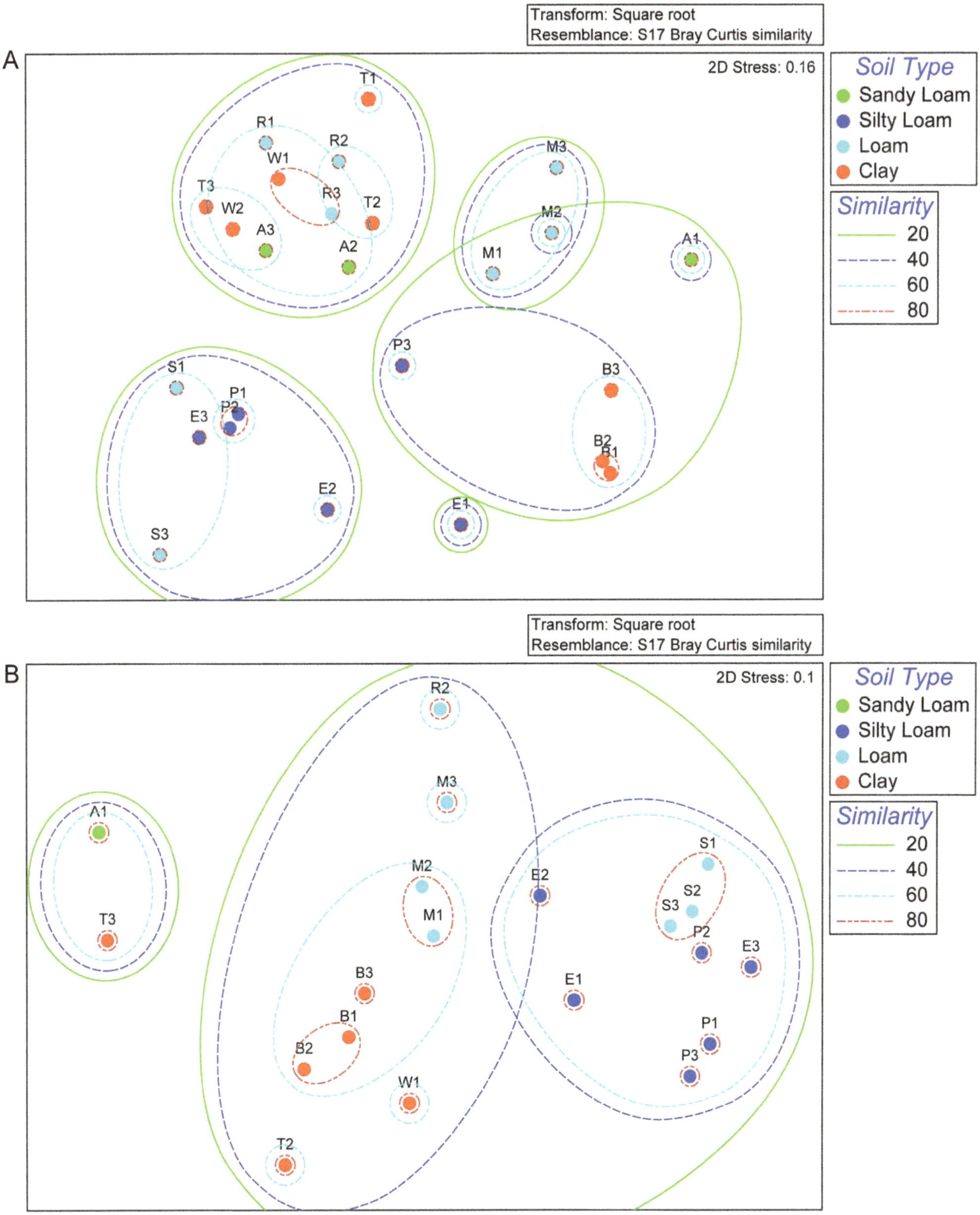

Figure 7. Non-metric multidimensional scaling plots of bacterial *amoA* terminal restriction fragment profiles generated by digestion with MspI (Panel A) or RsaI (Panel B). Symbol colors correspond to soil types, letters indicate soil names and values are the replicate number from the indicated soil. Soil name abbreviations: A, Arena; B, Brasso; E, Ecclesville; P, Piarco; R, River Estate; S, St. Augustine; T, Talparo; W, Princes Town.

Table 6. BEST Variables explaining variation between soils in bacterial *amoA* MspI terminal restriction fragment profiles[a].

No. Variables	Correlation	Variable[a]										
		P	BD	Fe	Al	Cu	Na	Clay	S	B	OC	K
3	0.648	X	X	X								
3	0.632	X		X	X							
4	0.623	X	X	X	X							
2	0.621	X	X									
6	0.619	X	X	X			X	X	X			
5	0.618	X	X	X	X	X						
5	0.617	X	X	X		X				X		
4	0.616	X	X		X	X						
5	0.612	X	X	X	X						X	
6	0.609	X	X	X			X	X				X

[a]Variables included in correlation indicated by "X". Abbreviations: P = Phosphorus; K = Potassium; S = Sulphur; B = Boron; Mn = Manganese; Fe = Iron, Cu = Copper; Al = Aluminium; Na, Sodium; BD, Bulk Density; OC, Organic Carbon.

(ammonium-N and organic N pools) were the major factor affecting AOA communities, but were not identified to affect those of AOB. This finding was significant as it provided field-based evidence that AOA ecophysiology was responsive to natural variation in N levels, data which is limited in the current literature. Also, absence of soil N as a factor affecting AOB would be consistent with adaptation of AOB to relatively high N inputs, which were lacking in the soils studied here. If so, our field-based data would support the concept that soil N levels are one factor important in affecting niche separation of AOA and AOB.

For AOB, the dominant edaphic factors affecting community structure were the combination of P, BD and Fe. Soil P level was the only variable identified to affect communities of both AOA and AOB, but it was much more significant for the AOB. Phosphorus levels have been previously identified as affecting AOB communities [19,20,24]. But, those effects emerged in response to P enrichment through exogenous P application. In contrast, in the present study, the effect of P was associated with variation in P levels endogenous to the soils. The eco-physiological significance of the P effect on AOB community structure is unclear. It may reflect a direct impact on some aspect of AOB physiology and/or indirect effects, such as competition for P with other organisms [19,20,24]. The effect of BD on AOB could have reflected one or more impacts of BD on soil characteristics, perhaps most importantly oxygen diffusion. The strong effect of iron on AOB community structure might have reflected the iron-heme dependent respiration of AOB [53,54], and possibly indicated iron acquisition was a limiting factor. The BEST analyses identified Cu and Al as community structure determinants for AOB. Both metals could have impacted community structure *via* toxicity effects, but the significance of Cu could also have reflected its core role in AOB biology as a component of ammonia mono-oxygenase and other enzymes in N metabolism [53,54]. However, since Cu and Al were intercorrelated with Fe levels, it's difficult to resolve metal-specific effects.

Conclusions

Detectable *amoA* genes of AOA predominated over detectable bacterial *amoA* genes in these tropical soils, and their community structure was affected primarily by soil N characteristics. In contrast, communities of the low abundance AOB were shaped by edaphic factors other than soil N. Contrary to our hypothesis, pH was not a major factor affecting abundance or structure of either AOA or AOB communities. Nitrification levels were strongly affected by soil pH and clay content, but not by the abundance of AOA or AOB, or aspects of AOA or AOB community structure. Thus, edaphic factors other than pH differentially affected evolution and niche differentiation of AOA and AOB communities. But, the impact of pH (and clay content) on nitrification levels appeared to more strongly reflect physico-chemical effects on AOA or AOB activity, rather than alterations in ammonia-oxidizer abundance or selection for AOA or AOB phylotypes differing in nitrifying capacity.

Supporting Information

Figure S1 Rarefaction curves for clone libraries of archaeal *amoA* (Panel A) and bacterial *amoA* (Panel B).

File S1 Supporting tables. Table S1. Soil Taxonomy and Soil Sites Sampled. **Table S2.** Amplification of Bacterial *amoA* by qPCR. **Table S3.** Distribution of archaeal *amoA* OTU by soil. **Table S4.** Distribution of bacterial *amoA* OTU by soil.

Acknowledgments

The authors thank Ronald Roopnarine and David Dyer for their assistance with this study. The authors are also grateful to Lisa Stein for many constructive comments that improved this manuscript.

Author Contributions

Conceived and designed the experiments: WJH VdG GE. Performed the experiments: VdG. Analyzed the data: WJH VdG GE. Contributed reagents/materials/analysis tools: WJH VdG GE. Wrote the paper: VdG WJH.

References

1. de Gannes V, Eudoxie G, Dyer DH, Hickey WJ (2012) Diversity and abundance of ammonia oxidizing archaea in tropical compost systems. Front Microbio 3: 1–16.
2. Di HJ, Cameron KC, Shen JP, Winefield CS, O'Callaghan M, et al. (2009) Nitrification driven by bacteria and not archaea in nitrogen-rich grassland soils. Nat Geosci 2: 621–624.
3. Gubry-Rangin C, Nicol GW, Prosser JI (2010) Archaea rather than bacteria control nitrification in two agricultural acidic soils. FEMS Microbiol Ecol 74: 566–574.
4. Treusch AH, Leininger S, Kletzin A, Schuster SC, Klenk H, et al. (2005) Novel genes for nitrite reductase and amo-related proteins indicate a role of uncultivated mesophilic crenarchaeota in nitrogen cycling. Environ Microbiol 7: 1985–1995.
5. He J-Z, Shen J-P, Zhang L-M, Zhu Y-M, Zheng M-G, et al. (2007) Quantitative analyses of the abundance and composition of ammonia-oxidizing archaea of a Chinese upland red soil under long term fertilization practices. Environ Microbiol 9: 2364–2374.
6. Taylor AE, Zeglin LH, Dooley S, Myrold DD, Bottomley PJ (2011) Evidence for different contributions of archaea and bacteria to the ammonia-oxidizing potential of diverse Oregon soils. Appl Environ Microbiol 76: 7691–7698.
7. Yao H, Gao Y, Nicol GW, Campbell CD, Prosser JI, et al. (2011) Links between ammonia oxidizer community structure, abundance and nitrification potential in acidic soils. Appl Environ Microbiol 77: 4618–4625.
8. Zeglin LH, Taylor AE, Myrold DD, Bottomley PJ (2011) Bacterial and archaeal amoA gene distribution covaries with soil nitrification properties across a range of land uses. Environ Microbiol Rps 3: 717–726.
9. Jia Z, Conrad R (2009) Bacteria rather than archaea dominate microbial ammonia oxidation in an agricultural soil. Environ Microbiol 11: 1658–1671.
10. Shen J-P, Zhang L-M, Zhu Y-G, Zhang J-B, He J-Z (2008) Abundance and composition of ammonia-oxidizing bacteria and ammonia-oxidizing archaea communities of an alkaline sandy loam. Environ Microbiol 10: 1601–1611.
11. Gubry-Rangin C, Hai B, Quince C, Engel M, Thomson BC, et al. (2011) Niche specialization of terrestrial archaeal ammonia oxidizers. Proc Natl Acad Sci (USA) 108: 21206–21211.
12. Levicnik-Hofferle S, Nicol GW, Ausec L, Mandic-Mulec I, Prosser JI (2012) Stimulation of thaumarchaeal ammonia oxidation by ammonia derived from organic nitrogen but not added inorganic nitrogen. FEMS Microbiol Ecol 80: 114–123.
13. Hatzenpichler R, Lebedeva EV, Spieck E, Stoecker K, Richter A, et al. (2008) A moderately thermophilic ammonia-oxidizing crenarchaeota from a hot spring. Proc Natl Acad Sci (USA) 105: 2134–2139.
14. Tourna M, Stieglmeier M, Spang A, Konneke M, Schintlmeister A, et al. (2011) *Nitrososphaera viennensis*, an ammonia oxidizing archaeon from soil. Proc Natl Acad Sci (USA) 108: 8420–8425.
15. Hatzenpichler R (2012) Diversity, physiology and niche differentation of ammonia-oxidizing archaea. Appl Environ Microbiol 78: 7501–7510.
16. Prosser JI, Nicol GW (2008) Relative contributions of archaea and bacteria to aerobic ammonia oxidation in the environment. Environ Microbiol 10: 2931–2941.
17. Yao H, Campbell CD, Chapman SJ, Freitag TE, Nicol GW, et al. (2013) Multi-factorial drivers of ammonia oxidizer communities: Evidence from a national soil survey. Environ Microbiol 15: 2545–2556.
18. Di HJ, Cameron KC, Shen JP, Winefield CS, O'Callaghan M, et al. (2010) Ammonia-oxidizing bacteria and archaea grow under contrasting soil nitrogen conditions. FEMS Microbiol Ecol 72: 386–394.
19. He JZ, Shen JP, Zhang LM, Zhu YG, Zheng YM, et al. (2007) Quantitative analyses of the abundance and composition of ammonia-oxidizing bacteria and ammonia-oxidizing archaea of a Chinese upland red soil under long-term fertilization practices. Environ Microbiol 9: 2364–2374.
20. Klumpp K (2008) Effects of aboveground grazing on coupling among nitrifier activity, abundance and community structure. ISME J 2: 221–232.
21. Wessen E, Soderstrom M, Stenberg M, Bru D, Hellman M, et al. (2011) Spatial distribution of ammonia-oxidizing bacteria and archaea across a 44-hectare farm related to ecosystem functioning. ISME J 5: 1213–1225.
22. Ying J-Y, Zhang L-m, He J-Z (2010) Putative ammonia-oxidizing bacteria and archaea in an acidic red soil with different land utilization patterns. Environ Microbiol Rps 2: 304–312.
23. O'Sullivan CA, Wakelin SA, Fillery RP, Roper MM (2013) Factors affecting ammonia-oxidizing microroganisms and potential nitrification rates in southern Australian agricultural soils. Soil Res 51: 240–252.
24. Fierer N, Carney KM, Horner-Devine MC, Megonigal JP (2009) The biogeography of ammonia-oxidizing bacterial communities in soil. Microb Ecol 58: 434–445.
25. Pester M, Rattei T, Flechl S, Gröngröft A, Richter A, et al. (2012) *AmoA*-based consensus phylogeny of ammonia-oxidizing archaea and deep sequencing of *amoA* genes from soils of four different geographic regions. Environ Microbiol 14: 525–539.
26. Hart SC, Stark JM, Davidson EA, Firestone MK (1994) Nitrogen mineralization, immobilization, and nitrification. In Methods of Soil Analysis, Part 2. Microbiological and Biochemical Properties; Weaver RW, Angle S, Bottomley P, Bezdicek D, Smith S et al., editors. Madison, WI: Soil Science Society of America. 985–1018 p.
27. Shand C, Williams B, Coutts G (2008) Determination of N-species in soil extracts using microplate techniques. Talanta 74: 648–654.
28. Karlsson AE, Johansson T, Per Bengtson P (2012) Archaeal abundance in relation to root and fungal exudation rates. FEMS Microbiol Ecol 80: 305–311.
29. Horz H-P, Rotthauwe J-H, Lukow T, Liesack W (2000) Identification of major subgroups of ammonia-oxidizing bacteria in environmental samples by T-RFLP analysis of *amoA* PCR products. J Microbiol Meth 39: 197–120.
30. Francis CA, Roberts KJ, Beman JM, Santoro AE, Oakley BB (2005) Ubiquity and diversity of ammonia-oxidizing archaea in water columns and sediments of the ocean. Proc Natl Acad Sci (USA) 102: 14683–14688.
31. Rees GN, Baldwin DS, Watson GO, Perryman S, Nielsen DL (2004) Ordination and significance testing of microbial community composition derived from terminal restriction fragment length polymorphisms: application of multivariate statistics. Antonie van Leeuwenhoek 86: 339–347.
32. Schloss PD, Westcott SL, Ryabin T, Hall JR, Hartmann M, et al. (2009) Introducing mothur: Open-source, platform-independent, community-supported software for describing and comparing microbial communities. Appl Environ Microbiol 75: 7537–7541.
33. de Boer W, Kowalchuk GA (2001) Nitrification in acid soils: Microorganisms and mechanisms. Soil Biol & Biochem 28: 203–211.
34. Sahrawat KL (1982) Nitrification in some tropical soils. Plant Soil 65: 281–286.
35. Sahrawat KL (2008) Factors affecting nitrification in soils. Commun Soil Sci Plan 39: 1436–1446.
36. Kyveryga PM, Blackmer AM, Ellsworth JW, Isla R (2004) Soil pH effects on nitrification of fall-applied anhydrous ammonia. Soil Sci Soc Am J 68: 545–551.
37. Kunc F, Stotzky(1980) Acceleration by montmorillonite of nitrification in soil. Folia Microbiol 25: 106–125.
38. Powell SJ, Prosser JI (1991) Protection of *Nitrosomonas europaea* colonizing clay minerals from inhibition by nitrapyrin. J Gen Microbiol 137: 1923–1929.
39. Jang X, Maa Y, Yuan J, Wright AL, Li H (2011) Soil particle surface electrochemical property effects on abundance of ammonia-oxidizing bacteria and ammonia-oxidizing archaea, NH_4 activity, and net nitrification in an acid soil. Soil Biol & Biochem 43: 2215–2221.
40. Hofferle S, Nicol GW, Pal L, Hacin J, Prosser JI, et al. (2010) Ammonium supply rate influences archaeal and bacterial ammonia oxidizers in a wetland soil vertical profile. FEMS Microbiol Ecol 74: 302–315.
41. Zhang LM, Offre PR, He J-Z, Verhamme DT, Nicol GW, et al. (2010) Autotrophic ammonia oxidation by soil thaumarchaea. Proc Natl Acad Sci (USA) 107: 17240–17245.
42. Verhamme DT, Prosser JI, Nicol GW (2011) Ammonia concentration determines differential growth of ammonia-oxidising archaea and bacteria in soil microcosms. ISME J 5: 1067–1071.
43. Stopnisek N, Gubry-Rangin C, Hofferle S, Nicol GW, Mandic-Mulec I, et al. (2010) Thaumarchaeal ammonia oxidation in an acidic forest peat soil is not influenced by ammonium amendment. Appl Environ Microbiol 76: 7626–7634.
44. Erguder TH, Boon N, Wittebolle L, Marzorati M, Verstraete W (2009) Environmental factors shaping the ecological niches of ammonia-oxidizing archaea. FEMS Microbiol Rev 33: 855–869.
45. Valentine DL (2007) Adaptations to energy stress dictate the ecology and evolution of the Archaea. Nat Rev Microbiol 5: 316–323.
46. Green CJ, Blackmer AM, Yang NC (1994) Release of fixed ammonium during nitrification in soils. Soil Sci Soc Am J 58: 1411–1415.
47. Pett-Ridge J, Petersen DG, Nuccio E, Firestone MK (2013) Influence of oxic/anoxic fluctuations on ammonia oxidizers and nitrification potential in a wet tropical soil. FEMS Microbiol Ecol 85: 179–194.
48. Avrahami S, Conrad R, Braker G (2002) Effect of soil ammonium concentration on N_2O release and on the community structure of ammonia oxidizers and denitrifiers. Appl Environ Microbiol 68: 5685–5692.
49. Boyle-Yarwood SA, Bottomley PJ, Myrold DD (2008) Community composition of ammonia-oxidizing archaea in soils under stands of red alder and Douglas fir in Oregon. Environ Microbiol 10: 2956–2965.
50. Nicol GW, Leininger S, Schleper C, Prosser JI (2008) The influence of soil pH on the diversity, abundance and transcriptional activity of ammonia oxidizing archaea and bacteria. Environ Microbiol 10: 2966–2978.
51. Webster G, Embley TM, Freitag TE, Smith Z, Prosser PI (2005) Links between ammonia oxidizer species composition, functional diversity and nitrification kinetics in grassland soils. Environ Microbiol 7: 676–684.
52. Wessen E, Hallin S, Philippot L (2010) Differential responses of bacterial and archaeal groups at high taxonomical ranks to soil management. Soil Biol & Biochem 42: 1759–1765.
53. Glass JB, Orphan VJ (2012) Trace metal requirements for microbial enzymes involved in the production and consumption of methane and nitrous oxide. Front Microbio 3:61.
54. Godfrey LV, Glass JB (2011) The geochemical record of the ancient nitrogen cycle, nitrogen isotopes, and metal cofactors. Methods Enzymol 486: 483–506.

PERMISSIONS

All chapters in this book were first published in PLOS ONE, by The Public Library of Science; hereby published with permission under the Creative Commons Attribution License or equivalent. Every chapter published in this book has been scrutinized by our experts. Their significance has been extensively debated. The topics covered herein carry significant findings which will fuel the growth of the discipline. They may even be implemented as practical applications or may be referred to as a beginning point for another development.

The contributors of this book come from diverse backgrounds, making this book a truly international effort. This book will bring forth new frontiers with its revolutionizing research information and detailed analysis of the nascent developments around the world.

We would like to thank all the contributing authors for lending their expertise to make the book truly unique. They have played a crucial role in the development of this book. Without their invaluable contributions this book wouldn't have been possible. They have made vital efforts to compile up to date information on the varied aspects of this subject to make this book a valuable addition to the collection of many professionals and students.

This book was conceptualized with the vision of imparting up-to-date information and advanced data in this field. To ensure the same, a matchless editorial board was set up. Every individual on the board went through rigorous rounds of assessment to prove their worth. After which they invested a large part of their time researching and compiling the most relevant data for our readers.

The editorial board has been involved in producing this book since its inception. They have spent rigorous hours researching and exploring the diverse topics which have resulted in the successful publishing of this book. They have passed on their knowledge of decades through this book. To expedite this challenging task, the publisher supported the team at every step. A small team of assistant editors was also appointed to further simplify the editing procedure and attain best results for the readers.

Apart from the editorial board, the designing team has also invested a significant amount of their time in understanding the subject and creating the most relevant covers. They scrutinized every image to scout for the most suitable representation of the subject and create an appropriate cover for the book.

The publishing team has been an ardent support to the editorial, designing and production team. Their endless efforts to recruit the best for this project, has resulted in the accomplishment of this book. They are a veteran in the field of academics and their pool of knowledge is as vast as their experience in printing. Their expertise and guidance has proved useful at every step. Their uncompromising quality standards have made this book an exceptional effort. Their encouragement from time to time has been an inspiration for everyone.

The publisher and the editorial board hope that this book will prove to be a valuable piece of knowledge for researchers, students, practitioners and scholars across the globe.

LIST OF CONTRIBUTORS

Naomi Kodama, Seiichiro Yonemura and Satoshi Kaneda
National Institute for Agro-Environmental Sciences (NIAES), 3-1-3 Kannondai, Tsukuba, Ibaraki, Japan,

Toshifumi Kimura, Mizue Ohashi and Hidetoshi Ikeno
School of Human Science and Environment, University of Hyogo, 1-1-12 Shinzaike-honcho, Himeji, Hyogo, Japan

Maria Julia de Lima Brossi, Lucas William Mendes, Amanda Barbosa Lima and Siu Mui Tsai
Cellular and Molecular Biology Laboratory, Center for Nuclear Energy in Agriculture, University of São Paulo, Piracicaba, SP, Brazil

Mariana Gomes Germano
Brazilian Agricultural Research Corporation, Embrapa Soybean, Londrina, PR, Brazil

Ricardo Luis Louro Berbara
Soil Science Department, Agronomy Institute, Federal Rural University of Rio de Janeiro, Seropèdica-RJ, Brazil

Francy Junio Gonçalves Lisboa
Soil Science Department, Agronomy Institute, Federal Rural University of Rio de Janeiro, Seropèdica-RJ, Brazil,
The James Hutton Institute, Craigiebuckler, Aberdeen, United Kingdom

Pedro R. Peres-Neto
Canada Research Chair in Spatial Modelling and Biodiversity; Université du Québec á Montréal, Département des sciences biologiques, Québec, Canada

Guilherme Montandon Chaer and Ederson da Conceição Jesus
Embrapa Agrobiologia, Seropédica-RJ, Brazil

Ruth Joy Mitchell and Stephen James Chapman
The James Hutton Institute, Craigiebuckler, Aberdeen, United Kingdom

Dongmei He and Honghua Ruan
Faculty of Forest Resources and Environmental Science, and Key Laboratory of Forestry and Ecological Engineering of Jiangsu Province, Nanjing Forestry University, Nanjing, Jiangsu, China

Leah C. Blasiak, Temesgen Mulaw and Russell T. Hill
Institute of Marine and Environmental Technology, University of Maryland Center for Environmental Science, Baltimore, Maryland, United States of America

Alex W. Schmidt and Thomas M. Schmidt
The Center for Microbial Sciences, University of Michigan, Ann Arbor, Michigan, United States of America

Honoré Andriamiarinoro, Wendy L. Applequist and Chris Birkinshaw
Missouri Botanical Garden, St. Louis, Missouri, United States of America

Rado Rasolomampianina and Félicitée Rejo-Fienena
Laboratoire de Microbiologie de Environnement, Centre National de Recherches sur l9Environnement, Antananarivo, Madagascar

Porter P. Lowry
Missouri Botanical Garden, St. Louis, Missouri, United States of America
Département Systématique et Èvolution (UMR 7205), Muséum National d'Histoire Naturelle, Paris, France

Lei Deng and Zhou-Ping Shangguan
State Key Laboratory of Soil Erosion and Dryland Farming on the Loess Plateau, Northwest A&F University, Yangling, Shaanxi, China

Sandra Sweeney
Institute of Environmental Sciences, University of the Bosphorus, Istanbul, Turkey

Fasheng Zhang
Institute of Applied Ecology, Chinese Academy of Sciences, Shenyang, China

Eastern Cereal and Oilseed Research Centre, Agriculture and Agri-Food Canada, Ottawa, Canada

Guanghua Yin, Zhenying Wang and Zuoxin Liu
Institute of Applied Ecology, Chinese Academy of Sciences, Shenyang, China,

Neil McLaughlin and Xiaoyuan Geng
Eastern Cereal and Oilseed Research Centre, Agriculture and Agri-Food Canada, Ottawa, Canada

Marianne Stenrød, Sonja S. Klemsdal, Hans Ragnar Norli and Ole Martin Eklo
Norwegian Institute for Agricultural and Environmental Research (Bioforsk), Ås, Norway

Zhaojun Li and Yongchao Liang
Institute of Agricultural Resources and Regional Planning, Chinese Academy of Agricultural Sciences, Key Laboratory of Plant Nutrition and Fertilizer, Ministry of Agriculture, Beijing, China

Hua Yang
Institute of Agricultural Resources and Regional Planning, Chinese Academy of Agricultural Sciences, Key Laboratory of Plant Nutrition and Fertilizer, Ministry of Agriculture, Beijing, China
Guizhou Key Laboratory of Mountain Environment, Guizhou Normal University, Guiyang, China

Yupeng Li
Institute of Agricultural Resources and Regional Planning, Chinese Academy of Agricultural Sciences, Key Laboratory of Plant Nutrition and Fertilizer, Ministry of Agriculture, Beijing, China
College of City and Environmental Sciences, Shanxi Normal University, Linfen, China

Jian Long
Guizhou Key Laboratory of Mountain Environment, Guizhou Normal University, Guiyang, China

Selene Muñiz, Juan Lacarta, María P. Pata, Juan José Jiménez and Enrique Navarro
Department of Biodiversity & Conservation and Ecosystem Restoration, Pyrenean Institute of Ecology, Spanish National Research Council, Zaragoza, Spain

Kara R. Lind
Department of Materials Science & Engineering, Iowa State University, Ames, Iowa, United States of America

Tom Sizmur and Saida Benomar
Department of Materials Science & Engineering, Iowa State University, Ames, Iowa, United States of America
Ames Laboratory, US Department of Energy, Iowa State University, Ames, Iowa, United States of America

Anthony Miller
Department of Materials Science & Engineering, Iowa State University, Ames, Iowa, United States of America
Department of Agronomy, Iowa State University, Ames, Iowa, United States of America

Kadri Koorem, Antonio Gazol, Maarja Öpik, Mari Moora, Ülle Saks, Annika Uibopuu and Martin Zobel
Department of Botany, Institute of Ecology and Earth Sciences, University of Tartu, Tartu, Estonia

Virve Sõber
Department of Zoology, Institute of Ecology and Earth Sciences, University of Tartu, Tartu, Estonia

Ludovico Cademartiri
Department of Materials Science & Engineering, Iowa State University, Ames, Iowa, United States of America
Ames Laboratory, US Department of Energy, Iowa State University, Ames, Iowa, United States of America
Department of Chemical & Biological Engineering, Iowa State University, Ames, Iowa, United States of America

Irene Criscuoli
FoxLab, Forest & Wood Science, Fondazione E.Mach, San Michele all'Adige (Trento), Italy
Biogéochimie et écologie des milieux continentaux, CNRS-INRA-ENS-Paris 6, Thiverval-Grignon, France

Giorgio Alberti
MountFor Project Centre, European Forest Institute, San Michele all'Adige (Trento), Italy
Department of Agricultural and Environmental Sciences, University of Udine, Udine, Italy

Silvia Baronti and Emanuela Pusceddu
Institute of Biometeorology, National Research Council, Firenze, Italy

Filippo Favilli
European Academy of Bolzano, Bolzano, Italy

Cristina Martinez and Franco Miglietta
FoxLab, Forest & Wood Science, Fondazione E.Mach, San Michele all'Adige (Trento), Italy
Institute of Biometeorology, National Research Council, Firenze, Italy

Costanza Calzolari
Institute of Research for Hydrogeological Protection, National Research Council, Sesto Fiorentino (FI), Italy

Cornelia Rumpel
Biogéochimie et écologie des milieux continentaux, CNRS-INRA-ENS-Paris 6, Thiverval-Grignon, France

Roberto Viola
Centre for Research and Innovation, Fondazione E.Mach, San Michele all'Adige (Trento), Italy

Chunye Lin, Ruiping Li, Hongguang Cheng and Xiao Shao
State Key Joint Laboratory of Environmental Simulation and Pollution Control, School of Environment, Beijing Normal University, Beijing, China

Jing Wang
Key Laboratory of Land Use, China Land Surveying and Planning Institute, Ministry of land and Resources, Beijing, China

Oxána Bánszegi and Miklós Dombos
Institute for Soil Sciences and Agricultural Chemistry, Centre for Agricultural Research, Hungarian Academy of Sciences, Budapest, Hungary

András Kosztolányi
MTA-DE "Lendület" Behavioural Ecology Research Group, Department of Evolutionary Zoology and Human Biology, University of Debrecen, Debrecen, Hungary

Gábor Bakonyi
Szent István University, Faculty of Agricultural and Environmental Science, Gödöllő, Hungary

Borbála Szabó
Szent István University, Faculty of Veterinary Science, Budapest, Hungary

Pimonrat Tiansawat
Department of Plant Biology, University of Illinois, Urbana, Illinois, United States of America

Adam S. Davis
Global Change and Photosynthesis Research Unit, United Department of Agricultural - Agricultural Research Service, Urbana, Illinois, United States of America

Mark A. Berhow
National Center of Agricultural Utilization Research, United Department of Agricultural - Agricultural Research Service, Peoria, Illinois, United States of America

Paul-Camilo Zalamea
Smithsonian Tropical Research Institute, Ancon, Republic of Panama

James W. Dalling
Department of Plant Biology, University of Illinois, Urbana, Illinois, United States of America Smithsonian Tropical Research Institute, Ancon, Republic of Panama

Jinman Wang and Zhongke Bai
College of Land Science and Technology, China University of Geosciences, Beijing, People's Republic of China
Key Laboratory of Land Consolidation and Rehabilitation, Ministry of Land and Resources, Beijing, People's Republic of China

Peiling Yan
College of Hydraulic and Civil Engineering, China Agricultural University, Beijing, People's Republic of China

Jade Mitchell
Biosystems & Agricultural Engineering, Michigan State University, East Lansing, Michigan, United States of America

Nicolas Pabon
Physics Department, Carnegie Mellon University, Pittsburgh, Pennsylvania, United States of America

Zachary A. Collier and Igor Linkov
Environmental Laboratory, Engineer Research and Development Center, United States Army Corps of Engineers, Concord, Massachusetts, United States of America

Peter P. Egeghy, Elaine Cohen-Hubal and Daniel A. Vallero
Office of Research and Development, United States Environmental Protection Agency, Research Triangle Park, North Carolina, United States of America

Lin Huang, Bi-Cheng Dong, Wei Xue, Yi-Ke Peng, Ming-Xiang Zhang and Fei-Hai Yu
School of Nature Conservation, Beijing Forestry University, Beijing, China

Brent F. Kim
Johns Hopkins Center for a Livable Future, Baltimore, Maryland, United States of America
Department of Environmental Health Sciences, Johns Hopkins Bloomberg School of Public Health, Baltimore, Maryland, United States of America

Melissa N. Poulsen
CLF-Lerner Fellow, Johns Hopkins Center for a Livable Future, Baltimore, Maryland, United States of America
Department of International Health, Johns Hopkins Bloomberg School of Public Health, Baltimore, Maryland, United States of America

Jared D. Margulies
Johns Hopkins Center for a Livable Future, Baltimore, Maryland, United States of America
Department of Environmental Health Sciences, Johns Hopkins Bloomberg School of Public Health, Baltimore, Maryland, United States of America
Department of Geography and Environmental Systems, University of Maryland, Baltimore, Maryland, United States of America

Katie L. Dix
Community Greening Resource Network, Parks & People Foundation, Baltimore, Maryland, United States of America

Anne M. Palmer
Johns Hopkins Center for a Livable Future, Baltimore, Maryland, United States of America
Department of Environmental Health Sciences, Johns Hopkins Bloomberg School of Public Health, Baltimore, Maryland, United States of America
Department of Health, Behavior and Society, Johns Hopkins Bloomberg School of Public Health, Baltimore, Maryland, United States of America

Keeve E. Nachman
Johns Hopkins Center for a Livable Future, Baltimore, Maryland, United States of America
Department of Environmental Health Sciences, Johns Hopkins Bloomberg School of Public Health, Baltimore, Maryland, United States of America
Department of Health Policy and Management, Johns Hopkins Bloomberg School of Public Health, Baltimore, Maryland, United States of America

Baoming Du
School of Agriculture and Biology and Research Center for Low-Carbon Agriculture, Shanghai Jiao Tong University, Shanghai, China

Chunjiang Liu, Hongzhang Kang, Penghua Zhu, Shan Yin and Guangrong Shen
School of Agriculture and Biology and Research Center for Low-Carbon Agriculture, Shanghai Jiao Tong University, Shanghai, China
Key Laboratory of Urban Agriculture (South), Ministry of Agriculture, People's Republic of China, Shanghai, China

Jingli Hou
Instrumental Analysis Center of SJTU, Shanghai Jiao Tong University, Shanghai, China
Hannu Ilvesniemi
Finnish Forest Research Institute, Vantaa, Finland

Weiping Zhou
Key laboratory of Vegetation Restoration and Management of Degraded Ecosystems, South China Botanical Garden, Chinese Academy of Sciences, Guangzhou, China
Graduate University of the Chinese Academy of Sciences, Beijing, PR China

Dafeng Hui
Department of Biological Sciences, Tennessee State University, Nashville, Tennessee, United States of America

Weijun Shen
Key laboratory of Vegetation Restoration and Management of Degraded Ecosystems, South China Botanical Garden, Chinese Academy of Sciences, Guangzhou, China

Yidong Wang, Changcheng Guo and Qing Chen
Tianjin Key Laboratory of Water Resources and Environment, Tianjin Normal University, Tianjin, China

Zhong-Liang Wang
Tianjin Key Laboratory of Water Resources and Environment, Tianjin Normal University, Tianjin, China
State Key Laboratory of Environmental Geochemistry, Institute of Geochemistry, Chinese Academy of Sciences, Guiyang, China

Xiaoping Feng
Tianjin Key Laboratory of Water Resources and Environment, Tianjin Normal University, Tianjin, China
College of Urban and Environmental Sciences, Tianjin Normal University, Tianjin, China

Vidya de Gannes and Gaius Eudoxie
Dept. Food Production, Faculty of Food and Agriculture, University of the West Indies, St. Augustine Campus, Republic of Trinidad and Tobago

William J. Hickey
O.N. Allen Laboratory for Soil Microbiology, Dept. Soil Science, University of Wisconsin-Madison, Madison, Wisconsin, United States of America

Index